Celestial Mechanics

Celestial Mechanics

A COMPUTATIONAL GUIDE
FOR THE PRACTITIONER

LAURENCE G. TAFF

A WILEY–INTERSCIENCE PUBLICATION

John Wiley & Sons

New York   Chichester   Brisbane   Toronto   Singapore

Library of Congress Cataloging in Publication Data:
Taff, Laurence G., 1947—
  Celestial mechanics.

  "A Wiley-Interscience publication."
  Bibliography: p.
  Includes index.
  1. Mechanics, Celestial. 2. Gravitation.
  3. Artificial satellites—Orbits—Measurement.
  4. Perturbation (Mathematics) I. Title.

QB351.T34   1985   521'.1   84-20989
ISBN 0-471-89316-1

Printed in the United States of America

10 9 8 7 6 5 4 3 2 1

For Lois and Matthew

Preface

The subject of this volume is Newtonian gravitation—how it manifests itself, how to calculate its effects on a variety of objects (e.g., artificial and natural satellites, planets, and stars), and how to utilize (passive or active) optical or radar data to predict the future locations and velocities of such objects. The physics involved is relatively simple. The associated practical mathematics is only slightly more complex. The theoretical mathematical basis is more difficult and the art of celestial mechanics is elusive. A thorough reading of this book will provide a complete basis for practicing celestial mechanics. But if I have chosen the illustrative topics well and written the text entertainingly, then you may be able to help invent new celestial mechanics.

While pursuing this goal many of the formal aspects and traditional presentations of the material integral to the standard texts (e.g., Plummer, Smart, Moulton, Brouwer and Clemence) have been slighted. Other texts (e.g., McCuskey, Danby, Battin, Escobal, Deutsch) are at a lower level of mathematical sophistication than is this work. Finally, many new and still active fields of research—formula manipulation on digital computers, Hamilton–Jacobi theory, topological methods and KAM theory, the nontrivial three-body problem, Lie series—are touched upon only briefly or in the references. My focus has been to provide a comprehensive grounding in all of the pertinent physics, the essential utilitarian and theoretical mathematics, and an entrée into areas of applied and potentially fruitful research. I hope that the numerical examples employed within the text and the problems I have collected at the end of the chapters will illuminate the subject, challenge the reader, and interest him or her into trying to develop the knack of celestial mechanics.

This book does cover Newtonian gravitation in great detail. Respective chapters are devoted to Newton's Laws of Motion and his law of gravity as applied to systems of particles and solids; motion in the gravitational field (especially a full discussion of the two-body problem); celestial and geographic coordinate systems and time systems; data reduction; initial orbit determination; perturbation theory; the differential correction of orbits; and astrophysical applications of gravity to binary stars, clusters of stars, galactic dynamics, and clusters of galaxies. The favored object of interest is the artificial satellite and the discussion on orbit determination and refinement is the most complete (theoretical and practical) extant. Similarly, the two chapters (of thirteen) devoted to perturbation theory are unique in scope and thrust. Finally, the discussion of astrophysical applications is the only complete presentation in the textbook or journal literature.

The reader is expected to have had at least one course in classical mechanics and to have a fair degree of mathematical sophistication (both theoretical and applied). Also presumed to be an integral part of the reader's repertoire are least squares, maximum likelihood, the numerical solution of ordinary differential equations, vector calculus, and analysis (especially as it pertains to infinite series). I have aimed for a graduate/senior-level audience, and a mix between a textbook and a reference work. My opinions on several areas of active research are clearly expressed. I hope that the result is a more relaxed text that involves the reader in learning, and communicates some of my excitement while encouraging the next generation to delve further into celestial mechanics. I have attempted to enlighten you on the computational aspects while persuading you to practice your new expertise.

My work at the University of Rochester and the University of Pittsburgh on large-scale gravitational problems (principally star clusters and clusters of galaxies) coupled with my last decade's involvement in M.I.T. Lincoln Laboratory's artificial satellite surveillance and tracking activities provide a part of my expertise in the subject matter. Lincoln Laboratory tracks and searches for artificial satellites, both passively and actively, at several wavelengths of the electromagnetic spectrum. I have called on this expertise and would like to thank Donald Batman, Robert J. Bergemann, V. Alexander Nedzel, William P. Seniw, Ramaswamy Sridharan, John M. Sorvari, and William J. Taylor for easing my passage. Many of the computations were carried out by Sharon A. Stansfield, Kathleen M. Sommerer, Patrice M. S. Randall, or Iva M. Poirier. Lynne M. Perry typed the first draft of the manuscript. Eleanor Harne drafted all of the astronomical symbols and complex mathematical expressions. As he did with my first book, William J. Finnegan, Jr. improved the final version of the text. And as they did with my first book, the Editorial Committee of the Massachusetts Institute of Technology's Lincoln Laboratory supported me through this venture. Finally, Patricia Kennedy Graham supervised the editing and additional typing by Jane Ellen Galus, Eleanor M. Harne, Catherine M. Martino, and Mary Ellen Shortsleeves. Pat's perseverance, dedication, and professionalism in difficult circumstances were critical in assisting me to make this book my best effort.

Before getting to the legal acknowledgments, I would like to mention my wife Lois's enthusiasm for this (and the next!) endeavor as well as my son Matt's fascination with an astronomer/father. Their important psychological aid and emotional comfort (or vice versa—they are not sure) have been extremely valuable.

I have quoted from or paraphrased from several copyrighted works. I wish to acknowledge the permission to do so from the *Astronomical Journal* from Herget (1965) and Kozai (1959); R. Sridharan and W. P. Seniw for permission to use parts of their work (1979, 1980); McGraw-Hill Book Co. to use a paraphrase of part of Reif (1965); J. Wiley & Sons, Inc. to paraphrase a section of Huang (1963); the University of Chicago Press for permission to quote from Spitzer and Hart (1971) as well as to Prof. L. Spitzer, Jr.; Springer-Verlag for permission to quote from Hori and Kozai (1975) as well as Profs. Hori and Kozai; Dover Publication for permission to use a figure from Moulton (1902) and one from Bate, Mueller, and White (1971); Bell Telephone Laboratories for permission to use a part of Geyling and Westerman (1971); Pergamon Press Ltd. for the use of a section of Kurth (1959) and a table in Batten (1973); Annual Reviews, Inc., for permission to quote from Toomre (1977); D. Reidel Publication Company for permission to quote from Hénon (1971) and Miller (1972); and to Batchworth Press Ltd. for the reproductions of several figures in Chapter 12.

LAURENCE G. TAFF

Lexington, Massachusetts
January 1985

Contents

Contents   **xix**

Celestial Mechanics

chapter one

The Basics

This introductory chapter lays the physical and mathematical foundations for the applications of classical mechanics to the problems of celestial mechanics— the study of the motions of natural and artificial bodies in space. Since the subject of this text is a branch of physics, mathematics is an integral part of the formulation of the basic physical principles and the operating tool utilized in the solution of the problems we will encounter. Mathematics, at varying levels from calculus to group theory, is the expressive language of physics. An effort has been made to keep the mathematical sophistication at the lowest possible level while not sacrificing completeness.

The first part of the chapter presents the four fundamental principles of celestial mechanics. They are then applied to individual particles, systems of particles, and lastly to continuous material distributions. Conservation theorems are discussed as well. The last part of this chapter develops the central questions pertinent to any mathematical problem:

1. Is there a solution?
2. Is the solution unique?
3. How can one calculate the solution?
4. If the problem is altered slightly, in what fashion does (do) the solution(s) (if any) change?

FUNDAMENTALS

Classical mechanics principally rests on Isaac Newton's three Laws of Motion:*

1. Every body continues in its state of rest, or of uniform motion in a right line, unless it is compelled to change that state by forces impressed upon it.

* From Cajori's (1934) translation. Details of the references are given separately at the end of the book.

2. The change of motion is proportional to the motive force impressed and is made in the direction of the right line in which that force is impressed.

3. To every action there is always opposed an equal reaction; or, the mutual actions of two bodies upon each other are always equal and directed to contrary parts.

To do celestial mechanics, one supplements these postulates with Newton's Law of Gravitation:

4. Every particle in the Universe attracts every other particle in the Universe with a force that varies directly as the product of their masses and inversely as the square of the distance between them; furthermore, this force acts along the line joining the two particles.

There is essentially no more physics in this book—the rest is mathematics.

The undefined concepts of classical mechanics—length, mass, and time—are not discussed herein. Similarly, the Newtonian concepts of uniform space and of uniform time, the foundations of special relativity or of general relativity, and the experimental equality of inertial and gravitational mass [to 1 part in 10^{11}; see Eötvös (1891, 1896) and Dicke (1961)] will be hardly mentioned. Also, note that the mathematical representations of statements 2 and 4 (see Eqs. 1.7 and 1.10) are invariant under transformations of the form (known as Galilean transformations)

$$\mathbf{r}' = R\mathbf{r} + \mathbf{r}_0 + \mathbf{v}_0(t - t_0) \tag{1.1}$$

where R is an arbitrary rotation matrix (see Chapter Two), $\mathbf{r}_0$ and $\mathbf{v}_0$ are arbitrary constant vectors, and t_0 is some arbitrary fixed epoch.

A particle's *location** relative to an arbitrary origin O can be uniquely specified by providing, for example, the three components of the vector $\mathbf{r} = (x, y, z)$. Here x, y, and z are understood to be the ordinary rectangular components of the vector $\mathbf{r}$, that is, the values of the projections of $\mathbf{r}$ onto the three mutually perpendicular axes OX, OY, and OZ (see Fig. 1) (Newtonian space is perforce Euclidean). An alternative I frequently employ is the spherical coordinates r, θ, and ϕ (see Fig. 1 again) where

$$\mathbf{r} = (x, y, z) = r(\cos \theta \cos \phi, \cos \theta \sin \phi, \sin \theta) \tag{1.2}$$

Here ϕ is the azimuth, θ is the latitude, and r is the distance, $r = +(x^2 + y^2 + z^2)^{1/2}$. A shorthand notation for Eq. 1.2 is

$$\mathbf{r} = r\mathbf{l}(\phi, \theta) \tag{1.3}$$

$\mathbf{l}(\phi, \theta)$ is the vector of direction cosines or *position* vector.

---

*The term *position* is reserved for the direction of a point. Location implies the full three-dimensional specification of placement.

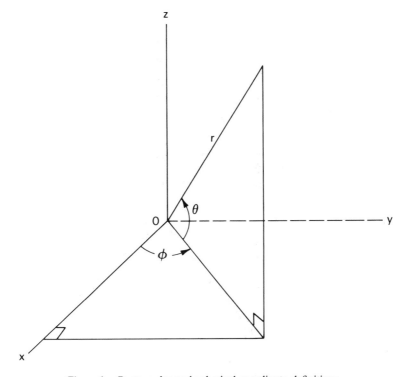

Figure 1. Rectangular and spherical coordinate definitions.

The *velocity* of this particle, relative to the origin O, is given by

$$\mathbf{v} = \frac{d\mathbf{r}}{dt} = \dot{\mathbf{r}} = (\dot{x}, \dot{y}, \dot{z})$$

In terms of the spherical coordinates r, θ, and ϕ

$$\dot{\mathbf{r}} = \dot{r}\mathbf{l}(\phi, \theta) + r\dot{\phi} \begin{pmatrix} -\cos\theta\sin\phi \\ \cos\theta\cos\phi \\ 0 \end{pmatrix} + r\dot{\theta} \begin{pmatrix} -\sin\theta\cos\phi \\ -\sin\theta\sin\phi \\ \cos\theta \end{pmatrix} \tag{1.4}$$

The *linear momentum* (Newton's "motion") is defined as

$$\mathbf{p} = m\mathbf{v} \tag{1.5}$$

and the *acceleration* of the particle, relative to O, is given by

$$\mathbf{a} = \dot{\mathbf{v}} = \ddot{\mathbf{r}} \tag{1.6}$$

The *speed* of the particle is $v = |\mathbf{v}|$. The *angular velocity* is equal to $\dot{\mathbf{l}}(\phi, \theta)$ and the *angular speed* is equal to ω; $\omega^2 = \dot{\mathbf{l}} \cdot \dot{\mathbf{l}} = \dot{\phi}^2 \cos^2\theta + \dot{\theta}^2$, $\omega \geqslant 0$.

Newton's second law of motion can be expressed as

$$\mathbf{F} = \dot{\mathbf{p}} \tag{1.7}$$

where **F** is the impressed motive force. For most problems in classical mechanics the mass m is a constant so that $\dot{\mathbf{p}} = m\dot{\mathbf{v}}$ whence, by Eq. 1.6,

$$\mathbf{F} = m\mathbf{a} \tag{1.8}$$

A notable exception is rockets wherein mass is lost in the act of propulsion. Note that Newton's first law of motion may be deduced from the second.

Newton's third law of motion can be formulated as

$$\mathbf{F}_{12} = -\mathbf{F}_{21} \tag{1.9}$$

where $\mathbf{F}_{12}$ is the force exerted on system 1 by system 2. Finally, Newton's law of gravitation between two particles with (gravitational) masses m_1, m_2 may be symbolized as

$$\mathbf{F}_{12} = -\frac{Gm_1 m_2 (\mathbf{r}_1 - \mathbf{r}_2)}{|\mathbf{r}_1 - \mathbf{r}_2|^3} \tag{1.10}$$

where $\mathbf{r}_1$, $\mathbf{r}_2$ are the locations of the two particles and G is a positive constant. G is known as Newton's (universal) constant of gravitation, $G = 6.672 \times 10^{-11}$ m^3/kg · sec^2 [the 1976 International Astronomical Union (IAU) system of constants is used; see the Appendix]. Equations 1.8–1.10 comprise *the two-body problem.*

Equations 1.8–1.10 are invariant under transformations of the form in Eq. 1.1. Hence, if Newton's laws hold in one coordinate system, then they hold in infinitely many coordinate systems connected by Galilean transformations. Such special coordinate systems are known as *inertial* coordinate (or reference) systems (or frames). In practice the reference frame of the "fixed" stars provides a very good approximation to an inertial reference frame. The misleading and incorrect phrase "Earth-centered inertial," or "ECI," reference frame shall nowhere else appear in this book.

Constants of the Motion

The central problem of celestial mechanics is the problem posed by the solution of Eqs. 1.8–1.10. This is called the two-body problem.* One can deduce several important properties of this mathematical system that have general applicability in much of physics and for the N-body problem in particular.

It should be clear to the reader that Eq. 1.10 can be written as

$$\mathbf{F}_{12} = -\nabla_{\mathbf{r}_1} \left(-\frac{Gm_1 m_2}{|\mathbf{r}_1 - \mathbf{r}_2|} \right) = -\nabla_{\mathbf{r}_1} U \tag{1.11a}$$

U is called the *potential energy.* Similarly, as $\mathbf{F}_{12} = -\mathbf{F}_{21}$ and $\nabla_{\mathbf{r}_1} U = -\nabla_{\mathbf{r}_2} U$,

---

* Nature is wonderfully perverse. Here, in the first rational formulation of gravity and mechanics one can solve, completely, the two-body problem. In the next step up (general relativity), one can solve, completely, the one-body problem. No doubt when the super grand unified theory is developed, one will only be able to solve, completely, the zero-body problem.

one can write that

$$\mathbf{F}_{21} = -\nabla_{\mathbf{r}_2} U \tag{1.11b}$$

But $\mathbf{F}_{12} = m_1\ddot{\mathbf{r}}_1$ and $\mathbf{F}_{21} = m_2\ddot{\mathbf{r}}_2$, so after these substitutions and the scalar multiplication of Eq. 1.11a by $\dot{\mathbf{r}}_1$ and of Eq. 1.11b by $\dot{\mathbf{r}}_2$, we have, adding the results,

$$\frac{d}{dt}\left(\frac{m_1}{2}|\dot{\mathbf{r}}_1|^2 + \frac{m_2}{2}|\dot{\mathbf{r}}_2|^2 + U\right) = \frac{d}{dt}(K+U) = \frac{d\mathscr{E}}{dt} = 0 \tag{1.12}$$

K is called the *kinetic energy* and $\mathscr{E}$ is called the (total) *energy*. Similarly, the quantity

$$\mathbf{J} = \mathbf{r}_1 \times \mathbf{p}_1 + \mathbf{r}_2 \times \mathbf{p}_2 \tag{1.13}$$

may be shown to have a vanishing first derivative in this instance. $\mathbf{J}$ is called the (total) *angular momentum*. Thus, both $\mathbf{J}$ and $\mathscr{E}$ are constants of the motion. These conserved quantities separately illustrate the facts that the problem under consideration is rotationally invariant and time reversible. [In the absence of an external force a particle's linear momentum is conserved and this (literally) reflects the translational invariance of the system. Thus the form of Eq. 1.1.]

In general there are six other constants of the motion. These are connected with the motion of the *center of mass*. Define

$$\mathbf{R} = \frac{m_1\mathbf{r}_1 + m_2\mathbf{r}_2}{m_1 + m_2} \quad \text{and} \quad \mathbf{r} = \mathbf{r}_1 - \mathbf{r}_2$$

Then Eqs. 1.11 are equivalent to the set

$$\ddot{\mathbf{R}} = 0$$

$$\ddot{\mathbf{r}} = -\frac{G(m_1 + m_2)\mathbf{r}}{|\mathbf{r}|^3} \tag{1.14}$$

Clearly the center of mass, whose location is given by $\mathbf{R}$, moves along a right (i.e., straight) line at a constant (i.e., uniform) speed. This implies the constancy of six quantities and simultaneously (no pun intended) proves that the center of mass reference frame is an inertial one. In terms of $\mathbf{R}$, $\mathbf{r}$, and their derivatives we may rewrite Eqs. 1.12 and 1.13 as

$$\frac{d}{dt}\left(\frac{\mu}{2}|\dot{\mathbf{r}}|^2 + \frac{M}{2}|\dot{\mathbf{R}}|^2 + U\right) = \frac{d\mathscr{E}}{dt} = 0$$

$$\mathbf{J} = \mathbf{r} \times \mathbf{p} + \mathbf{R} \times \mathbf{P}$$

where μ is the reduced mass, $\mu = m_1 m_2/M$, and M is the total mass, $M = m_1 + m_2$; $\mathbf{p} = \mu\dot{\mathbf{r}}$, $\mathbf{P} = M\dot{\mathbf{R}}$, and $\dot{\mathbf{J}} = 0$ still.

Systems of Particles

It is an experimental fact that the gravitational force exerted by a system of particles with masses $m_1, m_2, \ldots, m_N$ located at $\mathbf{r}_1, \mathbf{r}_2, \ldots, \mathbf{r}_N$ on a particle of mass m located at $\mathbf{r}$ is equal to

$$\mathbf{F} = \sum_{n=1}^{N} \mathbf{F}_n = -\sum_{n=1}^{N} \frac{Gmm_n(\mathbf{r}-\mathbf{r}_n)}{|\mathbf{r}-\mathbf{r}_n|^3} \tag{1.15}$$

Therefore, in a self-gravitating system of N particles the force exerted on particle i by all of the other $N-1$ particles is

$$\mathbf{F}_i = \sum_{\substack{j=1 \\ j \neq i}}^{N} \mathbf{F}_{ij}$$

where $\mathbf{F}_{ij}$ is given by Eq. 1.10 ($1 \to i, 2 \to j$). The potential energy of such a system is

$$U = -\sum_{N \geqslant j > i \geqslant 1} \frac{Gm_im_j}{|\mathbf{r}_i-\mathbf{r}_j|}$$

and the equations of motion of the ith particle may be written as

$$\mathbf{F}_i = -\nabla_{\mathbf{r}_i} U$$

The kinetic energy is $K = \sum_{i=1}^{N} m_i |\dot{\mathbf{r}}_i|^2/2$ and the total energy $\mathscr{E} = K + U$ is conserved. Similarly, the total angular momentum $\mathbf{J} = \sum_{i=1}^{N} \mathbf{r}_i \times \mathbf{p}_i$ is a constant vector. Lastly, the center of mass

$$\mathbf{R} = \sum_{i=1}^{N} \frac{m_i \mathbf{r}_i}{M}, \qquad M = \sum_{j=1}^{n} m_j$$

moves along a straight line (the geodesic of Euclidean space) at a constant speed. These 10 integrals of the motion are the only ones known in general.

CONTINUOUS DISTRIBUTIONS OF MATTER

The critical problem faced by Newton, after he convinced himself of the correctness of statements 1–4, was in applying them to the bodies of the solar system. Therefore, he needed to compute the total gravitational force exerted by a continuously distributed material system. The mathematical apparatus necessary to solve this problem did not exist (i.e., the integral calculus), so he invented it. He also made the reasonable approximation that the Sun, the planets, and their moons could be accurately represented by spherically symmetric distributions of mass. It turns out that the gravitational field exterior to a spherically symmetric distribution of matter is given by $-GM\mathbf{r}/|\mathbf{r}|^3$ where M is the total mass within the sphere and $\mathbf{r}$ is the vector from the sphere's center to the point in question. Hence, the spherically symmetric macroscopic two-body problem is identical to the problem posed above for two point masses.

From Eq. 1.15, as $m_n \to 0$, $N \to \infty \ni \sum_{n=1}^{N} m_n$ stays finite,

$$\mathbf{F} = -Gm \int_V \frac{(\mathbf{r}-\mathbf{r}') \, dm'}{|\mathbf{r}-\mathbf{r}'|^3}$$

where V is the volume enclosing the distributed mass. If dm' is replaced by $\rho(\mathbf{r}') \, d\mathbf{r}'$ (ρ is called the *mass density*), then the force can be written as

$$\mathbf{F} = -Gm \int_V \frac{(\mathbf{r}-\mathbf{r}')\rho(\mathbf{r}') \, d\mathbf{r}'}{|\mathbf{r}-\mathbf{r}'|^3} \tag{1.16}$$

In terms of the *potential* U (mU is now the potential energy) of this mass distribution, the force is given by

$$\mathbf{F} = -m\nabla U(\mathbf{r}) \tag{1.17a}$$

with the potential U defined in Eq. 1.17b:

$$U(\mathbf{r}) = -G \int_V \frac{\rho(\mathbf{r}') \, d\mathbf{r}'}{|\mathbf{r}-\mathbf{r}'|} \tag{1.17b}$$

Because the potential is a scalar, the system 1.17 is usually easier to deal with than is Eq. 1.16. The gradient of U is called the gravitational field of the mass distribution. As

$$\nabla^2 \left(\frac{1}{|\mathbf{r}-\mathbf{r}'|} \right) = -4\pi\delta(\mathbf{r}-\mathbf{r}')$$

where δ is Dirac's delta function, Poisson's equation can be derived:

$$\nabla^2 U(\mathbf{r}) = +4\pi G\rho(\mathbf{r})$$

Clearly, much of the formal development of the potential in electromagnetic theory can be taken over (but not the wave equation because Newtonian gravity has an infinite propagation speed). What is needed here deals with associated Legendre functions and trigonometric functions. If Eq. 1.21 makes sense to the reader, then he can skip directly to there.

Formal Potential Theory. Formal potential theory is usually developed in courses on electromagnetic theory rather than in courses (or texts) on gravitational theory. The reason is that classical gravitational theory is totally dominated by spherically symmetric distributions of matter and the two-body problem. Nonetheless, a brief aside here might be enriching reading for the reader.

Given that the force between two particles is given by Eq. 1.10, one can continue to focus on the force, the gravitational *field* (= the force per unit mass), the potential energy (whose negative gradient is equal to the force), or the potential (whose negative gradient is equal to the field). Herein the gravitational field **g** and the potential U shall be emphasized.

The force acting on a particle of mass m in the presence of the gravitational field **g** is

$$\mathbf{F} = m\mathbf{g} = -m\nabla U$$

The field due to a point mass m' situated at $\mathbf{r}'$ at the location $\mathbf{r}$ is (cf. Eq. 1.10)

$$\mathbf{g} = \frac{-Gm'(\mathbf{r} - \mathbf{r}')}{|\mathbf{r} - \mathbf{r}'|^3}$$

whereas that due to a system of N point masses $\{m_n\}$ at $\{\mathbf{r}_n\}$ is (cf. Eq. 1.15)

$$\mathbf{g} = -G \sum_{n=1}^{N} \frac{m_n(\mathbf{r} - \mathbf{r}_n)}{|\mathbf{r} - \mathbf{r}_n|^3}$$

If there are so many mass points per unit volume that one can speak of a macroscopic mass density $\rho(\mathbf{r})$, then (cf. Eq. 1.16)

$$\mathbf{g} = -G \int_V \frac{(\mathbf{r} - \mathbf{r}')\rho(\mathbf{r}')\, d\mathbf{r}'}{|\mathbf{r} - \mathbf{r}'|^3}$$

The volume V encloses the nonzero range of ρ.

Gauss's law relates the surface integral of the normal component of **g** to the total amount of matter enclosed. It is really a statement about the $1/r^2$ central nature of the force, the three dimensionality of Euclidean space, and the observed linear superposition of the gravitational effects of matter. Gauss's law may be written as

$$\oint_S \mathbf{g} \cdot d\mathbf{S} = -4\pi G \int_V \rho(\mathbf{r})\, d\mathbf{r}$$

for the surface S bounding the volume V. If the divergence theorem is used, this may be written as a differential relationship,

$$\nabla \cdot \mathbf{g} = -4\pi G\rho$$

Another differential relationship is

$$\nabla \times \mathbf{g} = 0$$

which follows directly from the generalized Newtonian law,

$$\mathbf{g} = -G \int_V \frac{(\mathbf{r} - \mathbf{r}')\rho(\mathbf{r}')\, d\mathbf{r}'}{|\mathbf{r} - \mathbf{r}'|^3} = \nabla \left[G \int_V \frac{\rho(\mathbf{r}')\, d\mathbf{r}'}{|\mathbf{r} - \mathbf{r}'|} \right]$$

since $(\mathbf{r} - \mathbf{r}')/|\mathbf{r} - \mathbf{r}'|^3 = -\nabla(1/|\mathbf{r} - \mathbf{r}'|)$ and $\nabla \times \nabla u = 0$ for any twice differentiable scalar u. This is how the potential (cf. Eq. 1.17b) is introduced,

$$\mathbf{g} = -\nabla U, \qquad U = -G \int \frac{\rho(\mathbf{r}')\, d\mathbf{r}'}{|\mathbf{r} - \mathbf{r}'|}$$

As $\nabla \cdot \mathbf{g} = -4\pi G\rho$ and $\mathbf{g} = -\nabla U$, it follows that

$$\nabla^2 U = 4\pi G\rho(\mathbf{r})$$

It is at this point that the introduction of the concept of the Dirac delta function makes sense.

$\delta(u)$ is an improper mathematical function [actually a generalized function; see Lighthill (1958)] which is zero everywhere except when its argument vanishes. There it is "infinite" in such a way that

$$\int \delta(x-a)\,dx = \begin{cases} 1 & a \text{ in the region of integration} \\ 0 & \text{otherwise} \end{cases}$$

For any function $f(x)$

$$\int_{-\infty}^{\infty} f(x)\,\delta(x-a)\,dx = f(a) \qquad \int_{-\infty}^{\infty} f(x)\,\delta'(x-a)\,dx = -f'(a)$$

where the prime means differentiation with respect to the argument. As an example of the use of the delta function consider

$$\rho(\mathbf{r}) = \sum_{n=1}^{N} m_n \delta(\mathbf{r}-\mathbf{r}_n)$$

This represents the mass density of a system of N point masses $\{m_n\}$ located at $\{\mathbf{r}_n\}$. In a symbolic notation one could write

$$\nabla^2 \left(\frac{1}{r}\right) = -4\pi\delta(\mathbf{r})$$

since $\nabla^2(1/r) = 0$ for $r \neq 0$ whereas the volume integral of $\nabla^2(1/r)$, if the volume includes the origin, is -4π. This kind of symbolism justifies the statement immediately below Eq. 1.17b.

Another example of the utility of the delta function concept is in the development of Green's functions for the solution of the field in more general matter distributions. See Jackson (1975). Of more importance are the boundary conditions that $\mathbf{g}$ or U have to satisfy. (Linear and surface distributions of mass will not be discussed.) Consider any finite amount of matter M contained in some finite volume V. Far enough away from V the internal structure of the mass distribution becomes invisible and, intuitively, infinitely far away the mass collection must appear as a mass point of total mass M. Thus,

$$U \xrightarrow[r \to \infty]{} -\frac{GM}{r}$$

Within the volume V there is the problem of an infinity in the integrand for both the potential and the force (see Eqs. 1.16 and 1.17b). Take the expression for the potential first and translate to a new coordinate system whose origin is at $\mathbf{r}$. In the infinitesimal volume element surrounding $\mathbf{r}$ the integrand for the potential is

$$\lim_{r \to 0} \frac{\rho(r) r^2 \, d\Omega}{r}$$

where $d\Omega$ is the element of solid angle. Clearly, as long as ρ itself is finite, this expression is too, and thence U. Thus, one can demand that U be finite everywhere within a continuous material distribution. The situation for the force is similar but a bit more complicated.

Spherical Harmonics

When $\rho = 0$ everywhere, Poisson's equation is known as Laplace's equation,

$$\nabla^2 U = 0$$

Coordinate systems in which this first-degree, second-order, three-dimensional partial differential equation separates are of special importance in mathematical physics. Of the 11 such systems the spherical coordinate system* (see Fig. 1) is one of them. Hence, assume that $U(\mathbf{r}) = U(r, \theta, \phi)$ can be written as $U = [R(r)/r]\Theta(\theta)\Phi(\phi)$. Then Laplace's equation

$$\frac{1}{r} \frac{\partial^2 (rU)}{\partial r^2} + \frac{\sec \theta}{r^2} \frac{\partial}{\partial \theta} \left(\cos \theta \frac{\partial U}{\partial \theta} \right) + \frac{\sec^2 \theta}{r^2} \frac{\partial^2 U}{\partial \phi^2} = 0$$

becomes

$$\Theta\Phi \frac{d^2 R}{dr^2} + R\Phi \frac{\sec \theta}{r^2} \frac{d}{d\theta} \left(\cos \theta \frac{d\Theta}{d\theta} \right) + R\Theta \frac{\sec^2 \theta}{r^2} \frac{d^2\Phi}{d\phi^2} = 0$$

Multiply through by $r^2 \cos^2 \theta / U$ to find that the ϕ dependence may be separated out:

$$r^2 \cos^2 \theta \left[\frac{1}{R} \frac{d^2 R}{dr^2} + \frac{\sec \theta}{r^2 \Theta} \frac{d}{d\theta} \left(\cos \theta \frac{d\Theta}{d\theta} \right) \right] = -\frac{1}{\Phi} \frac{d^2\Phi}{d\phi^2}$$

As the functional dependence of both sides appears to be different—but must be the same—they must both be equal to a constant. Call this separation constant m^2; then

$$\frac{d^2\Phi}{d\phi^2} = -m^2 \Phi$$

The two linearly independent solutions of this equation are

$$\Phi_+ e^{im\phi}, \qquad \Phi_- e^{-im\phi}$$

where $\Phi_\pm$ are the constants of integration. Note that the form of the separation constant is immaterial. A form that makes the final answer neater has been deliberately used.

---

* Note that many authors use the colatitude $\pi/2 - \theta$ instead of the latitude.

Continuing the analysis of Laplace's equation, the remaining terms are divided by $\cos^2 \theta$ and the result rearranged to obtain

$$-\frac{r^2}{R}\frac{d^2 R}{dr^2} = \frac{\sec \theta}{\Theta}\frac{d}{d\theta}\left(\cos \theta \frac{d\Theta}{d\theta}\right) - m^2 \sec^2 \theta$$

This time the separation constant is defined to be $-l(l+1)$ so that

$$\frac{d^2 R}{dr^2} - \frac{l(l+1)R}{r^2} = 0$$

$$\sec \theta \frac{d}{d\theta}\left(\cos \theta \frac{d\Theta}{d\theta}\right) + [l(l+1) - m^2 \sec^2 \theta]\Theta = 0$$

The linearly independent solutions for R are just powers,

$$Ar^{l+1}, \qquad Br^{-l}$$

Before solving the equation for Θ, the change of variable $u = \sin \theta$ is introduced and $P(u) = \Theta(\theta)$. P satisfies

$$\frac{d}{du}\left[(1-u^2)\frac{dP}{du}\right] + \left[l(l+1) - \frac{m^2}{1-u^2}\right]P = 0 \qquad (1.18)$$

Legendre Polynomials

Equation 1.18 is called the generalized Legendre equation and its solutions are called the associated Legendre functions. When $m = 0$ and l is an integer, the solutions are known as Legendre polynomials. Because a physical problem is being solved, the solution should possess certain properties: It must be finite in the range of interest, continuous, and single-valued. The well-behaved Legendre polynomials and functions are usually symbolized by P (with appropriate subscripts or superscripts and functional dependence). The Legendre polynomials and functions of the second kind are denoted by Q with its corresponding set of addenda.

The solution of Eq. 1.18 when $m = 0$ will now be derived using the method of Frobenius: that is, a series solution of the form

$$P(u) = u^a \sum_{j=0}^{\infty} a_j u^j \qquad (1.19)$$

will be attempted. After direct substitution into Eq. 1.18, this works if

$$a(a-1) = 0 \quad (a_0 \neq 0) \qquad \text{or} \qquad a(a+1) = 0 \quad (a_1 \neq 0)$$

and

$$a_{j+2} = \frac{(a+j)(a+j+1) - l(l+1)}{(a+j+1)(a+j+2)} a_j$$

Thus, only one of a_0, a_1 can be nonzero, and P contains only even or only odd powers of u. Moreover, for $a = 0$ or 1 the series 1.19 converges for $u^2 < 1$

and all real values of l, but it will diverge if $u^2 = 1$ unless it terminates. To terminate, a_j must be zero for all $j > J$. Because both a and j are integers, it is clear that this implies that l must be one too. The solution for a particular value of l is written as $P_l(u)$. Furthermore, the standard (arbitrary) normalization for the Legendre polynomials is $P_l(1) = 1$. The first few are

$$P_0(u) = 1, \qquad P_1(u) = u$$

$$P_2(u) = \frac{3u^2 - 1}{2}, \qquad P_3(u) = \frac{5u^3 - 3u}{2}$$

$$P_4(u) = \frac{35u^4 - 30u^2 + 3}{8}, \qquad P_5(u) = \frac{63u^5 - 70u^3 + 15u}{8}$$

In general, one has Rodrigues's formula for $P_l(u)$:

$$P_l(u) = \frac{1}{2^l l!} \frac{d^l}{du^l} (u^2 - 1)^l$$

The set of Legendre polynomials forms a complete orthogonal set on $u \in [-1, 1]$,

$$\int_{-1}^{+1} P_l(u) P_{l'}(u) \, du = \frac{2\delta_{ll'}}{2l + 1}$$

where δ_{mn} is the Kronecker delta function ($= 1$ if $m = n$, 0 otherwise). One can expand any appropriately smooth function $f(u)$ in a series

$$f(u) = \sum_{l=0}^{\infty} f_l P_l(u)$$

wherein the expansion coefficients are calculated via

$$f_l = \frac{2l + 1}{2} \int_{-1}^{+1} f(u) P_l(u) \, du$$

Finally, the Legendre functions satisfy a variety of recurrence relations:

$$(l+1) P_{l+1} - (2l+1) u P_l + l P_{l-1} = 0$$

$$P'_{l+1} - P'_{l-1} = (2l+1) P_l$$

$$P'_{l+1} - u P'_l = (l+1) P_l$$

$$(1 - u^2) P'_l = l P_{l-1} - l u P_l$$

$$= (l+1) u P_l - (l+1) P_{l+1}$$

The argument above is always u and $P'_l = dP_l(u)/du$.

Associated Legendre Functions

So far the regular solution to the generalized Legendre equation 1.18 has been obtained only when $m = 0$. Because it is necessary that Φ be single-valued, only integer values of m are interesting. The general regular solution of Eq. 1.18 is denoted by $P_l^m(u)$ and is called the associated Legendre function of order l, degree m. It is simply related to the ordinary Legendre polynomial of order l via

$$P_l^m(u) = (-1)^m (1-u^2)^{m/2} \frac{d^m P_l(u)}{du^m} \qquad \text{if } m \geq 0$$

For $m < 0$ one uses

$$P_l^{-m}(u) = (-1)^m \frac{(l-m)!}{(l+m)!} P_l^m(u)$$

Note that $P_l^0(u) \equiv P_l(u)$.

For a fixed value of degree m the associated Legendre functions form an orthogonal, complete set in the order l when $u \in [-1, 1]$:

$$\int_{-1}^{+1} P_l^m(u) P_{l'}^m(u)\, du = \frac{2}{2l+1} \frac{(l+m)!}{(l-m)!} \delta_{ll'}$$

There is an orthogonality condition on the degree m too:

$$\int_{-1}^{+1} P_l^m(u) P_l^{m'}(u) \frac{du}{1-u^2} = \frac{(l+m)!}{m(l-m)!} \delta_{mm'}$$

There is an assortment of recurrence relationships also:

$$(2l+1)u P_l^m = (l+m) P_{l-1}^m + (l-m+1) P_{l+1}^m$$

$$(2l+1)(1-u^2)^{1/2} P_l^m = P_{l+1}^{m+1} - P_{l-1}^{m+1}$$

$$2(1-u^2)^{1/2} \frac{dP_l^m}{du} = P_l^{m+1} - (l+m)(l-m+1) P_l^{m-1}$$

Lastly there is the parity relationship

$$P_l^m(-u) = (-1)^{l+m} P_l^m(u)$$

The zeros of P_l^m are at the ends of the interval $[-1, 1]$,

$$P_l^m(\pm 1) = 0 \qquad \text{for } m > 0$$

Spherical Harmonics. The *spherical harmonic* $Y_{lm}(\theta, \phi)$ is defined as

$$Y_{lm}(\theta, \phi) = \left[\frac{2l+1}{4\pi} \frac{(l-m)!}{(l+m)!} \right]^{1/2} P_l^m(\sin \theta)\, e^{im\phi} \qquad (1.20)$$

for $m > 0$ and for $m < 0$ by

$$Y_{l,-m}(\theta, \phi) = (-1)^m Y_{lm}^*(\theta, \phi)$$

where the asterisk denotes complex conjugation. The orthonormality relationship for these functions is

$$\int_0^{2\pi} d\phi \int_{-\pi/2}^{\pi/2} Y_{lm}(\theta, \phi) Y_{l'm'}^*(\theta, \phi) \cos \theta \, d\theta = \delta_{ll'} \delta_{mm'}$$

The completeness relationship for them is

$$\sum_{l=0}^{\infty} \sum_{m=-l}^{l} Y_{lm}(\theta, \phi) Y_{lm}^*(\theta', \phi') = \delta(\phi - \phi') \, \delta(\sin \theta - \sin \theta')$$

Finally, for any sufficiently smooth function $f(\theta, \phi)$ one can write

$$f(\theta, \phi) = \sum_{l=0}^{\infty} \sum_{m=-l}^{l} f_{lm} Y_{lm}(\theta, \phi)$$

wherein the expansion coefficients are computed from

$$f_{lm} = \int_0^{2\pi} d\phi \int_{-\pi/2}^{\pi/2} f(\theta, \phi) Y_{lm}^*(\theta, \phi) \cos \theta \, d\theta$$

If we pull together the three pieces that U was factored into, then the complete, general solution to Laplace's equation is obtained (expressed in spherical coordinates):

$$U(\mathbf{r}) = \sum_{l=0}^{\infty} \sum_{m=-l}^{\infty} [A_{lm} r^l + B_{lm} r^{-(l+1)}] Y_{lm}(\theta, \phi)$$

with Y_{lm} defined in Eq. 1.20.

Development of the Potential

There are several requirements that the potential must meet: It must be finite at $r = 0$, it must approach constant r as $|\mathbf{r}| \to \infty$, it must satisfy Poisson's equation (Laplace's equation wherever $\rho = 0$), and it must satisfy certain *boundary conditions*. In particular, over some closed surface either U (Dirichlet boundary conditions) or its gradient (Neumann boundary conditions) will be specified. Under these constraints the solution for U exists and is unique. With the mathematical underpinnings secure we return to the physical problem.

Among the myriad things one can prove about spherical harmonics is the result

$$\frac{1}{|\mathbf{r} - \mathbf{r}'|} = 4\pi \sum_{p=0}^{\infty} \sum_{q=-p}^{+p} \frac{r_<^p}{r_>^{p+1}} \frac{Y_{pq}^*(\theta', \phi') Y_{pq}(\theta, \phi)}{2p + 1} \tag{1.21}$$

where $r_<$ $(r_>)$ is the lesser (greater) of $|\mathbf{r}|$, $|\mathbf{r}'|$, θ and ϕ $(\theta'$ and $\phi')$ are the latitude and azimuth of $\mathbf{r}$ $(\mathbf{r}')$, and Y_{pq} is the spherical harmonic of order p, degree q. Utilizing the result in Eq. 1.21 the potential in Eq. 1.17b outside of the volume V can be written as

$$U(\mathbf{r}) = -G \sum_{p=0}^{\infty} \sum_{q=-p}^{+p} \frac{\rho_{pq} Y_{pq}(\theta, \phi)}{|\mathbf{r}|^{p+1}} \qquad (1.22a)$$

The moments of the density distribution are defined by

$$\rho_{pq} = \frac{4\pi}{2p+1} \int_V |\mathbf{r}'|^p Y_{pq}^*(\theta', \phi') \rho(\mathbf{r}') \, d\mathbf{r}' \qquad (1.22b)$$

The result in Eqs. 1.22 is completely general.

Consider the special case of a spherically symmetric mass distribution. This means that $\rho(\mathbf{r}') = \rho(|\mathbf{r}'|)$. As $Y_{00} = 1/\sqrt{4\pi}$, $d\mathbf{r}' = (r')^2 \cos\theta' \, d\theta' \, d\phi'$, and the spherical harmonics form a complete orthonormal system of basis functions on the unit sphere, only ρ_{00} is nonzero: $\rho_{00} = (4\pi)^{1/2} M$ where

$$M \equiv \int_V \rho(\mathbf{r}') \, d\mathbf{r}' = 4\pi \int_0^R (r')^2 \rho(r') \, dr'$$

is the total mass contained inside the sphere of radius R. Thus,

$$U(\mathbf{r}) = -\frac{GM}{|\mathbf{r}|} \qquad (1.23)$$

and obviously $-\nabla U = -GM\mathbf{r}/|\mathbf{r}|^3$. This result can also be obtained by a straightforward integration over spherical shells.

For nonspherically symmetric distributions of matter all of the terms in Eqs. 1.22 may be nonzero. The three $p = 1$ terms contribute

$$-\frac{GM\mathbf{r}}{|\mathbf{r}|^3} \cdot \int_V \frac{\mathbf{r}' \rho(\mathbf{r}') \, d\mathbf{r}'}{M}$$

to U. The above integral defines the location of the center of mass of the distributed system. If this is used as the origin of the coordinate system, then this term vanishes. Hence, gravitational dipole moments are eliminated by a proper choice of reference frame. The sum of the five $p = 2$ terms can be compactly written as

$$-\frac{G}{2|\mathbf{r}|^5} [\mathbf{r} \cdot \mathbf{r} \, \mathrm{Tr}(\mathbf{l}) - 3\mathbf{r} \cdot \mathbf{l} \cdot \mathbf{r}] \qquad (1.24)$$

where $\mathbf{l}$ is the *moment of inertia tensor* of the distribution,

$$\mathbf{l}_{\alpha\beta} = \int_V \rho(\mathbf{r}')[(r')^2 \delta_{\alpha\beta} - r'_\alpha r'_\beta] \, d\mathbf{r}', \qquad r_\alpha, r_\beta = x, y, \text{ or } z$$

Thus,

$$\mathbf{l} = \begin{pmatrix} \mathbf{l}_{xx} & -\mathbf{l}_{xy} & -\mathbf{l}_{xz} \\ -\mathbf{l}_{yx} & \mathbf{l}_{yy} & -\mathbf{l}_{yz} \\ -\mathbf{l}_{zx} & -\mathbf{l}_{zy} & \mathbf{l}_{zz} \end{pmatrix} = \mathbf{l}^T$$

and $I_{xx} = \int_V (y^2 + z^2)\rho(\mathbf{r})\, d\mathbf{r}$, $I_{yz} = \int_V yz\rho(\mathbf{r})\, d\mathbf{r}$, and so on. By at most three rotations one can always diagonalize $\mathbf{I}$ (because it is symmetric). The set of axes that does so is called the principal axes.

Once this is done, the five $p = 2$ terms can be written in three equivalent forms:

$$-\frac{G}{2r^3}\left[A + B + C - \left(\frac{3}{r^2}\right)(Ax^2 + By^2 + Cz^2)\right]$$

$$= -\left(\frac{G}{2r^3}\right)(A + B + C - 3I)$$

$$= -\left(\frac{G}{4r^3}\right)[(2C - A - B)(1 - 3\sin^2\theta) - 3(A - B)\cos^2\theta\cos 2\phi]$$

In all three forms $A = I_{xx}$, $B = I_{yy}$, $C = I_{zz}$. In the second form I is the moment of inertia of the distributed system along the line joining its center of mass with $\mathbf{r} = r\mathbf{l}(\phi, \theta)$ (see Eq. 1.3),

$$I = \int_V \left[(r')^2 - \frac{(\mathbf{r}\cdot\mathbf{r}'^2)}{r^2}\right]\rho(\mathbf{r}')\, d\mathbf{r}'$$

Even more special is the case of axial symmetry when $I_{xx} = I_{yy}$ (or $A = B$). The expression in Eq. 1.24 can now be written as

$$-\frac{G(C - A)}{2r^5}(3z^2 - r^2) = -\frac{G(C - A)}{r^3}P_2(\sin\theta)$$

A is called the equatorial moment of inertia and C is called the polar moment of inertia. Through the $p = 2$ terms of Eq. 1.22, in this case,

$$U = -\frac{GM}{r}\left[1 - J_2\frac{P_2(\sin\theta)}{r^2}\right], \qquad J_2 = \frac{A - C}{M} \tag{1.25}$$

We shall return to this approximation to the potential later.

Real Geopotential Models

Equation 1.23 represents the first approximation to the potential of the Earth. The next term, the J_2 term of Eq. 1.25, was poorly known before the launching of artificial satellites. This is because it is a very small quantity, $J_2/R_\oplus^2 = 1.08263 \times 10^{-3}$, with the higher-order coefficients at most equal (in absolute magnitude) to the square of this. With the large number of artificial satellites launched in the last 25 years, and the varied array of observing methods and techniques now available, the coefficients in the geopotential have been increasingly refined. Ten years ago terms with $l > 10$ were poorly known whereas today terms out to $l = 15$ are probably known within 10% (the $l = 14$ and 15 cases, as well as a few others, are especially well known because of resonance effects). Another by-product of this research as been the proliferation of geopotential models.

The Smithsonian Astrophysical Observatory (SAO) was given the job of optical data acquisition and reduction on artificial satellites in the 1950s, so it should not be surprising that as a part of this effort they produced models for the gravitational field of the Earth. *Smithsonian Standard Earth (II)* was published in 1969 (Gaposchkin and Lambeck 1970, 1971). The next in this series came out in 1973 (Gaposchkin 1973, 1974) and the most recent in 1977 (Gaposchkin 1977). Of course NASA creates models too, the most up to date of which are known as *GEM10A* and *GEM10B* (Lerch et al., 1977). *GEM* is an abbreviation for *G*oddard (Space Flight Center) *E*arth *M*odel. The *GEM10* models are based on a total of 840,000 observations (150,000 optical angles-only, 213,000 laser ranging, 270,000 radial velocity, and the rest radar distances and angles). These were used to determine the first 592 coefficients in the geopotential through $l = 22$. In addition to these two series of models the U.S. Navy publishes models through the Naval Research Laboratory (*NWL-10* is a recent one) and the U.S. Department of Defense publishes models. Its latest is called *WGS-72* and relies heavily on the *NWL-9D* and *NWL-10* versions.

All of these models are not only different for the geopotential but differ in their assumed constants ($GM_\oplus$, $R_\oplus$, etc.). Astronomers, not too concerned with the higher-order terms in Eq. 1.22, have their own system. (See the Appendix.)

A Simplified Model. It would be beneficial to the reader if all of this pure mathematics were made more concrete. Consider then the forces exerted on a point that is on the Earth's surface at a latitude θ. The principal ones, in a coordinate system rotating with the Earth's rotation rate $\omega_\oplus$, are the gravitational attraction of the Earth and the Coriolis force (due to the noninertial character of this reference frame). Now the Earth's surface is almost that of an oblate spheroid (e.g., an ellipsoid of revolution with the equatorial radius larger than the polar radius). Hence the equation of its nearly spherical surface can be written as

$$\frac{r}{R_\oplus} = \sum_{n=0}^{\infty} r_{2n} P_{2n}(\cos \theta)$$

where $r_0 \simeq 1$ and the other dimensionless parameters satisfy $|r_{2n}| \ll 1$ for $n \geq 1$. The departure of the Earth from spherical symmetry can be described in terms of the flattening of the ellipsoid of revolution,

$$f = \frac{a - b}{a}$$

where a is the Earth's equatorial radius and b is its polar radius. As $1/f = 298.257$, f is a very good quantity to use in power series expansions.

If the Earth were a fluid, then its surface would be an equipotential under the actions of its self-gravitation and rotational deformation (lunisolar perturbations are being neglected here). In units of $R_\oplus$ this potential may be written as

$$V = -GM_\oplus \left(\frac{R_\oplus}{r}\right)\left[1 - J_2\left(\frac{R_\oplus}{r}\right)^2 P_2 - J_4\left(\frac{R_\oplus}{r}\right)^4 P_4 - \cdots\right]$$
$$+ \frac{\omega_\oplus^2}{3}\left(\frac{r}{R_\oplus}\right)^2 (1 - P_2)$$

where the last term represents the potential of the Coriolis force. Note that there are no azimuth-dependent terms because of the assumed axial symmetry. The odd-numbered Legendre polynomials are absent because of the symmetry about the equatorial plane. If the equation of the Earth's surface is used for r in the expression for V and then the P_0 ($=1$), P_2, P_4, etc., terms isolated, an identity is obtained. In order to close this system, it is necessary that the gradient of V at $\theta = 0$ match the observed equatorial gravitational acceleration g_e. Through terms of order f this yields three equations for $GM_\oplus$, J_2, and J_4 in terms of r_2, the quantity $\omega_\oplus^2 R_\oplus / g_e$ (usually denoted by m), and g_e itself. The result is

$$GM_\oplus = R_\oplus^2 g_e \left(1 + \frac{3m}{2} + \frac{3r_2}{2}\right)$$

$$J_2 = -\frac{m}{3} - r_2$$

$$J_4 = 0$$

Remember that terms of order f^2 and higher have been dropped.

If we go back to the equation of the ellipsoid, then we discover that

$$r_0 = 1 - \frac{f}{3}$$

$$r_2 = -\frac{2f}{3}$$

$$r_4 = 0$$

so that the Earth's surface is approximately given by

$$\frac{r}{a} = 1 - f \sin^2 \theta$$

This type of analysis can be laboriously extended to include higher-order terms and provides a connection between the Earth-bound results of gravity measurements (gravimetry) and celestial mechanics as seen in artificial satellite orbits.

THE SOLUTION

Equation 1.14 represents the basic problem considered in this book. Several mathematical questions pertinent to this set of first-degree, second-order, three-dimensional ordinary differential equations should be addressed. For instance we might ask "Do they have a solution?" "Do they have a unique solution?" "Is the solution continuous?" "How does the solution change as I alter the problem specification?" In this section all of these questions will be answered with mathematical rigor. Ince's (1927) method of presentation will be followed because it is straightforward and to the point. First the problem of

$$\frac{dy}{dx} = f(x, y) \qquad (1.26)$$

will be treated and then the more complicated problems of $\mathbf{F} = m\mathbf{a}$.

Why is this subject here and why is it treated with such rigor? The subject is here because it is not enough to know how to solve a problem. One must also know *if* one can solve it and how many answers might turn up. In addition, one should understand the properties of any possible solutions and how the solution(s) varies (vary) as the parameters of the problem are changed. The discussion to be presented encompasses all of these topics. Moreover, the method of proof used provides a technique for actually solving the problem. The process is known as the method of successive approximations (developed especially by Picard).

Existence, Uniqueness, and Continuity

Let (x_0, y_0) be a pair of values assigned to the real variables (x, y) such that within the domain $D: |x - x_0| < a, |y - y_0| < b, f(x, y)$ is a single-valued continuous function of x and y. Let F be the maximum value of $|f|$ in D and let $h = \min(a, b/F)$. If $h < a$, then redefine D to be $|x - x_0| < h, |y - y_0| < b$. It is also necessary that f satisfy a Lipschitz condition in D. By this is meant that if $(x, y) \in D$ and $(x, Y) \in D$, then

$$|f(x, Y) - f(x, y)| < K|Y - y|$$

for some constant K. This assumption is necessary to prove uniqueness.

Theorem: Under the above conditions there exists a unique, continuous function of x, say $y(x)$, defined $\forall x \ni |x - x_0| < h$ which satisfies the differential Equation 1.26 and is equal to y_0 when $x = x_0$.

The proof is by the method of successive approximations. Define the sequence of functions $y_n(x)$ for $n \geqslant 1$ by

$$y_n(x) = y_0 + \int_{x_0}^{x} f[t, y_{n-1}(t)] \, dt$$

First, show that if $x \in D$, then $y_n(x) \in D$ too. Next, show that as $n \to \infty$, the sequence of functions has a limit and that the limit function is continuous for $x \in D$. The penultimate step shows that the limit function satisfies the differential equation and the initial conditions. Finally, show that it is unique. Many of the proofs rely on the principle of mathematical induction.*

Suppose that $|y_{n-1} - y_0| \leq b$. Then $|f[t, y_{n-1}(t)]| \leq F$ and

$$|y_n(x) - y_0| = \left| \int_{x_0}^x f[t, y_{n-1}(t)] \, dt \right| \leq \int_{x_0}^x |f[t, y_{n-1}(t)]| \, dt$$

$$\leq F(x - x_0) \leq Fh \leq b$$

Clearly $|y_1(x) - y_0| \leq b$ so $|y_n(x) - y_0| \leq b \ \forall n > 0$.

Now assume that for $x \in [x_0, x_0 + h]$ (remember F is the upper bound on $|f|$ and K is the Lipschitz constant)

$$|y_{n-1}(x) - y_{n-2}(x)| < \frac{FK^{n-2}(x - x_0)^{n-1}}{(n-1)!}$$

Then

$$|y_n(x) - y_{n-1}(x)| = \left| \int_{x_0}^x \{f[t, y_{n-1}(t)] - f[t, y_{n-2}(t)]\} \, dt \right|$$

$$\leq \int_{x_0}^x |f[t, y_{n-1}(t)] - f[t, y_{n-2}(t)]| \, dt$$

$$\leq \int_{x_0}^x K|y_{n-1}(t) - y_{n-2}(t)| \, dt$$

the last inequality following from the Lipschitz condition. Employing the above assumption in the integrand,

$$|y_n(x) - y_{n-1}(x)| < \frac{FK^{n-1}|x - x_0|^n}{n!}$$

which is certainly true for $n = 1$. The proof for $x \in [x_0 - h, x_0]$ is done similarly. Hence each term of the series

$$y_n(x) = y_0 + \sum_{m=1}^n [y_m(x) - y_{m-1}(x)]$$

is bounded for $x \in D$ and is a continuous function of x. Thus, this series converges absolutely and uniformly (since $|x - x_0| < h$) and the limit

$$y(x) = \lim_{n \to \infty} y_n(x)$$

exists and is a continuous function of $x \in D$.

---

* Let $A(n)$ denote a proposition (e.g., a formula) associated with the integer n. Suppose that it is possible to show that it is true for some finite value of $n = N$ and further suppose that for all $n > N$ the truth of $A(n+1)$ can be deduced if the truth of $A(n)$ is assured. Then $A(n)$ is true $\forall n \geq N$.

Consider at this point

$$\left| \int_{x_0}^x \{ f[t, y(t)] - f[t, y_n(t)] \} \, dt \right| \le \int_{x_0}^x \left| \{ f[t, y(t)] - f[t, y_n(t)] \} \right| \, dt$$

$$\le \int_{x_0}^x K |y(t) - y_n(t)| \, dt$$

$$\le K \varepsilon_n |x - x_0| < K \varepsilon_n h$$

Again the Lipschitz condition has been used and in the penultimate step the meaning of convergence has been utilized ($\varepsilon_n \to 0$ as $n \to \infty$ and is independent of x because the convergence is uniform). But it follows from this that

$$y(x) = \lim_{n \to \infty} y_n(x) = y_0 + \lim_{n \to \infty} \int_{x_0}^x f[t, y_{n-1}(t)] \, dt$$

$$= y_0 + \int_{x_0}^x \lim_{n \to \infty} f[t, y_{n-1}(t)] \, dt$$

$$= y_0 + \int_{x_0}^x f[t, y(t)] \, dt$$

The last step follows because f is continuous in D. Moreover, because f is continuous in D,

$$\frac{d}{dx} \left\{ y_0 + \int_{x_0}^x f[t, y(t)] \, dt \right\} = \frac{d}{dx} [y(x)]$$

exists and is equal to $f(x, y)$. But it is also equal to dy/dx. Hence the limit function not only exists and is continuous, but it satisfies the differential equation. Furthermore, from the equivalent integral equation above, one can see that $y(x_0) = y_0$. The only element of the theorem left to demonstrate is uniqueness.

Suppose that $Y(x) \ne y(x)$ satisfies $Y(x_0) = y_0$, for $|x - x_0| \le h' \le h |Y(x) - y_0| < b$, and also satisfies the differential equation 1.26. Then $Y(x)$ must satisfy the equivalent integral equation

$$Y(x) = y_0 + \int_{x_0}^x f[t, Y(t)] \, dt$$

Subtract $y_n(x)$. For $n = 1$ the Lipschitz condition is used to show that

$$|Y(x) - y_1(x)| < bK(x - x_0)$$

For $n = 2$ it is again used to show that

$$|Y(x) - y_2(x)| < \tfrac{1}{2} bK^2 (x - x_0)^2$$

and in general that

$$|Y(x) - y_n(x)| < \frac{bK^n (x - x_0)^n}{n!}$$

Therefore,

$$\lim_{n \to \infty} |Y(x) - y_n(x)| = 0$$

so

$$Y(x) = \lim_{n \to \infty} y_n(x) = y(x) \qquad \text{Q.E.D.}$$

The reader should study the statement of the theorem carefully, because much is demonstrated. We also have a procedure for constructing a solution. I shall not dwell on relaxing the various constraints on f except to remark that a Lipschitz condition is a weak restriction. Weaker restrictions yet will suffice; any function f for which $\partial f / \partial y$ exists will satisfy it too.

Variations

There are two quantities of interest that can be varied. The initial condition y_0 can be changed to (say) $y_0 + \delta y_0$ or the derivative of y, f, can be changed to (say) $f(x, y) + \delta f(x, y)$. We will now show that the solution $y(x)$ is a continuous function of the change (or *perturbation*) in either case. First suppose that the functional form of f changes to $f + \delta f$ where $|\delta f(x, y)| < \varepsilon$ for $(x, y) \in D$. Also assume that δf is continuous in D. The corresponding integral equation for the solution of the new differential equation is

$$Y(x) = y_0 + \int_{x_0}^{x} \{f[t, Y(t)] + \delta f[t, Y(t)]\} \, dt$$

where $y(x)$ satisfies the original one:

$$y(x) = y_0 + \int_{x_0}^{x} f[t, y(t)] \, dt$$

Subtract to find that (using Y as a first guess for y this time)

$$|y_1(x) - Y(x)| = \left| \int_{x_0}^{x} \delta f[t, Y(t)] \, dt \right| \leq \varepsilon |x - x_0|$$

It can also be deduced that

$$|y_n(x) - Y(x)| = \left| \int_{x_0}^{x} \{f[t, y_{n-1}(t)] - f[t, Y(t)] - \delta f[t, Y(t)] \, dt \right|$$

$$\leq K \int_{x_0}^{x} |y_{n-1}(t) - Y(t)| |dt| + \varepsilon |x - x_0|$$

Now we use this and the inequality already found for $|y_1 - Y|$ to deduce that

$$|y_2(x) - Y(x)| \leq \varepsilon |x - x_0| + \varepsilon K \int_{x_0}^{x} |t - x_0| |dt|$$

$$= \varepsilon |x - x_0| + \varepsilon K \frac{|x - x_0|^2}{2}$$

$$|y_3(x) - Y(x)| \leq \varepsilon |x - x_0| + \varepsilon K \frac{|x - x_0|^2}{2!} + \varepsilon K^2 \frac{|x - x_0|^3}{3!}$$

so, in general,

$$|y_n(x) - Y(x)| \leq \varepsilon|x - x_0| + \varepsilon|x - x_0| \sum_{m=1}^{n-1} \frac{K^m|x - x_0|^m}{(m+1)!}$$

Taking the limit as $n \to \infty$ and recognizing the power series expansion of the exponential function, we can rewrite this as

$$|y(x) - Y(x)| \leq (\varepsilon/K)\{\exp[K|x - x_0|] - 1\}$$
$$\leq (\varepsilon/K)[\exp(Kh) - 1] \tag{1.27}$$

Clearly as $\varepsilon \to 0$, $Y(x) \to y(x)$ showing that the solution of the altered equation is a continuous function of the alteration.

The result may be similarly demonstrated when a parameter λ is involved. That is, if $f = f(x, y:\lambda)$ where f is single-valued, continuous, and satisfies a Lipschitz condition uniformly in D for $\lambda \in [\Lambda_1, \Lambda_2]$, then the solution $y(x; \lambda)$ is a continuous function of λ and is uniformly differentiable with respect to λ on D. *This result is the mathematical foundation of perturbation theory.*

The last topic to consider is a change in the initial condition from $y(x_0) = y_0$ to $y_0 + \delta y_0$. It is simplest to make this case depend on the preceding one. We have the equivalent integral equation

$$z(x) = y_0 + \delta y_0 + \int_{x_0}^x f[t, z(t)]\, dt$$

Let $Y(x) = z(x) - \delta y_0$ so that the integral equation for Y is

$$Y(x) = y_0 + \int_{x_0}^x f[t, Y(t) + \delta y_0]\, dt$$

From the Lipschitz condition on f,

$$|f[x, Y(x) + \delta y_0] - f[x, Y(x)]| < K|\delta y_0|$$

Let $\varepsilon > |\delta y_0|$. Then

$$f[x, Y(x) + \delta y_0] = f(x, Y) + \delta f(x, Y)$$

where $|\delta f| < K\varepsilon$. Inequality 1.27 is now satisfied if ε is replaced by $K\varepsilon$, thereby proving the desired result.

The Problem of F = ma

The rigorous mathematical results on the solution of Eq. 1.26 need to be applied to the solution of 1.14. To do this the second-order differential equation is changed into twice as many first-order equations by the introduction of new variables, namely the velocity vector (or better the momentum **p**; see Eq. 1.5)

$$\frac{d\mathbf{r}}{dt} = \mathbf{v}, \qquad \frac{d\mathbf{v}}{dt} = \frac{\mathbf{F}}{m}$$

Next I assert that the results proved above are just as valid for a *system* of first-degree ordinary differential equations as they are for a single one. The proof offers no new results or insights. [It can be found in Ince's (1927) text.] The only points of interest are the new forms of the domain D and the Lipschitz condition. For the Nth-order system

$$\frac{d\mathbf{y}}{dx} = \mathbf{f}(\mathbf{y}, x), \qquad \mathbf{y} = (y_1, y_2, \ldots, y_N)$$

with initial conditions $\mathbf{y}(x_0) = \mathbf{y}_0$, one supposes that the vector function $\mathbf{f}$ is a single-valued and continuous function of its $N+1$ arguments for the domain

$$|x - x_0| \le a \qquad |y_n(x) - y_n(x_0)| < b_n, \qquad n = 1, 2, \ldots, N$$

Again if $F \ge |f_n| \ \forall n \in [1, N]$, $h = \min(a, b_n/F)$, then the range of x may be further restricted to $|x - x_0| \le h$. Finally the form of the N Lipschitz conditions is

$$|f_m(\mathbf{Y}, x) - f_m(\mathbf{y}, x)| < \sum_{n=1}^{N} K_n |Y_n - y_n|, \qquad m = 1, 2, \ldots, N$$

PROBLEMS

1. Suppose that an attractive force $F = -k/x^2$ acts on a particle of mass m initially a distance D from $x = 0$. Show that the time to reach $x = 0$ is given by $\pi(m/k)^{1/2}(D/2)^{3/2}$ if the particle is released from rest.

2. Suppose that a particle of mass m is acted on by both gravity (gravitational acceleration g) and air resistance. Consider both linear and quadratic air drag so that the total force acting on the particle is

    $$f = -mg - cv$$

    or

    $$F = -mg - Cv^2$$

    If the particle is dropped from rest at a height h, solve for the motions and show that

    $$x = \frac{m^2 g}{c^2}\left[1 - \frac{ct}{m} - \exp\left(\frac{-ct}{m}\right)\right] + h$$

    or

    $$X = -\frac{m}{C}\ln\left\{\cosh\left[\left(\frac{Cg}{m}\right)^{1/2} t\right]\right\} + h$$

    Further show that in each case there is a terminal speed $[= \lim_{t \to \infty} v(t)]$ given by mg/c or $(mg/C)^{1/2}$, respectively.
    Finally, if the air drag is quadratic, $h = 0$ but the initial speed is v_0 (upward), find the speed with which the particle strikes the ground. You should get $v_0 v_\infty/(v_0^2 + v_\infty^2)^{1/2}$ where v_∞ is the appropriate terminal speed.

3. An artillery piece has a muzzle speed of v_0. If it fires a shell up a mountain whose constant slope is θ, show that the maximum distance the shell can land up the mountain is $(v_0^2/g)/(1+\sin\theta)$.

4. Suppose that the coordinate system $\mathbf{r} = (x, y, z)$ is in rotation with angular velocity $\boldsymbol{\omega}$ with respect to an inertial reference frame. Further suppose that $d\boldsymbol{\omega}/dt \neq 0$ and that there is a translational acceleration $\mathbf{A}$. If the force exerted on a particle of mass m in an inertial reference frame is $\mathbf{F}$, then show that this particle's equation of motion in the noninertial reference frame is

$$m\ddot{\mathbf{r}} = \mathbf{F} - m\mathbf{A} - 2m\boldsymbol{\omega}\times\dot{\mathbf{r}} - m\dot{\boldsymbol{\omega}}\times\mathbf{r} - m\boldsymbol{\omega}\times(\boldsymbol{\omega}\times\mathbf{r})$$

5. The coefficient of restitution is defined as the ratio of the speed of separation to the speed of approach (during a collision). If the collision is perfectly elastic, then $\varepsilon = 1$. A value of $\varepsilon < 1$ implies a frictional loss during an inelastic collision. Suppose that for a ball hitting the pavement $\varepsilon < 1$. If such a ball is dropped from a height H, show that the total (vertical) distance the ball will travel before coming to rest is given by $H(1+\varepsilon^2)/(1-\varepsilon^2)$.

6. Assume that as a raindrop falls through the atmosphere it grows in a spherically symmetric fashion with an accretion rate proportional to its surface area. If it is dropped from rest and stays homogeneous, prove that there is a terminal acceleration if $m(0) = 0$.

7. A man of mass M (weight $= Mg$) is climbing up a ladder of length L, mass m. The ladder rests against a rough wall at an angle θ. If the coefficients of friction between the ladder and the wall and the ladder and the floor are the same ($= \mu$), calculate the maximum height of the man off the floor.

8. A rigid body's equations of motion in the absence of external torques is given by

$$\dot{\mathbf{L}} + \boldsymbol{\omega}\times\mathbf{L} = 0$$

where $\mathbf{L} = (I_x\omega_x, I_y\omega_y, I_z\omega_z)$ and I_x, I_y, I_z are the principal moments of inertia. Show that if these are unequal, then motion about the axis of intermediate moment of inertia is unstable whereas motion about an axis of minimum or maximum moment of inertia is stable.

9. A rocket is fired vertically upward with a mass loss of $\dot{m}$ due to exhaust. The exhaust leaves the rocket with a relative speed v'. Neglecting the variation of gravity with height, air resistance, winds, the Earth's rotation, and so on, show that the rocket's equation of motion is

$$m\frac{dv}{dt} = -v'\frac{dm}{dt} - mg$$

If $\dot{m}$ is a constant, solve for $v(m)$. Show that, after exhausting a fraction f of its initial mass m_0 its maximum height is larger for larger values of $-\dot{m}$ (≥ 0).

10. Show that the debris left over (in the forms of gas and dust, e.g., particles) from the formation of a spherical star must eventually coalesce into a ring after losing an amount of energy $\Delta\mathscr{E}$ due to internal friction (e.g., collisions). Use conservation arguments, assume that the mass of the debris is small compared to that of the star (e.g., ignore its self-gravity), and calculate the radius of the ring for a given initial mass, energy, and angular momentum of the debris.

11. A wheel of mass m and radius of gyration r spins on a fixed horizontal axle of radius a. If there is friction between the bearings (coefficient of friction $= \mu$) show that the wheel will spin through $r^2\Omega^2(1 + \mu^2)^{1/2}/(4\pi g\mu a)$ turns before coming to a stop. Ω is the initial angular speed.

12. Three particles, all of mass m, are constrained to move on a straight line. Their initial locations and velocities are $0, L, 2L$ and $2V, V$, and $V/2$. Assuming that the coefficient of restitution between them is unity, compute their final velocities.

13. Prove that the gravitational force anywhere inside a spherically symmetric distribution of matter vanishes.

14. Imagine a tunnel bored through a homogeneous Earth. Suppose a particle is dropped from the surface into the hole. Find its motion.

15. Prove the Parallel Axis Theorem: The moment of inertia of a body about any given axis is equal to the moment of inertia about a parallel axis through its center of mass plus Md^2 where M is the mass of the body and d is the separation between the axes.

16. Prove the Perpendicular Axis Theorem: The sum of the moments of inertia of any plane lamina about two mutually perpendicular axes in the plane of the lamina is equal to the moment of inertia about an axis through their point of intersection and normal to the lamina's plane.

17. Show that the moment of inertia of a solid uniform ellipsoid of mass m is given by $m(a^2 + b^2)/5$ where a and b are the semimajor axes perpendicular to the axis (which is along the third axis of symmetry and through the ellipsoid's center).

18. Use the preceding to show that the moment of inertia of a sphere about any axis through its center is $2MR^2/5$. Show further that the moment of inertia about the straight edge of an octant of a sphere of radius R and mass M is also $2MR^2/5$.

19. A ring of infinitesimal thickness has a total mass m and radius a. Show that the gravitational potential at any point in the plane of the ring is given by

$$U = -2\frac{GM}{\pi}\int_0^{\pi/2}(r^2 - a^2\sin^2 x)^{-1/2}\,dx = -2\frac{GM}{\pi r}K\left(\frac{a}{r}\right)$$

where r is the distance from the point in question to the origin and K is the complete elliptic integral of the first kind.

20. Show that if $a = 1$ but the point in question is on the axis of symmetry of the ring, for the situation in Problem 19,

$$U = -\frac{GM}{(1+r^2)^{1/2}}$$

Further show that for a point a distance r away from the ring at a latitude θ

$$U = -\frac{GM}{r}\sum_{n=0}^{\infty}\frac{(-1)^n(2n-1)!!}{(2n)!!}\frac{P_{2n}(\sin\theta)}{r^{2n}}$$

where $(2n)!! = (2n)(2n-2)(2n-4)\cdots 2$, and so on.

21. Show that the potential due to the ellipsoid

$$\frac{x^2}{a^2}+\frac{y^2}{b^2}+\frac{z^2}{c^2}=1$$

is given by

$$U = -G\pi abc\int_0^{\infty}[(a^2+s)(b^2+s)(c^2+s)]^{-1/2}S(s)\,ds$$

where

$$S(s) = \int_{f(s)}^{\infty}\rho(t)\,dt$$

Herein the mass density ρ is a function of $t = x^2/a^2 + y^2/b^2 + z^2/c^2$ and f is defined as

$$f(s) = \frac{x^2}{a^2+s}+\frac{y^2}{b^2+s}+\frac{z^2}{c^2+s}$$

If the ellipsoid is homogeneous ($\rho = \rho_0$), deduce that

$$U = -G\rho_0\pi abc\int_{f\leq1}^{\infty}[(a^2+s)(b^2+s)(c^2+s)]^{-1/2}[1-f(s)]\,ds$$

22. In terms of the distance r, longitude ϕ, and the colatitude θ show that

$$x = r \sin \theta \cos \phi$$

$$y = r \sin \theta \sin \phi$$

$$z = r \cos \theta$$

and that $\mathbf{v} = (\dot{x}, \dot{y}, \dot{z}) = (\dot{r}, r\dot{\theta}, r\dot{\phi} \sin \theta)$ as well as

$$\mathbf{a} = \dot{\mathbf{v}} = [\ddot{r} - r\dot{\phi}^2 \sin^2 \theta - r\dot{\theta}^2, \; r\ddot{\theta} + 2\dot{r}\dot{\theta} - r\dot{\phi}^2 \sin \theta \cos \theta,$$

$$(r\ddot{\phi} + 2\dot{r}\dot{\phi}) \sin \theta + 2r\dot{\phi}\dot{\theta} \cos \theta]$$

chapter two

The Two-Body Problem

FORMULATION

The quintessential problem of celestial mechanics is the two-body problem. The two-body (or Kepler) problem is to solve for the motions of two particles interacting through their mutual self-gravitation. As demonstrated in Chapter One (Eq. 1.14), the problem can be split into two mutually exclusive parts—the motion of the center of mass and the relative motion. The location of the center of mass $\mathbf{R}$ follows a trajectory determined by

$$\ddot{\mathbf{R}} = 0$$

whereas the relative location $\mathbf{r} = \mathbf{r}_1 - \mathbf{r}_2$ has its time development governed by

$$\ddot{\mathbf{r}} = -\frac{G(m_1 + m_2)\mathbf{r}}{|\mathbf{r}|^3} = -\frac{GM\mathbf{r}}{|\mathbf{r}|^3} = -\nabla U(\mathbf{r}) \qquad (2.1)$$

The solution of the former is trivial and the ramifications of the latter shall occupy our attention in the remainder of this chapter.

Before delving into the multiple facets of the two-body problem, I shall very briefly outline this Chapter. Immediately below the complete analytical solution to Eq. 2.1 is presented, illustrating both the time development and the form of the orbit. The next part of this chapter seeks to explain more general aspects of central force motion and expands on other than elliptic motion. The last section of the chapter addresses expansions, especially with regard to initial orbit determination and the solution of Kepler's equation.

Reduction

We should first notice that the force is *central*, that is, U depends on $|\mathbf{r}|$ alone. This fact implies that $\mathbf{L} = \mathbf{r} \times \dot{\mathbf{r}}$ is a constant of the motion. From the properties of the vector cross product $\mathbf{r} \cdot \mathbf{L} = 0 = \dot{\mathbf{r}} \cdot \mathbf{L}$ so that both $\mathbf{r}$ and $\dot{\mathbf{r}}$ lie in a plane (the *invariable plane*) perpendicular to $\mathbf{L}$. The constancy of $\mathbf{L}$ means that the problem is essentially two-dimensional, not three-dimensional. It requires two independent angles to fix the direction cosines of **L**. These will be defined later. For now assume that we are in the orbital plane and that the plane polar coordinates r, ϕ will be used. Equation 2.1 now takes the form (cf. Problem 1.22)

$$\ddot{r} - r\dot{\phi}^2 = -\frac{\partial U}{\partial r} = -\frac{GM}{r^2}$$

$$r\ddot{\phi} + 2\dot{r}\dot{\phi} = -\frac{1}{r}\frac{\partial U}{\partial \phi} = 0 \qquad (2.2)$$

The lower equation in 2.2 is immediately integrable to $d(r^2\dot{\phi})/dt = 0$. However, the norm of $\mathbf{L} = \mathbf{r} \times \dot{\mathbf{r}}$ is $r^2\dot{\phi}$; thus

$$L = r^2\dot{\phi}$$

and $\dot{\phi}$ can now be eliminated from the radial equation of motion,

$$\ddot{r} - \frac{L^2}{r^3} = -\frac{dU(r)}{dr}$$

This is simply integrable to

$$\frac{1}{2}\left(\dot{r}^2 + \frac{L^2}{r^2}\right) + U = \text{const}$$

Noticing that the sum of the first two terms is $|\dot{\mathbf{r}}|^2/2$, the entire expression is recognized as (cf. Eq. 1.12) the relative energy per unit mass. Call this quantity $\mathscr{E}$. Progress (because of the central and conservative nature of the force) so far can be summarized in

$$\dot{r}^2 = 2(\mathscr{E} - U) - \frac{L^2}{r^2} \qquad (2.3a)$$

$$r^2\dot{\phi} = L \qquad (2.3b)$$

The next steps are to obtain the orbit $r(\phi)$, then the time dependence of r and ϕ, and then recover the full three-dimensional aspects of the motion.

The Orbit

First the constants of motion L and $\mathscr{E}$ are parameterized in a different fashion. The quantities a and e are introduced via

$$\mathscr{E} = -\frac{GM}{2a}, \qquad L^2 = GMa(1 - e^2) \qquad (2.4)$$

and Eq. 2.3a is rewritten as

$$\dot{r}^2 = -\frac{GM}{ar^2}[r^2 - 2ar + a^2(1 - e^2)] = \frac{GM(r_+ - r)(r - r_-)}{ar^2} \qquad (2.5)$$

with $r_\pm = a(1 \pm e)$. Because $\dot{r}^2 \geqslant 0$, r must lie in $[r_-, r_+]$. The points where $\dot{r}$ vanishes $(r_\pm)$ are known as the *turning points* of the motion. In celestial mechanics these points have special names. The point of closest approach (r_-) is known as the *periapse* (perigee for Earth orbits, perihelion for solar orbits, peri-Jove for orbits about Jupiter, etc.). The point of maximum distance is known as *apoapse* (apogee, aphelion, apo-Jove, etc.). The line joining the apses is called the *line of apsides*. The time interval associated with two successive passages through periapse is termed the *anomalistic period*.

Back to applied mathematics. Use Eq. 2.3b to turn Eq. 2.5 from an equation for dr/dt into one for $r' = dr/d\phi$; $r^2 d\phi = L\, dt$ so $\dot{r} = Lr'/r^2$; namely,

$$(r')^2 = \frac{r^2(r_+ - r)(r - r_-)}{a^2(1 - e^2)}$$

The orientation of the r, ϕ coordinate system was arbitrary. Suppose that the line of apsides makes an angle ω (the *argument of periapse*) with respect to the $\phi = 0$ line. Set v (the *true anomaly*) equal to $\phi - \omega$; see Fig. 2. Then $r'(v = 0) = 0$, $r(v = 0) = r_-$. As v increases, r must, so $r' > 0$. This continues until v reaches (say) v_+ when $r' = 0$ again and $r = r_+$. As v increases further, r must decrease $(r' < 0)$ until r reaches r_- at some value of $v = v_-$. Thereafter this sequence repeats because r is a periodic function of v given implicitly by

$$\int_{r_-}^{r} \frac{ds}{s[(r_+ - s)(s - r_-)]^{1/2}} = \int_0^v \frac{dv}{(ap)^{1/2}}, \qquad v \in [0, v_+], \ r(v_+) = r_+$$

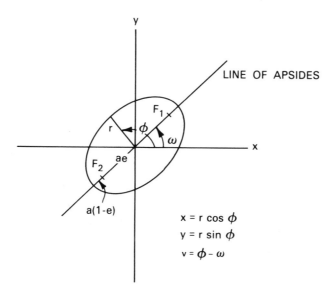

$$x = r \cos \phi$$
$$y = r \sin \phi$$
$$v = \phi - \omega$$

Figure 2. Definition of the semi-major axis (a), eccentricity (e), and argument of periapse (ω). The center of the ellipse is at the origin and its foci are at F_1 and F_2.

$$\int_{r_+}^{r} \frac{ds}{s[(r_+ - s)(s - r_-)]^{1/2}} = -\int_{v_+}^{v} \frac{dv}{(ap)^{1/2}}, \quad v \in [v_+, v_-], r(v_-) = r_-$$

After performing the integrations, we discover that $v_+ = \pi$, $v_- = 2\pi$, and for *both* branches of the orbit

$$r = \frac{a(1 - e^2)}{1 + e \cos v} \tag{2.6}$$

Equation 2.6 is the equation of a conic with one focus at the origin (see Fig. 3). The *semi-major axis* is a, the *eccentricity* is e, and $p = a(1 - e^2)$ is half the *semi-latus rectum* distance. When L is a maximum for a given energy $\mathcal{E}$ ($e = 0$), the orbit is a circle. For larger eccentricities and bound orbits ($0 < e < 1$, $\mathcal{E} < 0$) the orbit is an ellipse. As the energy tends toward zero, the orbit becomes parabolic ($a = \infty$, $e = 1$ such that p is finite). Finally, for unbound orbits ($\mathcal{E} > 0$) $a < 0$ with $e > 1$ or hyperbolic orbits.

Remembering that r is the relative distance between the two particles, applying this to the solar system, and assuming that the Sun is much more massive than is any of the planets, I've just derived Kepler's first law. If we also remind ourselves of the formula for the element of area in polar coordinates, then we see that L is twice the rate at which the radius vector sweeps

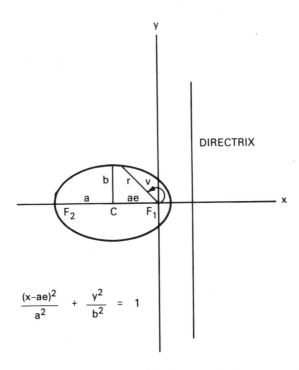

Figure 3. Geometry of an ellipse center C, foci F_1, F_2, semi-major axis a, and eccentricity e.

out area. But L is a constant, so equal areas are swept out in equal times. This statement is Kepler's second law. Kepler's third (and last law) will be derived below. Kepler deduced these laws by analyzing Tyco Brahe's naked-eye observations of Mars. He published the first two in 1609, the third in 1632—the year Napier invented logarithms. Imagine how much arithmetic Kepler would have saved if . . . .

The Time Dependence

Return to Eq. 2.5 and suppose that the origin of time is such that if $t = 0$, then $\dot{r} = 0$ and $r = r_-$. As reasoned above, as time marches on, $\dot{r}$ will be positive until $t = t_+$ when $\dot{r} = 0$ again and $r = r_+$. At still later moments $\dot{r}$ will be negative until $t = t_-$ when $r = r_-$. From this moment on the oscillation of r repeats with period t_-. So from Eq. 2.5 the (implicit) time dependence of r is given by (taking $0 \le e < 1$, $a > 0$ and concentrating on elliptic motion until further notice)

$$\int_{r_-}^r \frac{s\, ds}{[(r_+ - s)(s - r_-)]^{1/2}} = na \int_0^t d\tau, \qquad t \in [0, t_+], \, r(t_+) = r_+$$

$$\int_{r_+}^r \frac{s\, ds}{[(r_+ - s)(s - r_-)]^{1/2}} = -na \int_{t_+}^t d\tau, \qquad t \in [t_+, t_-], \, r(t_-) = r_-$$

The quantity $n[= (GM/a^3)^{1/2}]$ is called the *mean motion*. Performing the integrations,

$$\cos^{-1}\left(\frac{a - r}{ae}\right) - e\left[1 - \left(\frac{a - r}{ae}\right)^2\right]^{1/2} = nt, \qquad t \in [0, t_+]$$

$$\pi - \cos^{-1}\left(\frac{a - r}{ae}\right) + e\left[1 - \left(\frac{a - r}{ae}\right)^2\right]^{1/2} = n(t - t_+), \qquad t \in [t_+, t_-]$$

From these it is deduced that $nt_+ = \pi$, $nt_- = 2\pi$ so that the *period* of the motion P is given by

$$P = \frac{2\pi}{n} = 2\pi\left(\frac{a^3}{GM}\right)^{1/2}$$

In the solar system $M \simeq M_\odot$ and the fact that P^2 is proportional to a^3 is known as Kepler's third law.

The functional form of the r term and the presence of the inverse cosine suggests that the introduction of the auxiliary angle E by $\cos E = (a - r)/ae$ or

$$r = a(1 - e \cos E) \qquad (2.7)$$

would be useful. E is called the *eccentric anomaly*. If, upon making this substitution an arbitrary time scale is reintroduced wherein $\dot{r} = 0$, $r = r_-$ at $t = T$, the *time of periapse passage*, then the $r(t)$ relationship takes the implicit form

$$E - e \sin E = n(t - T) = M \qquad (2.8)$$

This is known as *Kepler's equation* and M is called the *mean anomaly.** The solution for $E(M)$ is

$$E = M + \sum_{k=1}^{\infty} \frac{2}{k} J_k(ke) \sin kM \tag{2.9}$$

where J_k is a Bessel function (see below). Bessel invented Bessel functions to solve Kepler's equation (but he didn't call them Bessel functions of course).

The last task here is to obtain the time dependence of the true anomaly $v(=\phi-\omega)$. If Eq. 2.6 is equated to Eq. 2.7, then the problem is (implicitly) solved:

$$
\begin{aligned}
\tan\left(\frac{v}{2}\right) &= \left(\frac{1+e}{1-e}\right)^{1/2} \tan\left(\frac{E}{2}\right) \\[2mm]
\sin v &= \frac{(1-e^2)^{1/2}\sin E}{1-e\cos E}, \qquad \cos v = \frac{\cos E - e}{1-e\cos E} \\[2mm]
\sin E &= \frac{(1-e^2)^{1/2}\sin v}{1+e\cos v}, \qquad \cos E = \frac{\cos v + e}{1+e\cos v} \\[2mm]
1\pm\cos v &= \frac{(1\mp e)(1\pm\cos E)}{1-e\cos E}, \qquad 1\pm\cos E = \frac{(1\pm e)(1\pm\cos v)}{1+e\cos v}
\end{aligned}
\tag{2.10}
$$

At periapse all of the anomalies vanish. As motion away from periapse occurs they all increase, but in such a fashion that $v \geq E \geq M$ until the other apse is reached when they are all equal to π. As the motion continues back to periapse, the inequalities are all reversed until each anomaly is equal to 2π again. Also if E and M correspond to v, then $2\pi - E$ and $2\pi - M$ correspond to $2\pi - v$. The quantity $v - M$ is known as the *equation of the center*. Finally $v/2$ and $E/2$ are always in the same quadrant. Another approach is to use $r(v)$ from Eq. 2.6 directly in Eq. 2.3b to obtain

$$n(t-T) = 2\tan^{-1}\left[\left(\frac{1-e}{1+e}\right)^{1/2}\tan\left(\frac{v}{2}\right)\right] - \frac{e(1-e^2)^{1/2}\sin v}{1+e\cos v}$$

Example 2.1. A Russian Molniya satellite passed perigee at $66^d15^h51^m51\overset{s}{.}809$ in 1981. Its eccentricity is $e = 0.7248470$ and its semimajor axis is $a = 26{,}561.955$ km. Compute n, M, E, v, and r at midnight on day 67 of 1981. For Earth (neglecting the mass of the artificial satellite), $GM = 3.986005 \times 10^5$ km^3/sec^2.

Since $n^2 = GM/a^3$, $n = 12.600640$ rad/day. The time interval $t-T$ is $08^h08^m08\overset{s}{.}191$ so $M = n(t-T) = 244\overset{\circ}{.}73385$. In general, the best way to solve Kepler's equation is to use Newton's false-root method. Here it takes the form

$$E_n = E_0 - \frac{(E_0 - e\sin E_0 - M)}{1 - e\cos E_0}$$

---

* I assume that you can keep the sum of the masses, which only appears herein multiplied by G, separate from the mean anomaly and shall continue to use the same symbol for both.

where E_0 is the old guess for E and E_n is the new one. If $|E_n - E_0|$ is too large, then one relabels E_n as E_0 and continues the iteration. The original value of E_0 can be M (see below for a better initial E_0). For this problem $E_0 = 244°73385$, $216°05036$, $218°72548$, $218°74286$, and $218°74286$, indicating convergence. Now from Eq. 2.10 $v = 195°98736$ so $r = 41578.848$ km as may be deduced from Eq. 2.6 or 2.7.

The Orbit in Space

The spherical symmetry of the potential implied that $\mathbf{L} = \mathbf{r} \times \dot{\mathbf{r}}$ would be a constant of the motion. Spherical symmetry also implies an invariance to rotations about the origin or two degrees of freedom, two in the orientation of the line of apsides (ω) and the time of periapse passage (T), and two in the size (a) and shape (e) of the conic section characterizing the orbit. These six parameters have replaced the initial conditions $\mathbf{r}(t_0)$, $\dot{\mathbf{r}}(t_0)$. This interrelationship is more fully discussed below and in Chapter Eight. What is needed now is to fix the last two orbital elements; the *longitude of the ascending node* Ω and the *inclination i*.

Before using the conservation of angular momentum to reduce an inherently three-dimensional problem to a two-dimensional one, one must imagine that we are discussing the two-body problem in some three-dimensional coordinate system that had a fundamental reference plane. The plane of the orbit is tilted relative to this plane by the inclination $i \in [0°, 180°]$. When i is in the first quadrant, one says that the motion is direct or prograde. When i is in the second quadrant, the motion is described as being retrograde. Orbits for which $i = 90°$ are called polar. Many low-altitude ($P \simeq 90$–180 min) artificial satellites are in nearly polar orbits (see Fig. 4). The angular momentum vector is given

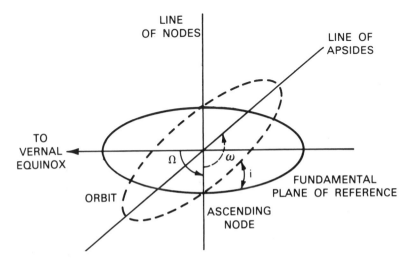

Figure 4. Euler angles argument of periapse (ω), longitude of the ascending node (Ω), and inclination (i) defining the orientation of an orbit in space.

by

$$L = L \begin{pmatrix} \sin i \sin \Omega \\ -\sin i \cos \Omega \\ \cos i \end{pmatrix} \tag{2.11}$$

Unless $i = 0°$ or $180°$, the orbital plane will cut the fundamental reference plane in two points or *nodes*. At one node the latitude is increasing, at the other it is decreasing. These points are known as the *ascending node* and the *descending node*. The line connecting the nodes is called the *line of nodes*. The specification of the orientation of the line of nodes in the fundamental reference plane is done via the *longitude of the ascending node* Ω. The time interval between successive passages through the ascending node is called the *draconitic period* (P is the *sidereal period* or just the period). Some other well-worn combinations of angles are the longitude of periapse $\tilde{\omega} = \omega + \Omega$, the mean longitude $M + \omega + \Omega$, the true longitude or the true orbital longitude $v + \omega + \Omega$, the longitude at epoch $\omega + \Omega - nT$, and the *argument of latitude*

$$u = v + \omega$$

The six quantities a, e, ω, i, Ω, and T are known as an *orbital element set*. The transformation from the orbital plane coordinate system, aligned with the line of apsides, back to the full three-dimensional space is accomplished via the rotation matrix S,

$$S = R_3(-\Omega) R_1(-i) R_3(-\omega) \tag{2.12a}$$

Thus, if

$$\mathbf{q} = r \begin{pmatrix} \cos v \\ \sin v \\ 0 \end{pmatrix} = a \begin{pmatrix} \cos E - e \\ (1 - e^2)^{1/2} \sin E \\ 0 \end{pmatrix}$$

so that

$$\dot{\mathbf{q}} = \frac{na}{(1 - e^2)^{1/2}} \begin{pmatrix} -\sin v \\ e + \cos v \\ 0 \end{pmatrix} = \frac{na^2}{r} \begin{pmatrix} -\sin E \\ (1 - e^2)^{1/2} \cos E \\ 0 \end{pmatrix}$$

then

$$\mathbf{r} = S\mathbf{q}, \qquad \dot{\mathbf{r}} = S\dot{\mathbf{q}} \tag{2.13a}$$

and

$$S = \begin{pmatrix} \cos \Omega \cos \omega - \sin \Omega \sin \omega \cos i \\ \sin \Omega \cos \omega + \cos \Omega \sin \omega \cos i \\ \sin \omega \sin i \end{pmatrix}$$

$$\begin{matrix} -\cos \Omega \sin \omega - \sin \Omega \cos \omega \cos i & \sin \Omega \sin i \\ -\sin \Omega \sin \omega + \cos \Omega \cos \omega \cos i & -\cos \Omega \sin i \\ \cos \omega \sin i & \cos i \end{matrix} \tag{2.12b}$$

Rotation Matrices. The three basic rotations about the x, y, or z axes by a total amount of α, β, or γ (all angles are in radians unless explicitly indicated otherwise) are equivalent to multiplication by the matrices

$$R_1(\alpha) \equiv \begin{pmatrix} 1 & 0 & 0 \\ 0 & \cos\alpha & \sin\alpha \\ 0 & -\sin\alpha & \cos\alpha \end{pmatrix}$$

$$R_2(\beta) \equiv \begin{pmatrix} \cos\beta & 0 & -\sin\beta \\ 0 & 1 & 0 \\ \sin\beta & 0 & \cos\beta \end{pmatrix}$$

$$R_3(\gamma) \equiv \begin{pmatrix} \cos\gamma & \sin\gamma & 0 \\ -\sin\gamma & \cos\gamma & 0 \\ 0 & 0 & 1 \end{pmatrix}$$

Note that these matrices are unitary ($M^{-1} = M^T$ for any one of them). Note too that we have been a bit cavalier about writing a vector as a column vector

$$\mathbf{r} = \begin{pmatrix} x \\ y \\ z \end{pmatrix}$$

and then using the same symbol for the corresponding row vector, $\mathbf{r} = (x, y, z)$. The transpose notation, $\mathbf{r}^T$, should be used for one of them. Neither the absence or presence of this notation can cause a problem, and I shall continue to be a bit loose about it.

Example 2.2. Suppose that the remaining orbital elements for the satellite in Example 2.1 are $\omega = 274°7383$, $i = 64°0265$, and $\Omega = 271°1076$. Compute, at midnight on day 67 of 1981, the rectangular and spherical geocentric location of the satellite.

From the results already obtained $\mathbf{q}$ is calculated in kilometers:

$$\mathbf{q} = \begin{pmatrix} -39970.681 \\ -11451.866 \\ 0 \end{pmatrix}$$

For the given angles

$$S = \begin{pmatrix} -0.4347803 & 0.0554344 & -0.8988287 \\ -0.0910260 & -0.9956969 & -0.0173777 \\ -0.8959243 & 0.0742614 & 0.4379554 \end{pmatrix}$$

so

$$\mathbf{r} = \begin{pmatrix} 16743.638 \\ 15040.961 \\ 34960.272 \end{pmatrix}$$

If **q** is multiplied by S analytically,

$$\mathbf{r} = S\mathbf{q} = r \begin{pmatrix} \cos \Omega \sin u - \sin \Omega \cos i \sin u \\ \sin \Omega \cos u + \cos \Omega \cos i \sin u \\ \sin i \sin u \end{pmatrix} \qquad (2.13b)$$

In this case $u = 110°72566$, so

$$\mathbf{r} = 41578.848 \begin{pmatrix} 0.4026961 \\ 0.3617455 \\ 0.8408187 \end{pmatrix} = \begin{pmatrix} 16743.638 \\ 15040.961 \\ 34960.272 \end{pmatrix}$$

For the spherical geocentric coordinates r, θ, ϕ it is best to equate the forms in Eqs. 1.2 and 2.13b for **r** and play with them; namely,

$$\phi = \Omega + \tan^{-1}(\cos i \tan u), \qquad \theta = \sin^{-1}(\sin i \sin u) \qquad (2.14)$$

Here, $\phi = 221°93364$ and $\theta = 57°22667$.

Example 2.3. Most asteroids are in nearly circular, 4.7 year period orbits about the Sun with very low inclinations (here the fundamental plane is called the ecliptic, see Chapter Three). An exception is the Earth-crossing minor planet 1865 Cerberus whose orbital element set is

$$a = 1.0800957 \text{ A.U.} \qquad i = 16°09211$$

$$e = 0.4669872 \qquad \Omega = 212°43247$$

$$\omega = 325°05162 \qquad T = 10^h 30^m 38.6 \quad \text{on March 28, 1982}$$

It will be brightest near noon, August 31, 1982. Compute its location then.

An A.U. is one astronomical unit, almost the average Earth–Sun distance. Units are discussed at the end of this chapter. For now accept that $n^2 a^3 = 0.9714224 \text{ (A.U.)}^3 (\text{deg/day})^2$. Whence, $n = 0°87803272/\text{day}$. $M = n(t - T)$ and $t - T = 156^d 062053$ or $M = 137°02759$. Using this as the first estimate for E_0, $E_n = 150°61987$, $150°28771$, $150°28756$, and $150°28756$. Next $v = 161°82997$ and $r = 1.5181128$ A.U. The argument of latitude $u = v + \omega = 126°88910$ so that the longitude is $160°41349$ and the latitude is $12°80752$.

The Orbital Elements and the Initial Conditions

Equations 2.6 or 2.7, 2.8, and 2.13 allow one to predict the location and velocity at any time from the orbital element set. One needs to be able to solve the inverse problem—given **r** and **ṙ** at some time $t = t_0$ to compute the orbital element set. To this end the conservation equations 2.3 and 2.11 are used to compute a, e, i, and Ω as follows: Equation 2.3a can be written as

$$\frac{1}{a} = \frac{2}{|\mathbf{r}|} - \frac{|\dot{\mathbf{r}}|^2}{GM}$$

Combining Eqs. 2.3b, 2.4, and 2.11 with $\mathbf{L} = \mathbf{r} \times \dot{\mathbf{r}}$ results in

$$\mathbf{L} = L \begin{pmatrix} \sin i \sin \Omega \\ -\sin i \cos \Omega \\ \cos i \end{pmatrix} = \begin{pmatrix} y\dot{z} - \dot{y}z \\ \dot{x}z - x\dot{z} \\ x\dot{y} - \dot{x}y \end{pmatrix}, \qquad L^2 = GMa(1 - e^2)$$

This fixes e, i, and Ω. For small values of e this will be inaccurate. If Eqs. 2.3a, 2.4, and 2.7 and the time derivative of Eq. 2.7 are manipulated,

$$e \cos E = 1 - \frac{|\mathbf{r}|}{a}$$

$$e \sin E = \frac{\mathbf{r} \cdot \dot{\mathbf{r}}}{na^2}, \qquad n^2 a^3 = GM$$

From this pair both the eccentricity and the eccentric anomaly can be accurately determined. Alternatively one can compute E from Kepler's equation now that a is known. Note that Kepler's equation takes the form

$$E - e \sin E = M = n(t - t_0) + M_0$$

as the moment corresponding to $t = t_0$ is not necessarily the instant of periapse passage $(T = t_0 - M_0/n)$. Lastly one needs to determine ω. This is indirectly computed from the true anomaly (obtained from E) and the argument of latitude $u(= v + \omega)$. The argument of latitude comes from (see Eq. 2.13b)

$$r \cos u = x \cos \Omega + y \sin \Omega$$

$$r \sin u = (-x \sin \Omega + y \cos \Omega) \cos i + z \sin i$$

These are intermediary forms in the derivation of Eq. 2.14 from Eqs. 2.12.

Finally, there are really seven orbital elements and not six. The seventh is the mass of the body that appears in $M = $ sum of the masses (cf. Eq. 1.14). In solar system problems we are so used to the primary (the Sun or the appropriate planet) dominating that we may tend to get sloppy and forget that $M \neq M_\odot$ or $M_\oplus$. Hence, as Gauss (1809) points out nicely (Section 48), one truly needs seven orbital elements.

MORE GENERAL ASPECTS

Solvability

The preceding sections introduced the reader to those practical aspects of the two-body problem that one needs in artificial satellite or solar system work. There are other facets of the problem, both formal and practical, which a complete discussion should contain. One of these is the more general question of the solvability of an equation of the form

$$\ddot{\mathbf{r}} = -\nabla U(\mathbf{r})$$

The conservation of the angular momentum $\mathbf{L} = \mathbf{r} \times \dot{\mathbf{r}}$ is due to the central nature (e.g., $|\mathbf{r}|$ dependence) of the force (i.e., it is an expression of the spherical symmetry of the force). We took advantage of this fact to reduce the dimensionality of the problem by one as well as to concentrate on the orbit $r(\phi)$ rather than on the time dependence $r(t)$. The questions to address here are "For what potentials U are Eqs. 2.3 solvable?" and "For what potentials U are Eqs. 2.3 solvable explicitly?"

The answer to the first question is "All continuous potentials" and this is easy to see. Equation 2.3a can be integrated in the form

$$\int_{t_0}^{t} d\tau = \pm \int_{r_0}^{r} \left\{ 2[\mathscr{E} - U(s)] - \frac{L^2}{s^2} \right\}^{-1/2} ds \qquad (2.15)$$

This gives $t(r)$ explicitly and $r(t)$ implicitly. Next Eq. 2.3b is integrated with $r(t)$ known,

$$\int_{\phi_0}^{\phi} d\psi = L \int_{t_0}^{t} \frac{d\tau}{r^2(\tau)}$$

Hence as long as the integral in Eq. 2.15 exists (which it certainly does when U is a continuous function of $r = |\mathbf{r}|$), the problem is solvable in the form of two quadratures.

More interesting are the instances when the integral in Eq. 2.15 can be explicitly evaluated in terms of circular or elliptic functions. The simple case of $U(s) = As^{m+1}$ where A and m are constants has long been dealt with (e.g., Whittaker 1937). The result is that the problem is solvable by circular or elliptic functions when m takes on one of the following 14 values:

$$m = -7, -5, -4, -3, -2, 0, 1, 3, 5 \quad \text{or} \quad -\tfrac{7}{3}, -\tfrac{5}{3}, -\tfrac{1}{3}, -\tfrac{5}{2}, -\tfrac{3}{2}$$

This result has been extended by Broucke (1980). He discusses potentials of the form $A/r^m + B/r^n$, for example. He also discusses the possibility of circular orbits under such potentials and the stability of such orbits.

Stability

The stability of a circular orbit can be examined without the mathematical superstructure one invokes for a generalized linear stability analysis in mathematical physics. This edifice will be utilized when the stability of the equilibrium solutions of the restricted three-body problem is discussed (see Chapter Six).

Let us go back to the derivative of Eq. 2.3a, which can be written in the form

$$\ddot{r} - \frac{L^2}{r^3} = f(r)$$

If a circular orbit of radius a is a solution, then $\dot{r} = \ddot{r} = 0$ and a satisfies

$$\frac{-L^2}{a^3} = f(a) \qquad (2.16)$$

Suppose that the motion is slightly perturbed and $R = r - a$ is used as a new variable. Assume that $|R|$ is small (in some sense). The equation for d^2R/dt^2 is

$$\frac{d^2R}{dt^2} - \frac{L^2}{(R+a)^3} = f(R+a)$$

Now decide that "$|R|$ is small" means that, using Taylor's theorem, the two terms that contain R in a nonlinear fashion can be replaced by their linear approximations. Thus, the equation of motion for R is simplified to

$$\ddot{R} = \left[\frac{3f(a)}{a} + f'(a)\right] R$$

where Eq. 2.16 has been used twice. The solutions for R are proportional to $\exp(\pm \lambda t)$ with

$$\lambda^2 = \frac{3f(a)}{a} + f'(a)$$

If $\lambda^2 < 0$ then R is an oscillatory function of t and the motion is stable (e.g., $|R|$ is bounded). If $\lambda^2 > 0$ then R is an exponential function of t and the motion is unstable (e.g., $|R|$ increases as t does). If λ^2 should vanish, then one needs to look at higher-order terms in the Taylor series to decide the issue.

Power Law Forces

Consider the special case of $f \propto r^m$, say $f = Fr^m$. Then

$$\lambda^2 = 3Fa^{m-1} + mFa^{m-1} = Fa^{m-1}(m+3)$$

Presumably we are only interested in attractive forces (after all it doesn't make sense to inquire concerning the stability, in the sense of bounded motion, of an orbit near a repelling center of force) so $F < 0$ whence $\lambda^2 < 0$ becomes equivalent to $m > -3$. In particular, the inverse square case ($m = -2$) is stable to small radial perturbations as is the linear case ($m = 1$) which corresponds to an harmonic oscillator. It turns out that the $m = -3$ circular orbits are unstable.

The Apsidal Angle. We already know that at an apse the radial coordinate is a maximum or a minimum. The difference in azimuth or longitude between consecutive apsides is called the *apsidal angle*. For an ellipse this is just π but it need not be in general. As an example, consider an orbit that represents a stable oscillation about a circular orbit. As discussed above, in such a case $\lambda^2 < 0$ and the solution for the perturbed motion is of the form $\exp(\pm i\Lambda t)$ where $\Lambda^2 = -\lambda^2$. What happens to ϕ? From Eq. 2.3b

$$\dot{\phi} = \frac{L}{r^2} = \frac{L}{(R+a)^2} \simeq \frac{L}{a^2} - \frac{2LR}{a^3}$$

During an anomalistic period ($=2\pi/\Lambda$) the longitude increases by the apsidal angle (Φ) twice,

$$2\Phi = \int_{t_0}^{t_0+2\pi/\Lambda} \dot{\phi}\, d\tau = \frac{2L}{a^2}\frac{\pi}{\Lambda}$$

for arbitrary t_0. Hence

$$\Phi = \frac{L\pi}{a^2\Lambda} = \pi\left[3 + \frac{d\ln f(r)}{d\ln r}\bigg|_{r=a}\right]^{-1/2}$$

In the case of the power law $f = Fr^m$

$$\Phi = \pi(3+m)^{-1/2}$$

Thus, for the inverse square force ($m = -2$) the apsidal angle is π. For values of m such that $3 + m$ is not a perfect square, Φ/π will be irrational and the motion, although periodic in time, will not repeat spatially. Figure 27 shows an example of this.

Another Solution

Let us return to Eq. 2.1. We know that the vector $\mathbf{L} = \mathbf{r} \times \mathbf{v}$ is a constant of the motion. It is possible to find another one, denoted by $\mathbf{e}$. Form the vector cross product of $\mathbf{L}$ with Eq. 2.1 (on the right),

$$\ddot{\mathbf{r}} \times \mathbf{L} = \frac{-GM\mathbf{r}}{r^3} \times \mathbf{L}$$

But $\ddot{\mathbf{r}} \times \mathbf{L} = \ddot{\mathbf{r}} \times (\mathbf{r} \times \dot{\mathbf{r}}) = (\ddot{\mathbf{r}} \cdot \dot{\mathbf{r}})\mathbf{r} - (\ddot{\mathbf{r}} \cdot \mathbf{r})\dot{\mathbf{r}}$ using the standard result from vector algebra on the triple vector product. Thus,

$$\ddot{\mathbf{r}} \times \mathbf{L} = \frac{-GM(\dot{r}\mathbf{r} - r\dot{\mathbf{r}})}{r^2} = GM\frac{d}{dt}\left(\frac{\mathbf{r}}{r}\right)$$

whence

$$\dot{\mathbf{r}} \times \mathbf{L} = \frac{GM\mathbf{r}}{r} + \text{constant vector}$$

Call the constant vector $GM\mathbf{e}$ and next form the scalar product with $\mathbf{r}$,

$$\mathbf{r} \cdot (\dot{\mathbf{r}} \times \mathbf{L}) = GM(r + \mathbf{r} \cdot \mathbf{e})$$

Now the triple scalar product is unchanged by a cyclic permutation of its factors so $\mathbf{r} \cdot (\dot{\mathbf{r}} \times \mathbf{L}) = (\mathbf{r} \times \dot{\mathbf{r}}) \cdot \mathbf{L} = \mathbf{L} \cdot \mathbf{L} = L^2$. Thus, we have rederived the equation of the orbit,

$$r = \frac{L^2/GM}{1 + \mathbf{e} \cdot \mathbf{r}/r}$$

or, if we denote the angle between **e** and **r** by v and the norm of **e** by e,

$$r = \frac{L^2/GM}{1 + e\cos v}$$

Hence the vector **e** points, in the orbital plane, to the periapse and, since this is the general form for a conic section, its magnitude is the eccentricity of the conic. One can also show that another form for **e** is

$$GM\mathbf{e} = \left(\dot{\mathbf{r}}\cdot\dot{\mathbf{r}} - \frac{GM}{r}\right)\mathbf{r} - (\mathbf{r}\cdot\dot{\mathbf{r}})\dot{\mathbf{r}}$$

The vector **e** is called the Laplace–Runge–Lenz vector (which shows how many times it has been rediscovered).

Constants of the Motion

The reader might be confused. It would appear that there are seven constants of the motion: three in **L**, three in **e**, and one in $\mathscr{E}$. Since we started with three second-order, ordinary differential equations, there can only be six constants. The paradox is resolved when one realizes that $\mathbf{L}\cdot\mathbf{e} = 0$, which provides a relationship between the apparently independent seven parameters. Of course, there is yet another one as $L^2 = GMa(1 - e^2)$. To derive it, we return to the original integration of $\dot{\mathbf{r}}\times\mathbf{L}$. We accomplished this in the form

$$\dot{\mathbf{r}}\times\mathbf{L} = GM\left(\frac{\mathbf{r}}{r} + \mathbf{e}\right)$$

We form the scalar product of this with itself and set $(\dot{\mathbf{r}}\times\mathbf{L})^2 = \dot{r}^2 L^2$. This follows from a laborious expansion of the scalar product of two vector products or the insight that $\dot{\mathbf{r}}\cdot\mathbf{L} = 0$. Thus,

$$\dot{r}^2 L^2 = G^2 M^2\left(1 + \frac{2\mathbf{e}\cdot\mathbf{r}}{r} + e^2\right)$$

We have already shown that $\mathbf{e}\cdot\mathbf{r} = L^2/GM - r$. Using this and the conservation of energy equation $\dot{\mathbf{r}}\cdot\dot{\mathbf{r}} = 2\mathscr{E} + 2GM/r$, the above is transformed into

$$(GM)^2(1 - e^2) = -2\mathscr{E}L^2$$

Compare with Eq. 2.4.

Lambert's Theorem. In 1761 Lambert developed a relationship between a chord of an elliptical orbit, the radius vectors to the ends of the chord, and the time difference needed to traverse the associated arc. The relationship has some value in certain types of initial orbit determination.

Consider two points on an elliptical orbit, say 1 and 2. Then

$$r_1 = a(1 - e\cos E_1) \quad\text{and}\quad r_2 = a(1 - e\cos E_2)$$

The sum of the distances can be written as

$$r_1 + r_2 = 2a \left[1 - e \cos\left(\frac{E_1 + E_2}{2}\right) \cos\left(\frac{E_1 - E_2}{2}\right) \right]$$

The length of the chord joining these two points, c, is given by

$$c^2 = (x_2 - x_1)^2 + (y_2 - y_1)^2$$
$$= [a(\cos E_2 - \cos E_1)]^2 + [b(\sin E_2 - \sin E_1)]^2$$
$$= 4a^2 \sin^2\left(\frac{E_1 - E_2}{2}\right)\left[1 - e^2 \cos^2\left(\frac{E_1 + E_2}{2}\right) \right]$$

where $b^2 = a^2(1 - e^2)$ as is usual for ellipses.

From this result it is straightforward to deduce that

$$\frac{r_1 + r_2 \pm c}{2a} = 1 - \cos\left(\frac{\pm[E_2 - E_1]}{2}\right) + \cos^{-1}\left[e \cos\left(\frac{E_1 + E_2}{2}\right) \right]$$

or

$$2 \sin^{-1}\left[\frac{1}{2}\left(\frac{r_1 + r_2 \pm c}{a}\right)^{1/2} \right] = \pm\frac{E_2 - E_1}{2} + \cos^{-1}\left[e \cos\left(\frac{E_1 + E_2}{2}\right) \right]$$

The quantities p and q are defined by

$$\sin\left(\frac{p}{2}\right) = \frac{1}{2}\left(\frac{r_1 + r_2 + c}{a}\right)^{1/2}, \qquad \sin\left(\frac{q}{2}\right) = \frac{1}{2}\left(\frac{r_1 + r_2 - c}{a}\right)^{1/2}$$

Finally, it is deduced from Kepler's equation that

$$n(t_1 - t_2) = E_1 - E_2 - 2e \sin\left(\frac{E_1 - E_2}{2}\right) \cos\left(\frac{E_1 + E_2}{2}\right)$$

$$n(t_2 - t_1) = p - q - 2 \cos\left(\frac{p+q}{2}\right) \sin\left(\frac{p-q}{2}\right)$$

This relates r_1, r_2, and c to the time difference $t_1 - t_2$.

Other Conics

In the discussion of the time dependence we parenthetically remarked that the general solution was now limited to ellipses, that is, $0 \leq e < 1$, $a > 0$. The conic sections include the circle ($e = 0$, $a > 0$), which is a special case of an ellipse; the parabola [$e = 1$, $a = \infty$ but $p = a(1 - e^2)$ finite]; and the hyperbola ($e > 1$, $a < 0$). Here I turn to motion on a parabola or hyperbola.

Parabolic Motion

Motion along a parabola is especially simple. Analytically take the limit as a approaches infinity and e approaches unity such that the quantity $p = a(1 - e^2)$

stays finite. If

$$p = \frac{L^2}{GM} = 2q$$

then, for a parabola,

$$r = \frac{2q}{1 + \cos v} = q \sec^2\left(\frac{v}{2}\right)$$

From Eq. 2.3b ($\phi = v - \omega$ so $\dot\phi = \dot v$)

$$q^2 \dot v \sec^4\left(\frac{v}{2}\right) = (2GMq)^{1/2}$$

since $\mathscr{E} = 0$, $L^2 = 2GMq$. After integration we have Kepler's equation for parabolic motion,

$$\tan\left(\frac{v}{2}\right) + \tfrac{1}{3}\tan^3\left(\frac{v}{2}\right) = \left(\frac{GM}{2q^3}\right)^{1/2}(t - T)$$

This can be solved analytically through the following set of intermediaries

$$2 \cot 2\alpha = \tan\left(\frac{v}{2}\right)$$

$$\cot^{1/3}\left(\frac{\beta}{2}\right) = \cot \alpha$$

$$\cot \beta = 3\left(\frac{GM}{2q^3}\right)^{1/2}(t - T)$$

Finally, for a parabolic orbit, the speed is always given by $(2GM/r)^{1/2}$. This is the speed a particle initially at rest at infinity has at a distance r and, therefore, is the escape speed from distance r. It is also $\sqrt{2}$ times the circular orbit speed of an orbit of size r. Also note that for a parabola Lambert's theorem assumes the especially simple form

$$t_2 - t_1 = \frac{(r_1 + r_2 + c)^{3/2} - (r_1 + r_2 - c)^{3/2}}{6(GM)^{1/2}}$$

Hyperbolic Motion

Formally, the appropriate formulas can be obtained by regarding e as greater than 1, a as less than 0, and replacing circular functions of the eccentric anomaly with hyperbolic ones. Thus,

$$r = \frac{a(e^2 - 1)}{1 + e \cos v} = a(e \cosh E - 1)$$

$$(\dot r)^2 = GM\left(\frac{2}{r} + \frac{1}{a}\right)$$

$$e \sinh E - E = M = n(t - T)$$

$$\cos v = \frac{e - \cosh E}{e \cosh E - 1}, \qquad \sin v = \frac{(e^2 - 1)^{1/2} \sinh E}{e \cosh E - 1}$$

$$\sinh E = \frac{(e^2 - 1)^{1/2} \sin v}{1 + e \cos v}, \qquad \cosh E = \frac{e + \cos v}{1 + e \cos v}$$

and so on. If ε is the Gudermannian function of E,

$$E = \ln \left[\tan \left(\frac{\varepsilon}{2} + \frac{\pi}{4} \right) \right]$$

then it too can be used as an auxiliary since

$$\sinh E = \tan \varepsilon, \qquad \cosh E = \sec \varepsilon, \qquad \tanh \left(\frac{E}{2} \right) = \tan \left(\frac{\varepsilon}{2} \right)$$

Finally note that the true anomaly is restricted to the range interior to the roots of $1 + e \cos v = 0$, that is,

$$-\pi + \cos^{-1} \left(\frac{1}{e} \right) \leq v \leq \pi - \cos^{-1} \left(\frac{1}{e} \right)$$

Rectilinear Motion. Until now it has been assumed that $L > 0$. If $L = 0$, then the solution is that of a degenerate conic, namely a straight line. Obviously if the particles are bound they will collide, so this represents an atypical case.* It makes sense to reinitiate the problem on the assumption that the motion is rectilinear. The alternative is to obtain these results as limits of the general solution.

The equations of motion reduce to

$$\frac{d^2 x}{dt^2} = -\frac{GM}{x^2}$$

This can be integrated ($\ddot{x} = \dot{x} \, d\dot{x}/dx$) to give

$$\dot{x}^2 = GM \left(\frac{2}{x} - \frac{1}{a} \right)$$

where the constant of integration is $-GM/(2a)$. Assume that the initial conditions are $x(t = T) = x(T)$ and $\dot{x}(t = T) = \dot{x}(T) \leq 0$ (The case $\dot{x}(T) \geq 0$ will be treated momentarily). Then

$$\int_{x(T)}^{x(t)} \frac{dz}{[2/z - 1/a]^{1/2}} = -(GM)^{1/2} \int_T^t d\tau$$

---

* There exists a group of Sun-grazing comets known as Kreutz comets. As I write this (behind schedule in the Fall of 1981), recent artificial satellite observations have shown that some members of this cometary family do collide with the Sun. Once again idle speculation meets reality.

There are three cases, depending on the sign of a. If $\infty > a > 0$, then $x \leq 2a (\dot{x} = 0$ when $x = 2a$, the turning point) and

$$t - T = \left(\frac{a^3}{GM}\right)^{1/2} \left\{ \sin^{-1}\left(\frac{[z(2a-z)]^{1/2}}{a}\right) - \frac{[z(2a-z)]^{1/2}}{a} \right\} \Bigg|_{x(t)}^{x(T)}$$

If $x(T) = 2a$ and $\dot{x}(T) = 0$ (e.g., a body free falling to the center of attraction), then the time interval required to reach the origin is $\pi a^{3/2}/(GM)^{1/2}$ or half of the periodic time for a particle moving on an elliptical orbit with semimajor axis a. If $|a|$ is infinite, then the integration is especially easy,

$$t - T = \frac{2}{3(2GM)^{1/2}} [x^{3/2}(T) - x^{3/2}(t)]$$

Finally, if $0 > a > -\infty$, then the result is

$$t - T = -\left(\frac{-a^3}{GM}\right)^{1/2} \left\{ \sinh^{-1}\left(\frac{[z(z-2a)]^{1/2}}{-a}\right) + \frac{[z(z-2a)]^{1/2}}{a} \right\} \Bigg|_{x(t)}^{x(T)}$$

If instead $\dot{x}(t) > 0$ and $\infty > a > 0$, then once $x = 2a$, $\dot{x}$ will vanish. The object turns back (is bound), eventually colliding with the center of force. If $|a| = \infty$, then $\dot{x} (t = 0)$ is infinite so escape to infinity is possible where $\dot{x} = 0$. Lastly, if $0 > a > -\infty$, then escape occurs with a nonzero speed, $\dot{x} (x = \infty) = (GM/-a)^{1/2}$.

SERIES EXPANSIONS

There is a lot of nice mathematics associated with Kepler's equation, Eq. 2.8, and various expansions of r, $r \cos v$, and so on in powers of the eccentric or mean anomaly. Most of these series are convergent for $0 \leq e < 1$ and most practical applications had been to planetary orbits (i.e., low-eccentricity ellipses). Hence, this section will be limited to nonhyperbolic motion and when a result is true for parabolic ($e = 1$) motion that fact will be explicitly indicated.

Motion under Newtonian gravitation in ellipical orbits is periodic. One should not be surprised, therefore, to learn that series expansions in elliptic motion involve Fourier series. A simplified introduction to Fourier series is given below. Such problems also lead to Bessel functions as it turns out that Bessel functions are the coefficients of the trigonometric terms in the Fourier series. A brief discussion of these functions (lack of space prohibits excessive motivation of examples of utilizations) follows too. Then analytical, numerical, and geometrical aspects of Kepler's equation, the radius of convergence of the f and g series (which are critical for Gaussian initial orbit determination), generalized expansions in elliptic motion, and examples thereof will be discussed. As I don't believe such series are of much utility, the latter sections are very brief. The classical texts contain much more on the subject—little of it relevant to real problems.

Fourier Series

A Fourier series is a series of the form

$$\frac{a_0}{2} + \sum_{n=1}^{\infty} a_n \cos\left(\frac{n\pi x}{L}\right) + \sum_{n=1}^{\infty} b_n \sin\left(\frac{n\pi x}{L}\right) \qquad (2.17a)$$

One becomes interested in such series because of their ability to approximate periodic functions $f(x)$. Here the period is $2L$ that is, $f(x+2L)=f(x) \forall x$. If we exploit the results of Euler and DeMoivre, then the series can be written in the more compact form ($i^2 = -1$)

$$\sum_{n=-\infty}^{\infty} c_n \exp\left(\frac{in\pi x}{L}\right) \qquad (2.17b)$$

where $c_n = (a_n - ib_n)/2$ and $c_{-n} = (a_n + ib_n)/2$. Half-range series or odd and even series come about if all of the a_n vanish or if all of the b_n vanish. One needs to know several things about such series if one is to utilize them:

1. Under what circumstances do they converge?
2. For a given function $f(x)$ how does one relate (say) c_n to $f(x)$?
3. Given 1 and 2, does the series converge (and if so how) to $f(x)$?

Answers to these questions are:

1. A sufficient condition for the absolute and uniform convergence of the series in Eq. 2.17b is that $|c_n|$ be of the order $|n|^{-1-\varepsilon}$ as $|n| \to \infty$ for some $\varepsilon > 0$.
2. The formula for the expansion coefficients is

$$a_n = \frac{1}{L} \int_{-L}^{L} f(x) \cos\left(\frac{n\pi x}{L}\right) dx, \qquad b_n = \frac{1}{L} \int_{-L}^{L} f(x) \sin\left(\frac{n\pi x}{L}\right) dx \quad (2.18a)$$

or

$$c_n = \frac{1}{2L} \int_{-L}^{L} f(x) \exp\left(\frac{-in\pi x}{L}\right) dx \qquad (2.18b)$$

Note that if f is an odd function of x so that $f(x) = -f(-x)$, then

$$a_n = 0, \qquad b_n = \frac{2}{L} \int_{0}^{L} f(x) \sin\left(\frac{n\pi x}{L}\right) dx$$

whereas if $f(x) = f(-x)$ (i.e., f is an even function of x),

$$b_n = 0, \qquad a_n = \frac{2}{L} \int_{0}^{L} f(x) \cos\left(\frac{n\pi x}{L}\right) dx$$

3. If $f(x)$ is a periodic function of x and piecewise differentiable, then the
    partial sums

$$S_N(x) = \sum_{n=-N}^{N} c_n \exp\left(\frac{in\pi x}{L}\right) \qquad (2.19)$$

converge to $\frac{1}{2}[f(x+) + f(x-)]$. Hence, if f is continuous at x, they converge
to $f(x)$.

There are a few other results concerning Fourier series worth mentioning
here. The first concerns the least squares trigonometric approximation to $f(x)$.
Basically, if one is to represent $f(x)$ by a finite number of terms in the form
of a trigonometric series, then the representation is best (in the least squares
sense) if the coefficients are the Fourier coefficients of f: If the $\{c_n\}$ are related
to a continuous periodic function $f(x)$ by Eq. 2.18b and $S_N(x)$ is as defined
by Eq. 2.19, then

$$\lim_{N\to\infty} \int_{-L}^{L} |S_N(x) - f(x)|^2 \, dx = 0$$

Moreover, if

$$R_N(x) = \sum_{n=-N}^{N} C_n \exp\left(\frac{in\pi x}{L}\right)$$

for arbitrary coefficients C_n, then

$$\int_{-L}^{L} |R_N(x) - f(x)|^2 \, dx > \int_{-L}^{L} |S_N(x) - f(x)|^2 \, dx$$

unless $C_n = c_n \forall n \in [-N, N]$.

Two more results of some importance are Bessel's inequality

$$2L \sum_{n=-N}^{N} |c_n|^2 \le \int_{-L}^{L} [f(x)]^2 \, dx$$

and Parseval's equality

$$2L \sum_{n=-\infty}^{\infty} |c_n|^2 = \int_{-L}^{L} |f(x)|^2 \, dx$$

In each case the $\{c_n\}$ are the Fourier coefficients of f (i.e., related to f as in
Eq. 2.18b).

Bessel Functions

Consider expanding the function $\exp[(z/2)(t - t^{-1})]$ in a Laurent series as a
function of t. Clearly the coefficient of t^n will depend on both n and z. Call

it $J_n(z)$ and write

$$\exp\left[\left(\frac{z}{2}\right)\left(\frac{t-1}{t}\right)\right] = \sum_{n=-\infty}^{\infty} J_n(z)t^n$$

$J_n(z)$ is known as the *Bessel function* of argument z and order n. By direct expansion of the exponential one can show that

$$J_n(z) = \sum_{m=0}^{\infty} \frac{(-1)^m(z/2)^{n+2m}}{m!(n+m)!} \qquad (n \geq 0)$$

This result can be extended to complex z and nonintegral values of n. Note that for integral n, $J_{-n}(z) = (-1)^n J_n(z)$ and that

$$J_n(z) = \frac{(z/2)^n}{n!}\left[1 - \frac{z^2}{4(n+1)} + \frac{z^4}{32(n+2)(n+1)} - \cdots\right]$$

By manipulating the various partial derivatives of the generating function and the corresponding series, one can derive assorted recurrence relationships:

$$\left(\frac{2n}{z}\right)J_n(z) = J_{n-1}(z) + J_{n+1}(z)$$

$$2\frac{dJ_n(z)}{dz} = J_{n-1}(z) - J_{n+1}(z)$$

$$\frac{d[z^{\pm n}J_n(z)]}{dz} = \pm z^{\pm n}J_{n\mp 1}(z)$$

Further manipulations demonstrate that $J_n(z)$ satisfies the linear, first-degree, second-order, homogeneous, ordinary differential equation

$$z^2 y'' + zy' + (z^2 - n^2)y = 0, \qquad y = J_n(z)$$

One also has the integral representation

$$J_n(z) = \frac{1}{2\pi}\int_0^{2\pi} \cos(n\theta - z\sin\theta)\, d\theta \tag{2.20}$$

See Watson's (1922) reference work for much much more. A necessary result is that as long as $z^2 - 1$ is not a real positive number, then

$$|J_n(nz)| \leq \left|\frac{z^n \exp[n(1-z^2)^{1/2}]}{[1 + (1-z^2)^{n/2}]}\right| \tag{2.21}$$

Kepler's Equation

Kepler's equation

$$E - e\sin E = M \tag{2.8}$$

expresses, implicitly, the time dependence of the motion on an ellipse. The mean anomaly is related to the time via

$$M = n(t - T) \tag{2.22a}$$

or

$$M = n(t - t_0) + M_0 \tag{2.22b}$$

The representation in Eq. 2.22a is standard astronomy. M is measured from periapse (where all of the anomalies are equal to zero) and T is the time of periapse passage. The form in Eq. 2.22b is common in artificial satellite work. Here t_0 is some convenient epoch and M_0 is the corresponding value of M. The eccentric anomaly E is related to place in the orbit via $E = \cos^{-1}(1 - r/a)$ where r is the distance between the particle and the force center (which is the occupied focus of the conic section). The connection with the true anomaly v was given in Eq. 2.10, of which the most commonly used formula is

$$\tan\left(\frac{v}{2}\right) = \left(\frac{1+e}{1-e}\right)^{1/2} \tan\left(\frac{E}{2}\right) \tag{2.23}$$

The Solution of Kepler's Equation

A numerical method of solving Kepler's equation was presented in Example 2.1, a more detailed discussion is given below. Here the analytical basis for the solution of Kepler's equation will be discussed. Denote the difference between E and M by χ so that

$$\chi(M) = E - M = e \sin E \tag{2.24}$$

First it will be shown that M is a strictly monotonic increasing function of E. To see this, Eq. 2.8 is differentiated with respect to E and we deduce that

$$\frac{dM}{dE} = 1 - e \cos E$$

whence (for elliptic motion)

$$\frac{dM}{dE} \geqslant 1 - e > 0$$

This implies that E is uniquely determined by M (the eccentricity e is a fixed parameter $0 \leqslant e < 1$). It is also clear from Eq. 2.8 that $M(E)$ is an odd function and thus this is also true of both $E(M)$ and $\chi(M)$. Equally clear is that M is an analytic function of E with continuous derivatives of all orders. Therefore one can develop $M(E)$ in a power series in E. As the derivative $dM/dE \neq 0$, one can invert the $M(E)$ relationship and develop $E(M)$ in a power series in $M - M'$ in some sufficiently small neighborhood about $M' \forall M'$. You can further see from Eq. 2.24 that

$$\chi = e \sin E = e \sin(\chi + M) \tag{2.25}$$

so $\chi(M)$ is unchanged if M changes by 2π. This is the definition of periodicity.

For real χ and M it also follows that

$$|\chi| \le e$$

A result concerning Fourier series that went unmentioned above is that such series exist for continuous, periodic functions of bounded variation. Furthermore, because $\chi(M)$ is odd, if expressed in a series of the form Eq. 2.17a, by Eq. 2.18a,

$$a_n = 0, \qquad b_n = \frac{2}{\pi} \int_0^\pi \chi(M) \sin nM \, dM = \frac{1}{\pi} \int_0^{2\pi} \chi(M) \sin nM \, dM$$

Since χ is an analytic function of M, it may be differentiated. Knowing this we integrate by parts using the second form for b_n,

$$b_n = \frac{1}{\pi} \left[\frac{-\chi(M)\cos nM}{n} \Big|_0^{2\pi} + \frac{1}{n} \int_0^{2\pi} \chi'(M) \cos nM \, dM \right]$$

$$= -\frac{1}{\pi n^2} \int_0^{2\pi} \chi''(M) \sin nM \, dM$$

As $|\chi''(M)| \le e/(1-e)^3$ for $e \in [0, 1)$ [this is left as an exercise for the reader (I had to say that at least once!)], $|b_n| \le 2e/[n^2(1-e)^3]$, hence the Fourier series for $\chi(M)$ converges (uniformly and absolutely). We can even calculate the $\{b_n\}$ for

$$b_n = \frac{2}{\pi} \int_0^\pi \chi(M) \sin nM \, dM$$

which upon change of variable from M to E becomes

$$b_n = \frac{2e}{\pi} \int_0^\pi \sin E \, \sin[n(E - e \sin E)](1 - e \cos E) \, dE$$

We can say even more from this. The right-hand side is obviously an analytic function of e for all real e and therefore, b_n can be expanded in a convergent power series in e. One might think that this yields a power series in e for χ and hence E. This is true, but not for all values of e, not even for $e < 1$. See below for the exact radius of convergence of such a series in connection with the f and g series. It is a historical note that the discussion of this question motivated Cauchy's fundamental work on the convergence (in general) of infinite series. Obviously Kepler's equation has been the stimulus for much of pure mathematics (as has astronomy in general).

The point of all this is to show how to analytically express $\chi = E - M$ as a function of M, namely in a Fourier sine series. Thus, we have E explicitly as a function of M, which is what solving Kepler's equation means. Solving Kepler's equation numerically means something else entirely. The next two sections will deal with that and then, to lighten the analytic load, a geometrical interpretation of the problem will be provided. Finally we shall return to the essential end result, Eq. 2.29, which is the radius of convergence formula for the f and g series. This result is central to initial orbit determination by Gauss's technique (or modifications thereof).

Solving for χ. The last few paragraphs have followed the development in Kurth's (1959) text and we will continue to do so here. Kurth investigates in detail the successive substitution method for χ from Eq. 2.25. In particular, he shows that if $\chi_0 = 0$ and

$$\chi_{n+1} = e \sin(M + \chi_n), \qquad n = 0, 1, 2, \ldots$$

then (1) this series converges absolutely and uniformly for all M, (2) this series converges to a solution of Eq. 2.25 and it is unique, and (3) the difference between the value of χ that satisfies Eq. 2.25 differs from χ_n by at most $e^{n+1}|\sin M|/(1-e)$.

In order to see that the series of successive substitutions converges, observe that

$$|\chi_{n+1} - \chi_n| = e|\sin(M + \chi_n) - \sin(M + \chi_{n-1})|$$

$$= e|\cos(M + \psi_n)||\chi_n - \chi_{n-1}| \leqslant e|\chi_n - \chi_{n-1}|$$

where $\chi_{n-1} \leqslant \psi_n \leqslant \chi_n$ or $\chi_n \leqslant \psi_n \leqslant \chi_{n-1}$. This last result follows after an application of the mean value theorem of differential calculus. A straightforward induction yields

$$|\chi_2 - \chi_1| \leqslant e|\chi_1| = e^2|\sin M|$$

$$|\chi_3 - \chi_2| \leqslant e|\chi_2 - \chi_1| \leqslant e^3|\sin M|$$

so that

$$|\chi_{n+1} - \chi_n| \leqslant e^{n+1}|\sin M|$$

Hence the series $(\chi_1 - \chi_0) + (\chi_2 - \chi_1) + \cdots$, which has partial sums equal to $\chi_1, \chi_2, \ldots (\chi_0 = 0)$, converges absolutely and uniformly. Call the limit X.

To demonstrate that X satisfies Eq. 2.25, observe that

$$|X - e \sin(M + X)| = |(X - \chi_{n+1}) - e[\sin(M + X) - \sin(M + \chi_n)]|$$

$$\leqslant |X - \chi_{n+1}| + e|X - \chi_n|$$

(Show this.) Since the series for X converges, as $n \to \infty$, each term on the right-hand side approaches zero. Thus, X satisfies Eq. 2.25. Uniqueness is proved in a similar fashion. Suppose that Y satisfies Eq. 2.25 too and that $X \neq Y$. Then

$$|X - Y| = e|\sin(M + X) - \sin(M + Y)| \leqslant e|X - Y|$$

But $e < 1$ so X must equal Y.

The last point is to construct the error estimate for χ_n. Now

$$|X - \chi_n| = \left| \sum_{m=1}^{\infty} (\chi_{m+n} - \chi_{n+m-1}) \right|$$

$$\leq \sum_{m=1}^{\infty} |\chi_{m+n} - \chi_{n+m-1}|$$

$$\leq \sum_{m=1}^{\infty} e^{n+m} |\sin M| = \frac{e^n}{1-e} |\sin M|$$

Obviously convergence can be improved upon if one starts with $\chi = e \sin M$.

The Numerical Solution of Kepler's Equation. There is more literature on the numerical solution of Kepler's equation than I would want to read. I have not examined any of it (almost) and strongly suggest that you follow my lead. A study of Newton's method of solution applied to Kepler's equation (which is my all-purpose recommendation) is of value though.

While designing real-time asteroid pointing and observation scheduling software (Taff 1980a) a thorough test of four different methods of solving Kepler's equation was undertaken. It was, however, restricted to small eccentricities ($e \leq 0.3$). Four different methods were tested. Over all mean anomalies [$M = 0(15)360°$, $e = 0(0.05)0.30$] Newton's method (see Example 2.1) was fastest. If the starting value for E was M, then three iterations were necessary for $0\overset{''}{.}1$ convergence (on the average). If the starting value $M + e \sin M$ was used, then the average number of iterations decreased to 2.5.

An improvement for larger values of the eccentricity was suggested by Smith (1979). He noted that if we rewrite Kepler's equation as

$$f(E) = E - e \sin E - M$$

then $f(M) \leq 0$ while $f(M + e) \geq 0$. Since $f'(E) > 0, f$ must vanish for $E \in [M, M + e]$. We are now specifically considering M between 0 and π. The eccentric anomaly E' corresponding to a mean anomaly M' between π and 2π is equal to that eccentric anomaly corresponding to $2\pi - M'$.) By using a linear interpolation formula, we can calculate that a better value of E is

$$M + \frac{e \sin M}{1 - \sin(M + e) + \sin M}$$

Now this can be used to commence a sequence of iterations via Newton's method.

As an illustration, consider $e = 0.995$ and $M = 0.1$. The table below shows the successive values of E for the four starting values: (1) M, (2) $M + e \sin M$, (3) $M + e \sin M + (e^2/2) \sin 2M$, and (4) Smith's starting value.

(1)	(2)	(3)	(4)
05° 43′ 46″481	11° 25′ 15″640	17° 03′ 20″506	32° 42′ 55″166
216 32 11.350	238 00 47.917	127 41 42.260	56 12 34.235
80 31 31.570	54 14 19.280	79 54 40.914	49 16 00.071
58 19 30.285	48 51 58.576	58 02 47.174	48 18 10.440
49 47 04.985	48 17 28.629	49 42 42.297	48 17 05.687
48 19 34.713	48 17 05.667	48 19 20.830	48 17 05.665
48 17 05.783	48 17 05.664	48 17 05.762	48 17 05.664
48 17 05.665	48 17 05.664	48 17 05.665	48 17 05.664
48 17 05.665		48 17 05.664	
48 17 05.665		48 17 05.664	
48 17 05.664			

A Geometrical Perspective

Consider Fig. 5, which shows an elliptical orbit. The semi-major axis is a and the eccentricity is e. Also shown is a circle concentric with the ellipse and tangent to it at the ends of the major axis. This circle is known as the auxiliary circle. The source of gravitational attraction is at the occupied focus F and not at the center C. If the particle is at point P, then the angle AFP is the true anomaly v (in the figure A is the periapse). The angle ACQ is the eccentric anomaly E where the point Q is obtained by dropping a perpendicular from P to the major axis of the ellipse and then extending it upward to meet the

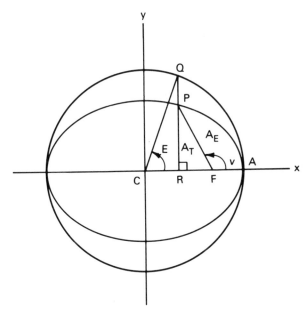

Figure 5.   Geometrical derivation of Kepler's equation with an ellipse and its external auxiliary circle.

auxiliary circle. Kepler's equation can be derived by considering the relationship between the area of the triangle PRF and the elliptical sector bounded by the line segments FP, AF, and the arc of the ellipse AP. Call these areas A_T and A_E. The time aspect is introduced via Kepler's second law, Eq. 2.3b.

The area swept out by the radius vector while opening the true anomaly from 0 to v is A_E. Call α the area bounded by the line segments AR, PR, and the arc of the ellipse between A and P. Then

$$A_E = \alpha - A_T$$

It is simple to compute A_T if we remember that $AC = a$, $CF = ae$ so that $FP = a(1 - \cos E)$. To compute PR, observe the general relationship between y values relative to the ellipse and its auxiliary circle:

$$\frac{y_{\text{ellipse}}}{y_{\text{circle}}} = \frac{b}{a}$$

To see that this is valid, remember that the equation of the ellipse in the coordinate system of Fig. 5 is

$$\frac{x^2}{a^2} + \frac{y^2}{b^2} = 1$$

whereas that of the auxiliary circle is just

$$x^2 + y^2 = a^2$$

Now $QR = e \sin E$, so $PR = a(b/a) \sin E$ and, therefore,

$$A_T = \frac{a}{2}\left(\frac{b}{a}\right) \sin E (ae - a \cos E)$$

Similar reasoning shows that $\alpha = (b/a)\beta$ where β is the area bounded by the line segments AR, RQ, and the arc of the auxiliary circle from A to Q. But E is the central angle of the circular sector ACQ so its area is $a^2E/2$, and $\beta = a^2E/2$ minus the area of triangle RCQ or

$$\beta = \frac{a^2 E}{2} - \frac{(a \cos E)(a \sin E)}{2}$$

Whence

$$\alpha = \left(\frac{ab}{2}\right)(E - \sin E \cos E)$$

so

$$A_E = \left(\frac{ab}{2}\right)(E - e \sin E)$$

Now invoke Kepler's law of equal areas swept out in equal times in the form

$$\frac{t - T}{P} = \frac{A_E}{\pi ab}$$

where P is the period and t is the time corresponding to v or E or $M =$ $n(t-T) = 2\pi(t-T)/P$. Clearly,

$$E - e \sin E = M = n(t-T)$$

This geometrical construction can also be used to derive Eq. 2.23. We saw that

$$r \cos v = a(e - \cos E)$$

Exploiting the relationship between auxiliary circle and ellipse y values we can also derive the formula

$$r \sin v = b \sin E$$

whence, after squaring, adding, and taking the positive square root,

$$r = a(1 - e \cos E)$$

Using these three formulas and the half-angle formulas for the trigonometric functions,

$$r \cos^2\left(\frac{v}{2}\right) = a(1-e) \cos^2\left(\frac{E}{2}\right)$$

$$r \sin^2\left(\frac{v}{2}\right) = a(1+e) \sin^2\left(\frac{E}{2}\right)$$

and thence Eq. 2.23.

The f and g Series

Recall that the initial conditions of the problem consist of the specification of $\mathbf{r}$ and $\dot{\mathbf{r}} = d\mathbf{r}/dt$ at some time $t = t_0$. Denote these values by $\mathbf{r}_0 = \mathbf{r}(t_0)$ and $\dot{\mathbf{r}}_0 = \dot{\mathbf{r}}(t_0)$. Then the solution of the equations of motion, Eq. 2.1, can be written as, for example,

$$\mathbf{r}(t) = f(\mathbf{r}_0, \dot{\mathbf{r}}_0, t - t_0)\mathbf{r}_0 + g(\mathbf{r}_0, \dot{\mathbf{r}}_0, t - t_0)\dot{\mathbf{r}}_0 \qquad (2.26)$$

Explicit formulas for f and g can be derived by remembering that $\mathbf{L} = \mathbf{r} \times \dot{\mathbf{r}} = \mathbf{r}_0 \times \dot{\mathbf{r}}_0$, whence

$$f = (\mathbf{r} \times \dot{\mathbf{r}}_0) \cdot \mathbf{L}/L^2, \qquad g = -(\mathbf{r} \times \mathbf{r}_0) \cdot \mathbf{L}/L^2$$

In terms of the true anomaly v and the distance r

$$f = \left(\frac{r}{L}\right)[r_0 \dot{v}_0 \cos(v - v_0) - \dot{r}_0 \sin(v - v_0)]$$

$$\qquad\qquad (2.27)$$

$$g = \left(\frac{r}{L}\right) r_0 \sin(v - v_0)$$

where $r_0 = r(t_0)$, $\dot{v}_0 = dv/dt$ at $t = t_0$, and so on. The corresponding functions expressed in terms of the eccentric anomaly are messier-looking:

$$f = \left(\frac{a}{Lr_0}\right)(r_0\dot{r}_0 \sin v_0 + L \cos v_0)(\cos E - e)$$

$$+ \left[\frac{a(1-e^2)^{1/2}}{r_0 L}\right](L \sin v_0 - r_0\dot{r}_0 \cos v_0) \sin E \qquad (2.28)$$

$$g = -\left(\frac{ar_0}{L}\right)\sin v_0(\cos E - e) + \left[\frac{ar_0(1-e^2)^{1/2}}{L}\right]\cos v_0 \sin E$$

When one analyzes Gauss's method of orbit determination, one needs f and g expressed as power series in the time or equivalently in the mean anomaly $M = n(t - T)$. This turns out to be the equivalent to requiring the $E(M)$ expansion. The relevant theorem, first proved by Moulton (1903) [and independently by Taff (1979a) using a different technique], is described next.

The f and g series defined by Eq. 2.26 have the following joint radius of convergence in $t - t_0$: (1) If $e = 0$, then the radius of convergence is infinite; (2) if $e = 1$, then the radius of convergence is $(8Q^3/9GM)^{1/2}$ where Q is the distance from the focus of the parabola to its directrix; and (3) if $e \in (0, 1)$, then the radius of convergence is given by $P\imath/(2\pi)$ where P is the period and

$$\imath = [M_0^2 + \{\ln[1 + (1 - e^2)^{1/2}] - \ln e - (1 - e^2)^{1/2}\}^2]^{1/2} \qquad (2.29)$$

Here M_0 is the value of the mean anomaly corresponding to $t = t_0$, $M_0 \in [-\pi, \pi]$. This theorem will be proved in Chapter Eight. Also in that chapter a detailed discussion of the implications of this theorem for Gaussian-type initial orbit determination will be presented. Basically, Gauss did not solve the initial orbit determination problem at all.

Expansions In Elliptic Motion

The classical celestial mechanics texts usually contain several pages devoted to the evaluation of Fourier–Bessel functions for such things as $r^p \cos qv$. Some of the more common forms are exhibited below. The first set of formulas lists complete analytical results. The second set gives explicit expansions in powers of the eccentricity. The highest order noted by this author have been included. Such expansions are not empty for several reasons. First try to derive them. This will give you a good practical introduction into Bessel function manipulation. Second they are of real use if the eccentricity is small. [See Taff and Sorvari (1982) or Chapter Eleven for a fully analytical theory of differential correction for a near-stationary artificial satellite based on power series.] Unless explicitly indicated otherwise all sums are from $n = 1$ to $n = \infty$ and $J'_n(ne)$ means $\partial J_n(ne)/\partial e$.

$$\frac{r}{a} = 1 + \frac{e^2}{2} - \sum \left(\frac{2e}{n^2}\right) J'_n(ne) \cos nM$$

$$\left(\frac{r}{a}\right)\cos v = \cos E - e = -\frac{3e}{2} + \sum \left(\frac{2}{n^2}\right) J_n'(ne) \cos nM$$

$$\left(\frac{r}{a}\right)\sin v = (1-e^2)^{1/2}\sin E = (1-e^2)^{1/2}\sum \left(\frac{2}{ne}\right) J_n(ne) \sin nM$$

$$\left(\frac{r}{a}\right)^2 = 1 + \frac{3e^2}{2} - \sum \left(\frac{4}{n^2}\right) J_n(ne) \cos nM$$

$$\frac{a}{r} = 1 + 2\sum J_n(ne) \cos nM$$

$$\left(\frac{a}{r}\right)^2 \cos v = 2\sum J_n'(ne) \cos nM$$

$$\left(\frac{a}{r}\right)^2 \sin v = \frac{1-e^2}{e}\sum 2nJ_n(ne) \sin nM$$

$$\cos v = -e + \frac{2(1-e^2)}{e}\sum J_n(ne) \cos nM$$

$$\sin v = (1-e^2)^{1/2}\sum \left(\frac{2}{n}\right) J_n'(ne) \sin nM$$

$$v - M = \sum \left\{ \left(\frac{2}{n}\right) J_n(ne) + \sum_{m=1}^{\infty} g^m[J_{n-m}(ne) + J_{n+m}(ne)] \right\} \sin nM$$

$$\text{with } g = e/[1+(1-e^2)^{1/2}]$$

$$\cos mE = \sum \left(\frac{m}{n}\right)[J_{n-m}(ne) - J_{n+m}(ne)] \cos nM \qquad m \geqslant 1, \text{ integral}$$

$$\sin mE = \sum \left(\frac{m}{n}\right)[J_{n-m}(ne) + J_{n+m}(ne)] \sin nM \qquad m \geqslant 1, \text{ integral}$$

$$\left(\frac{r}{a}\right)^2 = (1-e^2)^{1/2}\left\{ 1 + 2\sum (-1)^n[1 + n(1-e^2)^{1/2}] \right.$$

$$\left. \times \left[\frac{1-(1-e^2)^{1/2}}{1+(1-e^2)^{1/2}}\right]^{n/2} \cos nv \right\}$$

$$M = v + 2\sum (-1)^n[n^{-1} + (1-e^2)^{1/2}]\left[\frac{1-(1-e^2)^{1/2}}{1+(1-e^2)^{1/2}}\right]^{n/2} \sin nv$$

$$\sin(v - M) = \left[e + \frac{1-e^2}{e}J_2(2e) + \frac{(1-e^2)^{1/2}}{2e}J_2'(2e)\right]\sin M$$

$$+ \sum_{n=2}^{\infty}\left(\frac{1-e^2}{e}\{J_{n+1}[(n+1)e] - J_{n-1}[(n-1)e]\}\right.$$

$$\left. + \frac{(1-e^2)^{1/2}}{n^2-1}\{(n+1)J_{n-1}'[(n-1)e] + (n-1)J_{n+1}'[(n+1)e]\}\right)\sin M$$

$$\ln\left(\frac{r}{a}\right) = eg - \ln(1+g^2) - \sum \frac{2}{n}\left(\sum_{m=1}^{\infty} g^m[J_{n-m}(ne) - J_{n+m}(ne)]\right)\cos nM$$

$$\text{with } g = e/[1+(1-e^2)^{1/2}]$$

$$\frac{r}{a} = 1 + \frac{1}{2}e^2 + [-e + \frac{3}{8}e^3 - \frac{5}{192}e^5 + \frac{7}{9216}e^7]\cos M$$

$$+ [-\frac{1}{2}e^2 + \frac{1}{3}e^4 - \frac{1}{16}e^6]\cos 2M + [-\frac{3}{8}e^3 + \frac{45}{128}e^5 - \frac{567}{5120}e^7]\cos 3M$$

$$+ [-\frac{1}{3}e^4 + \frac{2}{5}e^6]\cos 4M + [-\frac{125}{384}e^5 + \frac{4375}{9216}e^7]\cos 5M$$

$$- \frac{27}{80}e^6 \cos 6M - \frac{16807}{46080}e^7 \cos 7M$$

$$\frac{r}{a}\cos v = \cos E - e = -\frac{3}{2}e + (1 - \frac{3}{8}e^2 + \frac{5}{192}e^4 - \frac{7}{9216}e^6)\cos M$$

$$+ (\frac{1}{2}e - \frac{1}{3}e^3 + \frac{1}{16}e^5 - \frac{1}{180}e^7)\cos 2M$$

$$+ (\frac{3}{8}e^2 - \frac{45}{128}e^4 + \frac{567}{5120}e^6)\cos 3M$$

$$+ (\frac{1}{3}e^3 - \frac{2}{5}e^5 + \frac{8}{45}e^7)\cos 4M + (\frac{125}{384}e^4 - \frac{4375}{9216}e^6)\cos 5M$$

$$+ (\frac{27}{80}e^5 - \frac{81}{140}e^7)\cos 6M + \frac{16807}{46080}e^6 \cos 7M$$

$$\frac{r}{a}\sin v = (1-e^2)^{1/2}\sin E = (1 - \frac{5}{8}e^2 - \frac{11}{192}e^4 - \frac{457}{9216}e^6)\sin M$$

$$+ (\frac{1}{2}e - \frac{5}{12}e^3 + \frac{1}{24}e^5 - \frac{1}{45}e^7)\sin 2M$$

$$+ (\frac{3}{8}e^2 - \frac{51}{128}e^4 + \frac{543}{5120}e^6)\sin 3M$$

$$+ (\frac{1}{3}e^3 - \frac{13}{30}e^5 + \frac{13}{72}e^7)\sin 4M + (\frac{125}{384}e^4 - \frac{4625}{9216}e^6)\sin 5M$$

$$+ (\frac{27}{80}e^5 - \frac{135}{224}e^7)\sin 6M + \frac{16807}{46080}e^6 \sin 7M$$

$$\frac{(r/a)\sin v}{(1-e^2)^{1/2}} = \sin E = \sin M + \left(\frac{e}{2}\right)\sin 2M + \frac{e^2(3\sin 3M - M)}{8}$$

$$+ \frac{e^3(2\sin 4M - \sin 2M)}{6}$$

$$+ \frac{e^4(125\sin 5M - 81\sin 3M + 2\sin M)}{384}$$

$$+ \frac{e^5(81\sin 6M - 64\sin 4M + 5\sin 2M)}{240}$$

$$+ \frac{e^6(16807\sin 7M - 15625\sin 5M + 2187\sin 3M - 5\sin M)}{46080}$$

$$\frac{a}{r} = 1 + \left(e - \frac{e^3}{8} + \frac{e^5}{192} - \frac{e^7}{9216}\right) \cos M + \left(e^2 - \frac{e^4}{3} + \frac{e^6}{24}\right) \cos 2M$$

$$+ \left(\frac{9e^3}{8} - \frac{81e^5}{128} + \frac{729e^7}{5120}\right) \cos 3M + \left(\frac{4e^4}{3} - \frac{16e^6}{15}\right) \cos 4M$$

$$+ \left(\frac{625e^5}{384} - \frac{15625e^7}{9216}\right) \cos 5M + \left(\frac{81e^6}{40}\right) \cos 6M$$

$$+ \left(\frac{117649e^7}{46080}\right) \cos 7M$$

$$v = M + (2e - \tfrac{1}{4}e^3 + \tfrac{5}{96}e^5 + \tfrac{107}{4608}e^7) \sin M$$

$$+ (\tfrac{5}{4}e^2 - \tfrac{11}{24}e^4 + \tfrac{17}{192}e^6) \sin 2M + (\tfrac{13}{12}e^3 - \tfrac{43}{64}e^5 + \tfrac{95}{512}e^7) \sin 3M$$

$$+ (\tfrac{103}{96}e^4 - \tfrac{451}{480}e^6) \sin 4M + (\tfrac{1097}{960}e^5 - \tfrac{5957}{4608}e^7) \sin 5M$$

$$+ \tfrac{1223}{960}e^6 \sin 6M + \tfrac{47273}{32256}e^7 \sin 7M$$

$$E = M + (e - \tfrac{1}{8}e^3 + \tfrac{1}{192}e^5 - \tfrac{1}{9216}e^7) \sin M$$

$$+ (\tfrac{1}{2}e^2 - \tfrac{1}{6}e^4 + \tfrac{1}{48}e^6) \sin 2M + (\tfrac{3}{8}e^3 - \tfrac{27}{128}e^5 + \tfrac{243}{5120}e^7) \sin 3M$$

$$+ (\tfrac{1}{3}e^4 - \tfrac{4}{15}e^6) \sin 4M + (\tfrac{125}{384}e^5 - \tfrac{3125}{9216}e^7) \sin 5M$$

$$+ \tfrac{27}{80}e^6 \sin 6M + \tfrac{16807}{46080}e^7 \sin 7M$$

$$M = v - 2e \sin v + \left\{\frac{3e^2}{4} + \frac{e^4}{8} + \frac{3e^6}{64}\right\} \sin 2v$$

$$- \left\{\frac{e^3}{3} + \frac{e^5}{8} + \frac{e^7}{16}\right\} \sin 3v + \left\{\frac{5e^4}{32} + \frac{3e^6}{32}\right\} \sin 4v$$

$$- \left\{\frac{3e^5}{40} + \frac{e^7}{16}\right\} \sin 5v + \left\{\frac{7e^6}{192}\right\} \cos 6v - \left\{\frac{e^7}{56}\right\} \sin 7v$$

$$\sin(v - M) = e(2 - \tfrac{5}{4}e^2) \sin M + \tfrac{5}{4}e^2 \sin 2M + \tfrac{17}{12}e^3 \sin 3M$$

$$E = v - e \sin v + \frac{e^2}{4} \sin 2v - \frac{e^3}{12} (\sin 3v + e \sin v) + \frac{e^4}{32} (\sin 4v + 4 \sin 2v)$$

$$v = E + e \sin E + \frac{e^2}{4} \sin 2E + \frac{e^3}{12} (\sin 3E + 3 \sin E) + \frac{e^4}{96} (\sin 4E + 4 \sin 2E)$$

$$\ln \frac{r}{a} = \tfrac{1}{4}e^2 + \tfrac{1}{32}e^4 + \tfrac{1}{96}e^6 + (-e + \tfrac{3}{8}e^3 + \tfrac{1}{64}e^5 + \tfrac{127}{9216}e^7) \cos M$$

$$+ (-\tfrac{3}{4}e^2 + \tfrac{11}{24}e^4 - \tfrac{3}{64}e^6) \cos 2M + (-\tfrac{17}{24}e^3 + \tfrac{77}{128}e^5 - \tfrac{743}{5120}e^7) \cos 3M$$

$$+ (-\tfrac{71}{96}e^4 + \tfrac{129}{160}e^6) \cos 4M + (-\tfrac{523}{640}e^5 + \tfrac{10039}{9216}e^7) \cos 5M$$

$$- \tfrac{899}{960}e^6 \cos 6M - \tfrac{355081}{322560}e^7 \cos 7M$$

PROBLEMS

1. Prove that

$$\tan\left(\frac{v-E}{2}\right) = \frac{\sin v}{1/g + \cos v} = \frac{\sin E}{1/g - \cos E}$$

and

$$v - M = 2\tan^{-1}\left(\frac{\sin E}{1/g - \cos E}\right) + e\sin E$$

$$= 2\tan^{-1}\left(\frac{\sin v}{1/g + \cos v}\right) + \frac{e(1-e^2)^{1/2}\sin v}{1 + e\cos v}$$

   where $g = e/[1 + (1-e^2)^{1/2}]$ (Broucke and Cefola 1973).

2. Use the results obtained above to prove that $v - E$ is a maximum when $r = b$, $E - M$ is a maximum when $r = a$, and $v - M$ is a maximum when $r = (ab)^{1/2}$. Also show that if $v - E$ is a maximum, $E + v = \pi$ with $E < \pi/2$ and $v > \pi/2$, whereas when $E - M$ is a maximum, $E = \pi/2$, $M = \pi/2 - E$ (Broucke and Cefola 1973).

3. Show that if $Ax^2 + 2Bxy + Cy^2 + Dx + Ey + F = 0$ represents an ellipse of semimajor axis a, eccentricity e, argument of periapse ω, and with focus at $(0, 0)$, then

$$A = 1 - e^2\cos^2\omega, \qquad B = -e^2\sin\omega\cos\omega, \qquad C = 1 - e^2\sin^2\omega$$

$$D = 2ae(1 - e^2)\cos\omega, \qquad E = 2ae(1 - e^2)\sin\omega$$

$$F = -a^2(1 - e^2)^2$$

   Generalize to three dimensions.

4. Set up the two-body problem in spherical coordinates. Show that if the equatorial plane contains both $\mathbf{r}$ and $\dot{\mathbf{r}}$ at some instant, then it does so for all times. Hence deduce Eqs. 2.2.

5. Derive an equation for $d^2r/d\phi^2$ from Eqs. 2.2 and 2.3b. Set $r = 1/u$ (Binet's transformation) and solve. Compare with Eq. 2.6.

6. Show that the relationship between the speed $|\dot{\mathbf{r}}|$ and the distance $r = |\mathbf{r}|$ is

$$|\dot{\mathbf{r}}|^2 = GM\left(\frac{2}{r} - \frac{1}{a}\right)$$

   Is this true for all conic sections?

7. Derive Newton's law of gravitation from his three laws of motion and Kepler's three laws.

8. Suppose that two particles are revolving about each other under the action of their mutual gravitational attraction. Let the period of this

motion be P. Suppose that their motion is instantaneously arrested and then they are allowed to fall toward their common center of mass. Prove that they will collide in a time equal to $P/4\sqrt{2}$.

9. Show that S in Eq. 2.12a is unitary.

10. Show that the average values of r, r^2 $1/r$, and $1/r^2$ with respect to E, v, and M(in elliptic motion) are given as below $[b^2 = a^2(1-e^2)$, $p = a(1-e^2)]$:

	$\langle r \rangle$	$\langle r^2 \rangle$	$\langle 1/r \rangle$	$\langle 1/r^2 \rangle$
E	a	$a^2\left(1+\dfrac{e^2}{2}\right)$	$\dfrac{1}{b}$	$\dfrac{1}{pb}$
v	b	ab	$\dfrac{1}{p}$	$\dfrac{(1+e^2/2)}{p^2}$
M	$a\left(1+\dfrac{e^2}{2}\right)$	$a^2\left(1+\dfrac{3e^2}{2}\right)$	$\dfrac{1}{a}$	$[a^2(1-e^2)^{1/2}]^{-1}$

(Serafin 1980: this paper contains several typographical errors).

11. Repeat Problem 10 for $|\dot{r}|$ and $|\dot{r}|^2$ (Serafin 1980)

	$\langle |\dot{r}| \rangle$	$\langle |\dot{r}|^2 \rangle$
E	$\dfrac{2}{\pi}\left(\dfrac{GM}{a}\right)^{1/2}K(e)$	$\dfrac{GM}{a}\left(\dfrac{2a}{b}-1\right)$
v	$\dfrac{2}{\pi}\left(\dfrac{GM}{a}\dfrac{1+e}{1-e}\right)^{1/2}E\left(\dfrac{2\sqrt{e}}{1+e}\right)$	$\dfrac{GM}{a}\left(\dfrac{2a}{p}-1\right)$
M	$\dfrac{2}{\pi}\left(\dfrac{GM}{a}\right)^{1/2}E(e)$	$\dfrac{GM}{a}$

where K and E are the complete elliptic integrals of the first and second kinds.

12. Let a particle be in ordinary Keplerian motion on an ellipse with eccentricity e. Suppose that at a periapse passage the other body is instantaneously moved from its current location to the other focus of the ellipse. Show that the particle's new eccentricity is given by $e(e+3)/(1-e)$.

13. Show that if the motion is hyperbolic, then the angle between the asymptotes is $2\tan^{-1}[(2\mathscr{E})^{1/2}L/GM]$.

14. Show that for small inclinations

$$u \approx \lambda - \Omega + \sin 2(\lambda - \Omega) \tan^2\left(\frac{i}{2}\right)$$

where λ is the longitude.

15. Show that the circular and parabolic speeds at a given point represent the lower and upper limits to the elliptic motion through that point. (No calculation is necessary!)

16. Deduce that for any three longitudes (λ_j) and latitudes (β_j)

$$\tan \beta_1 \sin(\lambda_2 - \lambda_3) + \tan \beta_2 \sin(\lambda_3 - \lambda_1) + \tan \beta_3 \sin(\lambda_1 - \lambda_2) = 0$$

under central force motion.

17. Show that a bound orbit under the potential $-k/r + C/2r^2$ is a precessing ellipse

$$r = \frac{a(1-e^2)}{1 + e \cos \gamma v}$$

If $|\gamma - 1|$ is small, derive an approximate formula for the precession rate of periapse.

18. Let a particle move in a circular orbit of radius R. Let the center of force be interior to the orbit. Suppose that v is the minimum speed and that V is the maximum speed of the particle. Prove that its period is given by $\pi(v + V)R/(vV)$.

19. Regard the Earth's orbit as circular. Suppose that a comet on a parabolic orbit about the Sun crosses the Earth's orbit moving toward perihelion. Derive a general expression for the length of time that the comet spends interior to the Earth's orbit in terms of the comet's perihelion distance. Show that this time is bounded above by $2/(3\pi)$ years.

20. Show that motion under the potential $kr^2/2$ is on an ellipse centered at the origin.

21. Show that for a polar orbit $(i = \pi/2)$ the absolute value of the Jacobian $\partial(\mathbf{r}, \dot{\mathbf{r}})/\partial(\mathbf{a})$ is equal to (use spherical coordinates)

$$\frac{e(GM)^2 \sec^2 u}{2ar^4}$$

22. Show that if $M \in [0, \pi/2 - e]$, then

$$M + e \geqslant E \geqslant \frac{\pi M}{\pi - 2e}$$

and if $M \in [\pi/2 - e, \pi]$, then

$$M + e \geqslant E \geqslant \frac{\pi(M + 2e)}{\pi + 2e}$$

(Smith 1979).

23. Derive

$$|\dot{\mathbf{r}}|^2 = n^2 a^2 \left[1 + \sum_{m=1}^{\infty} J_m(me) \cos mM \right]$$

$$\frac{|\dot{\mathbf{r}}|}{an} = P_{1/2}(z) + 2 \sum_{m=1}^{\infty} \frac{2^m}{(2m+1)!!} P_{1/2}^m(z) \cos mv$$

$$\frac{an}{|\dot{\mathbf{r}}|} = P_{-1/2}(z) + 2 \sum_{m=1}^{\infty} \frac{2^m}{(2m-1)!!} P_{-1/2}^m(z) \cos mv$$

$$\left(\frac{an}{|\dot{\mathbf{r}}|} \right)^2 = 1 + 2 \sum_{m=1}^{\infty} (-e)^m \cos mv$$

(Kinoshita 1977) where the P's are Legendre polynomials and

$$z = \frac{1 + e^2}{1 - e^2}$$

24. Show that

$$E = M + e \sin M + \frac{e^2}{2!2} 2 \sin 2M + \frac{e^3}{3!2^2} (3^2 \sin 3M - 3 \sin M)$$

$$+ \frac{e^4}{4!2^3} (4^3 \sin 4M - 4 \times 2^3 \sin 2M)$$

$$+ \frac{e^5}{5!2^4} (5^4 \sin 5M - 5 \times 3^4 \sin 3M + 10 \sin M)$$

$$+ \frac{e^6}{6!2^5} (6^5 \sin 6M - 6 \times 4^5 \sin 4M + 15 \times 2^5 \sin 2M)$$

$$\frac{r}{a} = 1 - e \cos M + \frac{e^2}{2} (1 - \cos 2M) - \frac{e^3}{2!2^2} (3 \cos 3M - 3 \cos M)$$

$$- \frac{e^4}{3!2^3} (4^2 \cos 4M - 4 \times 2^2 \cos 2M)$$

$$- \frac{e^5}{4!2^4} (5^3 \cos 5M - 5 \times 3^3 \cos 3M + 10 \cos M)$$

$$- \frac{e^6}{5!2^5} (6^4 \cos 6M - 6 \times 4^4 \cos 4M + 15 \times 2^4 \cos 2M)$$

$$v = M + 2e \sin M + \tfrac{5}{4}e^2 \sin 2M + e^3(\tfrac{13}{12} \sin 3M - \tfrac{1}{4} \sin M)$$
$$+ e^4(\tfrac{103}{96} \sin 4M - \tfrac{11}{24} \sin 2M)$$
$$+ e^5(\tfrac{1097}{960} \sin 5M - \tfrac{43}{64} \sin 3M + \tfrac{5}{96} \sin M)$$
$$+ e^6(\tfrac{1223}{960} \sin 6M - \tfrac{451}{480} \sin 4M + \tfrac{17}{192} \sin 2M)$$

25. Show that the first few terms of the f and g series are given by (see Eqs. 2.27 and 2.28; $GM = 1$)

$$f = 1 - \frac{at^2}{2!} + \frac{3abt^3}{3!} + \frac{(-15ab^2 + 3ac - 2a^2)t^4}{4!} + \cdots$$

$$g = t - \frac{at^3}{3!} + \frac{6abt^4}{4!} - \cdots$$

where

$$a = \frac{1}{r^3}, \qquad b = \frac{\dot{r}}{r}, \qquad c = \left(\frac{\dot{r}}{r}\right)^2 + \frac{\ddot{r}}{r}$$

chapter three

Coordinate Systems and Time Systems

TYPES OF COORDINATE SYSTEMS

We need precisely defined coordinate systems on the Earth and on the sky. The former are generically known as geographic coordinate systems, the latter as celestial coordinate systems. Three types of celestial coordinate systems—*horizon*, *equatorial*, and *ecliptic*—will be introduced in the next section. As they evolved from the apparent, Earth-centered view that mankind has of the Universe, they are all spherical coordinate systems in which the distance to an object plays a subordinate role. With the advent of radars and lasers, the refinement of trigonometric parallaxes, and always within the solar system, distance plays a more substantive role. After the sundry coordinate systems are introduced, the subjects of rotational and translational transformations between them are discussed. This involves the concept of parallax. Different formulas are routinely used in the artificial satellite, solar system, and stellar cases—all three are treated herein.

Likewise, three types of geographic coordinate systems—*astronomical*, *geodetic*, and *geocentric*—will be presented. A spherical coordinate system may be characterized by a fundamental or reference (great) circle, a particular point on this circle (the origin of the longitudinal or azimuthal coordinate), one of the poles of this circle (the origin of the colatitude), and a sense of helicity. The origin of the coordinate system itself may be observer centered (*topocentric*), Earth centered (*geocentric*), or Sun centered (*heliocentric*). It is usual for the positive z axis to point toward the selected pole of the reference circle and for the positive x axis to point toward the origin of longitude. The y axis lies in the plane of the fundamental circle and both right-handed and left-handed coordinate systems are in common usage.

Certainly when discussing dynamics the ideal would be to use an inertial coordinate system. Within the solar system the closest commonly used approximation to an inertial frame is the solar system barycentric reference frame. It is based on the ecliptic, the vernal equinox, and the North Celestial Pole. However, the ecliptic is in motion and the vernal equinox moves on a fixed ecliptic. Therefore, one needs to specify an epoch for such a coordinate system. This topic (precession and nutation) is treated in the next chapter. For now it is of no consequence.

After concluding this overview of coordinate systems, a brief overview of time systems is provided. Then with both **r** and t thoroughly explored, we can get back to dynamics [but not until Chapter Five—the next chapter deals with data reduction].

Celestial Coordinate Systems

The *celestial sphere* is the imaginary spherical surface, centered on the observer (or Earth), on which the stars and planets have apparently been placed. Its radius is infinite. The boundary between the visible and invisible portions of the celestial sphere is called the *horizon*. The poles of the horizon, those points directly overhead and beneath, are called the zenith and the nadir.

The celestial sphere appears to rotate about a fixed point. This point is known as the *North (South) Celestial Pole* in the northern (southern) terrestrial

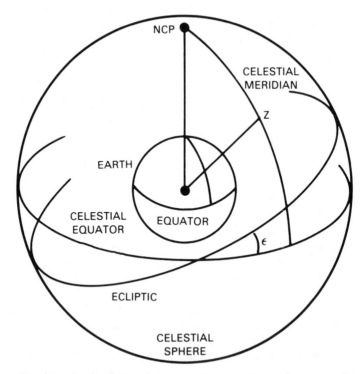

Figure 6. Celestial sphere with principal great circles (ecliptic, celestial equator, celestial meridian) indicated. NCP marks the North Celestial Pole, Z the astronomical zenith, and ε is the obliquity of the ecliptic.

hemisphere. The axis of the earth's rotation pierces the celestial sphere in these two points. The great circle passing through the celestial poles and the (astronomical) zenith is called the *celestial meridian*. Clearly it also passes through the nadir. The celestial meridian intersects the (astronomical) horizon at the north and south points. The great circle passing through the astronomical zenith and the nadir, and orthogonal to the celestial meridian at the zenith, is called the prime vertical. It intersects the astronomical horizon at the east and west points. See Fig. 6.

The Sun and planets move in a nearly coplanar fashion on the celestial sphere. This plane is approximately that of the *ecliptic*—the intersection of the instantaneous mean orbital plane of the Earth with the celestial sphere. The angle between the ecliptic and the celestial equator (the great circle cut on the celestial sphere whose normal is the Earth's instantaneous axis of rotation) is known as the *obliquity of the ecliptic*. It is about 23°.5. The two intersections of the ecliptic with the celestial equator are known as the equinoctial points (they are the nodes of the Earth's orbit). The Sun appears to pass through the vernal equinox (or spring equinox or the First Point of Aries) when moving northward. This usually occurs near March 21. Six months (roughly) later the sun appears to pass through the autumnal equinox moving southward. The Sun is farthest from the celestial equator at the summer and winter solstices (roughly June 21 and December 22).

The Horizon System

The visible horizon is the projection onto the celestial sphere of the local surface of the Earth as seen by the observer. This is rarely a great circle. The *astronomical horizon* is the great circle defined by the intersection of the celestial sphere and a plane whose normal is given by the direction of the local pseudo-gravitational field (i.e., it includes the Coriolis acceleration). This direction is known as the *astronomical vertical*. The line defining the astronomical vertical pierces the celestial sphere in the *astronomical zenith* (overhead) and in the nadir. These points are, therefore, the poles of the astronomical horizon, and the astronomical zenith is the special one chosen. The origin of longitudes (in this book) will be the south point with negative helicity.

Consider a point P on the celestial sphere (see Fig. 7). The *altitude a* of the point P is the angular distance measured positive (negative) towards the astronomical zenith (nadir) from the astronomical horizon along the great circle passing through the point P and the astronomical zenith (Z). See Fig. 7. The complement of the altitude is called the zenith distance, $z = 90° - a$. If $a < 0$ ($z > 90°$), the quantity $-a$ is called the depression. Altitude is sometimes incorrectly termed elevation. The altitude of the North Celestial Pole is the observer's *astronomical latitude* Φ. The *azimuth A* of the point P is the angular distance measured toward the west, from the south, along the astronomical horizon to the intersection of the great circle passing through the points P and Z with the astronomical horizon. A common feature of all spherical coordinate systems is the ambiguity of the longitude coordinate (azimuth here) at the poles of the fundamental circle.

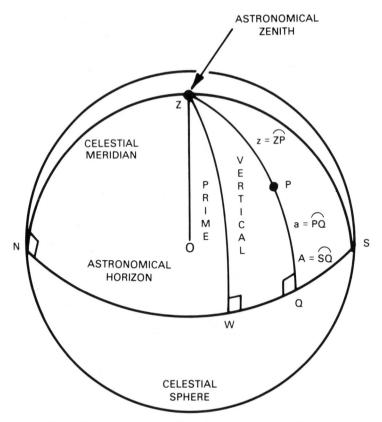

Figure 7. Definition of horizon system coordinates z (or a) and A. N, W, and S mark the north, west, and south points. The object of interest is at P.

The position vector of a point with altitude a (or zenith distance $z = 90° - a$) and azimuth A is

$$\mathbf{l}(A, a) = \begin{pmatrix} \cos a \cos A \\ \cos a \sin A \\ \sin a \end{pmatrix} = \begin{pmatrix} \sin z \cos A \\ \sin z \sin A \\ \cos z \end{pmatrix} = \mathbf{l}(A, 90° - z) \qquad (3.1)$$

and $\mathbf{l} \cdot \mathbf{l} = 1$. The location of P, when its distance is r, is given by $\mathbf{r} = r\mathbf{l}$. Note that the horizon system is left handed.

As the horizon system is topocentric, its theoretical usefulness (due both to being noninertial and dependent on one's geographic location) is limited.

Example 3.1. Three systems of units are used for angular measure—radians, decimal degrees, and sexagesimal (degrees, minutes, seconds, and fractions thereof). Different subfields have different traditions and use different systems. Express the position $A = 123°45'67''.890$, $a = 13°57'24''.680$ in the other two systems.

It is useful to remember two general conversion factors. One second of arc is $\pi/(180 \times 60 \times 60) = \pi/6.48 \times 10^5 = 1/206264.8062 = 4.848136812 \times 10^{-6}$ in radians or about $5\,\mu\text{rad}$. Also $1\,\text{mdeg} = 3''6$, $1\,\mu\text{deg} = 0''004$. Finally $A = 123°45'67''890$ means $123° + (45 + 67.890/60)°/60$ or $123°7688583$ but neither $123°768858$ nor $123°76885833$. The reason is that the original azimuth was given to the nearest milli-arc second, that is to $\pm 5 \times 10^{-4}$ arc seconds. Now this, in decimal degrees, is 1.4×10^{-7} $(= 5 \times 10^{-4}/3600)$ so that seven decimal places (no more, no less) are appropriate. Thus, $a = 13°9568556$. Lastly, in radian measure, $A = 2.160174089$ and $a = 0.243593083$.

Example 3.2.   Compute the vector $l(A, a)$ for the above position.

Here the question of precision is not so easily dealt with because, depending on the value of the angle itself and the trigonometric function in question, more or less digits are required. One cannot treat each case separately and a uniform point of view—eight decimal digits—will be adopted. So, from Eq. 3.1

$$l = \begin{pmatrix} -0.53943404 \\ 0.80674515 \\ 0.24119118 \end{pmatrix}$$

The Equatorial Systems

The fundamental circle of the two equatorial systems is the celestial equator. The particular pole of the celestial equator that is used is the North Celestial Pole. To fix a particular place on the celestial equator, the vernal equinox is chosen. It is symbolized by Υ and positive helicity is used.

The *declination* δ of the point P (see Fig. 8) is the angular distance measured positive (negative) toward the North (South) Celestial Pole from the celestial equator along the great circle passing through the point P and the North Celestial Pole (NCP). The complement of the declination is called the north polar distance. The *right ascension* α of the point P is the angular distance measured toward the east, from the vernal equinox, along the celestial equator to the intersection of the great circle passing through the points P and the NCP with the celestial equator.

The position vector of a point with declination δ and right ascension α is

$$l(\alpha, \delta) = \begin{pmatrix} \cos\delta \cos\alpha \\ \cos\delta \sin\alpha \\ \sin\delta \end{pmatrix} \tag{3.2}$$

Declination is usually reckoned in sexagesimal arc measure whereas right ascension is usually reckoned in sexagesimal time measure. The connection between the two systems of units is based on the fact that a complete revolution of the celestial sphere covers 360° and requires 24^h to complete. Thus, $1^h = 15°$, $1^m = 15'$, and $1^s = 15''$. The location of P, when its distance is r, is given by $\mathbf{r} = rl(\alpha, \delta)$.

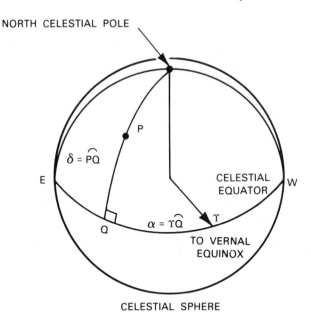

NORTH CELESTIAL POLE

Figure 8.   Definition of equatorial system coordinates δ and α. E is the east point and the object of interest is at P.

As will be discussed in Chapter Four in depth, at a fixed epoch (which really specifies the exact position of the North Celestial Pole and the vernal equinox) the right ascension/declination system is an inertial one. Common epochs have been 1900.0 and 1950.0. The future standard epoch will be 2000.0.

The close connection between right ascension and time becomes even more apparent if one considers the angular distance, measured positive westward along the celestial equator, from the intersection of the celestial meridian with the celestial equator to the intersection of the great circle through the points P and the NCP with the celestial equator. This measure, for the stars, directly gives the time until meridian passage. This quantity is called the *hour angle* h. Hour angle is usually reckoned positive westward (0^h to $+12^h$) of the celestial meridian and negative eastward (0^h to -12^h) of the celestial meridian. Obviously, when an object possesses a large intrinsic or proper motion, this simple interpretation of hour angle is no longer valid. The relationship between hour angle and right ascension is simply

$$h = \tau - \alpha \qquad\qquad (3.3)$$

where τ is the hour angle of the vernal equinox. This quantity is also the right ascension of a point on the celestial meridian or the *sidereal time*. Note that the hour angle/declination coordinate system is not an inertial one.

Example 3.3.   In the use of hour angles we see yet a fourth system of measurement—sexagesimal time measure. The factor of 15 difference between a second of time and a second of arc ($1^s = 15''$) causes no end of difficulty. It is, however, historically embedded in the subject. Given $\alpha = 123^\circ\!.456789$ and $\delta = -0^\circ\!.9876543$, convert them into normal astronomical usage.

The right ascension, after division by 15, is $8^h\!.230453$ or $8^h13^m49^s\!.63$. That it should be this rather than $8^h\!.2304526$ and $8^h13^m49^s\!.629$ can be seen by examining the number of truly significant digits in each instance. Nor can it be that the declination is equivalent to $-0°59'15''\!.555$, which shows the nature of the problem.

Example 3.4. Suppose that the hour angle for the above position is $20^h20^m07^s\!.00$. What is the sidereal time?

Since $\tau = h + \alpha = 28^h33^m56^s\!.63$ or, as times are normally given modulo 24, $\tau = 4^h33^m56^s\!.63$.

The Ecliptic System

The ecliptic or celestial latitude β of the point P is the angular distance measured positive (negative) toward the north (south) pole of the ecliptic from the ecliptic along the great circle passing through the points P and the north pole of the ecliptic. The ecliptic or celestial longitude λ of the point P is the angular distance measured toward the east, from the vernal equinox, along the ecliptic to the intersection of the great circle passing through the points P and the north pole of the ecliptic with the celestial equator. The position vector of a point with ecliptic latitude β and ecliptic longitude λ is

$$\mathbf{l}(\lambda, \beta) = \begin{pmatrix} \cos \beta \cos \lambda \\ \cos \beta \sin \lambda \\ \sin \beta \end{pmatrix} \tag{3.4}$$

Again, if the distance of P is r, its location is $\mathbf{r} = r\mathbf{l}(\lambda, \beta)$.

Rotational Transformations of Celestial Coordinates

Transformations between the celestial coordinate systems can be accomplished by the compounding of elementary rotations or by solving the appropriate spherical triangles. As mentioned above, in going to or from the horizon system, the observer's astronomical latitude (Φ) and the sidereal time (τ) are needed (because it is noninertial). The obliquity of the ecliptic (ε) is necessary when going to or from the ecliptic system. The results are:

1. From the horizon system to the hour angle and declination system

$$\mathbf{l}(h, \delta) = R_2(\Phi - 90°)\mathbf{l}(A, a)$$
$$= R_2(\Phi - 90°)\mathbf{l}(A, 90° - z) \tag{3.5}$$

or

$$\cos \delta \cos h = \cos \Phi \sin a + \sin \Phi \cos a \cos A$$
$$\cos \delta \sin h = \cos a \sin A \qquad \qquad (3.6)$$
$$\sin \delta = \sin \Phi \sin a - \cos \Phi \cos a \cos A$$

2. From the hour angle and declination system to the horizon system

$$l(A, a) = R_2(90° - \Phi)l(h, \delta) \qquad \qquad (3.7)$$

or

$$\cos a \cos A = -\cos \Phi \sin \delta + \sin \Phi \cos \delta \cos h$$
$$\cos a \sin A = \cos \delta \sin h \qquad \qquad (3.8)$$
$$\sin a = \sin \Phi \sin \delta + \cos \Phi \cos \delta \cos h$$

3. From the ecliptic system to the equatorial system

$$l(\alpha, \delta) = R_1(-\varepsilon)l(\lambda, \beta) \qquad \qquad (3.9)$$

or

$$\cos \delta \cos \alpha = \cos \beta \cos \lambda$$
$$\cos \delta \sin \alpha = -\sin \varepsilon \sin \beta + \cos \varepsilon \cos \beta \sin \lambda \qquad \qquad (3.10)$$
$$\sin \delta = \cos \varepsilon \sin \beta + \sin \varepsilon \cos \beta \sin \lambda$$

4. From the equatorial system to the ecliptic system

$$l(\lambda, \beta) = R_1(\varepsilon)l(\alpha, \delta) \qquad \qquad (3.11)$$

or

$$\cos \beta \cos \lambda = \cos \delta \cos \alpha$$
$$\cos \beta \sin \lambda = \sin \varepsilon \sin \delta + \cos \varepsilon \cos \delta \sin \alpha \qquad \qquad (3.12)$$
$$\sin \beta = \cos \varepsilon \sin \delta - \sin \varepsilon \cos \delta \sin \alpha$$

The interrelationship between the second and the first equatorial systems is given in Eq. 3.3.

Example 3.5.   My wife's favorite star is Alpha Orionis, which is located at $\alpha = 05^h54^m15\overset{s}{.}020$, $\delta = +07°24'17\overset{''}{.}41$ (1983.0). Her birthday is on May 26. Will she be able to see Betelgeuse from the Anglo–Australian Observatory at local midnight on her birthday? (If the answer is yes, she wants to know does this mean she gets to go!) The location of the 3.9-m reflector is $\lambda = 210°56'02\overset{''}{.}10$ (W), $\Phi = -31°16'37\overset{''}{.}3$, height above sea level $H = 1164$ m. The local sidereal time at 0^h local time May 26, 1983 is $\tau = 02^h08^m16\overset{s}{.}319$.

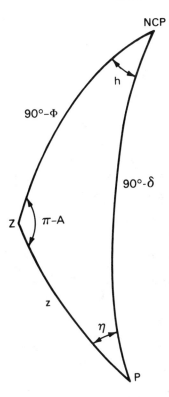

Figure 9. The astronomical triangle and the parallactic angle η.

simplifies the formula for the astronomical refraction correction in equatorial coordinates. The complete solution of the astronomical triangle in most applications is given by Eqs. 3.5–3.8 and 3.13 and 3.14:

$$\text{sgn}(\Phi) \sin z \cos \eta = \sin \Phi \cos \delta - \cos \Phi \sin \delta \cos h \qquad (3.13)$$

$$\sin z \sin \eta = \cos \Phi |\sin h|$$

$$\text{sgn}(\Phi) \cos \delta \cos \eta = \sin \Phi \sin z + \cos \Phi \cos z \cos A \qquad (3.14)$$

$$\cos \delta \sin \eta = \text{sgn}(h) \cos \Phi \sin A$$

Classical Formula

A problem that arises in solar system dynamics is the change of the orbital elements from one reference plane (say the ecliptic) to another (say Jupiter's orbital plane). A straightforward method of handling such problems is to use D'Alembert's analogies for a spherical triangle. Let the orbital elements inclination i, argument of periapse ω, and longitude of the ascending node Ω be symbolized by unprimed symbols relative to one reference frame and by the corresponding primed symbols when referred to the other. Note that there is

The hour angle is $h = \tau - \alpha = 20^h14^m01\overset{s}{.}299 = -03^h45^m58\overset{s}{.}701$, which does not look good. From the last part of Eqs. 3.8, $\sin a = 0.40094399$, or $a = 23\overset{\circ}{.}63721$, so that the star will be up $(a > 0)$, but low in the (obviously) northern sky. From the first two of Eqs. 3.8

$$\cos a \cos A = -0.39435238, \qquad \cos a \sin A = -0.82687975$$

whence $A = 244\overset{\circ}{.}50277$, which implies that it is almost due northeast (which the hour angle told us). Had we used the matrix form in Eq. 3.7, then

$$
\mathbf{l}(A, a) = \begin{pmatrix} -0.51917648 & 0 & -0.85466706 \\ 0 & 1 & 0 \\ 0.85466706 & 0 & -0.51917648 \end{pmatrix} \begin{pmatrix} 0.54741210 \\ -0.82687975 \\ 0.12887930 \end{pmatrix}
$$

$$
= \begin{pmatrix} -0.39435238 \\ -0.82687975 \\ 0.40094399 \end{pmatrix}
$$

yielding the same altitude and azimuth. To complete the numerical exercise, suppose that $\varepsilon = 23°26'29''\!.38$. What are Betelgeuse's ecliptic coordinates? From either Eqs. 3.11 or 3.12

$$
\mathbf{l}(\lambda, \beta) = \begin{pmatrix} 0.02487582 \\ 0.96079884 \\ -0.27612818 \end{pmatrix}
$$

Whence (α and λ are always in the same quadrant!)

$$\lambda = 88°31'00''\!.84, \qquad \beta = -16°01'45''\!.33$$

It is sometimes desirable to transform between the ecliptic coordinate system and the horizon coordinate system. The trick is to remember that the declination of your astronomical zenith is equal to your astronomical latitude Φ and that the right ascension of your astronomical zenith is equal to the local sidereal time τ. Thus, the ecliptic coordinate position vector of the astronomical zenith is

$$
\mathbf{l}_z = \begin{pmatrix} \cos \Phi \cos \tau \\ \sin \varepsilon \sin \Phi + \cos \varepsilon \cos \Phi \sin \tau \\ \cos \varepsilon \sin \Phi - \sin \varepsilon \cos \Phi \sin \tau \end{pmatrix}
$$

The Astronomical Triangle

The astronomical triangle is that spherical triangle which has vertices at the North Celestial Pole, the astronomical zenith, and the point in question; see Fig. 9. The lengths of the sides are the north polar distance $(90° - \delta)$, the zenith distance (z), and the complement of the astronomical latitude $(90° - \Phi)$. The corresponding angles are the supplement of the azimuth $(180° - A)$, the hour angle (h), and the *parallactic angle* (η). The use of the parallactic angle no translation involved, only rotations. The inclination and longitude of the ascending node of the primed reference frame upon the unprimed one are I, L. Finally, allow the origins of longitude in the two systems to differ by L_0 (=the longitude of the primed zero longitude in the unprimed system). After substitution into the aforementioned standard relationships, one can see that

$$\sin\left[\frac{(\Omega' - L_0) + (\omega - \omega)}{2}\right] \sin\left(\frac{i'}{2}\right) = \sin\left(\frac{\Omega - L}{2}\right) \sin\left(\frac{i + I}{2}\right)$$

$$\cos\left[\frac{(\Omega' - L_0) + (\omega - \omega')}{2}\right] \sin\left(\frac{i'}{2}\right) = \cos\left(\frac{\Omega - L}{2}\right) \sin\left(\frac{i - I}{2}\right)$$

$$\sin\left[\frac{(\Omega' - L_0) - (\omega - \omega')}{2}\right] \cos\left(\frac{i'}{2}\right) = \sin\left(\frac{\Omega - L}{2}\right) \cos\left(\frac{i + I}{2}\right)$$

$$\cos\left[\frac{(\Omega' - L_0) - (\omega - \omega')}{2}\right] \cos\left(\frac{i'}{2}\right) = \cos\left(\frac{\Omega - L}{2}\right) \cos\left(\frac{i - I}{2}\right)$$

For the special case of ecliptic–equatorial transformations, let the primed quantities be relative to the celestial equator and the unprimed relative to the ecliptic. Then, with ε equal to the obliquity of the ecliptic,

$$\sin i \sin \Omega = \sin i' \sin \Omega'$$

$$\sin i \cos \Omega = -\sin \varepsilon \cos i' + \cos \varepsilon \sin i' \cos \Omega'$$

$$\cos i = \cos \varepsilon \cos i' + \sin \varepsilon \sin i' \cos \Omega'$$

$$\sin i \sin(\omega' - \omega) = \sin \varepsilon \sin \Omega'$$

$$\sin i \cos(\omega' - \omega) = \cos \varepsilon \sin i' - \sin \varepsilon \cos i' \cos \Omega'$$

and inversely

$$\sin i' \sin \Omega' = \sin i \sin \Omega$$

$$\sin i' \cos \Omega' = \sin \varepsilon \cos i + \cos \varepsilon \sin i \cos \Omega$$

$$\cos i' = \cos \varepsilon \cos i - \sin \varepsilon \sin i \cos \Omega$$

$$\sin i' \sin(\omega' - \omega) = \sin \varepsilon \sin \Omega$$

$$\sin i' \cos(\omega' - \omega) = \cos \varepsilon \sin i + \sin \varepsilon \cos i \cos \Omega$$

Geographic Coordinate Systems

There are three commonly used coordinate systems associated with the Earth. One is the astronomical system of geographic coordinates. It is based on the direction to the astronomical vertical and the direction to the elevated celestial pole. This system is, therefore, directly realizable and independent of the form and figure of the Earth. Another system, the geodetic system of geographic coordinates, is based on a model for the size and shape of the Earth. The third coordinate system, with its origin at the center of the Earth, is called the geocentric coordinate system. In the next three sections these systems and their interrelationships are described.

The Astronomical System

The geographical poles of rotation are the two points on the Earth's surface where the axis of rotation pierces the surface. The prolongation of the Earth's axis of rotation meets the celestial sphere in the celestial poles. A plane through the Earth's center, perpendicular to the Earth's axis of rotation, intersects the Earth's surface in the geographical equator of rotation. This plane intersects the celestial sphere along the *celestial equator*. In addition to the direction to the elevated celestial pole, the other immediately realizable direction is that of the *astronomical vertical*. This is determined by the direction of the local pseudo-gravitational field. The extension of a plumb line intersects the celestial sphere in the astronomical zenith (above) and the nadir (below). The customary zero point of the longitude coordinate is the site of the Airy transit circle at the Old Royal Greenwich Observatory.

The *astronomical latitude* Φ of a point is the complement of the acute angle between the elevated celestial pole and the direction of the astronomical vertical and is positive (negative) in the northern (southern) hemisphere. The astronomical equator is the locus of points on the surface of the Earth such that $\Phi = 0°$. The astronomical equator is not a plane curve, nor are the astronomical latitudes of the geographical poles of rotation necessarily ±90°. The altitude of the elevated celestial pole is numerically equal to the observer's astronomical latitude. The declination of the astronomical zenith is algebraically equal to the astronomical latitude. The astronomical longitude Λ is the dihedral angle between the plane of the local astronomical meridian and the plane of the astronomical meridian through the site of the Airy transit circle at the Old Royal Greenwich Observatory. The reader will remember that the plane of the local astronomical meridian includes the celestial poles, the astronomical zenith, and the nadir. Hence it intersects the celestial sphere in the celestial meridian. The astronomical meridian is the locus of points on the surface of the Earth such that $\Lambda = $ constant. This too is not necessarily a plane curve. There are many conventions for giving longitudes (e.g., continuously eastward from 0° to 360°, continuously westward from 0° to 360°, both eastward and westward from 0° to 180°). The exact meaning of a particular longitude should always be provided.

The Geodetic System

The Earth's surface is nearly that of an oblate spheroid. Hence modern models use such a figure of revolution as a reference surface. Two independent parameters uniquely specify such an ellipsoid. It is important to realize that the definitions of the geodetic coordinates are independent of the numerical values of the parameters of the ellipsoid.

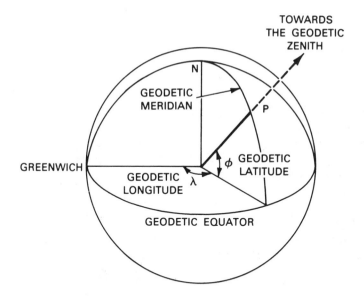

Figure 10. Geodetic latitude (ϕ) and longitude (λ).

The geodetic poles are the intersection of the axis of revolution of the generating ellipse with the surface of the ellipsoid. The geodetic equator is the intersection of the plane swept out by the major axis of the generating ellipse and the surface of the ellipsoid. The direction of the geodetic vertical is given by the local outward normal to the ellipsoid. A line along the direction of the geodetic vertical intersects the celestial sphere in the geodetic zenith.

The *geodetic latitude* ϕ of a point is the acute angle between the direction of the geodetic vertical and the plane of the geodetic equator. This is shown in Fig. 10. The *geodetic longitude* λ is the dihedral angle between the plane of the local geodetic meridian and the plane of the geodetic meridian through the site of the Airy transit circle at the Old Royal Greenwich Observatory. The angle between the astronomical and geodetic verticals is known as the deflection of the vertical (and is usually $\leqslant 20''$).

The Geocentric System

It is customary to specify the reference ellipsoid by the size of the semi-major axis a and its flattening* f. These are related to the semi-minor axis c and the eccentricity e by

$$f = \frac{a-c}{a} = 1 - (1-e^2)^{1/2}$$

---

* The quantity f is a purely geometrical one. The dynamical flattening (or mechanical ellipticity) of the Earth is the quantity $(C-A)/C$ where A is the equatorial moment of inertia and C is the polar moment of inertia. If the Earth were an oblate spheroid of uniform density, then $(C-A)/C$ would be equal to $f - f^2/2 = e^2/2$.

$$e = \left(1 - \frac{c^2}{a^2}\right)^{1/2} = (2f - f^2)^{1/2}$$

The *geocentric latitude* ϕ' of a point is the acute angle between the geocentric radius vector to that point and the geodetic equator. See Fig. 10. The *geocentric longitude* is equal to the geodetic longitude λ. The *geocentric distance* ρ is the magnitude of the geocentric radius vector to the point in question. The difference between geodetic and geocentric latitudes is called the angle of the vertical $(= \phi - \phi')$.

From the geometry of a meridian section of the reference ellipsoid it follows that (see Fig. 11)

$$\rho \cos \phi' \equiv C \cos \phi, \qquad \rho \sin \phi' \equiv S \sin \phi \qquad (3.15)$$

where

$$S = \frac{a(1 - e^2)}{(1 - e^2 \sin^2 \phi)^{1/2}} = (1 - e^2)C$$

$$C = \frac{a}{[\cos^2 \phi + (1 - f)^2 \sin^2 \phi]^{1/2}} \qquad (3.16)$$

$$= \frac{S}{(1 - f)^2}$$

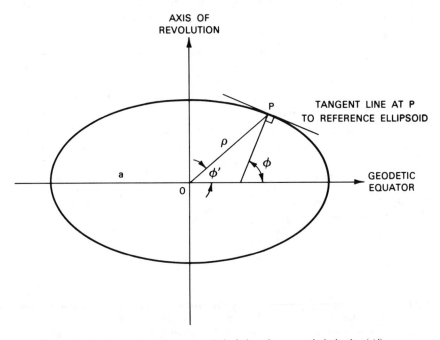

Figure 11. Relationship between geodetic (ϕ) and geocentric latitudes (ϕ').

Hence

$$\tan \phi' = (1-f)^2 \tan \phi$$

$$\rho^2 = C^2[\cos^2 \phi + (1-f)^4 \sin^2 \phi]$$

$$= \frac{C^2 + S^2}{2} + \frac{C^2 - S^2}{2} \cos 2\phi$$

$$= a^2 \frac{\cos^2 \phi + (1-f)^4 \sin^2 \phi}{\cos^2 \phi + (1-f)^2 \sin^2 \phi}$$

and, with $g = e^2/(2 - e^2)$,

$$\tan(\phi - \phi') = \frac{e^2 \sin \phi \cos \phi}{1 - e^2 \sin \phi} = \frac{g \sin 2\phi}{1 + g \cos 2\phi} = \frac{g \sin 2\phi'}{1 - g \cos 2\phi'}$$

The maximum departure of ϕ' from ϕ is $\tan^{-1}\{e^2/[2(1-e^2)^{1/2}]\}$ when $\sin \phi = 1/(2-e^2)^{1/2}$. This is $11'32''.7$ at $\phi = 45°05'46''.4$ for $1/f = 298.257$. Although the reference ellipsoid is a close approximation to the Earth's surface, local topographical irregularities abound on land. Hence the height H above the surface of the ellipsoid is necessary to complete the specification on one's geodetic location. With ρ, S, C, and H all measured in Earth radii the correction for $H \neq 0$ is adequately represented by modifying Eqs. 3.15 to

$$\rho \cos \phi' = (C + H) \cos \phi, \qquad \rho \sin \phi' = (S + H) \sin \phi \qquad (3.17)$$

with no significant change in the angle of the vertical.

Example 3.6. M.I.T.'s Wallace Observatory is located at $\lambda = 71°29'05''$ (W), $\phi = +42°36'37''$, $H = 107$ m. Use $1/f = 298.257$ and compute its geocentric location.

The computational sequence is straightforward; first calculate S and C from Eqs. 3.16:

$$\frac{S}{a} = 0.99483301, \qquad \frac{C}{a} = 1.00153769$$

Then compute ρ and ϕ' from Eq. 3.17 using $a = 6378.140$ km,

$$\rho = 0.99848703, \qquad \phi' = 42°25'06''.89$$

Had we not made the correction for $H \neq 0$, we would have found that $\phi' = 42°25'06''.88$ (Wallace is relatively low).

Variations of the Geographic Coordinates. From the definition of the direction of the astronomical vertical it should be clear that in addition to the gravitational attraction of the Earth and the Coriolis force due to the Earth's rotation, the gravitational attractions of the Moon and Sun contribute to the orientation of a plumb line. Thus, there are monthly and annual periodic variations of any geographically fixed observer's astronomical latitude. Separate from these small effects the geographical poles of rotation are not fixed to the solid body of the Earth. This polar motion (or polar wandering or Chandler wobble) is caused by a combination of the free precession of the axis of rotation, lunisolar forces, and geophysical rearrangements of the mass distribution of the Earth. They are irregularly periodic, with the principal periods being 12 and 14 months. The net displacement of the pole during the interval 1900–1950 is less than 0″4. Hence, only in extremely precise work need these effects be taken into account.

THREE-DIMENSIONAL TRANSFORMATIONS

The principal changes of origin that one needs to study are from the observer to the center of the Earth and from the Earth's center to the Sun's center. In all cases the quantity (origin to origin distance)/(celestial object to origin distance) plays a crucial role. This quantity is generically referred to as *parallax*. More accurately the geocentric or diurnal parallax is the difference between the topocentric zenith distance and the geocentric zenith distance—both zenith distances being measured from the geocentric zenith. The situation is illustrated in Fig. 12. Using primes to denote the topocentric value of a quantity and p for the parallax,

$$p = z' - z$$

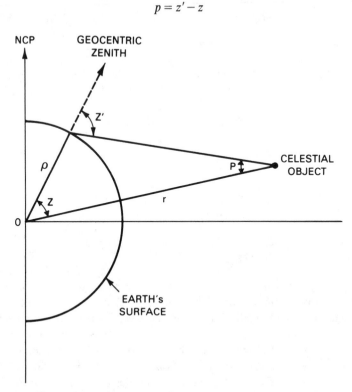

Figure 12.   The illustration of diurnal or geocentric parallax p.

From the geometry and the law of sines one can verify that

$$\sin p = \left(\frac{\rho}{r}\right) \sin z'$$

where ρ is the geocentric distance of the observer and r is the geocentric distance of the celestial object. The value of p when $z' = 90°$ is known as the *horizontal parallax*. When, in addition, ρ is equal to the equatorial radius of the Earth, this particular value of the horizontal parallax is called the *equatorial horizontal parallax*. Finally, the value of the equatorial horizontal parallax when the object is at its mean geocentric distance is called the *mean equatorial horizontal parallax*. When the approximation $p = \sin p$ is valid, then the parallax in azimuth may be neglected and the relation between the topocentric zenith distance measured from the geocentric zenith (z') and the topocentric zenith distance measured from the goedetic zenith ($\jmath'$) is

$$z' = \jmath' - (\phi - \phi') \cos A$$

Within errors introduced by the neglect of the deflection of the vertical, $\jmath'$ is also the topocentric zenith distance measured from the astronomical zenith.

There is no name for the corresponding parallax due to a heliocentric/geocentric change of origin. Although annual parallax might seem appropriate, this term is reserved for the parallactic displacement of stars as the Earth revolves about the Sun in its yearly journey. There is also a quantitative difference between the sizes of the heliocentric/geocentric parallax and the geocentric/topocentric parallax. Thus, in some cases approximate formulas are all that is needed whereas in other cases only the rigorous equations derived below will be sufficient.

Geocenter–Topocenter Parallax

Horizon System Diurnal Parallax

Let the topocentric location of a celestial object be $\mathbf{r}' = r'\mathbf{l}(A', 90° - z')$ where r' is the topocentric distance. The geocentric location is $r\mathbf{l}(A, 90° - z)$ where z is measured from the geodetic zenith. The observer's geocentric location is $\boldsymbol{\rho} = \rho[\sin(\phi - \phi'), 0, \cos(\phi - \phi')]$, and

$$\mathbf{r} = \mathbf{r}' + \boldsymbol{\rho} \tag{3.18}$$

Trigonometric manipulations of this fundamental relationship yield

$$\sin(z' - z) = \left(\frac{\rho}{r}\right) \cos(\phi - \phi') \sec \Gamma \sin(z' - \Gamma)$$

$$\sin(A' - A) = \left(\frac{\rho}{r}\right) \sin(\phi - \phi') \csc z \sin A' \tag{3.19}$$

$$\frac{r'}{r} = \sin(z - \Gamma) \csc(z' - \Gamma)$$

where

$$\tan \Gamma = \tan(\phi - \phi') \cos\left(\frac{A' + A}{2}\right) \sec\left(\frac{A' - A}{2}\right) \qquad (3.20)$$

Useful forms for the inverse relationships are

$$\tan(A' - A) = \frac{P \sin A}{1 - P \cos A}$$

$$\tan(z' - z) = \frac{Q \sin(z - \Gamma)}{1 - Q \cos(z - \Gamma)} \qquad (3.21)$$

$$\left(\frac{r'}{r}\right)^2 = 1 - 2Q \cos(z - \Gamma) + Q^2$$

with P and Q defined by

$$P = \left(\frac{\rho}{r}\right) \csc z \sin(\phi - \phi'), \qquad Q = \left(\frac{\rho}{r}\right) \sec \Gamma \cos(\phi - \phi')$$

When P and Q are both small, these reduce to (approximately),

$$\Gamma = (\phi - \phi') \cos A$$

$$A' - A = P \sin A$$

$$z' - z = Q \sin(z - \Gamma)$$

Equatorial System Diurnal Parallax

With $\mathbf{r}' = r'\mathbf{l}(\alpha', \delta)$, $\mathbf{r} = r\mathbf{l}(\alpha, \delta)$, and $\boldsymbol{\rho} = \rho\mathbf{l}(\tau, \phi')$ where τ is the local sidereal time, the implications of Eq. 3.18 are

$$\sin(\alpha - \alpha') = p \sin h'$$

$$\sin(\delta - \delta') = q \sin(\gamma - \delta') \qquad (3.22)$$

$$\frac{r'}{r} = \sin(\delta - \gamma) \csc(\delta' - \gamma)$$

where

$$\tan \gamma = \tan \phi' \cos\left(\frac{\alpha - \alpha'}{2}\right) \sec\left(h + \frac{\alpha - \alpha'}{2}\right) \qquad (3.23)$$

Useful forms for the inverse relationships are

$$\tan(\alpha - \alpha') = \frac{p \sin h}{1 - p \cos h}$$

$$\tan(\delta - \delta') = \frac{q \sin(\gamma - \delta)}{1 - q \cos(\gamma - \delta)} \qquad (3.24)$$

Here $h(h')$ is the geocentric (topocentric) hour angle, $h = \tau - \alpha$ $(h' = \tau - \alpha')$, and

$$p = \left(\frac{\rho}{r}\right) \cos \phi' \sec \delta, \qquad q = \left(\frac{\rho}{r}\right) \sin \phi' \csc \gamma \qquad (3.25)$$

When p and q are both small, these formulas may be approximated by

$$\alpha - \alpha' = p \sin h$$
$$\delta - \delta' = q \sin(\gamma - \delta)$$
$$\tan \gamma = \tan \phi' \sec h$$

The reader might be interested in seeing these computations in detail. Let us start with $\mathbf{r} = \mathbf{r'} + \boldsymbol{\rho}$ in component form; namely,

$$r \cos \delta \cos \alpha = r' \cos \delta' \cos \alpha' + \rho \cos \phi' \cos \tau \qquad (3.26a)$$

$$r \cos \delta \sin \alpha = r' \cos \delta' \sin \alpha' + \rho \cos \phi' \sin \tau \qquad (3.26b)$$

$$r \sin \delta = r' \sin \delta' + \rho \sin \phi' \qquad (3.26c)$$

Equation 3.26a is multiplied by $\sin \alpha'$, Eq. 3.26b by $\cos \alpha'$, and the results are subtracted. With $h' = \tau - \alpha'$

$$r \cos \delta \sin(\alpha - \alpha') = \rho \cos \phi' \sin h'$$

which, considering the definition of p in Eq. 3.25, is the top line of Eq. 3.22. The declination result is a bit more involved. This time Eq. 3.26a is multiplied by $\sin \alpha$ instead of $\sin \alpha'$ and Eq. 3.26b is multiplied by $\cos \alpha$ instead of $\cos \alpha'$. Then we subtract to derive, with $h = \tau - \alpha$,

$$r' \cos \delta' \sin(\alpha - \alpha') = \rho \cos \phi' \sin h$$

Returning to Eqs. 3.26 and multiplying Eq. 3.26a by $\cos \alpha$ and Eq. 3.26b by $\sin \alpha$ and adding,

$$r' \cos \delta' \cos(\alpha - \alpha') = r \cos \delta - \rho \cos \phi' \cos h$$

The first intermediate result is then multiplied by $\sin[(\alpha - \alpha')/2]$ and the second intermediate result by $\cos[(\alpha - \alpha')/2]$. These two are summed; the trigonometric addition theorems are used and the result divided by $\cos[(\alpha - \alpha')/2]$. We derive

$$r' \cos \delta' = r \cos \delta - \rho \cos \phi' \sec\left(\frac{\alpha - \alpha'}{2}\right) \cos\left(h + \frac{\alpha - \alpha'}{2}\right)$$

From Eq. 3.23 γ is introduced and this becomes

$$r' \cos \delta' = r \cos \delta - \rho \sin \phi' \cot \gamma$$

This is then multiplied by $\sin \delta'$, Eq. 3.26c repeated,

$$r' \sin \delta' = r \sin \delta - \rho \sin \phi'$$

and multiplied by $\cos \delta'$. Subtracting yields

$$r \sin(\delta - \delta') = \rho \sin \phi' \csc \gamma \sin(\gamma - \delta')$$

Remembering the definition of q in Eq. 3.25, one can see that this is the middle line of Eqs. 3.22. Had, instead, the last two intermediaries been solved for $\rho \sin \phi'$, we would have discovered that

$$\rho \sin \phi' = r \sin \delta - r' \sin \delta' = (r \cos \delta - r' \cos \delta') \tan \gamma$$

which is an unreduced form of the bottom line of Eq. 3.22.

To deduce Eqs. 3.24, two of the intermediate results obtained above will be repeated:

$$r \cos \delta' \sin(\alpha - \alpha') = \rho \cos \phi' \sin h$$

$$r \cos \delta' \cos(\alpha - \alpha') = r \cos \delta - \rho \cos \phi' \cos h$$

Division yields the $\tan(\alpha - \alpha')$ formula. The last result is obtained as follows; reiterate the r', δ' equations,

$$r' \cos \delta' = r \cos \delta - \rho \sin \phi' \cot \gamma$$

$$r' \sin \delta' = r \sin \delta - \rho \sin \phi'$$

The first equation is multiplied by $\cos \delta$ and the second by $\sin \delta$ and then added to deduce

$$r' \cos(\delta - \delta') = r - \rho \sin \phi' \csc \gamma \cos(\delta - \gamma)$$

The same pair are multiplied by $\sin \delta$, $\cos \delta$ and then subtracted to get

$$r' \sin(\delta - \delta') = -\rho \sin \phi' \csc \gamma \sin(\delta - \gamma)$$

Division will now yield the lower line of Eq. 3.24. The derivations of Eqs. 3.19–3.21 and the ecliptic coordinate formulas below are similar.

Example 3.7. An unusual orbit for an artificial satellite is

$$n = 2.28829615 \text{ rev/day} \qquad i = 47°3894$$

$$e = 0.7234293 \qquad\qquad \Omega = 283°3360$$

$$\omega = 359°8897 \qquad\qquad M_0 = 0°4279$$

The epoch is $36^d87440145$, 1982. M.I.T.'s artificial satellite observatory is at $\lambda = 106°30'33''786$ (W), $\phi = +33°49'03''854$, $H = 1502.42$ m. On my birthday (Feb. 5) this satellite is closest near 0^h40^m U.T. Compute the topocentric location in the horizon system of coordinates.

This will be done the long way around for practice. 0^h40^m U.T. on Feb. 5, $1982 = 36^d0^h40^m$, 1982 so $M = 22°99065$. Solving Kepler's equation starting with $E = M + e \sin M + (e^2/2) \sin 2M + (e^3/8)(3 \sin 3M - \sin M)$ yields $56°49548$, $58°25598$, $58°22947$, $58°22946$, and $58°22946$. From this $v = 108°54558$ whence

$u = 108°43528$. From Eqs. 2.13b, $\mathbf{r} = S\mathbf{q}$, and in units of the Earth's radius,

$$\mathbf{r} = 2.36117286 \begin{pmatrix} 0.55987847 \\ 0.44614891 \\ 0.69820287 \end{pmatrix}$$

so $\alpha = 2^h34^m12^s048$, $\delta = +44°16'58''79$. Using the equatorial coordinates and the fact that $\tau = 2^h32^m44^s979$, we then get $h = \tau - \alpha = -0^h01^m27^s069$. For the location of interest $\phi' = +33°38'24''20$, $\rho = 0.99920303$, so, following Eqs. 3.24, $p = 0.49212529$, $\alpha' = 2^h35^m36^s414$, $\gamma = 33°64110$, $q = 0.43216923$, $\delta' = +51°54'11''62$, and $r' = 1.39147080$ or $r' = 8875.00$ km. We transform α', δ' to A', a' using Eqs. 3.8 with the approximation that Φ is equal to ϕ:

$$A' = 181°41925, \qquad a' = 71°90720$$

Example 3.8. Given that artificial satellite 4632, as viewed from M.I.T.'s artificial satellite observatory, had $r' = 6.168415$ (in Earth radii), $\alpha' = 06^h50^m03^s915$, and $\delta' = +0°0'01''82$ (I was aiming for equator crossing) when the local mean sidereal time was $\tau = 02^h51^m12^s235$. Compute $rl(\alpha, \delta)$.

From Eqs. 3.22 and 3.25, it is clear that one does not have (apparently) enough information to solve the problem (obviously one would use ρ, ϕ' from Example 3.7). In this situation one needs to run through the following cycle; (1) get p, (2) get α, (3) get γ, (4) get q, (5) get δ, (6) get r, and (7) check for convergence. Starting with $\delta = \delta'$ and $r = r'$, the following sequence were derived for α, δ, and r (the initial values for p, γ, and q were $p = 0.13485971$, $\gamma = 50°18202$, and $q = 0.11683188$; the final values were $p = 0.12552838$, $\gamma = 50°36092$, and $q = 0.10808969$):

α	δ	$r/R_\oplus$
$06^h23^m18^s895$	$05°08'56''11$	6.696418
06 25 19.948	04 44 31.20	6.688073
06 25 19.006	04 44 52.56	6.646722
06 25 09.719	04 46 39.13	6.650347
06 25 10.472	04 46 29.74	6.650326
06 25 10.469	04 46 29.79	6.650031
06 25 10.406	05 46 30.56	6.650033
06 25 10.407	04 46 30.55	6.650033
06 25 10.407	04 46 30.55	6.650033

Solar System Form. Maxwell (1932) has put the parallax equations into a slightly different form particularly applicable for solar system use. Let

$$j = \left(\frac{\pi_\odot \rho}{r} \right) \cos \phi' \sec \delta \sin h$$

$$k = \left(\frac{\pi_\odot \rho}{r}\right) \sin \phi' \csc c \sin(c - \delta)$$

$$\tan c = \tan \phi' \sec h$$

where $\pi_\odot$ is the mean equatorial horizontal parallax of the Sun ($=8\overset{''}{.}794148$), r is measured in astronomical units, and ρ is measured in units of a. Also let j', k' be j and k when $\alpha \rightarrow \alpha'$, $\delta \rightarrow \delta'$. Then, to first order,

$$\alpha - \alpha' = j$$
$$\delta - \delta' = k$$

and

$$\alpha - \alpha' = j'$$
$$\delta - \delta' = k'$$

Heliocentric–Geocentric Parallax

A common problem is to transform the heliocentric location $\mathbf{r} = (x, y, z) = r\mathbf{l}(\alpha, \delta)$ of a solar system object to its geocentric location $\mathbf{R} = R\mathbf{l}(A, \Delta)$. Since the geocentric coordinates of the Sun, $\mathbf{R}_\odot = (X_\odot, Y_\odot, Z_\odot)$, are tabulated for every day of the year in *The Astronomical Almanac*, this translation is easily accomplished via

$$\mathbf{r} + \mathbf{R}_\odot = \mathbf{R} \tag{3.27}$$

or

$$x + X_\odot = R \cos \Delta \cos A = r \cos \delta \cos \alpha + X_\odot$$
$$y + Y_\odot = R \cos \Delta \sin A = r \cos \delta \sin \alpha + Y_\odot$$
$$z + Z_\odot = R \sin \Delta = r \sin \delta + Z_\odot$$

In the ecliptic system $\mathbf{r} = r\mathbf{l}(\lambda, \beta)$, $\mathbf{R} = R\mathbf{l}(\Lambda, B)$, and $\mathbf{R}_\odot = R_\odot \mathbf{l}(\Lambda_\odot, B_\odot)$, so Eq. 3.27 becomes

$$x + X_\odot = R \cos B \cos \Lambda = r \cos \beta \cos \lambda + X_\odot$$
$$y + Y_\odot = R \cos B \sin \Lambda = r \cos \beta \sin \lambda + Y_\odot$$
$$z + Z_\odot = R \sin B = r \sin \beta + Z_\odot$$

As $|B_\odot|$ never exceeds $1\overset{''}{.}2$ the approximation $\mathbf{R}_\odot = R_\odot \mathbf{l}(\Lambda_\odot, 0)$ is frequently adequate.

Example 3.9. Let us consider the full computation of the geocentric right ascension and declination of the minor planet 1620 Geographos near its closest approach, 12^h March 16, 1983. Its orbital element set for 0^h September 23,

1983, is

$$a = 1.2445845 \text{ A.U.} \quad \text{or} \quad n = 0°70985147/\text{day}$$

$$e = 0.3354146 \qquad\qquad i = 13°32071$$

$$\tilde{\omega} = 276°53348 \qquad\qquad \Omega = 336°74956$$

$$M_0 = 90°94616$$

as taken from the 1983 edition of the *Ephemerides of Minor Planets*.

The first thing needed is M at $t = 12^h$ March 16, $1983 = 75\overset{d}{.}5$, 1983. Since 1983 is not a leap year 0^h September 23, $1983 = 266\overset{d}{.}0$, 1983. Hence, as $M = M_0 + n(t - t_0)$, here $M = 315°71946$. Solving Kepler's equation with the starting value $M + e \sin M + (e^2/2) \sin 2M = 299°08014$, one finds that the successive iterates on E via Newton's method are $298°89383$ and $298°89393$. The corresponding true anomaly is $v = 280°15764$ so that the argument of latitude $u = 196°69112$. From Eqs. 2.13b the heliocentric, spherical, equatorial, location vector $\mathbf{r} = r\mathbf{l}(\alpha, \delta)$ is

$$r = 1.0428761 \text{ A.U.}$$

$$\alpha = 11^h 28^m 21\overset{s}{.}091$$

$$\delta = -00°42'44\overset{''}{.}98$$

As an aside the corresponding ecliptic location has direction cosines $\mathbf{l}(\lambda, \beta) = (-0.99040378, 0.12133151, -0.06617409)$ since $\varepsilon = 23°26'44\overset{''}{.}836$ in this case. So (λ is always in the same quadrant as is α)

$$\lambda = 173°01566, \qquad \beta = -3°79427$$

r being unchanged by a rotation. In order to transform to the Earth's center we need $\mathbf{R}_\odot = (X_\odot, Y_\odot, Z_\odot)$, the geocentric, rectangular, equatorial, location vector of the Sun. Using four-point Lagrangian interpolation among the days surrounding March $16\overset{d}{.}5$ (from page C20 of *The Astronomical Almanac* for 1983),

$$\mathbf{R}_\odot = \begin{pmatrix} 0.9908314 \\ -0.0814966 \\ -0.0353476 \end{pmatrix}$$

As $\mathbf{r} = (-1.0328684, 0.1435451, -0.0129682)$,

$$\mathbf{R} = (X, Y, Z) = \begin{pmatrix} -0.0420371 \\ 0.0620485 \\ -0.0483158 \end{pmatrix} = R\mathbf{l}(A, \Delta)$$

Whence $R = 0.0891715$ A.U., $A = 8^h 16^m 28\overset{s}{.}113$, and $\Delta = -32°48'30\overset{''}{.}02$.

Ecliptic Coordinate System Parallax

Formulas analogous to Eqs. 3.22–3.25 for ecliptic coordinates and the heliocentric $rl(\lambda, \beta)$ to geocentric $Rl(\Lambda, B)$ coordinate translation are

$$\sin(\lambda - \Lambda) = m \sin(\Lambda_\odot - \Lambda)$$

$$\sin(\beta - B) = n \sin(\theta - B)$$

$$\frac{R}{r} = \sin(\beta - \theta) \csc(B - \theta)$$

$$\tan \theta = \tan B_\odot \cos\left(\frac{\lambda - \Lambda}{2}\right) \sec\left(\Lambda_\odot - \lambda + \frac{\lambda - \Lambda}{2}\right)$$

$$\tan(\lambda - \Lambda) = \frac{m \sin(\Lambda_\odot - \lambda)}{1 - m \cos(\Lambda_\odot - \lambda)}$$

$$\tan(\beta - B) = \frac{n \sin(\theta - B)}{1 - n \cos(\theta - B)}$$

$$m = -\left(\frac{R_\odot}{r}\right) \cos B_\odot \sec \beta$$

$$n = -\left(\frac{R_\odot}{r}\right) \sin B_\odot \csc \theta$$

For small $|m|$ and $|n|$

$$\lambda - \Lambda = n \sin(\Lambda_\odot - \lambda)$$

$$\beta - B = n \sin(\theta - \beta)$$

$$\tan \theta = \tan B \sec(\Lambda_\odot - \lambda)$$

    Annual Parallax.   The smallness of the largest annual parallax of any star, coupled with the comparable size of $|B_\odot|$, means that the rigorous formulas for the annual parallax of stars are never needed. The approximate ecliptic rectangular transformation is ($\mathbf{r}$ is the heliocentric equatorial location, $\imath$ is the heliocentric ecliptic location, and $\mathbf{R}$ is the geocentric location)

$$X - x = R_\odot \cos \Lambda_\odot$$

$$Y - y = R_\odot \sin \Lambda_\odot$$

$$Z - z = 0$$

whereas the approximate equatorial rectangular transformation is

$$X - x = R_\odot \cos \Lambda_\odot$$

$$Y - y = R_\odot \sin \Lambda_\odot \cos \varepsilon$$

$$Z - z = R_\odot \sin \Lambda_\odot \sin \varepsilon$$

For spherical ecliptic coordinates this can be written in the form

$$\lambda - \Lambda = \pi \sin(\lambda - \Lambda_\odot) \sec \beta, \qquad \beta - B = \pi \cos(\lambda - \Lambda_\odot) \sin \beta$$

where λ, β are heliocentric ecliptic latitude and longitude of the star, Λ, B are its geocentric ecliptic latitude and longitude, and π is the star's annual parallax,

$$\sin \pi = 1 \text{ A.U.}/r$$

The semi-major axis of the Earth's orbit is actually 1.00000003 A.U. but the difference is negligible.

On the celestial sphere the annual displacement of a star due to the revolution of the Earth about the Sun takes place on an ellipse. The semi-major axis of the ellipse is π, the semi-minor axis is $\pi \sin \beta$, and the ellipse is called the parallactic ellipse. Finally, for equatorial spherical coordinates

$$\alpha - A = \pi[\sin \alpha \cos \Lambda_\odot - \cos \alpha \cos \varepsilon \sin \Lambda_\odot] \sec \delta$$

$$\delta - \Delta = \pi\{[\sin \alpha \sin \delta \cos \varepsilon - \cos \delta \sin \varepsilon] \sin \Lambda_\odot$$

$$+ \cos \alpha \sin \delta \cos \Lambda_\odot\}$$

In terms of the geocentric equatorial rectangular coordinates of the Sun these may be written as

$$\alpha - A = \pi(X_\odot \sin \alpha - Y_\odot \cos \alpha) \sec \delta$$

$$\delta - \Delta = \pi(X_\odot \cos \alpha \sin \delta + Y_\odot \sin \alpha \sin \delta - Z_\odot \cos \delta)$$

If the approximation $Z_\odot = Y_\odot \tan \varepsilon$ is used and the eccentricity of the Earth's orbit is neglected, then the parallax correction takes the simple form

$$\Delta \alpha = -\left(\frac{\pi}{\kappa}\right)(Cd \sec \varepsilon - Dc \cos \varepsilon)$$

$$\Delta \delta = -\left(\frac{\pi}{\kappa}\right)(Cd' \sec \varepsilon - Dc' \cos \varepsilon) \qquad (3.28)$$

The quantity κ is the constant of aberration and C, D are known as aberrational day numbers (see Chapter Four). The annual parallax correction is usually applied only if $\pi \geq 0\rlap{.}''01$. The maximum error of Eqs. 3.28 is $0\rlap{.}''01$.

SOLAR SYSTEM FORMULAS

Planets, Comets, and Asteroids

For a long time the most significant computational advance (over logarithms) was a table of proportional parts. Thus, analytical rearrangements of fundamental results were much desired if they shortened labor. A good computational trick, even if invented during the days of logarithms, is still a good computational trick.

Return to Eq. 2.13a, $\mathbf{r} = S\mathbf{q}$, and write it out fully,

$$\frac{x}{r} = \cos \Omega \cos(\omega + v) - \sin \Omega \sin(\omega + v) \cos i$$

$$\frac{y}{r} = \sin \Omega \cos(\omega + v) + \cos \Omega \sin(\omega + v) \cos i \qquad (3.29)$$

$$\frac{z}{r} = \sin(\omega + v) \sin i$$

If these orbital elements (e.g., the angles) are relative to the ecliptic and equatorial values desired, then one has to rotate about the x axis by the negative of the obliquity of the ecliptic, (see Eq. 3.9 or 3.10). Using uppercase letters to denote the rectangular equatorial coordinates,

$$X = x$$

$$Y = y \cos \varepsilon - z \sin \varepsilon$$

$$Z = y \sin \varepsilon + z \cos \varepsilon$$

In terms of the orbital elements,

$$\frac{X}{r} = \cos \Omega \cos(\omega + v) - \sin \Omega \sin(\omega + v) \cos i$$

$$\frac{Y}{r} = \cos \varepsilon \sin \Omega \cos(\omega + v)$$

$$\qquad + (\cos \varepsilon \cos \Omega \cos i - \sin \varepsilon \sin i) \sin(\omega + v) \qquad (3.30)$$

$$\frac{Z}{r} = \sin \varepsilon \sin \Omega \cos(\omega + v)$$

$$\qquad + (\sin \varepsilon \cos \Omega \cos i + \cos \varepsilon \sin i) \sin(\omega + v)$$

Looking at the combinations of i, Ω, ω, and ε that appear in Eq. 3.30, Gauss introduced the six auxiliary angles a, b, c, A, B, and C:

$$\sin a \sin A = \cos \Omega$$

$$\sin a \cos A = -\cos i \sin \Omega$$

$$\sin b \sin B = \cos \varepsilon \sin \Omega$$

$$\sin b \cos B = \cos \varepsilon \cos \Omega \cos i - \sin \varepsilon \sin i$$

$$\sin c \sin C = \sin \varepsilon \sin \Omega$$

$$\sin c \cos C = \sin \varepsilon \cos \Omega \cos i + \cos \varepsilon \sin i$$

These allowed him to rewrite Eqs. 3.30 as

$$X = r \sin a \sin(A + \omega + v)$$
$$Y = r \sin b \sin(B + \omega + v)$$
$$Z = r \sin c \sin(C + \omega + v)$$

This clearly separates the time-varying from the constant quantities. In general one chooses the sign convention so that $\sin a$, $\sin b$, and $\sin c$ are all non-negative.

Yet another useful form can be derived from Eq. 3.30. Define two vectors **A, B** by (a is the semi-major axis again)

$$\mathbf{A} = a \begin{pmatrix} \cos \Omega \cos \omega - \cos i \sin \Omega \sin \omega \\ (\sin \Omega \cos \omega + \cos i \cos \Omega \sin \omega) \cos \varepsilon - \sin \varepsilon \sin i \sin \omega \\ (\sin \Omega \cos \omega + \cos i \cos \Omega \sin \omega) \sin \varepsilon + \cos \varepsilon \sin i \sin \omega \end{pmatrix}$$

$$\mathbf{B} = a(1-e^2)^{1/2} \begin{pmatrix} -\cos \Omega \sin \omega - \cos i \sin \Omega \cos \omega \\ (-\sin \Omega \sin \omega + \cos i \cos \Omega \cos \omega) \cos \varepsilon - \sin \varepsilon \sin i \cos \omega \\ (-\sin \Omega \sin \omega + \cos i \cos \Omega \cos \omega) \sin \varepsilon + \cos \varepsilon \sin i \cos \omega \end{pmatrix}$$

Then [note that $\mathbf{A} \cdot \mathbf{B} = 0$, $\mathbf{A} \cdot \mathbf{A} = a^2$, and $\mathbf{B} \cdot \mathbf{B} = a^2(1-e^2) = b^2$]

$$\mathbf{R} = (\cos E - e)\mathbf{A} + \sin E \mathbf{B}$$

and with n the mean motion,

$$\dot{\mathbf{R}} = \left(\frac{na}{r}\right)(-\sin E \mathbf{A} + \cos E \mathbf{B})$$

Return to Eq. 3.29 and the form for $\mathbf{r} = r\mathbf{l}(\lambda, \beta)$ in Eq. 3.4. some straightforward trigonometry will yield ($u = \omega + v$ as usual)

$$\sin(u - \lambda + \Omega) = 2 \sin^2\left(\frac{i}{2}\right) \sin u \cos(\lambda - \Omega)$$

$$= \tan\left(\frac{i}{2}\right) \sin \beta \cos(\lambda - \Omega)$$

$$= \tan\left(\frac{i}{2}\right) \tan \beta \cos u$$

$$\sin(u + \lambda - \Omega) = 2 \cos^2\left(\frac{i}{2}\right) \sin u \cos(\lambda - \Omega)$$

$$= \cot\left(\frac{i}{2}\right) \sin \beta \cos(\lambda - \Omega)$$

$$= \cot\left(\frac{i}{2}\right) \tan \beta \cos u$$

For direct orbits $(i < 90°)$ the quantity $u - \lambda + \Omega$ is called the reduction to the ecliptic. For retrograde orbits it is $u + \lambda - \Omega$. The quantity $r \cos \beta$ is known as the curtate distance—a quaint term that is rarely used. Preliminaries to the above formulas are

$$\tan(\lambda - \Omega) = \cos i \tan u$$

$$\sin(\lambda - \Omega) = \cot i \tan \beta$$

$$\cos(\lambda - \Omega) = \sec \beta \cos u$$

$$\sin \beta = \sin i \sin u$$

If you look back to the approximate forms of the equatorial parallax correction, then you will see that the quantities

$$r\pi_\alpha \equiv \left(\frac{\rho\pi_\odot}{15}\right) \cos \phi' \sec \delta \sin h$$

$$r\pi_\delta \equiv (\rho\pi_\odot) \sin \phi' \csc \gamma \sin(\gamma - \delta)$$

are important. These are known as the *parallax factors* in right ascension ($r\pi_\alpha$ in seconds of time) and in declination ($r\pi_\delta$ in seconds of arc). The equatorial horizontal parallax of the Sun is $\pi_\odot = 8\rlap{.}''794148$.

Units

The history of units in astronomy is long and inglorious. Physics has been slightly more rational so that the common areas of the two are a mess. The physicists would have had us use cgs (or mks) units. The essential parameter of gravity, G, is hard to come by in these units (I don't even want to bring up SI units!) Cavendish tried long ago. Gauss took the point of view that the sidereal year should be the fundamental unit of time, the solar mass should be unity, and the semi-major axis of the Earth's (actually the Earth–Moon barycenter) orbit also should be one. This too was another quantity difficult to ascertain (until recently) in centimeters. Modern radars, instrumental refinements leading to a precise value of the speed of light, and spacecraft have given us the scale of the solar system in light time and thence centimeters. Anyway, Gauss looked at Kepler's third law for the Earth,

$$\left(\frac{2\pi}{P}\right)^2 a^3 = k^2(M_\odot + M_\oplus)$$

and fixed k such that $M_\odot = 1$, $a = 1$ A.U. (or astronomical unit), and $P = 1$ sidereal year, which Gauss took to be 365.2563835 days. The resulting value of k is ($M_\oplus = 1/354710$)

$$k = 0.01720209895$$

or $\log k = 8.2355814414 - 10$. This value became so ingrained in extensive numerical tables of Newcomb and others that we have kept k but changed

a! This decision dates back to 1939 [*Transactions of the International Astronomical Union* **6**, 20, 336, 357 (1939)]. According to modern values of the length of the sidereal year ($365^d06^h09^m10\overset{s}{.}5$ at 2000.0 or $365\overset{d}{.}25637$) and the mass of the Earth–Moon system ($=1/328900.5$), the semi-major axis of the Earth–Moon barycenter's orbit is 1.00000003 A.U. Of course, one could use $1/k = 58.132441$ as a unit of time, in which case k itself would be unity.

Artificial Satellites

Since artificial satellite astronomy is coeval with the computer age, analytical dexterity is rapidly departing the subject. This trend will be advanced as digital algebraic manipulators become more sophisticated. Nonetheless there are some (old) tricks worth being aware of. We start from Eqs. 3.29 and use the half-angle formulas for $\sin i$ and $\cos i$. One should be able to deduce that

$$\frac{x}{r} = \cos^2\left(\frac{i}{2}\right)\cos(\omega +\Omega +v) + \sin^2\left(\frac{i}{2}\right)\cos(\omega -\Omega +v)$$

$$\frac{y}{r} = \cos^2\left(\frac{i}{2}\right)\sin(\omega +\Omega +v) - \sin^2\left(\frac{1}{2}\right)\sin(\omega -\Omega +v)$$

$$\frac{z}{r} = \sin i \sin(\omega +v)$$

We find the equations in 2.14 so useful that we gave them earlier. Their derivation is straightforward from $\mathbf{r} = r\mathbf{l}(\phi, \theta) = S\mathbf{q} = Sr\mathbf{l}(v, 0)$. In terms of right ascension and declination they are

$$\alpha = \Omega +\tan^{-1}(\cos i \tan u), \qquad \delta = \sin^{-1}(\sin i \sin u)$$

If the Euler angles ω, Ω, i refer to the ecliptic instead of the celestial equator, then

$$\lambda = \Omega +\tan^{-1}(\cos i \tan u), \qquad \beta = \sin^{-1}(\sin i \sin u)$$

If i is small, then $\lambda \approx \Omega +u - (i^2/2)\tan u \sec^2 u$ and the latter term is known as the reduction of the orbit or the reduction to the orbit.

Similar manipulations yield the equation of the plane of the orbit (i.e., $\mathbf{r} \cdot \mathbf{L} = 0$)

$$x \sin i \sin \Omega - y \sin i \cos \Omega +z \cos i = 0$$

or

$$\tan \delta = \tan i \sin(\alpha - \Omega)$$

Other useful forms are

$$r \cos u = x \cos \Omega +y \sin \Omega$$

$$r \sin u = (-x \sin \Omega +y \cos \Omega) \cos i +z \sin i$$

$$\cos u = \cos \delta \cos(\alpha - \Omega)$$

$$\sin u = \cos \delta \cos i \sin(\alpha - \Omega) + \sin i \sin \delta$$

Units

The rational system of units for artificial satellites uses the mean solar day as the unit of time, the equatorial radius of the Earth as the unit of distance, and the mass of the Earth as the unit of mass. Then $GM_{\oplus} = 3.986005 \times 10^{20}$ cm^3/sec^2 becomes ($R_{\oplus} = 6.378140 \times 10^8$ cm, 1 mean   solar day $= 86400$ sec) 11,467.88. To give scale to this, consider a geosynchronous satellite. Its period is 1 mean sidereal day $= 23^h56^m04\!\!.\!\!^s09054$ and the size of its orbit is 6.6107.

TYPES OF TIME SYSTEMS

There are four different time systems in use in astronomy: sidereal, Universal (or solar), Ephemeris, and atomic. The situation gets worse in 1984 (see below). Each time system has a fundamental epoch and a fundamental measure of duration. Each one also uses days of 24 equal hours, hours of 60 equal minutes, and minutes of 60 equal seconds; $1^d = 24^h = 1440^m = 86400^s$. Moreover, each time system also uses sexagesimal arc measure to represent times according to $1^h = 15°$. Lastly, each time system shares the idea of a Julian century. A Julian century is a unit of exactly 36525 days. It does not signify an interval of time equal to 36,525 days in some time system, merely a count of that many days.

Sidereal Time

Sidereal time is the time system derived from the Earth's rotation with respect to the stars. Apart from the small effects of rapid fluctuations in the Earth's rotation rate and polar motion, local sidereal time is the hour angle of the vernal equinox. Local apparent sidereal time is the hour angle of the true equinox of date. Local mean sidereal time is the hour angle of the mean equinox of date. Sidereal time on the geographic Greenwich meridian (i.e., where $\lambda = 0$) is known as Greenwich sidereal time (apparent or mean as appropriate). The relationship between local sidereal time, apparent or mean, and the corresponding Greenwich sidereal time is simply

$$(\text{Local sidereal time}) = (\text{Greenwich sidereal time}) - \lambda \qquad (3.31)$$

Here λ is the observer's longitude measured positively to the west.

The difference between apparent and mean sidereal time (in the sense apparent minus mean) is known as the *equation of the equinoxes*. Its magnitude is $<1^s$ and it is tabulated for each day of the year in *The Astronomical Almanac*. One mean sidereal day is the duration of the time interval between two successive upper transits of the mean equinox of date relative to a fixed meridian. The instant of transit is called sidereal noon. An apparent sidereal day is similarly defined with respect to transits of the true equinox of date. It follows from its definition that a mean sidereal day is shorter than the Earth's sidereal period of rotation ($P_\oplus$) by the daily precession in right ascension, $0^s\!.008412 + 5^s\!.1 \times 10^{-6} T_E$. Or

$$(\text{1 mean sidereal day})/P_\oplus = 0.999999902907 - 5.9 \times 10^{-11} T_E$$

$$= (1.000000097093 + 5.9 \times 10^{-11} T_E)^{-1}$$

Here T_E is the time interval measured in Julian centuries since 12^h Jan. 0, 1900 E.T. = Jan. $0^d\!.5$, 1900 E.T. Ephemeris Time is discussed below. The fundamental epoch of sidereal time is the instant 12^h Jan. 0, 1900 = Jan. $0^d\!.5$, 1900, where the hours are measured in mean sidereal units.

Universal Time

Universal Time is an outgrowth of the time system derived from the Earth's rotation with respect to the Sun. That this is different from its rotation with respect to the stars becomes clear if one remembers that the Earth is revolving about the Sun in its annual orbit as it rotates about its axis. The net result is that a *tropical year* (the time interval from equinox to equinox) contains one more apparent sidereal day than it does apparent solar days (e.g., 366.2422 vs. 365.2422). Mean solar time, as formerly defined, was in terms of the hour angle of the fictitious mean Sun. Mean solar time in this sense no longer exists. It has been replaced by the concept of Universal Time. Since Universal Time is very close to the now defunct mean solar time, both the mean solar day and the mean solar second remain the phrases used to describe the measures of Universal Time.

Local apparent solar time is 12^h plus the hour angle of the Sun (i.e., of the center of the actual solar disk). This is the only definition of apparent solar time; it is the time kept by a sundial and is no longer used in astronomy (post-1965). The fictitious mean Sun is the name for the formulas that describe the position of a point on the celestial sphere whose average motion on the celestial equator closely approximates the projection of the motion of the real body (i.e., the Sun) from the ecliptic to the equator during its apparent annual orbit. The position of the fictitious mean Sun is given by

$$\delta_{\text{FMS}} = 0$$

$$\alpha_{\text{FMS}} = 18^h 38^m 45^s\!.836 + 8640184^s\!.542 T_U + 0^s\!.0929 T_U^2 \tag{3.32}$$

where T_U is the time interval measured in Julian centuries of 36525 mean solar days since 12^h Jan. 0, 1900 U.T. = Jan. $0^d\!.5$, 1900 U.T. Just as with the sidereal times, the relationship between Greenwich solar time* and local solar

---

* Greenwich mean solar time as just defined is the post-1925 definition. The abbreviation GMT should only be used in conjunction with the civil time of the standard meridian $\lambda = 0$ (i.e., only in the United Kingdom).

time is given by (cf. Eq. 3.31)

$$\text{(Local solar time)} = \text{(Greenwich solar time)} - \lambda$$

where λ is the observer's west longitude.

The difference between apparent and mean solar time (in the sense apparent minus mean) is known as the equation of time. Its magnitude is $< 16^m$ and it is a concept no longer used in astronomy. It is still tabulated for the use of navigators and surveyors. Sometimes the difference mean minus apparent has been given.

With the recognition of the departure of the Earth's rotation rate from a uniform rate, the concept of mean solar time, as just given, was abandoned. In its place Universal Time was introduced. Both sidereal and solar times depend on the Earth's rotation rate, so when Universal Time was invented a strict numerical relationship between it and sidereal time served as its definition. Universal Time defined below is very close to the concept of Greenwich mean solar time.

The instant 0^h U.T. corresponds to that instant when the Greenwich mean sidereal time is $12^h + \alpha_{FMS}$ or

$$6^h38^m45\overset{s}{.}836 + 8640184\overset{s}{.}542 T_U + 0\overset{s}{.}0929 T_U^2 \qquad (3.33)$$

This quantity is tabulated for every day in *The Astronomical Almanac*. At any other instant Universal Time is defined to be 12^h plus the hour angle at Greenwich of the fictitious mean Sun (with α_{FMS} measured relative to the mean equinox of date). Alternatively,

$$\text{U.T.} + 12^h = \text{(Greenwich mean sidereal time)} - \alpha_{FMS} \qquad (3.34)$$

The local mean solar time can now be defined as (cf. Eq. 3.31)

$$\text{Local mean solar time} = \text{(Universal Time)} - \lambda$$

where λ is the west longitude of the observer.

One mean solar day used to be the duration of the time interval between two successive upper transits of the point represented by α_{FMS}, δ_{FMS} relative to a fixed meridian. Now, from the formulas, the interval of mean sidereal time in 1 mean solar day is $24^h + (1/36525)\, d\alpha_{FMS}/dT_U$. Hence the ratio of a mean solar day to a mean sidereal day is

$$\frac{1 \text{ mean solar day}}{1 \text{ mean sidereal day}} = \frac{24^h + (1/36525)(d\alpha_{FMS}/dT_U)}{24^h}$$

$$= \frac{86636\overset{s}{.}5553605 + 5\overset{s}{.}087 \times 10^{-6} T_U}{86400^s}$$

$$= 1.002737909265 + 5.89 \times 10^{-11} T_U$$

$$= (0.997269566414 - 5.86 \times 10^{-11} T_U)^{-1} \qquad (3.35)$$

Alternatively, apart from the negligible secular term,

$$1 \text{ mean sidereal day} = 23^h56^m04\overset{s}{.}09054 \text{ of mean solar measure}$$

$$1 \text{ mean solar day} = 24^h03^m56\overset{s}{.}55536 \text{ of mean sidereal measure}$$

The extra $3^m56\overset{s}{.}55536$/day is known as the acceleration of the equinox or the acceleration of the fixed stars. The relationship between the Earth's sidereal period of rotation $P_\oplus$ and the mean solar day is

$$1 \text{ mean solar day} = 1.002737811906 P_\oplus$$

$$P_\oplus = 23^h56^m04\overset{s}{.}098904 \text{ of mean solar measure}$$

$$\frac{2\pi}{P_\oplus} = 15\overset{''}{.}04106718/\text{second of mean solar measure}$$

The measure of sidereal (hence Universal) time at any particular location on the Earth is represented by the sum of (1) the measure at some epoch, (2) the integral of the true angular speed of rotation of the Earth with respect to a uniform time scale from the initial epoch to the current instant, (3) the general precession in right ascension, (4) the equation of the equinoxes, (5) the motion of the pole (maximum $\sim 0\overset{s}{.}035$), and (6) the variations in the astronomical vertical. Therefore, sidereal time as determined by star transits (for the local sidereal time is equal to the right ascension of any star on the celestial meridian) is nonuniform and peculiar to the observer's geographic location. The Universal Time derived from this time by the fixed numerical relationships of Eq. 3.35 is designated UT0. When UT0 is corrected for polar motion, it is called UT1. This is the time implicit in the definition of Eq. 3.33. UT1 is available with good accuracy on a real-time basis and to excellent accuracy from the national time services after a short delay. It is UT1 that enters into computations such as hour angle and the topocentric place of an object. When UT1 is corrected for seasonal inequalities (maximum $\sim 0\overset{s}{.}035$) in the Earth's rotation rate, it is called UT2. UT2 is a nearly uniform measure of time. UT0 is not immediately available and neither UT0 nor UT2 is for general usage. Instead, the national time services provide yet another Universal Time, UTC (Coordinated Universal Time). In the U.S. this is the time broadcast by WWV. UTC is a smoothed version of UT2 designed such that $|\text{UT1} - \text{UTC}| <$ 1^s. The correction $\Delta \text{UT1} = \text{UT1} - \text{UTC}$ is available after the fact. All of these corrections have been made since 1956.

In fact, since 1972 UTC is based on International Atomic Time (TAI, see below) and not UT2. This change was made without a change in nomenclature. Post-1972 UTC runs at the TAI rate and on January 1, 1972 the difference between TAI and UTC was exactly 10^s. Additional discontinuities in UTC, required to keep it near UT1, will be made in steps of exactly 1^s on either January 1 or July 1 but at no other instants. Finally, the SI second is nearly equal to an Ephemeris Second but is shorter than a second of Universal Time by about one-ten millionth.

Example 3.10.   Compute the mean sidereal times used in Examples 3.7 and 3.8.

The local mean sidereal time at 0^h40^m U.T. on February 5, 1982, for Example 3.7 and at $358^d03^h47^m28\overset{s}{.}342$, 1981, for Example 3.8 is needed. In each case the west longitude is $\lambda = 106°39'33\overset{''}{.}786 = 07^h06^m38\overset{s}{.}2524$. This author prefers to solve problems such as these directly from the definition, Eq. 3.33. In the first instance $T_U = (0.5 + 365 \times 82 + 20)/36525$, representing the half day from the fundamental epoch of Universal Time to January 1.0, 1900; 82 years of 365 days each; and 20 intercalary days for the leap years. Thus, after reducing the second two terms modulo 86400 in Eq. 3.33, we get $06^h38^m45\overset{s}{.}836 + 151\overset{s}{.}3870 = 06^h41^m17\overset{s}{.}2230$ for the mean sidereal time on the prime meridian (e.g., Greenwich mean sidereal time) at January 1.0, 1982. On page B8 of the 1982 *Astronomical Almanac* $06^h41^m17\overset{s}{.}2229$ is found. Next the west longitude is subtracted to get the local mean sidereal time at January 1.0, 1982: $23^h34^m38\overset{s}{.}9706$. Since February 5 = day 36, to this is added 35 days worth of acceleration of the equinox, which is $35 \times (03^m56\overset{s}{.}55536) = 02^h17^m59\overset{s}{.}4376$. Therefore, the local mean sidereal time at February 5.0 is $01^h52^m38\overset{s}{.}4082$. Finally, the interval of Universal Time since midnight must be converted to a sidereal interval. This just means multiplying by 1.0027379093 or $0^h40^m06\overset{s}{.}5710$ in this instance. Hence, $\tau = 02^h32^m44\overset{s}{.}9742$.

For the other date begin in the same fashion, $T_U = (0.5 + 365 \times 81 + 20)/36525$ to get GMST at January 1.0, 1981: $06^h38^m45\overset{s}{.}836 + 208\overset{s}{.}6780 = 06^h42^m14\overset{s}{.}5140$ (p. B8 of the 1981 *Astronomical Almanac* has $14\overset{s}{.}5148$). Next 357 days worth of acceleration of the equinox ($= 23^h27^m30\overset{s}{.}2635$) is summed to get GMST on December 24.0, 1981, that is, $06^h09^m44\overset{s}{.}7775$ (p. B15 of *The Astronomical Almanac* has $44\overset{s}{.}7800$; these small differences are in my hand calculator). Now adjust for the observer's longitude to get the local mean sidereal time at midnight, December 24: $23^h03^m06\overset{s}{.}5251$. Finally the interval of Universal Time past midnight is converted to one of sidereal measure by multiplying by 1.0027379093; $03^h48^m05\overset{s}{.}7099$, then added to obtain $\tau = 02^h51^m12\overset{s}{.}7350$.

Ephemeris Time

The irregularities in the Earth's rate of rotation make sidereal time (or Universal Time) unsuitable for the comparison of theory with observation. The reason is clear—Newton's laws of motion and his theory of gravitation are represented mathematically by differential equations wherein uniform time is the independent variable. Hence, to determine the accuracy of the theory, the times of observation of the Sun, Moon, and planets must be given on a uniform time scale. Ephemeris time (E.T.) was invented to fulfill this need. As the Moon has the largest geocentric motion of any natural body in the solar system, in practice it is more accurate and expeditious to use observations of the Moon rather than those of the Sun or the planets. Using a particular analytical or numerical expression for the position of the Moon as a function of Ephemeris Time is *not* the same as saying, "Ephemeris Time is the independent variable in the equations of motion of the bodies of the solar system." As our approximation techniques improve, astronomers approach this simple definition but they have not yet reached it. In fact there are three different Ephemeris Times (ET0, ET1, and ET2), each one obtained by comparing observations of the Moon with three different, and increasingly complex and sophisticated, models of the motion of the Moon. Indeed, lunar ranging by lasers is now good enough (~ 10 cm) that general relativistic effects must be taken into account.

The fundamental epoch of Ephemeris Time is 12^h Jan. 0, 1900 E.T. = Jan. $0^d.5$, 1990 E.T. = JED 2415020 (see below for a discussion of Julian Ephemeris Dates), when the geometric mean longitude of the Sun, affected by aberration and measured relative to the mean equinox of date, was exactly $279°41'48''.04$. The fundamental epoch of Ephemeris Time depends on the numerical value of the constant of aberration and the value $20''.47$ was used in the definition. The fundamental epoch of Ephemeris Time is about 4^s later than the fundamental epoch of Universal Time.

The unit of Ephemeris Time is the ephemeris second. It is defined in terms of the length of the tropical year (see below) at the fundamental Ephemeris Time epoch. The formula for the mean longitude of the Sun is

$$L_\odot = 279°41'48''.04 + 129,602,768''.13\, T_E + 1''.089\, T_E^2$$

where T_E is measured in Julian centuries of 36525 days of Ephemeris Time measured from January $0^d.5$, 1900 E.T. The time for $L_\odot$ to increase by $360°$ (i.e., a tropical year) at $T_E = 0$ is

$$360 \times 60 \times 60 \times 36,525 \times 86,400 \left/ \frac{dL_\odot}{dT_E}\right|_{T_E=0} = 31,556,925^s.9747$$

Therefore, one second of ephemeris measure is exactly $1/31,556,925.9747$ of the tropical year for 1900.

For some purposes it is useful to have the concept of an ephemeris transit. The ephemeris transit of a celestial object occurs when it crosses the ephemeris meridian. The geographic location of the ephemeris meridian is at $\lambda = 1.002738\, \Delta T$ east of $\lambda = 0$ where ΔT is the difference between Ephemeris Time and Universal Time and the numerical factor represents the mean solar to mean sidereal ratio. Ephemeris longitude is measured relative to the ephemeris meridian, as is ephemeris hour angle. The ephemeris hour angle of the vernal equinox is called the ephemeris sidereal time. Ephemeris Time is related to ephemeris sidereal time via (cf. Eq. 3.34)

$$\text{E.T.} + 12^h = (\text{ephemeris sidereal time}) - \alpha_{\text{FMS}}(T_E)$$

where $\alpha_{\text{FMS}}(T_E)$ means the expression in Eqs. 3.32 with T_E replacing T_U. The angular speed of rotation of the ephemeris meridian is such that it makes one complete revolution with respect to the mean equinox of date in $23^h56^m04^s.098904$ of ephemeris time measure.

The quantity $\Delta T = \text{E.T.} - \text{U.T.}$ can only be determined after observations of the Moon have been performed and reduced. Thus, there is a lag of about

1 year before it is known with precision. The value of ΔT is currently $\sim 50^s$ and vanished sometime between 1900 and 1905. Extrapolations of ΔT to $\pm 0\overset{s}{.}1$ for a year or two are published in *The Astronomical Almanac* (as well as past values).

Atomic Time

One second of *atomic time* is the time duration associated with 9,192,631,770 cycles of the hyperfine ground state transition of cesium 133. The fundamental epoch of atomic time is different for A.1 (U.S. Naval Observatory Atomic Time) and TAI (International Atomic Time). Both are simply related to Ephemeris Time via

$$\text{E.T.} = \text{A.1} + 32\overset{s}{.}15 \qquad \text{pre-1972}$$

$$\text{E.T.} = \text{TAI} + 32\overset{s}{.}18 \qquad \text{post-1972}$$

The difference between Ephemeris Time and atomic time $\Delta T(A)$ provides a first approximation to $\Delta T = \text{E.T.} - \text{U.T.}$,

$$\Delta T(A) = \text{A.1} + 32\overset{s}{.}15 - \text{UTC} \qquad \text{pre-1972}$$

$$= \text{TAI} + 32\overset{s}{.}18 - \text{UT1} \qquad \text{post-1972}$$

The fundamental epoch of A.1 was $0^h0^m0^s$ January 1, 1958 UT2, when A.1 was exactly 0. The epoch of TAI differs by ~ 30 msec and $\text{TAI} - \text{UTC} = 10^s$ exactly at January 1.0, 1972. The duration of 1 sec of atomic time is within 2 parts in a billion of the duration of 1 sec of Ephemeris Time.

Atomic time, in the general relativistic sense, probably keeps the proper time of a moving observer in a gravitational field. Within a microsecond it is related to coordinate time via (Mulholland 1972)

$$\text{Coordinate time} = \text{A.1} + 32\overset{s}{.}15 + 1\overset{s}{.}658 \times 10^{-3} (\sin E_\odot + 0.0368)$$

$$+ 2\overset{s}{.}03 \times 10^{-6} \cos \phi' [\sin(\text{U.T.} + \lambda) - \sin \lambda]$$

where $E_\odot$ is the eccentric anomaly of the Sun, λ is the observer's geocentric longitude measured eastward, and ϕ' is the observer's geocentric latitude. Moyer (1981a,b) has dealt with the Earth proper time to solar system barycentric coordinate time transformation.

Atomic clocks provide a unit of frequency and, therefore, allow the specification of a unit of time. The counting of such units, as seen in the continuous repetitive phenomena of the seasons or lit and dark portions of the day, determines a time interval for which astronomical observation will always be necessary.

Years

There are five types of years in astronomical use. The immediately observable ones are the tropical year (the time interval from equinox to equinox), the sidereal year (the time interval for one complete revolution of the Earth about the Sun relative to the stars), and the anomalistic year (the time interval from perihelion to perihelion). With T_E measured in Julian centuries of 36,525 days from Jan. $0\overset{d}{.}5$, 1900 E.T., their lengths (in Ephemeris Time measure) are

$$\text{Tropical year} = 365^d05^h48^m46\overset{s}{.}0 - 0\overset{s}{.}530\,T_E$$

$$= 365\overset{d}{.}24219878 \quad -6\overset{d}{.}14 \times 10^{-6}\,T_E$$

$$\text{Sidereal year} = 365^d06^h09^m09\overset{s}{.}5 + 0\overset{s}{.}01\,T_E$$

$$= 365\overset{d}{.}25636042 \quad +1\overset{d}{.}1 \times 10^{-7}\,T_E$$

$$\text{Anomalistic year} = 365^d06^h13^m53\overset{s}{.}0 + 0\overset{s}{.}26\,T_E$$

$$= 365\overset{d}{.}25964134 \quad +3\overset{d}{.}04 \times 10^{-6}\,T_E$$

The Besselian solar year is shorter than the tropical year by the secular acceleration of the fictitious mean Sun over the actual mean longitude of the Sun. This amounts to $0\overset{s}{.}148\,T_E$ and is usually ignored. The start of a Besselian solar year is used as the epoch of star catalogs. The notation .0 after a year, as in 1950.0, is used to denote this. The previous fundamental epochs have been 1900.0 and 1950.0. We will switch in 1984 to 2000.0. (See Table 1.)

The last type of year is the eclipse year—the time interval from lunar node to lunar node:

$$\text{Eclipse year} = 346^d14^h52^m50\overset{s}{.}7 + 2\overset{s}{.}8\,T_E$$

$$= 346\overset{d}{.}620031 + 3\overset{d}{.}2 \times 10^{-5}\,T_E$$

Julian Dates

Julian Day Numbers and Julian Dates are a simple means of continuously counting the number of days that have elapsed since a fundamental epoch. This epoch was chosen to be sufficiently far in the historical past that negative Julian Dates would not occur when reducing astronomical observations. The epoch is 12^h Jan. 1, 4713 B.C., where the hours are in mean solar time measure. The Julian Date (JD) at this instant was exactly 0. For the fundamental epoch of Universal Time, 12^h Jan. 1, 1900 U.T., JD = 2,415,020 exactly. The Modified Julian Date (MJD) sometimes appears in modern work,

$$\text{MJD} = \text{JD} - 2400000.5$$

Modified Julian Date is reckoned from 0^h Nov. 17, 1858 U.T. The relationship between JD and the time of the start of the Besselian solar year (B) is

$$\text{JD} = 2433282.423 + 365.2422\,(B - 1950.0)$$

A table (14.15) of Julian Day Numbers is given in *The Explanatory Supplement to The Astronomical Ephemeris and The American Ephemeris and Nautical Almanac.*

There is a completely analogous system of Julian Dates and Julian Day Numbers for Ephemeris Time. The fundamental epoch is 12^h Jan. 1, 4713 B.C., where the hours are in Ephemeris Time measure. The Julian Ephemeris Date (JED) at this instant was exactly 0. At the fundamental epoch of Ephemeris Time JED = 2,415,020. At the fundamental epoch of Universal time JED = 2,415,020.31352. The value of JED at any epoch T tropical centuries from 1900.0 can be computed from

$$JED(1900.0 + 100\,T) = JED(1900.0) + 100(365.2421988 - 8.56 \times 10^{-5}\,T)\,T$$

The neglect of the secular term amounts to $1\overset{s}{.}85$ at 1950.0.

An analogous system of Greenwich Sidereal Dates (GSD = 1.0027379093 × JD + 0.671, JD = 0.9972695664 × GSD − 0.669) also exists.

Changes in 1984.   The International Astronomical Union has adopted a series of resolutions in 1976, 1979, and 1982 that affects the reduction of all astronomical data from January 1, 1984 onward. This includes a new fundamental epoch designated J2000.0 = Jan. 1.5, 2000, with Julian Day Number 2,451,545. The precessional parameters were changed (see Chapter Four), the location of the vernal equinox was moved, a new time system was introduced, and several other changes were made. The reader is referred to Kaplan (1981) for more details.

The new expression for Greenwich mean sidereal time at 0^h UT1 is

$$06^h41^m50\overset{s}{.}54841 + 8640184\overset{s}{.}812866\,T_U + 9\overset{s}{.}3104 \times 10^{-2}\,T_U^2 - 6\overset{s}{.}2 \times 10^{-6}\,T_U^3$$

where T_U is the number of Julian centuries of 36525 days of Universal Time measure, elapsed since J2000.0.

The new Julian epoch for Julian Date JD is given by

$$J2000.0 + \frac{JD - 2451545}{365.25}$$

The corresponding Besselian epoch is given by

$$B1900.0 + \frac{JD - 2415020.31352}{365.242198781}$$

The new solar to sidereal ratio (at J2000.0) is 0.997269566329084 = 1/1.002737909350795 and the angular speed of the Earth's rotation is $15\overset{''}{.}04106718/\text{sec}$.

PROBLEMS

1. Suppose that a star's zenith distance when on the celestial meridian is z and when crossing the prime vertical it is Z. Prove that its declination is

given by

$$\cot \delta = \csc z \sec Z - \cot z$$

and that the observer's latitude is equal to

$$\cot^{-1}(\cot z - \csc z \cos Z)$$

2. Suppose that upon rising, the diurnal path of a star makes an angle ψ with the horizon. If its declination is δ and the astronomical latitude is Φ, show that

$$\cos \psi = \sin \Phi \sec \delta$$

3. Show that if $\delta > \Phi$, then the maximum azimuth is

$$\pi + \sin^{-1}(\sec \Phi \cos \delta)$$

4. Suppose that a vertical pole at astronomical latitude Φ casts a shadow of length s at noon on an equinox and length S on the day of the summer solstice as the Sun crosses the prime vertical. Prove that

$$s = S \tan \Phi \tan \psi, \qquad \sin \psi = \sin \varepsilon \csc \Phi$$

5. Two stars, whose equatorial positions are (α, δ) and (A, Δ), have the same celestial longitude. Prove that

$$\sin(\alpha - A) = (\cos \alpha \tan \Delta - \cos A \tan \delta) \tan \varepsilon$$

6. Show that if $|\Phi| < \varepsilon$, then the hour angle of the sun when the ecliptic is vertical is given by

$$h = \sin^{-1}(\sin \alpha \cot \delta \tan \Phi) - \alpha$$

where (α, δ) are the solar position.

7. If the eccentricity of the Earth's orbit were zero, prove that, in minutes, the equation of time would be given by

$$\frac{720}{\pi} \tan^{-1} \left[\frac{(1 - \cos \varepsilon) \tan \Lambda_\odot}{1 + \cos \varepsilon \tan^2 \Lambda_\odot} \right]$$

8. Show that the parallax in declination vanishes if $f = 0$ and

$$\tan \Phi = \tan \delta \cos h$$

9. Show that the true angular radius S of a planetary disc is related to its apparent angular radius S' by

$$\sin S' = \sin(\delta' - \gamma) \csc(\delta - \gamma) \sin S$$

where γ is defined in Eq. 3.23.

10. Show that the geocentric distance ρ, geocentric latitude ϕ', geodetic latitude ϕ, and height above (or below) sea level H (in units of the

Earth's equatorial radius) are given, approximately by (Long 1975)

$$\rho = (H+1) + \left[-\frac{1}{2}(1-\cos 2\phi) \right] f + \left\{ \left[\frac{1}{4(H+1)} + \frac{1}{16} \right] (1-\cos 4\phi) \right\} f^2$$

$$\phi' = \phi - \frac{\sin 2\phi}{H+1} f + \left(\frac{-\sin 2\phi}{2(H+1)^2} + \left[\frac{1}{4(H+1)^2} + \frac{1}{4(H+1)} \right] \sin 4\phi \right) f^2$$

or

$$\phi = \phi' + \left(\frac{\sin 2\phi'}{\rho} \right) f + \left[\left(\frac{1}{\rho^2} - \frac{1}{4\rho} \right) \sin 4\phi' \right] f^2$$

$$H = (\rho-1) + \left[\frac{1}{2}(1-\cos 2\phi') \right] f + \left[\left(\frac{1}{4\rho} - \frac{1}{16} \right)(1-\cos 4\phi') \right] f^2$$

where f is the flattening of the Earth.

11. Derive

$$\sin\left(\frac{\lambda-\alpha}{2} \right) = \tan\left(\frac{\varepsilon}{2} \right) \cos\left(\frac{\lambda+\alpha}{2} \right) \tan\left(\frac{\delta+\beta}{2} \right)$$

$$\tan\left(\frac{\delta-\beta}{2} \right) = \tan\left(\frac{\varepsilon}{2} \right) \sin\left(\frac{\lambda+\alpha}{2} \right) \sec\left(\frac{\lambda-\alpha}{2} \right)$$

12. The plane of the Galaxy is inclined to the celestial equator by 62°.6. The coordinates of the North Galactic Pole are $\alpha = 12^h 49^m = 192°15'$, $\delta = +27°24'$ (1950.0). The ascending node of the galactic plane is at a galactic longitude of $+33°$. Let (l, b) be the galactic position corresponding to (α, δ). Show that

$$\sin b = \sin(27°.4) \sin \delta + \cos(27°.4) \cos \delta \cos(\alpha - 192°.25)$$

$$\tan(l-33°) = [\sin \delta - \sin(27°.4) \sin b] \sec(27°.4) \sec \delta \csc(\alpha - 192°.25)$$

Also prove that

$$\sin \delta = \sin(27°.4) \sin b + \cos(27°.4) \cos b \sin(l-33°)$$

$$\tan(\alpha - 192°.25) = \frac{\cos b \cos(l-33°)}{\cos(27°.4) \sin b - \sin(27°.4) \cos b \sin(l-33°)}$$

13. Show that, if for artificial satellite problems we use a unit of length L and of time T,

$$L = 42164.172 \text{ km}, \qquad T = 13713.441 \text{ sec}$$

the mean speed of geosynchronous satellites is 1 as is $GM_\oplus$.

chapter four

Corrections to Coordinates

The celestial coordinate system used to catalog the stars is the equatorial system of right ascension and declination. Over the course of many years the right ascension and declination of a star change very little. The position of a star does change, however, and for two different reasons. One is that the stars are not fixed; many stars exhibit an intrinsic or proper motion of their own. All stars possess this characteristic but, because its magnitude is very small for most stars ($1''$/yr is large in this context), it is usually unobservable in practice. Proper motion is treated in the second section of this chapter. The other effect, general precession, is much larger ($\sim 50''$/yr on the celestial equator). It is a consequence of choosing reference planes (the planes of the celestial equator and of the ecliptic) that themselves are not fixed in inertial space. The discussion of general precession for equatorial coordinates occupies the first section of this chapter. Precession interacts with proper motion in a complicated fashion. The updating of a star's position for both is discussed in the third section. Reductions for ecliptic latitude and longitude, the (1976) International Astronomical Union precessional quantities, and the difference between mean place and mean catalog place are then discussed.

In addition to the above-mentioned year-to-year corrections a discussion of the intrayear adjustments for optical observations is also included herein. This encompasses such topics as general precession (again), nutation, proper motion, annual aberration, annual parallax, and astronomical refraction. Finally, I switch wavelengths and very briefly discuss the major components of radar and laser radar data reductions. These subjects are too complex to really deal with herein, but radial velocity reductions are treated in depth.

Although the presentation in this chapter is rigorous, it is not exhaustive. The reader should consult (at least) three other volumes for optical data reduction: Taff (1981a), Woolard and Clemence (1966), and *The Explanatory Supplement to The Astronomical Ephemeris and The American Ephemeris and Nautical Almanac* (HMSO, London, 1974). The classic radar reference work is Skolnik (1970).

In the terminology of astrometry the adjective *mean* refers to the orientation of the celestial equator, the ecliptic, or the equatorial reference frame (as in mean pole, mean ecliptic, or mean equator and equinox) caused by the secular changes brought about by general precession and the adjective *true* refers to the orientation caused by the additional changes brought about by nutation. The standard computational practice is to first correct a mean position or place (i.e., the position of a star referred to a particular mean equator and equinox) from one epoch (e.g., 1900.0) to another (e.g., 1987.0) for general precession and then to obtain the true place of date (e.g., April 1, 1987) by applying the intrayear general precession and then the corrections for nutation.

The terms *mean place* and *true place* for stars not only indicate the nature of the noninertial coordinate system utilized but also imply that the origin of the coordinate system is at the solar system barycenter (i.e., nearly heliocentric). The specification of the epoch for the mean place, as in 1900.0, defines a precise instant of time known as the start of the Besselian solar year. It was very near January 1, 1900 (as, e.g., 1988.0 is near January 1, 1988).

Star catalogs list mean positions. The standard modern epochs are 1900.0, 1950.0, and 2000.0. Actually star catalogs list what is known as mean catalog place. This differs slightly from mean place and is discussed after the discussions of general precession and proper motion. Catalogs with epochs of 2000.0 will use mean place.

GENERAL PRECESSION

The effects of general precession on the right ascension and declination may be described by three rotations. The Euler angles of these are (1) θ, the inclination of the mean equator at the new epoch to the mean equator of the old epoch; (2) $90° - \zeta_0$, the right ascension of the ascending node of the mean equator at the new epoch relative to the mean equator of the old epoch measured relative to the old equinox; and (3) $90° + z$, the right ascension of the ascending node of the mean equator at the new epoch measured relative to the new equinox. Clearly these angles depend on both epochs. The standard old epoch is 1900.0. Frequently we want to precess from the epoch of the star catalog (usually 1950.0) to an epoch near the date of observation (say 1986.0). For this purpose it is convenient to explicitly introduce the intermediate epoch T_i and the final epoch T_f. Then, with t_i and t_f measured in tropical centuries (or fractions thereof) and related to T_i and T_f by

$$T_i = 1900.0 + 100 t_i, \qquad T_f = 1900.0 + 100 t_i + 100 t_f \qquad (4.1)$$

the expressions for ζ_0, z, and θ are

$$\zeta_0(t_i, t_f) = (2304''.253 + 1''.3975 t_i + 0''.00006 t_i^2) t_f$$
$$+ (0''.3023 - 0''.00027 t_i) t_f^2 + 0''.01800 t_f^3$$

$$z(t_i, t_f) = \zeta_0(t_i, t_f) + (0''.7927 + 0''.00066 t_i) t_f^2 + 0''.00032 t_f^3 \qquad (4.2)$$

$$\theta(t_i, t_f) = (2004''.685 - 0''.8533 t_i - 0''.00037 t_i^2) t_f$$
$$- (0''.4267 + 0''.00037 t_i) t_f^2 - 0''.04180 t_f^3$$

The effect of general precession on right ascension and declination is given by compounding the three rotations $R_3[90° - \zeta_0(t_i, t_f)]$, $R_1[\theta(t_i, t_f)]$, and $R_3[-90° - z(t_i, t_f)]$ upon the mean position vector at T_i to obtain the mean position vector at T_f. (Rotation matrices were discussed in Chapter Two.) This can be written as

$$\mathbf{l}[\alpha(T_f), \delta(T_f)] = P(t_i, t_f)\mathbf{l}[\alpha(T_i), \delta(T_i)] \qquad (4.3)$$

where, dropping the explicit time dependence momentarily,

$$P = R_3(-90° - z)R_1(\theta)R_3(90° - \zeta_0) = R_3(-z)R_2(\theta)R_3(-\zeta_0) \qquad (4.4)$$

$$= \begin{pmatrix} \cos\zeta_0\cos z\cos\theta - \sin\zeta_0\sin z & -\sin\zeta_0\cos z\cos\theta - \cos\zeta_0\sin z & -\cos z\sin\theta \\ \cos\zeta_0\sin z\cos\theta + \sin\zeta_0\cos z & -\sin\zeta_0\sin z\cos\theta + \cos\zeta_0\cos z & -\sin z\sin\theta \\ \cos\xi_0\sin\theta & -\sin\zeta_0\sin\theta & \cos\theta \end{pmatrix}$$

A convenient component form of Eq. 4.3 is $[\delta_f = \delta(T_f),\ \alpha_i = \alpha(T_i),\ \text{etc.}]$,

$$\cos\delta_f\sin(\alpha_f - z) = \cos\delta_i\sin(\alpha_i + \zeta_0)$$
$$\cos\delta_f\cos(\alpha_f - z) = \cos\delta_i\cos(\alpha_i + \zeta_0)\cos\theta - \sin\delta_i\sin\theta \qquad (4.5)$$
$$\sin\delta_f = \cos\delta_i\cos(\alpha_i + \zeta_0)\sin\theta + \sin\delta_i\cos\theta$$

An analytical alternative of Eqs. 4.5 is

$$\tan(\alpha_f - \alpha_i - \zeta_0 - z) = \frac{q\sin(\alpha_i + \zeta_0)}{1 - q\cos(\alpha_i + \zeta_0)}$$

$$\tan\left(\frac{\delta_f - \delta_i}{2}\right) = \tan\left(\frac{\theta}{2}\right)\left[\cos(\alpha_i + \zeta_0) - \sin(\alpha_i + \zeta_0)\tan\left(\frac{\alpha_f - \alpha_i - \zeta_0 - z}{2}\right)\right]$$

$$= \tan\left(\frac{\theta}{2}\right)\sec\left(\frac{\alpha_f - \alpha_i - \zeta_0 - z}{2}\right)\cos\left(\alpha_i + \zeta_0 + \frac{\alpha_f - \alpha_i - \zeta_0 - z}{2}\right)$$

where

$$q = \left[\tan\delta_i + \tan\left(\frac{\theta}{2}\right)\cos(\alpha_i + \zeta_0)\right]\sin\theta$$

Example 4.1. The star numbered 22024 in the *Smithsonian Astrophysical Observatory Star Catalogue* (Staff of the SAO 1966) is listed as having the following 1950.0 position:

$$\alpha = 01^h04^m55^s\!.686, \qquad \delta = +54°40'32''\!.96$$

This happens to be incorrect [the star is μ Cassiopeia = *FK4* 1030 = 30 Cas = *GC* 1360 = *N30* 220 = *HR* (or *BS*) 321 = *HD* 6582 = *DM* + 54°0223] and the right numbers are

$$\alpha = 01^h04^m55^s\!.680, \qquad \delta = +54°40'32''\!.95$$

This star was selected for illustrative purposes not because of this error but because of its large annual parallax and radial velocity. The *SAOC* is used as an example of a catalog because it is the one the nonspecialist user will go to first. The other star catalogs mentioned are the *Fourth Fundamental Catalogue* (Fricke and Kopff 1963) from which the correct numbers are taken (the *FK4* defines the astronomical reference frame); the *General Catalogue* (Boss 1936), which lists its position as $01^h04^m55^s661$, $+54°40'33''03$; the *N30* (for normal system of 1930) *Catalog* (Morgan 1952), which has it at $01^h04^m55^s700$, $+54°40'32''88$; the *Yale Bright Star Catalog* (HR = *Harvard Revised Photometry*); the *Henry Draper Catalog*; and the *Bonner Durchmusterang Catalogue*. (The latter three are not positional catalogs.)

Compute its 1950.0 position in the 1985.0 coordinate system.

Be aware that the problem specifically requires only the changes on position wrought by the effects of general precession. This author prefers to use Eqs. 4.5 to acquire the answers. First, though, $T_i = 1950.0$ so $t_i = 0.5$ and $T_f = 1985.0$ so $t_f = 0.35$ in Eqs. 4.1. Therefore, from Eqs. 4.2

$$\xi_0 = 806''7709, \qquad z = 806''8681, \qquad \theta = 701''4363$$

Then

$$\cos \delta_f \sin(\alpha_f - z) = 0.16379345$$

$$\cos \delta_f \cos(\alpha_f - z) = 0.55173937$$

$$\sin \delta_f = 0.81777465$$

or

$$\alpha_f = 01^h07^m02^s057, \qquad \delta_f = +54°51'45''52$$

The numerical result from Eq. 4.3 is

$$l(\alpha_f, \delta_f) = (0.55109442, 0.16595049, 0.81777465)$$

and α_f, δ_f are as above.

Approximate Methods for Precession

An Iterative Method

The limits of the quotients $(\alpha_f - \alpha_i)/t_f$, $(\delta_f - \delta_i)/t_f$ as $t_f \to 0$ can be obtained directly from Eq. 4.5. The results are

$$p = \frac{d\alpha}{dt} = m + n \sin \alpha \tan \delta, \qquad p' = \frac{d\delta}{dt} = n \cos \alpha \qquad (4.6)$$

The quantities m and n are called the *speeds of general precession in right ascension and declination.* Since Eq. 4.6 is only first order, it is rarely used beyond a few years. As such their most frequent application is to update (or backdate) equatorial coordinates from one epoch T tropical centuries after (or before) 1900.0 to another epoch t tropical centuries from T. The appropriate equations then are

$$\alpha(T+t) = \alpha(T) + \{m(T) + n(T)\sin[\alpha(T)]\tan[\delta(T)]\}t$$
$$\delta(T+t) = \delta(T) + \{n(T)\cos[\alpha(T)]\}t \tag{4.7}$$

where

$$m(T) = (4608\overset{''}{.}506 + 2\overset{''}{.}7950\,T + 0\overset{''}{.}00012\,T^2)/\text{cent}$$

$$= (307\overset{s}{.}2337 + 0\overset{s}{.}18633\,T + 0\overset{s}{.}000008\,T^2)/\text{cent}$$

$$n(T) = (2004\overset{''}{.}685 - 0\overset{''}{.}8533\,T - 0\overset{''}{.}00037\,T^2)/\text{cent} \tag{4.8}$$

$$= (133\overset{s}{.}6457 - 0\overset{s}{.}05689\,T^2 - 0\overset{s}{.}000025\,T^2)/\text{cent}$$

A much more accurate procedure is to extend this idea one step further. Equations 4.7 are replaced by

$$\alpha(T+t) = \alpha(T) + \left\{ m(T+t/2) + n(T+t/2)\sin\left[\frac{\alpha(T+t)+\alpha(T)}{2}\right]\right.$$

$$\left. \times\tan\left[\frac{\delta(T+t)+\delta(T)}{2}\right]\right\}t$$

$$\delta(T+t) = \delta(T) + \left\{ n(T+t/2)\cos\left[\frac{\alpha(T+t)+\alpha(T)}{2}\right]\right\}t \tag{4.9}$$

One starts with an approximation for $\alpha(T+t)$ and $\delta(T+t)$ (from Eqs. 4.7, for example) and then iterates through Eqs. 4.9 by the method successive substitutions. Convergence will occur in two or three iterations.

Example 4.2. The problem posed in the previous Example will be solved by both of the new methods just outlined. *Note*: all approximate procedures for computing the effects of general precession degrade rapidly in accuracy as the time span increases (beyond ~25 years) or the absolute value of the declination approaches 90° (especially for $|\delta| > 80°$).

In order to use Eqs. 4.7, $T = 0.5$ in Eqs. 4.8. This yields

$$m = 4609\overset{''}{.}904/\text{cent}, \qquad n = 2004\overset{''}{.}258/\text{cent}$$

As $t = 0.35$

$$\alpha(1985.0) = 01^h07^m01\overset{s}{.}691, \qquad \delta(1985.0) = +54°51'46\overset{''}{.}48$$

In order to employ Eqs. 4.9, m and n must be at 1967.5,

$$m = 4610\overset{''}{.}393/\text{cent}, \qquad n = 2004\overset{''}{.}109/\text{cent}$$

and then we iterate through Eqs. 4.9. As the starting values for α and δ at 1985.0, those obtained from Eqs. 4.7 are used. The next set is

$$\alpha = 01^h07^m02\overset{s}{.}056, \qquad \delta = +54°51'45\overset{''}{.}52$$

and the one after that is

$$01^h07^m02{.}^s057, \qquad +54°51'45{.}''52$$

which is identical (to this number of digits) to the next set. Compare with Example 4.1.

The Power Series Method

If we go beyond the first-order expansion to include higher-order terms in the time, we would expect better results. Straightforward differentiation of Eqs. 4.6 yields

$$\frac{d^2\alpha}{dt^2} = \dot{m} + \dot{n}\sin\alpha\tan\delta + np\cos\alpha\tan\delta + np'\sin\alpha\sec^2\delta$$

$$(4.10)$$

$$\frac{d^2\delta}{dt^2} = \dot{n}\cos\alpha - np\sin\alpha$$

where, by differentiating Eqs. 4.8,

$$\dot{m}(T) = (2{.}''7950 + 0{.}''00024\,T)/\text{cent}^2$$

$$= (0{.}^s18633 + 0{.}^s000016\,T)/\text{cent}^2$$

$$\dot{n}(T) = -(0{.}''8533 + 0{.}''00074\,T)/\text{cent}^2 \qquad (4.11)$$

$$= -(0{.}^s05689 + 0{.}^s000049\,T)/\text{cent}^2$$

The analog of Eqs. 4.7 is now

$$\alpha(T+t) = \alpha(T) + \left.\frac{d\alpha}{dt}\right|_T t + \frac{1}{2}\left.\frac{d^2\alpha}{dt^2}\right|_T t^2$$

$$(4.12)$$

$$\delta(T+t) = \delta(T) + \left.\frac{d\delta}{dt}\right|_T t + \frac{1}{2}\left.\frac{d^2\delta}{dt^2}\right|_T t^2$$

Example 4.3. We close this discussion by computing the 1950.0 place of μ Cas in the 1985.0 coordinate system using Eqs. 4.12. We already have

$$\alpha(1950.0) = 01^h04^m55{.}^s680, \qquad \delta(1950.0) = +54°40'32{.}''95$$

$$m(1950.0) = 4609{.}''904/\text{cent}, \qquad n(1950.0) = 2004{.}''258/\text{cent}$$

so that (at 1950.0)

$$p = \frac{d\alpha}{dt} = 5400{.}''458/\text{cent}, \qquad p' = \frac{d\delta}{dt} = 1924{.}''364/\text{cent}$$

We also need from Eqs. 4.11

$$\dot{m}(1950.0) = 2{.}''7951/\text{cent}^2, \qquad \dot{n}(1950.0) = -0{.}''8537/\text{cent}^2$$

Thus, at 1950.0, using Eqs. 4.10,

$$\ddot{\alpha} = 89\rlap{.}''1890/\text{cent}^2, \qquad \ddot{\delta} = -15\rlap{.}''4881/\text{cent}^2$$

Whence by Eqs. 4.12

$$\alpha(1985.0) = 01^h07^m02\rlap{.}^s055, \qquad \delta(1985.0) = +54°51'45\rlap{.}''53$$

Elliptic Aberration and Mean Catalog Place

The *mean place* of a star is its solar system barycentric position referred to a specified mean equator and equinox. The epoch of the mean equator and equinox is generally the beginning of a Besselian solar year. For the Besselian solar year commencing near the beginning of the calendar year 1950 the notation 1950.0 is used. Star catalogs list *mean catalog place*. The difference between mean catalog place and mean place (for the same epoch of place and the same orientation of the coordinate system) is very small. The reason for the difference is that part of the aberration of light due to the ellipticity of the Earth's orbit has been left in the cataloged position.

Let (α, δ) be the mean position of a star at some epoch (of place and orientation). Let (A, Δ) be the corresponding mean catalog place. Then

$$\alpha - A = (e_\oplus \kappa \cos \omega_\odot \cos \varepsilon)c + (e_\oplus \kappa \sin \omega_\odot)d$$
$$\delta - \Delta = (e_\oplus \kappa \cos \omega_\odot \cos \varepsilon)c' + (e_\oplus \kappa \sin \omega_\odot)d' \qquad (4.13)$$

where $e_\oplus$ is the eccentricity of the Earth's orbit, $\omega_\odot$ is the longitude of perigee of the Sun, ε is the obliquity of the ecliptic, κ is the constant of aberration $(20\rlap{.}''49552)$, and c, d, c', and d' are known as *star constants*,

$$c = \cos \alpha \sec \delta, \qquad c' = \tan \varepsilon \cos \delta - \sin \alpha \sin \delta$$
$$d = \sin \alpha \sec \delta, \qquad d' = \cos \alpha \sin \delta$$

Formulas for $e_\oplus$, ε, and $\omega_\odot$ at T tropical centuries from 1900.0 are

$$e_\oplus = 0.01675104 - 4.180 \times 10^{-5} T - 1.26 \times 10^{-7} T^2$$
$$\varepsilon = 23°27'08\rlap{.}''26 - 46\rlap{.}''845 T - 0\rlap{.}''0059 T^2 + 0\rlap{.}''00181 T^3 \qquad (4.14)$$
$$\omega_\odot = 281°13'15\rlap{.}''04 + 6189\rlap{.}''03 T + 1\rlap{.}''63 T^2 + 0\rlap{.}''012 T^3$$

Frequently one sees ΔC and ΔD defined by

$$\Delta C = e_\oplus \kappa \cos \omega_\odot \cos \varepsilon, \qquad \Delta D = e_\oplus \kappa \sin \omega_\odot$$

and then instead of Eq. 4.13,

$$\alpha = A + c \,\Delta C + d \,\Delta D$$
$$\delta = \Delta + c' \,\Delta C + d' \,\Delta D$$

General precession and proper motion affect the mean place of a star, not its mean catalog place. Hence the rigorous reduction of the position of a star from one epoch to another consists of (1) removing the elliptic aberration from the mean catalog place to obtain the mean place, (2) updating for general precession and proper motion (see below) to obtain the new mean place, and (3) adding back in the elliptic aberration computed at the new epoch to obtain the new mean catalog place. This procedure is only necessary for stars near the celestial poles when the updating is to be over long periods of time.

If one ignores elliptic aberration, two types of errors are introduced. One arises from the time-varying nature of ΔC and ΔD. This, as can be seen from Eqs. 4.14, is very small. The other is the interaction between the elliptic aberration terms left in and those of general precession. Both of these are small, of similar form, and of opposite sign. In the region of a celestial pole the maximum centennial errors arising from the neglect of the rigorous procedure is $0\rlap{.}^{s}0001$ in $|\Delta\alpha\cos\delta|$ and $0\rlap{.}''002$ in $|\Delta\delta|$. At its 1976 meeting the IAU recommended that all future (e.g., epoch 2000.0) catalogs give mean place and not mean catalog place.

The New (1976) IAU Precessional Parameters

The formulas given above depend on the parameters of the solar system (planetary masses, distances, etc.), a fundamental epoch (1900.0), and a measure of time (the tropical year). A more attractive time system is called Ephemeris Time. One method of counting using this system is to keep a continuous count of the days from some fixed epoch. This is the system of Julian Day Numbers; when the measure of the day is Ephemeris Time, we speak of the Julian Ephemeris Date (JED). The JED of 1900.0 is 2415020.31352. The *International Astronomical Union* (IAU), the governing body of astronomy, recommended in 1976 that the measure of time to be used for computing mean places of stars be changed to the Julian century of 36,525 ephemeris days (that is 36,525 days where 1 day is a time interval of 24 ephemeris hours). They also recommended using J2000.0 (as opposed to what will now be denoted as B1900.0) as the new fundamental epoch. Finally, due to an improved system of masses for the planets and the inclusion of the geodesic precession, the values of the equatorial precessional quantities were slightly changed.

As above we introduce an intermediate epoch T_i and a final epoch T_f. These are now Julian epochs (e.g., J1950.0 and J1975.0). We define t_i and t_f, measured in Julian centuries of 36,525 ephemeris days, by

$$T_i = J2000.0 + 100t_i$$

$$T_f = J2000.0 + 100t_i + 100t_f$$

If the Julian Ephemeris Date of an epoch T is symbolized by JED(T), then

$$t_i = [\text{JED}(T_i) - \text{JED}(2000.0)]/36,525$$

$$t_f = [\text{JED}(T_f) - \text{JED}(T_i)]/36,525$$

The JED values of several epochs are given in Table 1. The new values of ζ_0,

Table 1. The Julian Ephemeris Dates of Some Important Epochs

Besselian Epoch	Julian Epoch	JED
1900.0	1900.000858	2415020.31352
1950.0	1949.999790	2433282.42345905
2000.0	1999.998722	2451544.5333981
1899.999142	1900.0	2415020
1950.000210	1950.0	2433282.5
2000.001278	2000.0	2451545

z, and θ are

$$\zeta_0(t_i, t_f) = (2306\rlap{.}''2181 + 1\rlap{.}''39656\,t_i - 0\rlap{.}''000139\,t_i^2)\,t_f$$
$$+ (0\rlap{.}''30188 - 0\rlap{.}''000344\,t_i)\,t_f^2 + 0\rlap{.}''017998\,t_f^3$$

$$z(t_i, t_f) = \zeta_0(t_i, t_f) + (0\rlap{.}''79280 + 0\rlap{.}''000410\,t_i)\,t_f^2 + 0\rlap{.}''000205\,t_f^3$$

$$\theta(t_i, t_f) = (2004\rlap{.}''3109 - 0\rlap{.}''85330\,t_i - 0\rlap{.}''000217\,t_i^2)\,t_f$$
$$- (0\rlap{.}''42665 + 0\rlap{.}''000217\,t_i)\,t_f^2 - 0\rlap{.}''041833\,t_f^3$$

For the speeds of general precession in right ascension and declination

$$m(T) = (4612\rlap{.}''4362 + 2\rlap{.}''79312\,T - 0\rlap{.}''000278\,T^2)/\mathrm{cent}$$

$$n(T) = (2004\rlap{.}''3109 - 0\rlap{.}''85330\,T - 0\rlap{.}''000217\,T^2)/\mathrm{cent}$$

where T is measured in Julian centuries of 36,525 ephemeris days from J2000.0. Also their new derivatives are given by

$$\dot{m}(T) = (2\rlap{.}''79312 - 0\rlap{.}''000556\,T)/\mathrm{cent}^2$$

$$\dot{n}(T) = -(0\rlap{.}''85330 + 0\rlap{.}''000434\,T)/\mathrm{cent}^2$$

The Besselian epoch B of an instant t whose Julian Ephemeris Date is $\mathrm{JED}(t)$ may be computed from (see also the last section in Chapter Three)

$$B = \mathrm{B}1900.0 + \frac{\mathrm{JED}(t) - \mathrm{JED}(\mathrm{B}1900.0)}{365.242198781}$$

Ecliptic Coordinate General Precession

The rigorous equation for computing the effects of general precession on the ecliptic coordinates λ and β is

$$\mathbf{l}[\lambda(T_f), \beta(T_f)] = Q(t_i, t_f)\mathbf{l}[\lambda(T_i), \beta(T_i)]$$

where the matrix Q is given by

$$Q(t_i, t_f) = R_1(\varepsilon_f)P(t_i, t_f)R_1(-\varepsilon_i)$$

The notation is that of Eqs. 4.1 and 4.3 with ζ_0, z, and θ given in Eqs. 4.2 and ε given in Eq. 4.14. To lowest order Q is antisymmetric and, therefore, approximate formulas can be written as

$$\lambda_f = \lambda_i + a - b \cos(\lambda_i + c) \tan \beta_i$$
$$\beta_f = \beta_i + b \sin(\lambda_i + c) \tag{4.15}$$

where

$$a = (5025\rlap{.}''6 + 2\rlap{.}''22 t_i) t_f$$
$$b \sin c = (4\rlap{.}''96 - 0\rlap{.}''75 t_i) t_f$$
$$b \cos c = (46\rlap{.}''8 + 0\rlap{.}''01 t_i) t_f$$

or

$$b = (47\rlap{.}''1 - 0\rlap{.}''07 t_i) t_f$$
$$\tan c = 0.1060 - 0.016 t_i$$

The New (1976) IAU Obliquity of the Ecliptic

Along with the new equatorial precessional quantities, the IAU adopted a new formula for the obliquity of the ecliptic relative to J2000.0:

$$\varepsilon = 23°26'21\rlap{.}''448 - 46\rlap{.}''8150 T - 0\rlap{.}''0059 T^2 + 0\rlap{.}''001813 T^3$$

where T is measured in Julian centuries of 36,525 ephemeris days from J2000.0. The corresponding values of a, b, and c in Eqs. 4.15 are now

$$a = (5029\rlap{.}''1 + 2\rlap{.}''22 t_i) t_f$$
$$b \sin c = (4\rlap{.}''20 - 0\rlap{.}''75 t_i) t_f$$
$$b \cos c = (46\rlap{.}''8 + 0\rlap{.}''01 t_i) t_f$$
$$b = (47\rlap{.}''0 - 0\rlap{.}''06 t_i) t_f$$
$$\tan c = 0.0897 - 0.016 t_f$$

The Precession of Orbital Planes

The rigorous formulas for the reduction of the elements of an orbit due to general precession are rarely required in practice. Instead approximate formulas based on the differential relationships of the appropriate spherical triangles will suffice. If i, Ω, and ω are the inclination, longitude of the ascending node, and argument of perihelion with respect to the ecliptic, then

$$\Omega_f = \Omega_i + a - b \sin(\Omega_i + c) \cot i_i$$
$$\omega_f = \omega_i + b \sin(\Omega_i + c) \csc i_i$$
$$i_f = i_i + b \cos(\Omega_i + c)$$

govern their changes due to general precession. The quantities a, b, and c were introduced in Eq. 4.15.

If the fundamental reference plane is the celestial equator instead of the ecliptic, then the rates of change of these three angles are given by

$$\frac{d\Omega}{dt} = m - n \cos \Omega \cot i$$

$$\frac{d\omega}{dt} = -n \cos \Omega \csc i$$

$$\frac{di}{dt} = -n \sin \Omega$$

where m and n are the speeds of general precession in right ascension and declination (see Eq. 4.8). In both cases higher accuracy can be achieved by using the intermediate values of the quantities on the right-hand sides and iterating (as in Eq. 4.9, for example).

PROPER MOTION

Relative to a fixed solar system barycentric coordinate system the stars are in motion. This intrinsic motion of the stars is called *proper motion*. The radial component of this motion is named the *radial velocity* $v_r = \dot{r}$. It is a signed quantity, measured with a spectrograph, whose units are kilometers per second. Positive values indicate recession. The tangential component of the motion (i.e., the component of the motion projected onto the plane of the sky) is known as *the* proper motion.

The equatorial solar system barycentric location vector of a star is given by $\mathbf{r} = r\mathbf{l}$ where r is its distance and $\mathbf{l}$ is given in Eq. 3.2. Differentiating this (compare with Eq. 1.4),

$$\dot{\mathbf{r}} = \dot{r} \begin{pmatrix} \cos \delta \cos \alpha \\ \cos \delta \sin \alpha \\ \sin \delta \end{pmatrix} + r\dot{\delta} \begin{pmatrix} -\sin \delta \cos \alpha \\ -\sin \delta \sin \alpha \\ \cos \delta \end{pmatrix} + r\dot{\alpha} \begin{pmatrix} -\cos \delta \sin \alpha \\ \cos \delta \cos \alpha \\ 0 \end{pmatrix} \qquad (4.16)$$

The distance r is a difficult quantity to measure and it can only be done for the closest stars. Its reciprocal, expressed in seconds of arc, is known as the *annual parallax* π. For the relationship $\pi r = 1$ to hold, the unit of distance must be the *parsec* (pc). At a distance of 1 pc from a line 1 A.U. (*astronomical unit*, essentially the mean distance of the Earth from the Sun) in length, that line subtends, or has a parallax of, one second of arc. Since 1 A.U. = 1.495978×10^8 km, 1 pc = $\csc 1'' \cdot 1$ A.U. = 3.08568×10^{13} km.

The quantity $\dot{\delta} = \mu'$ is known as *the proper motion in declination* and it is commonly expressed in seconds of arc per year or seconds of arc per century. The quantity $\dot{\alpha} = \mu$ is known as *the proper motion in right ascension* and it is commonly expressed in seconds of time per year or seconds of time per century. The quantity $\omega = [\mu^2 \cos^2 \delta + (\mu')^2]^{1/2}$ is known as *the proper motion*. Sometimes the direction of motion and ω are given instead of the components μ, μ'. If P is the *position angle* of the motion, then

$$\mu \cos \delta = \omega \sin P$$
$$\mu' = \omega \cos P \tag{4.17}$$

Position angle is measured from the North point positive to the East point.

Example 4.4. The values in the *SAOC* for the proper motions of μ Cas are $\mu = 0\overset{s}{.}3947/\text{yr}$, $\mu' = -1\overset{''}{.}575/\text{yr}$. A better set from the *FK4* is (1950.0) $\mu = 39\overset{s}{.}472/\text{cent}$, $\mu' = -157\overset{''}{.}53/\text{cent}$. Rigorous treatment is rarely necessary for proper motion updates and power series are usually used,

$$\alpha(T+t) = \alpha(T) + \mu t, \qquad \delta(T+t) = \delta(T) + \mu' t$$

Compute the 1985.0 position of μ Cas in the 1950.0 coordinate system. Make sure that you understand the difference between this calculation and that of Example 4.1.

Since $T = 1950.0$ and $t = 0.35$ cent, you should find that

$$\alpha = 01^h 05^m 09\overset{s}{.}495, \qquad \delta = +54°39'37\overset{''}{.}81$$

Approximate Matrix Formulation

If the star's motion was uniform on a great circle, then the effects of proper motion could be rigorously dealt with via

$$\mathbf{l} = R_3(-\alpha_0) R_2(\delta_0 - 90°) R_3(P_0) R_2(\omega_0 T) R_3(-P_0) R_2(90° - \delta_0) R_3(\alpha_0) \mathbf{l}_0 \tag{4.18}$$

where $T = t - t_0$, P_0 is the position angle defined in Eq. 4.17, and $\mathbf{l} = \mathbf{l}[\alpha(t), \delta(t)]$. This can also be written as

$$\begin{pmatrix} \cos \delta \cos \alpha \\ \cos \delta \sin \alpha \\ \sin \delta \end{pmatrix} = \begin{pmatrix} \sin \delta_0 \cos \alpha_0 & -\sin \alpha_0 & \cos \delta_0 \cos \alpha_0 \\ \sin \delta_0 \sin \alpha_0 & \cos \alpha_0 & \cos \delta_0 \sin \alpha_0 \\ -\cos \delta_0 & 0 & \sin \delta_0 \end{pmatrix}$$

$$\times \begin{pmatrix} -\cos P_0 \sin \omega_0 T \\ \sin P_0 \sin \omega_0 T \\ \cos \omega_0 T \end{pmatrix} \tag{4.19}$$

Equation 4.18 ignores the cross effects between radial and tangential motion. These are known as *foreshortening terms* and are proportional to $\pi v_r \mu$ and $\pi v_r \mu'$ in right ascension and declination. An approximate correction for the foreshortening term (i.e., the $v = \pi v_r$ factor) is to use the above formulas but with ω_0 replaced by $\omega = \omega_0(1 - vT)$. Over long time spans this is adequate for all but the closest and fastest moving stars.

Example 4.5. Something that you can't find out from the *SAOC* is that the parallax of μ Cas is $0''.136$ and that its radial velocity is -97.2 km/sec. When we look at the time derivatives of the proper motions, we find

$$\frac{d\mu}{dt} = 2\mu\mu' \tan \delta - 2\mu\nu, \qquad \frac{d\mu'}{dt} = -\mu^2 \sin \delta \cos \delta - 2\mu'\nu$$

where $\nu = \pi v_r$ (in appropriate units). This is the source of the correction to ω_0 noted above. Both with and without this term, compute α and δ at 1985.0 in the 1950.0 coordinate system.

From Eq. 4.17

$$\omega = 376''.85/\text{cent}, \qquad P = 114°.710$$

As $T = 0.35$, Eq. 4.19 implies that

$$\mathbf{l} = (0.55520046, 0.16224197, 0.81573892)$$

or

$$\alpha = 01^h 05^m 09^s.490, \qquad \delta = +54°39'37''.77$$

With π in seconds of arc and v_r in kilometers per second, $\nu = 1.0227 \times 10^{-4} \pi v_r$ in radians per century. Hence, here $\nu = -278''.86/\text{cent}$ so $\omega = \omega_0(1 - \nu T) = 377''.03/\text{cent}$ and we now derive, again from Eqs. 4.19,

$$\mathbf{l} = (0.55520049, 0.16224227, 0.81573885)$$

whence

$$\alpha = 01^h 05^m 09^s.497, \qquad \delta = +54°39'37''.74$$

A second-order series applied to proper motions, such as Eqs. 4.12, yields

$$\alpha = 01^h 05^m 09^s.497, \qquad \delta = +54°39'37''.74$$

whereas the answer is (from the rigorous equations)

$$\alpha = 01^h 05^m 09^s.497, \qquad \delta = +54°39'37''.74$$

The Precessional Effects on Proper Motions

Before presenting the solution to the complete problem of updating a star's mean position from one epoch to another epoch (i.e., correcting simultaneously for the effects of general precession and proper motion), the effects of general precession on the proper motions must be evaluated. Since the effects on the coordinate system due to general precession may be described by a series of rotations, neither the distance r nor the radial velocity v_r can be affected by it. This rigorous treatment for the proper motion is handled by the matrix P of Eq. 4.4 in a form analogous to that of Eq. 4.3,

$$\dot{\mathbf{r}}_f = P(t_i, t_f)\dot{\mathbf{r}}_i$$

Since $\dot{\mathbf{r}} = \dot{r}\mathbf{l} + r\dot{\mathbf{l}}$, it is clear that the rigorous procedure requires knowledge of $\mathbf{l}_i, \mathbf{l}_f,$ and $\dot{\mathbf{l}}_i$ in order to compute μ and μ' at T_f. This is cumbersome and the effects so small that power series expansions in t_f are generally used.

Approximate formulas for correcting μ and μ' at some epoch T tropical centuries from 1900.0 to some other epoch $T + t$ tropical centuries from 1900.0 for general precession are

$$\mu(T+t) = \mu(T) + \dot{\mu}(T)t$$

$$\mu'(T+t) = \mu'(T) + \dot{\mu}'(T)t$$

where

$$\dot{\mu} = n\mu \cos \alpha \tan \delta + n\mu' \sin \alpha \sec^2 \delta, \qquad \dot{\mu} = -n\mu \sin \alpha$$

THE COMBINED EFFECTS OF GENERAL PRECESSION AND PROPER MOTION

The problem is to update a star's position at the instant of some epoch represented by T_i in the coordinate system whose epoch of orientation is also T_i to some other instant of time represented by T_f in the coordinate system whose epoch of orientation is also T_f. This can be performed in two logically equivalent ways. One procedure is to first apply proper motion from T_i to T_f and then to correct the orientation of the coordinate system due to general precession from T_i to T_f. This is represented by

$$\mathbf{l}(\alpha_f, \delta_f) = P(t_i, t_f)\left[\mathbf{l}(\alpha_i, \delta_i) + \frac{t_f \dot{\mathbf{r}}(t_i)}{r_i}\right] \qquad (4.20)$$

The second rigorous procedure is to first apply the effects of general precession (on *both* the position and proper motion) from T_i to T_f and then adjust the epoch of place from T_i to T_f by applying the new proper motions; namely,

$$\mathbf{l}(\alpha_f, \delta_f) = P(t_i, t_f)\mathbf{l}(\alpha_i, \delta_i) + t_f\left[\frac{P(t_i, t_f)\dot{\mathbf{r}}(t_i)}{r_i}\right]$$

This formula is a special case of the more general one wherein one updates a position whose epoch of place is t_1 but whose equator and equinox have an epoch T_1 to an epoch of place t_2 in a coordinate system whose epoch of orientation is T_2; namely,

$$\mathbf{l}[\alpha(t_2, T_2), \delta(t_2, T_2)] = P(T_1, T_2)\left\{\mathbf{l}[\alpha(t_1, T_1), \delta(t_1, T_1)] + \frac{(t_2 - t_1)\dot{\mathbf{r}}(t_1, T_1)}{r_1}\right\}$$

Example 4.6. Compute the 1985.0 position of μ Cas in the coordinate system whose epoch is 1985.0 too.

We do this via Eq. 4.20 with $\dot{\mathbf{r}}$ given in Eq. 4.16, and start with

$$\alpha_i = 01^\mathrm{h}04^\mathrm{m}55\overset{s}{.}680, \qquad \delta_i = +54°40'32''\!.95$$

from Example 4.1 and

$$\mu = 39\overset{s}{.}472/\text{cent}, \qquad \mu' = -157\overset{"}{.}53/\text{cent}$$

from Example 4.4. We also need

$$\pi = 0\overset{"}{.}136, \qquad v_r = -97.2 \text{ km/sec}$$

from Example 4.5. With these values one can calculate

$$\mathbf{l}(\alpha_i, \delta_i) = (0.55515356, 0.16162330, 0.81589364)$$

$$\frac{t_f \dot{\mathbf{r}}(t_i)}{r_i} = (-2.1566427 \times 10^{-4}, 5.4223259 \times 10^{-4}, -5.4061616 \times 10^{-4})$$

Since $t_f = 0.35$ cent. Hence P operates on

$$(0.55493789, 0.16216553, 0.81535303)$$

Note that this is the position vector at epoch of place 1985.0 in the 1950.0 epoch of orientation coordinate system (the right ascension and declination are $01^h05^m09\overset{s}{.}497$, $+54°39'37\overset{"}{.}74$ as mentioned at the end of the preceding example). To perform the general precession, we need the equatorial precessional quantities ζ_0, z, and θ already calculated in Example 4.1:

$$\zeta_0 = 806\overset{"}{.}7709, \qquad z = 806\overset{"}{.}8681, \qquad \theta = 701\overset{"}{.}4363$$

From these results and the definition in Eq. 4.4

$$P = \begin{pmatrix} 9.9996362 \times 10^{-1} & -7.8230399 \times 10^{-3} & -3.4006265 \times 10^{-3} \\ 7.8230399 \times 10^{-3} & 9.9996940 \times 10^{-1} & -1.3302662 \times 10^{-5} \\ 3.4006265 \times 10^{-3} & -1.3301059 \times 10^{-5} & 9.9999422 \times 10^{-1} \end{pmatrix}$$

Thus

$$\mathbf{l}(\alpha_f, \delta_f) = (0.55087636, 0.16649102, 0.81723329)$$

whence

$$\alpha_f = 01^h07^m15\overset{s}{.}926, \qquad \delta_f = +54°50'50\overset{"}{.}11$$

The Power Series Method

The total rate of change of an equatorial coordinate because of the combination of general precession and proper motion is known as the *annual variation*. The rates of change of the annual variations are known as the *secular variations*. The annual variations for right ascension and declination are

$$\frac{d\alpha}{dt} = p + \mu, \qquad \frac{d\delta}{dt} = p' + \mu'$$

where p and p' were given in Eqs. 4.6. The rates of change of μ and μ' due to the combined effects of general precession, proper motion, and foreshortening are

$$\frac{d\mu}{dt} = n\mu \cos \alpha \tan \delta + n\mu' \sin \alpha \sec^2 \delta + 2\mu\mu' \tan \delta - 2\mu\nu$$

$$\frac{d\mu'}{dt} = -n\mu \sin \alpha - \left(\frac{\mu^2}{2}\right) \sin 2\delta - 2\mu'\nu$$

The secular variations of right ascension and declination may be expressed as

$$\frac{d^2\alpha}{dt^2} = \dot{m} + \dot{n} \sin \alpha \tan \delta + n(p + 2\mu) \cos \alpha \tan \delta$$

$$+ n(p' + 2\mu') \sin \alpha \sec^2 \delta + 2\mu\mu' \tan \delta - 2\mu\nu$$

$$\frac{d^2\delta}{dt^2} = \dot{n} \cos \alpha - n(p + 2\mu) \sin \alpha - \left(\frac{\mu^2}{2}\right) \sin 2\delta - 2\mu'\nu$$

Then

$$\alpha(T + t) = \alpha(T) + \dot{\alpha}(T)t + \frac{\ddot{\alpha}(T)}{2}t^2$$

$$\delta(T = t) = \delta(T) + \dot{\delta}(T)t + \frac{\ddot{\delta}(T)}{2}t^2$$

$$\mu(T + t) = \mu(T) + \dot{\mu}(T)t$$

$$\mu'(T + t) = \mu'(T) + \dot{\mu}'(T)t$$

There is one last point. It may happen that over long time intervals for stars whose right ascension is near 0^h (or 24^h), one obtains negative right ascensions (or values $> 24^h$). If this is the case 24^h is added (subtracted) from the computed right ascension. Similarly, over long time intervals for stars whose declination is near $+90°$ (or $-90°$), one can obtain declinations larger than $+90°$ (or smaller than $-90°$). If this is the case, $180° - \delta$ (or $-180° - \delta$) is used for the declination and 12^h is added to the right ascension.

Example 4.7.   Redo the 1985.0 computation of the mean place of μ Cas using the second-order power series method.

The information gathered in the preceding example will not be repeated. The 1950.0 values of m and n were obtained in Example 4.2,

$$m = 4609''904/\text{cent}, \qquad n = 2004''258/\text{cent}$$

and their first derivatives were derived in Example 4.3,

$$\dot{m} = 2''7951/\text{cent}^2, \qquad \dot{n} = 0''8537/\text{cent}^2$$

The required computations are straightforward and yield

$$\dot{\alpha} = 399\overset{s}{.}503/\text{cent}, \qquad \dot{\delta} = 1766\overset{''}{.}83/\text{cent}$$

$$\ddot{\alpha} = 6\overset{s}{.}836/\text{cent}^2, \qquad \ddot{\delta} = -19\overset{''}{.}94/\text{cent}^2$$

$$\dot{\mu} = 0\overset{s}{.}456/\text{cent}^2, \qquad \mu' = -2\overset{''}{.}84/\text{cent}^2$$

which results in

$$\alpha = 01^{h}07^{m}15\overset{s}{.}925, \qquad \delta = +54°50'50\overset{''}{.}12$$

$$\mu = 39\overset{s}{.}632/\text{cent}, \qquad \mu' = -158\overset{''}{.}52/\text{cent}$$

COMPUTING THE TOPOCENTRIC PLACE

The Places of a Star

The *mean position* (or place) of a star is that point on the celestial sphere at which it would be seen from the solar system barycenter when referred to the mean equator and equinox at the beginning of a Besselian solar year. The length of a Besselian solar year is almost equal to the duration of a tropical (or seasonal) year, about $365\overset{d}{.}2422$. It commences near the beginning of the corresponding calendar year. The notation .0 after a year (as in 1900.0 or 1983.0) signifies the start of the Besselian solar year.

The *mean catalog place* is the mean position with the *e* terms of aberration (or elliptic aberration) left in. In general, the *e* terms of aberration can be ignored.

The *true position* of a star is that point on the celestial sphere at which it would be seen from the solar system barycenter when referred to the true equator and equinox of date. To obtain the true place from the mean place, one adds to the mean place the effects of general precession, nutation, and proper motion.

The *apparent position* of a star is that point on the celestial sphere at which it would be seen from the center of the Earth when referred to the true equator and equinox of date. To obtain the apparent place from the true place, one adds to the true place the effects of annual aberration, annual parallax, and, if the star is a member of a multiple system, orbital motion.

The *topocentric position* of a star is that point on the celestial sphere at which it would be seen through a perfect optical instrument from a point on the surface of the Earth when referred to the true equator and equinox of date. To obtain the topocentric place from the apparent place, one adds to the apparent place the effects of diurnal aberration, diurnal parallax, and astronomical refraction.

The Computation of the Apparent Place

Let the mean place for the beginning of a Besselian solar year be (α_0, δ_0). A simple, accurate method of computing the apparent place (α, δ) is

$$\alpha = a_0 + \tau\mu + Aa + Bb + Cc + Dd + E + J\tan^2\delta_0$$

$$+ \left(\frac{\pi}{\kappa}\right)(Cd\sec\varepsilon_0 - Dc\cos\varepsilon_0)$$

$$\delta = \delta_0 + \tau\mu' + Aa' + Bb' + Cc' + Dd' + J'\tan\delta_0 \tag{4.21}$$

$$+ \left(\frac{\pi}{\kappa}\right)(Cd'\sec\varepsilon_0 - Dc'\cos\varepsilon_0)$$

This corrects for proper motion (the μ and μ' terms), general precession and nutation (the A, B, and E terms), annual aberration (the C and D terms), and annual parallax (the π/κ terms; cf. Eqs. 3.28) to first order and includes a general second-order term (the J and J' terms). When Eqs. 4.21 are used, the maximum error in $|(\alpha - \alpha_0)\cos\delta_0|$ or $|\delta - \delta_0|$ is $0\rlap{.}''003$ at all declinations. For $|\delta_0| \le 60°$ if the J, J' terms are ignored, then the maximum errors are less than $0\rlap{.}''010$. At $|\delta_0| = 80°$ the errors introduced by the neglect of the second-order terms can rise to $0\rlap{.}''02$.

The generic term for A, B, C, D, E, J, and J' is *Besselian day number*. The day numbers A–E are tabulated at 0^h E.T., for every day of the year, in *The Astronomical Almanac*. Their computation is explained in *The Explanatory Supplement*. Tables of J and J' are given therein too for every tenth day of the year for each hour of right ascension. Alternatively, they may be computed directly from

$$J = [(A \pm D)\sin\alpha_0 + (B \pm C)\cos\alpha_0][(A \pm D)\cos\alpha_0 - (B \pm C)\sin\alpha_0]$$
$$J' = -\tfrac{1}{2}[(A \pm D)\sin\alpha_0 + (B \pm C)\cos\alpha_0]^2 \tag{4.22}$$

where the upper (lower) sign is for $\delta_0 > 0$ $(\delta_0 < 0)$.

The quantities a, b, c, d, a', b', c', and d' are collectively known as (*Besselian*) *star constants*. They depend only on the mean place and its epoch;

$$a = \frac{m}{n} + \sin\alpha_0\tan\delta_0, \qquad a' = \cos\alpha_0$$

$$b = \cos\alpha_0\tan\delta_0, \qquad\qquad b' = -\sin\alpha_0$$

$$\tag{4.23}$$

$$c = \cos\alpha_0\sec\delta_0, \qquad\qquad c' = \tan\varepsilon_0\cos\delta_0 - \sin\alpha_0\sin\delta_0$$

$$d = \sin\alpha_0\sec\delta_0, \qquad\qquad d' = \cos\alpha_0\sin\delta_0$$

As earlier in this chapter (Eqs. 4.8), m and n are the speeds of general precession in right ascension and declination. A formula for the quotient m/n is given in Example 4.8.

It is modern (post-1960) practice to use Eqs. 4.21 from or to the beginning of the nearest Besselian solar year. This keeps the fraction of the year, τ, from (to) that epoch to less than one-half year (in absolute value). In particular, for the time interval from July 1 or 2, 1979, to June 30 or July 1, 1980, the preferred epoch for the mean equator and equinox is 1980.0. Between July 1, 1979 and 1980.0 ($=$January $1\overset{d}{.}189$, 1980), τ is negative and between 1980.0 and July 1, 1980, τ is positive. The practical effect of switching epochs at mid-calendar year is to improve the accuracy of Eqs. 4.21–4.23.

In the annual parallax correction π is the annual parallax, κ is the constant of aberration, and the approximation has been made that the geocentric distance of the Sun is exactly 1 A.U. The maximum error of this approximation is $0\overset{''}{.}013$ for α Cen, the star with the largest annual parallax. The annual parallax correction itself is only incorporated in the mean place to apparent place reduction when $\pi \geqslant 0\overset{''}{.}01$. The epoch of place and orientation of the proper motions should be that of the position. As a practical matter, proper motions whose epoch of orientation is the catalog epoch but whose epoch of place is that of α_0 and δ_0 may be used in all but the most precise work.

Example 4.8. The star 82 G. Eri [$=FK4\ 119 = SAOC\ 216263 = HR(BS)\ 1008 = HD\ 20794 = DM - 43°1028 = N30\ 684 = GC\ 4000$] has the following location and velocity at 2000.0 in the coordinate system mean equator and equinox 2000.0.

$$\alpha = 03^h 19^m 55\overset{s}{.}617, \qquad \delta = -43°04'11\overset{''}{.}06$$

$$\mu = 27\overset{s}{.}731/\text{cent}, \qquad \mu' = 73\overset{''}{.}32/\text{cent}$$

$$\pi = 0\overset{''}{.}156, \qquad v_r = 86.8 \text{ km/sec}$$

Compute its apparent place at $03^h 33^m 33\overset{s}{.}33$ Mountain Standard Time (MST) on July 4, 1983.

Since the required date is in the month of July, the 1984.0 position is needed. Following Eqs. 4.1 $T_i = 2000.0$ so $t_i = 1$, $T_f = 1984.0$ so $t_f = -0.16$. From Eqs. 4.2

$$\zeta_0 = -368\overset{''}{.}8964, \qquad z = -368\overset{''}{.}8761, \qquad \theta = -320\overset{''}{.}6238$$

and, consequently, from Eq. 4.4

$$P = \begin{pmatrix} 9.9999240 \times 10^{-1} & 3.5768127 \times 10^{-3} & 1.5544250 \times 10^{-3} \\ -3.5768127 \times 10^{-3} & 9.9999360 \times 10^{-1} & -2.7798776 \times 10^{-6} \\ -1.5544250 \times 10^{-3} & -2.7800306 \times 10^{-6} & 9.9999879 \times 10^{-1} \end{pmatrix}$$

Next $\mathbf{l}$ is computed at 1984.0 in the 1984.0 coordinate system via Eq. 4.20:

$$\mathbf{l}(\alpha_f, \delta_f) = (0.47073608, 0.55747608, -0.68350928)$$

or

$$\alpha_0 = \alpha(1984.0) = 03^h 19^m 17\overset{s}{.}293, \qquad \delta_0 = \delta(1984.0) = -43°07'49\overset{''}{.}34$$

Note that $\mathbf{l}$ at 1984.0 in the 2000.0 coordinate system is $\mathbf{l} =$ (0.46980097, 0.55915815, −0.68277829). We also need 1984.0 proper motions. These are (approximately) computed via the power series method. As at 2000.0

$$m = 4611\overset{\prime\prime}{.}3011/\text{cent}, \qquad n = 2003\overset{\prime\prime}{.}8313/\text{cent}$$

it follows that

$$\dot{\mu} = -0\overset{s}{.}1890/\text{cent}^2, \qquad \dot{\mu}' = -2\overset{\prime\prime}{.}879/\text{cent}^2$$

so 1984.0 values for the proper motions are

$$\mu = 27\overset{s}{.}761/\text{cent}, \qquad \mu' = 73\overset{\prime\prime}{.}78/\text{cent}$$

So far this problem has been nothing unusual. In order to compute the apparent place at a particular time, values of the Besselian day numbers are needed at that instant. A–E are tabulated in Section B of *The Astronomical Almanac* (pp. B24–31 in the 1983 edition; tables of J, J' are on pp. B32–35) as a function of Ephemeris Time. We obtain from the asked for Mountain Standard Time an approximate Ephemeris Time by the following steps: First MST is converted to GMT by applying the longitude correction for the standard meridian of MST (i.e., $105° = 7^h$). Thus, $03^h33^m33\overset{s}{.}33$ MST on July 4, 1983 = $10^h33^m33\overset{s}{.}33$ GMT on July 4, 1983. Having *no* choice (cf. pp. B4–5 of *The Astronomical Almanac*) we make the approximation that U.T. = U.T.C. = GMT. In terms of pointing a telescope, the error is trivial. Therefore $\Delta TA =$ TAI $+ 32\overset{s}{.}184 −$ U.T. can be used as an approximation to $\Delta ET = $ E.T. − U.T.C. On July 1, 1983, this is estimated to be $+53\overset{s}{.}5$ so that the value of E.T. for the Besselian day number interpolation is $10^h34^m26\overset{s}{.}8$ on July 4, 1983. Using cubic interpolation in the tables for this instant

$$A = -16\overset{\prime\prime}{.}116, \qquad C = +3\overset{\prime\prime}{.}8779, \qquad E = -0\overset{s}{.}0024$$
$$B = -0\overset{\prime\prime}{.}6132, \qquad D = -20\overset{\prime\prime}{.}0493$$

Also, from the tables on pages B34–5 (or Eqs. 4.22),

$$J = -0\overset{s}{.}00000, \qquad J' = -0\overset{\prime\prime}{.}0000$$

From page B1 of *The Astronomical Almanac* 1984.0 = January 1.158, 1984, so since 1983 was not a leap year, the fraction of year $\tau =$ $-(366.158 - 185.441)/365.2422 = -0.4948$. The obliquity of the ecliptic ε_0 is calculated from Eq. 4.14 with $T = 0.84$, $\varepsilon_0 = 23°26'28\overset{\prime\prime}{.}907$. The constant of aberration $\kappa = 20\overset{\prime\prime}{.}49552$, and we are now in a position to compute each term in Eq. 4.21 once m/n is evaluated. A convenient formula for the useful life of this text is (I should be more ambitious!)

$$\frac{m}{n} = 2.298868 + 2.3728 \times 10^{-3} T$$

Here $T = 0.84$ so $m/n = 2.300861$ (it is 2.300862, directly from Eqs. 4.8).

	Right Ascension	Declination
Start	$03^h19^m17\overset{s}{.}293$	$-43°07'49\overset{''}{.}34$
Proper motion	$-0\overset{s}{.}1374$	$-0\overset{''}{.}365$
Aa or Aa'	$-1\overset{s}{.}7554$	$-10\overset{''}{.}717$
Bb or Bb'	$+0\overset{s}{.}0247$	$+0\overset{''}{.}469$
Cc or Cc'	$+0\overset{s}{.}2285$	$+3\overset{''}{.}253$
Dd or Dd'	$-1\overset{s}{.}3993$	$+8\overset{''}{.}843$
E	$-0\overset{s}{.}0024$	0
Second order	0	0
Parallax	$+0\overset{s}{.}0105$	$+0\overset{''}{.}103$

$$\alpha = 03^h19^m14\overset{s}{.}262, \qquad \delta = -43°07'47\overset{''}{.}75$$

A Matrix Reduction

Some readers may prefer to reduce the mean place to the apparent place via matrices. Following Scott and Hughes (1964) one can accomplish this by using

$$l(\alpha, \delta) = R_3(-f)R_2(A)R_1(B)[l(\alpha_0, \delta_0) + \mathbf{a}] + \left(\frac{\pi}{\kappa}\right)\mathbf{p}$$

where

$$\mathbf{a} = \begin{pmatrix} -D \\ C \\ C \tan \varepsilon_0 \end{pmatrix}, \qquad \mathbf{p} = \begin{pmatrix} -C \sec \varepsilon_0 \\ -D \cos \varepsilon_0 \\ -D \sin \varepsilon_0 \end{pmatrix}, \qquad f = \left(\frac{m}{n}\right)A + E$$

The Computation of the Topocentric Place

The three remaining corrections are those of diurnal aberration, diurnal parallax, and astronomical refraction. The effects of diurnal parallax (for a star!) are less than those of annual parallax by (the equatorial radius of the Earth)/1 A.U. $= 4.3 \times 10^{-5}$. Thus diurnal parallax is universally ignored for stars. Starting from the apparent place (α, δ), the correction for diurnal aberration is

$$\alpha' = \alpha + 0\overset{s}{.}02133\rho \cos \phi' \cos h \sec \delta$$

$$\delta' = \delta + 0\overset{''}{.}32000\rho \cos \phi' \sin h \sin \delta$$

where h is the hour angle $[=$(local apparent sidereal time)$- \alpha]$, ρ is the observer's geocentric distance in units of the equatorial radius of the Earth, and ϕ' is the observer's geocentric latitude. From (a', δ') the computation of the topocentric place (α'', δ'') follows via

$$\alpha'' = \alpha' + \mathcal{R} \sec \delta'' \csc z' \cos \Phi \sin h'$$

$$\delta'' = \delta' + \mathcal{R} \sec \delta' \csc z' (\sin \Phi - \sin \delta' \cos z')$$

(4.24)

where $h' = $ (local apparent sidereal time) $- \alpha'$, z' is the zenith distance corresponding to h' and δ', and Φ is the observer's astronomical latitude. The quantity $\mathscr{R}$ is the value of the astronomical refraction R corrected for meteorological variations from the standard conditions of dry air at 50°F, 30 in. Hg. For most purposes, at visible wavelengths

$$\mathscr{R} = \frac{17P}{460 + T_F} R \qquad (4.25)$$

(where P is the atmospheric pressure in inches of Mercury and T_F is the atmospheric temperature in degrees Fahrenheit) is sufficient. When the amount of water vapor varies appreciably, a correction for the humidity must be included. The astronomical refraction R is related to z' (for $z' \leqslant 75°$, at larger zenith distances no reliable formula is available) through

$$R = R_1 \tan(z' - R) + R_2 \tan^3(z' - R) \qquad (4.26)$$

where $R_1 = 58\!''\!.294$, $R_2 = -0\!''\!.0668$. If R_2 is set equal to 0 in Eq. 4.26 (a reasonable approximation for altitudes $> 30°$), then the value of $58\!''\!.2$ should be used for R_1. Equations 4.24–4.26 represent mean or Besselian astronomical refraction. This only affects zenith distance or altitude, not azimuth. The refraction correction is prone to error on account of varying atmospheric conditions and topographical peculiarities. Finally, a refinement of the parameters in Eq. 4.25 and 4.26 for wavelength dependence may be necessary.

Planetary Aberration

That part of aberration, relative to the solar system barycentric reference frame, produced by the motion of the light source is called *planetary aberration*. Other names are the parallax of light and the correction for light time. Its computation requires a knowledge of the distance and motion of the light source. When the source of light is a star, this information is either unknown or poorly known. Hence, for stars planetary aberration is universally ignored. Because of the indeterminacy of who is in uniform rectilinear motion in special relativity, this neglect need not introduce any error.

When the motion of the light source is well known, the planetary aberration can and should be computed. This is the case for solar system objects. One can correct for planetary aberration in two different ways; one can either backdate the time of observation by $\Delta t = R/c$ where $\mathbf{R} = R\mathbf{l}(A, \Delta)$ is the topocentric location of the object or directly correct the topocentric position. If $\mathbf{V} = \dot{\mathbf{R}} = (\dot{X}, \dot{Y}, \dot{Z})$ is the topocentric rectangular velocity vector of the planet, these corrections take the form

$$(A - \alpha)\cos\delta = \frac{\dot{X}\sin\alpha - \dot{Y}\cos\alpha}{c}$$

$$\Delta - \delta = \frac{\dot{X}\cos\alpha\sin\delta + \dot{Y}\sin\alpha\sin\delta - \dot{Z}\cos\delta}{c}$$

where $\mathbf{l}(\alpha, \delta)$ is the geometric position and c is the speed of light.

Parallactic Refraction

Astronomical refraction is computed using the implicit assumption that the distance to the celestial object is infinite. When this is not the case, principally for artificial satellites but sometimes for the Moon too, a correction is necessary to *R*. If the actual refraction correction is $R + r$, then, to lowest order,

$$r = -\left(\frac{426''}{D}\right) \sec z' \tan z'$$

where *D* is the topocentric distance in kilometers and the notation is that of Eq. 4.26. More complicated formulas are necessary for near-Earth satellites.

Going Backwards. When one has an observed position and wants the mean place one simply needs to go backward through the sequence planetary aberration, refraction, diurnal aberration, general precession, nutation, annual aberration, and so on. To first order it is permissible to reverse all of the equations given above when the quantities on the right-hand sides are known to obtain those on the left-hand sides. Only Eqs. 4.21 really contains second-order terms, whose importance were separately discussed.

RADAR REDUCTIONS

Radars

Celestial mechanics mostly deals with nonluminous bodies. Moreover, the source of light is not under our control, is incoherent, unpolarized, and relatively steady. When the source of illumination is controlled, both electric bills and powers of observation increase. It is not misleading to write down a few formulas to describe normal (optical) astronomical practice, but it would be so for the radar case. Radio refraction alone is a whole subfield. For information on refraction, planetary aberration, frequency shifts, and polarization problems the reader is referred to the radar (i.e., engineering) literature and the radio astronomy literature. In particular, see Skolnik (1970), Meeks (1976), and Bean and Dutton (1966).

Laser Radars

Laser radars represent a more recent development in satellite and rocket observation. Traditionally, passive optical detection yielded accurate angles-only information with an associated time. Modern passive optical detection can also yield the angular velocity. The analysis necessary for this is given in the last section of Chapter Seven. Active optical detection [e.g., laser radars or LIDAR (for LIght Detection And Ranging) or LADAR (for LAser Detection And Ranging)] also provides distance and radial velocity information. This is obtained, as with an ordinary radar, using time of flight to deduce the distance and the Doppler shift of the returned signal to derive the radial velocity. The reductions necessary to treat laser radar observations are not easily accessible (presumably because of the potential military applications). Two references in textbook form are Bachman (1979) and Kingston (1978).

Radial Velocity Reductions

Radial velocity corrections for the motion of the Earth are necessary when reducing spectroscopic observations of stellar spectra, when correcting apparent pulsar periods to their true value, and when correcting the apparent direction of a source of electromagnetic radiation (i.e., aberration). All of these computations share the necessity to (1) compute the motion of the Earth about the Earth–Moon barycenter, (2) compute the motion of the Earth–Moon barycenter about the Sun, and (3) compute the motion of the Sun about the center of mass of the solar system. For stellar spectroscopy an additional correction to the local standard of rest is necessary. This aspect is not treated here. The accuracy with which all of this is to be done obviously depends on the application. The development given here, based on Ball (1969), is good to ±0.015 km/sec. A more modern treatment is Stumpff (1977).

As in Eqs. 3.15–3.17 the observer's geocentric location is given by a geocentric distance ρ, a geocentric latitude ϕ', and a geocentric longitude λ. At any instant of time t there exists a corresponding sidereal time τ so the observer's geocentric location $\boldsymbol{\rho} = \rho \mathbf{l}(\tau, \phi')$. If the external source has position $\mathbf{l}(\alpha, \delta)$, then the projection onto the line of sight (to the source) of the observer's geocentric velocity is

$$v_{\text{geo}} = 2\pi\rho \cos\phi' \cos\delta \sin(\tau - \alpha)/\text{sidereal day in mean solar measure}$$

The next step is to calculate the heliocentric component of the Earth–Moon barycentric motion. This computation is facilitated by recognizing the small eccentricity of the solar orbit. First the mean anomaly of the Sun is obtained (the symbol for this is g in *The Explanatory Supplement*, p. 98):

$$g = 358°28'33''04 + 129596579''10\,T_{\text{E}} - 0''54\,T_{\text{E}}^2 - 0''012\,T_{\text{E}}^3$$

where T_{E} is the time measured in Julian centuries from Jan. $0^{\text{d}}5$, 1900 ET. Next $e_{\oplus}$, ε, and $\omega_{\odot}$ are computed from Eqs. 4.14, and the true anomaly of the Sun $v_{\odot}$ is calculated. A cubic (in $e_{\oplus}$) approximation to the equation of the center is sufficient. Next $\mathbf{l}(\lambda, \beta)$ is obtained for the source in question from ε and Eq. 3.11. The penultimate step is the computation of the true longitude of the Sun, $\omega_{\odot} + v_{\odot} = \lambda_{\odot}$. The projection onto the line of sight (of the source) of the Earth–Moon barycenter's heliocentric motion is

$$v_{\text{hel}} = -\frac{n_{\oplus} a_{\oplus} \cos\beta}{(1 - e_{\oplus}^2)^{1/2}} [\sin(\lambda_{\odot} - \lambda) - e_{\oplus} \sin(\pi + \omega_{\odot} - \lambda)]$$

where $n_\oplus$ is the sidereal revolution rate of the Earth–Moon barycenter (i.e., 2π radians/tropical year), and $a_\oplus$ is 1 A.U.

The last step is to correct for the motion of the Earth about the Earth–Moon barycenter (planetary perturbations can amount to 0.015 km/sec).

The Lunar Orbit

The lunar orbit is expressed in terms of the eccentricity $e_{\mathbb{C}}$ ($=0.054900489$), the constant of inclination $\gamma = 0.044886967$ (which is the sine of half the inclination of the lunar orbit to the ecliptic, $i_{\mathbb{C}} = 5°1453964$), the semi-major axis of the lunar orbit $a_{\mathbb{C}}$ ($=60.2665$ equatorial Earth radii), the sidereal period $P_{\mathbb{C}}$ ($=27^d321661 = 27^d07^h43^m11^s5$), and three other angles. These are denoted by (1) $\mathbb{C}$, the mean longitude of the Moon, measured in the ecliptic from the mean equinox of date to the mean ascending node of the lunar orbit, and then along the orbit; (2) Γ', the mean longitude of the lunar perigee measured in the ecliptic from the mean equinox of date to the mean ascending node of the lunar orbit, and then along the orbit; and (3) Ω, the longitude of the ascending node of the lunar orbit on the ecliptic, measured from the mean equinox of date. In the usual notation equations for these three quantities are

$$\mathbb{C} = 270°26'02''99 + 1336^r307°52'59''31\,T_E - 4''08\,T_E^2 + 0''0068\,T_E^3$$

$$\Gamma' = 334°19'46''40 + 11^r109°02'02''52\,T_E - 37''17\,T_E^2 - 0''045\,T_E^3$$

$$\Omega = 259°10'59''79 - 5^r134°08'31''23\,T_E + 7''48\,T_E^2 + 0''008\,T_E^3$$

See p. 107 of *The Explanatory Supplement.*

If the external source has selenocentric latitude and longitude (Λ, B), then the projection onto the line of sight (to the source) of the observer's Earth–Moon barycentric motion is

$$v_{\text{Moon}} = -\frac{n_{\mathbb{C}} a_{\mathbb{C}}}{(1 - e_{\mathbb{C}}^2)^{1/2}}\left(\frac{M_{\mathbb{C}}}{M_\oplus}\right)[\sin(\Gamma' - \Omega + v_{\mathbb{C}} - \Lambda) - e_{\mathbb{C}}\sin(\Gamma' - \Omega - \Lambda)]\cos B$$

where $M_{\mathbb{C}}$ is the mass of the Moon, $M_\oplus$ that of the Earth (their ratio $= 1/81.30$), $n_{\mathbb{C}} = 2\pi/P_{\mathbb{C}}$, and $v_{\mathbb{C}}$ is the true anomaly computed from the equation of the center using $\mathbb{C} - \Gamma'$ as the pseudo-mean anomaly. Again a third-order expansion in $e_{\mathbb{C}}$ is adequate.

The total correction applied to the instrumental radial velocity is

$$v_{\text{geo}} + v_{\text{hel}} + v_{\text{moon}}$$

Because of the neglect of planetary perturbations and the less than rigorous treatment of the other effects, the net result is probably good to ±0.005 km/sec.

PROBLEMS

1. Let the astronomical refraction be given by $R = k\tan z$ where z is the observed zenith distance. Show that refraction makes the circular disc of

the Sun appear to be elliptical with semi-major axis $(1-k)S_\odot$ and semi-minor axis $(1 - k \sec^2 Z)S_\odot$. Here $S_\odot$ is the (circular) semi-diameter of the solar disk and Z is the true zenith distance corresponding to z.

2. Show that in terms of the parallactic angle η the astronomical refraction correction can be written as

$$\Delta\alpha = -R \sec \delta' \sin \eta$$

$$\Delta\delta = -R \cos \eta$$

chapter five

Odds and Ends

So far I have explained the solution of the two-body problem, shown how to juggle astronomical coordinate systems, and outlined data reduction. If this knowledge is going to be applied by the reader then the most likely object of interest is an artificial satellite. Thus, it seems appropriate to review the current (early 1984) artificial satellite population, to examine ground-based related satellite motions, observability, eclipses, and so on, and to investigate orbit-based satellite transfers, maneuvers, and interplanetary topics. This is a large amount of material, not necessarily closely related. Hence the title of the chapter.

Owing to space limitations and the lack of anything inherently illuminating in the analysis of these subjects, they have all been lightly treated. However, after having perused this chapter, the reader can at least claim an acquaintance with the most common satellite-related problems (except for orbit determination—see Chapters Seven and Eight). Moreover, when it comes to the nitty-gritty problem of when to fire a retrorocket or an attitude stabilizer or when to realign a telescope (or camera or antenna structure), the answer will come from a detailed numerical analysis and not a pretty, analytically tractable problem.

ARTIFICIAL SATELLITES

Artificial satellites are basically of three types; those close to the Earth (known as low-altitude* or near-Earth satellites), those more distant but still bound to the Earth (known as high-altitude or deep-space satellites), and the interplanetary (including lunar) ones. The last group is small in number, all different, and are best discussed in a text on planetary physics. The Air Force distinguishes between low- and high-altitude satellites by using a period of about

---

* Somehow the radar people got altitude and elevation confused long ago. Elevation to a radar person means altitude and vice versa. Those who know the long-standing astronomical terminology must be able to translate mentally. Low-altitude, deep-space, and so on, are too ingrained to challenge now.

220 min. The reason is that an artificial satellite in a circular orbit with this period (elevation ≈ 5700 km) can still be detected by the various ballistic missile warning radars. Artificial satellites farther out (obviously this depends in detail on the object's cross section, the characteristics of the radar, etc.) can not be routinely observed by these radars. Historically the U.S. Air Force has tracked these satellites by passive optical means. The first system was known as the Baker–Nunn network. The replacement system, which started to go on-line in the Summer of 1982, is known as the *GEODSS* network. *GEODSS* stands for *G*round-based *E*lectro-*O*ptical *D*eep *S*pace *S*urveillance [see *Sky and Telescope* **63**, 469 (1982) for a description of these observatories].

Low–Altitude Satellites

Some 5000 objects are now (or have been) in low-altitude orbits. These artificial satellites have been sent into orbit to serve a wide variety of functions— scientific, military, communications, meteorological, navigational, and Earth resources. The purely scientific satellites have done medical research, studied the Sun and solar–planetary phenomena, probed the Earth's atmosphere, and opened up observations in wavelength bands not available on the ground. The latter include ultraviolet, gamma ray, infrared, and x-ray observations. The meteorological satellites use some of these nonvisual wavelength bands to survey cloud cover, probe the vertical profile of temperature and relative humidity in the atmosphere, and to do multi-wavelength band radiometry. Using similar remote sensing techniques, the Earth Resources Technology Satellites look for mineral deposits, crop infestation, schools of fish, the spread of air and water pollution, forest fire or disease damage, and the detailed structure necessary for cartographic and geodetic purposes.

All of the uses mentioned above also have some military or intelligence value. A common type of purely military low-altitude satellite is the photo-graphic reconnaissance satellite. These usually stay up for only a few weeks and then their film packages are recovered. In addition the military also uses satellites for navigational and communication purposes.

There are two special low-altitude satellite orbits. The first is a polar orbit ($i = 90°$). In this case the longitude of the ascending node will not change. Hence the satellite's orbital plane remains fixed in inertial space while the Earth rotates underneath and revolves about the Sun. (The oblateness of the Earth causes second-order changes to this simplified description.) This causes the nature of the illumination at the sub-satellite point to vary, through its range, over a 3-month period. On the other hand, for certain types of missions it is desirable to have the illumination remain constant but to view different parts of the Earth. In order to accomplish this, the satellite's orbital plane must always include the Earth–Sun line. Such an orbit is termed Sun-synchronous. Since the Sun appears to move 360°/tropical year = 0°.986/day eastward, for these orbits the orbital element set is adjusted so that the precession in the longitude of the ascending node, due to the Earth's oblateness, is exactly this amount. It turns out that such orbits are nearly polar but retrograde ($i > 90°$). As an example, for a circular orbit, 98° at 500 km or 102° at 1700 km will do. Actually, since the ecliptic is not coincident with the celestial equator, the orientation of a Sun-synchronous orbital plane oscillates about the Earth–Sun line.

High-Altitude Satellites

Tracking

One of the missions of NORAD is keeping track of all artificial satellites, low- or high-altitude, that are in orbit about the Earth. Another mission that NORAD (the *North* American Aerospace *Defense* Command) is charged with includes providing terminal impact prediction on satellites that are decaying into the atmosphere (such as Cosmos 1402). Where possible NORAD tries to simplify this task by accepting data from other agencies such as NASA, other parts of the Air Force, commercial satellite operators like COMSAT, and so on. However, for noncooperating satellites NORAD has to maintain an extensive network of artificial satellite tracking sensors around the globe. Some of these are run by the military and some by civilians. The two types of sensors are radars and optical sensors. Together they constitute the SPACETRACK system. These sensors are distributed worldwide and are linked to NORAD by message and telephone circuits.

There are several types of radars that are part of SPACETRACK. The phased array radars are capable of illuminating and detecting many targets simultaneously in a large volume of space. The array limits are usually quoted in degrees in azimuth and altitude. The FPS-85 in Florida is an example of such a radar. It is capable of surveying ±60° in azimuth and altitude around a line pointing south at 45° altitude. Its distance limit is ~8000 km. Other examples of phased array radars are the PAVE PAWS (east and west) radars, COBRA DANE (Shemya, Alaska), and COBRA JUDY (a ship-borne radar). The primary missions of the phased array radars are missile warning or the monitoring of foreign ballistic missile tests. Fortunately few incoming missiles have been detected (so far) and the result is that some of these radars track some low- and high-altitude satellites. Because of their distance limitation, though, they can track deep-space satellites only near the perigee parts of their orbits. It is important to remember that a phased array radar has enormous search capability and the ability to track several objects simultaneously. The phased array radars provide distance, azimuth, altitude, and time in their data reports.

A second type of radar in the SPACETRACK network is the *Naval Space Surveillance* Radar. NAVSPASUR is a zenith fence stretching across the continent at ~31° latitude. It has three transmitters and six receiving stations. The detection of satellites is by an interferometric technique. Because it is a thin zenith fence, NAVSPASUR does not detect satellites whose inclinations (and hence their northernmost excursion in declination) are below the latitude of the fence. Hence it does not produce any data on satellites with orbital inclinations below 30°. Satellites in higher-inclination orbits are detected routinely as they pass through the fence. The distance limitation of the fence is approximately 12,000 km.

NAVSPASUR excels in several areas. Being a thin zenith fence, it can count pieces of breakups as they pass through. Since the fence is invariant in location, it can determine orbital periods accurately by timing successive penetrations of the fence by the same satellite. Thus, it can accurately monitor all maneuvers that affect the orbital period. It has been in business since the beginning of the satellite age and therefore provides a good historical data base. NAVSPASUR publishes an extensive satellite catalog every 6 months, with updates every month. It also provides excellent support for terminal impact prediction when satellites are in rapid decay (and their periods are changing fast).

A third type of radar consists of those like the Millstone Hill Radar run by M.I.T. There are several in the SPACETRACK system—the BMEWS (*B*allistic *M*issile *E*arly *W*arning *S*ystem) radars (at Clear in Alaska, Thule in Greenland, and Fylingdales in England), Diyarbakir in Turkey, ALTAIR (*A*RPA *L*ong-range *T*racking *a*nd *I*nstrumentation *R*adar) on Kwajalein Island, the Haystack radar in Massachusetts, and so on. These are all narrow-beam trackers operating at UHF or higher frequency and capable of tracking only one satellite at a time. Millstone and Haystack (also run by M.I.T.) are the only radars in the system capable of tracking in deep space out to synchronous distances. ALTAIR and Diyarbakir are being modified to provide deep-space observing capability.

Characteristics of the third type of radars are the ability to track anywhere in azimuth or altitude; the ability to provide good azimuth, altitude, distance, and occasionally radial velocity data; and the narrow pencil beam which precludes rapid searches.

The SPACETRACK system utilizes two types of optical sensors for deep-space tracking. The Baker–Nunn cameras are wide-angle ($5° \times 20°$) sensors that take photographs of the sky. The camera is pointed to the location of a satellite and an exposure is made. The exposure lasts from 1 to 10 sec. Frequently, several satellites are detected in a field of view and their positions are measured by using a star chart overlay. The time is automatically clocked on the film. Given the wide field of view, the Baker–Nunn cameras are efficient at detecting satellites, but the cameras are not real-time instruments—the film has to be developed, which is typically done at the end of the night. The precision of the data reduced in the field is a 1 minute of arc, although with special reduction techniques it can be as good as 2 seconds of arc. The Baker–Nunn system has been almost completely disbanded today.

The other type of optical sensor is the GEODSS network. The prototype is near Socorro, New Mexico (run by M.I.T.). There will be five GEODSS sites in the future. The first three are in New Mexico, South Korea, and Maui on Hawaii. The GEODSS sites have three telescopes per site. The detection of satellites and stars is by photoelectric means with several stages of amplification. The resulting picture is displayed, real time, on a television monitor. Thus satellites are detected in real time by their motion against the stellar background. The two main telescopes have a 2° field of view with a 2:1 zoom capability; the auxiliary telescope has a 6° field of view. The astrometric data produced consists of time, azimuth, and altitude (or time, right ascension, and declination). No distance information is available. The precision of the data is 5 seconds of arc. When the installation is completed, the GEODSS sites will be the major suppliers of deep-space tracking data to SPACETRACK. Finally the GEODSS network is capable of very rapid searches, but of course it will work only on clear nights.

The Satellite Population

A high-altitude or deep-space satellite is one whose period is in excess of $\sim$220 min ($n \simeq 6.5$ rev/day). The huge majority of the $\sim$600 routinely tracked deep-space artificial satellites can be grouped into three classes. One group (31%) is the near-stationary satellites. These objects are in nearly circular ($e \leq 0.1$ usually), nearly equatorial ($i \leq 10°$ usually), orbits with a period close to 1 mean sidereal day ($\tau = 1.0027379093$ rev/day; 0.9 rev/day $\leq n \leq 1.1$ rev/day usually). Most of these satellites are communication satellites but a few are spy (primarily missile launch detection) satellites. The second group (16%) are in $n \simeq 2$ rev/day, high-eccentricity ($e \simeq 0.7$), 28° inclination orbits. Almost all of these objects are rocket bodies from Cape Canaveral launches. The third group (40%) differs from the second only in inclination. Theirs is $\sim$63° and they are also tightly grouped in the argument of perigee ($\omega \simeq 293°$, giving a southern hemisphere perigee). Most of these are Russian Molniya (domestic communication) satellites and their associated rocket bodies. The remaining deep-space satellites (13%) range from the fast-moving retrograde LAGEOS, to the slow-moving, circular Vela's ($\sim$0.25 rev/day—these are nuclear explosion detection satellites), to the very high eccentricity ($e = 0.93$) Russian Prognoz satellites. The satellites in the Global Positioning System constellation are in high-inclination, half-synchronous ($P \simeq 12$ hr) orbits. The exceptions are numerous and diverse.

Because deep-space satellites are so high up, atmospheric drag is usually not a problem. However, some of the ones closer in or with relatively close in perigee distances do feel atmospheric effects. Appreciable decay is that which affects the mean motion in excess of $dn/dt > 0.00001$ rev/day^2.

In addition to the objects that can be identified as active payloads, inactive payloads, and rocket bodies, there is a large amount of debris up there. Explosive bolts, hatch covers, shrouds, upper truss assemblies, and so on, are also in orbit and are tracked (especially by the optical sensors). Some discrimination among these types is possible if one looks at the signal variation (radar or optical) of the object. Most active payloads are stable and their attitude is maintained. Rocket bodies and debris tumble about out of control. Rocket bodies are usually larger (and hence brighter) than debris.

Information on artificial satellites can be obtained from NASA (Office of Public Affairs, Goddard Space Flight Center, Greenbelt, Maryland—ask for the *Satellite Situation Report*); from the U.S. Navy (Commanding Officer, Naval Space Surveillance System, Dahlgren, Virginia—ask for the *Satellite Situation Summary*); from the Royal (Great Britain) Aircraft Establishment [Procurement Executive, Ministry of Defence, Farnborough, Hants., G.B.—ask for the *Table of Space Vehicles* (also available from the U.S. Defense Documentation Center, Cameron Station, Alexandria, Virginia 22314—ask for document AD-A109363)]; and from NORAD (Peterson AFB, Colorado 80914—ask for a copy of *Satellite Catalog Compilations*). The contents, accuracy, and precision of these catalogs vary and no overall statement would represent them fairly.

PLANET-RELATED TOPICS

In the next few pages several planet-related topics will be discussed, such as ground traces, observability including eclipses, the rise and set of artificial satellites, and searches.

Ground Trace

Imagine a line drawn from the center of the Earth to the satellite. As the satellite revolves about the Earth in its orbit, this line will cut the Earth's surface in a continuous sequence of sub-satellite points (each of which may be located by specifying the sub-satellite latitude and longitude—*not* the sub-latitude and sub-longitude). The locus of these points is called the *ground trace* of the satellite. In the mission planning stage of a satellite knowledge of the ground trace can be critical. Some obvious examples are for meteorological satellites, LANDSAT satellites, Earth Resources Technology Satellites, and so on. The ground trace is controlled by the inclination and the period. Variations in the other orbital elements either shift the ground trace east or west or alter the time when the satellite is over a particular point of the ground trace. Some ground traces are shown in Figs. 13–15 (for illustrative purposes).

Many low-altitude satellites are in nearly polar orbits and the perturbation of the orbit due to the departures of the Earth from sphericity (perturbations due to oblateness) cause the longitude of the ascending node to continuously increase (see Chapters Nine and Ten). Thus, from the point of view of the artificial satellite, the Earth turns underneath it (in addition to the Earth's usual 24-hr rotation about its axis) thereby allowing a simple mechanism for global longitudinal coverage via a polar orbit. In fact, the nonperturbed longitudinal displacement per sidereal day is

$$\Delta\lambda = 2\pi P / 1 \text{ sidereal day}$$

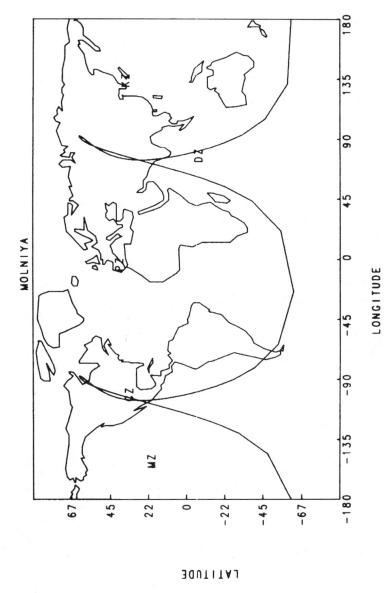

Figure 13. Ground trace of a Molniya satellite ($n = 2.01$ rev/day, $i = 62°8$, $\Omega = 254°8$, $e = 0.74$, $\omega = 280°2$) over 24 hours. The five GEODSS site locations are marked too (KZ, MZ, CZ, PZ, and DZ).

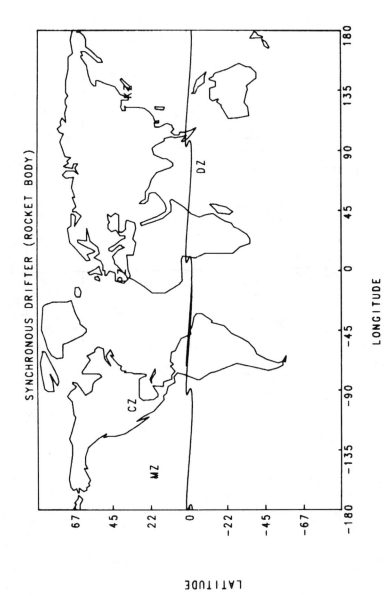

Figure 14. Ground trace of a drifting near-stationary satellite ($n = 1.32$ rev/day, $i = 2°$, $e = 0.2$) over $3\frac{1}{2}$ days.

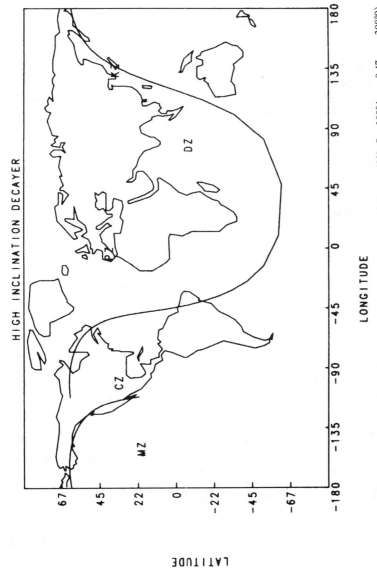

Figure 15. Ground trace of a decaying, high inclination satellite ($n = 5.72$ rev/day, $i = 62°.9$, $\Omega = 309°.4$, $e = 0.47$, $\omega = 308°.9$) over 6 hours.

141

where P is the satellite's period of revolution. After

$$N \simeq \frac{2\pi}{\Delta\lambda} \quad \text{orbits}$$

1 day has elapsed. Hence N is the number of orbits per day. For low-altitude satellites this might be as large as 16 whereas for most deep-space satellites it is 1 or 2. If $2\pi/\Delta\lambda$ is a rational fraction, then the orbit repeats (this ignores the perturbing effects of oblateness!). In general $2\pi/\Delta\lambda$ is not rational and the ground trace fills all longitudes within the latitude range $|\theta| <$ inclination (for prograde orbits; should the orbit be retrograde, then $i - \pi \leqslant \theta \leqslant \pi - i$). A detailed, numerical, ground-trace computation that included atmospheric drag, oblateness effects, and luni-solar perturbations was published by Kent and Betz (1969).

Observability

The topic of the observability of an artificial satellite has to include whether or not it is above the horizon, when it is above the horizon, and eclipses. In all of the discussion below, perturbations of the satellite's orbit are ignored, the (geometrical) oblateness of the Earth is neglected, and the motion of the Earth through space during the time interval of interest is disregarded.

Eclipses

Consider the Earth-centered coordinate system shown in Fig. 16. The equation of the projected surface is just $x^2 + y^2 = R_\oplus^2$, that of the Sun is $(x - D_\odot)^2 + y^2 = R_\odot^2$, and that of the circular satellite orbit is $x^2 + y^2 = R_s^2$. Tangents external to the Sun and Earth define the umbral shadow cone limits and the internal tangents generate the penumbral shadow cone limits. The umbral region is completely dark, the penumbral region partially so. We want to find e_u and e_p, the half angles of the umbral and penumbral shadow cones. The value of e_u will be obtained first.

The line L_u is tangent to the Earth's surface at $(x_\oplus, y_\oplus)$. Hence the normal form of its equation is

$$L_u: \quad xx_\oplus + yy_\oplus = R_\oplus^2$$

Similarly, because L_u is also tangent to the Sun at $(x_\odot, y_\odot)$

$$L_u: \quad (x - D_\odot)(x_\odot - D_\odot) + yy_\odot = R_\odot^2$$

Note that $x_\oplus \leqslant 0$, $y_\oplus \leqslant R_\oplus$, $x_\odot \leqslant D_\odot$, and $y_\odot \leqslant R_\odot$. As the line L_u has one and only one equation representing, it can be deduced that

$$\frac{x_\oplus}{R_\oplus^2} = \frac{x_\odot - D_\odot}{R_\odot^2 + D_\odot(x_\odot - D_\odot)}$$

$$\frac{y_\oplus}{R_\oplus^2} = \frac{y_\odot}{R_\odot^2 + D_\odot(x_\odot - D_\odot)}$$

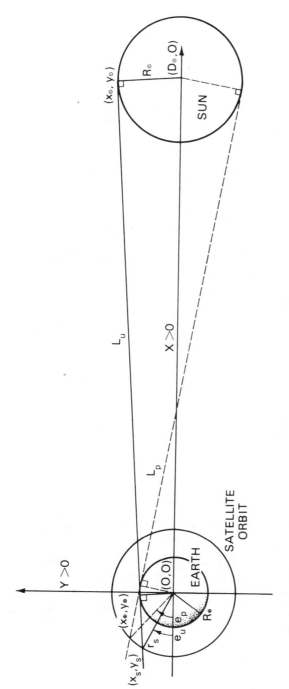

Figure 16. The geometry of eclipses (not drawn to scale) showing umbral (solid) and penumbral (dotted) cone generators.

143

Solving for $x_\odot$, we find that

$$x_\odot - D_\odot = -\frac{x_\oplus R_\odot^2}{x_\oplus D_\odot - R_\oplus^2} < 0$$

If this result and the equation of the solar surface is used,

$$y_\odot^2 = R_\odot^2 \frac{(x_\oplus D_\odot - R_\oplus^2)^2 - x_\oplus^2 R_\odot^2}{(x_\oplus D_\odot - R_\oplus^2)^2}$$

Using the latter two results in the equation for $y_\oplus$ and remembering that $y_\odot > 0$, we next obtain

$$y_\oplus = \frac{|(x_\oplus D_\odot - R_\oplus^2)^2 - x_\oplus^2 R_\odot^2|^{1/2}}{R_\odot}$$

Squaring this, using it in the equation for the surface of the Earth, and realizing that the $|x_\oplus D_\odot|$ term is dominant, we are left with, after reduction,

$$x_\oplus = -\frac{R_\oplus(R_\odot - R_\oplus)}{D_\odot}, \qquad y_\oplus = R_\oplus \left[1 - \frac{(R_\odot - R_\oplus)^2}{D_\odot^2} \right]^{1/2}$$

$$x_\odot = D_\odot - \frac{R_\odot(R_\odot - R_\oplus)}{D_\odot}, \qquad y_\odot = \frac{R_\odot y_\oplus}{R_\oplus}$$

For the next step the fact that L_u passes through both $(x_\oplus, y_\oplus)$ and $(x_\odot, y_\odot)$ will be utilized. Hence the two-point form of its equation is

$$L_u: \quad \frac{y - y_\oplus}{x - x_\oplus} = \frac{y_\odot - y_\oplus}{x_\odot - x_\oplus}$$

Eliminating the solar tangential point through the use of the above formulas, this is equivalent to

$$D_\odot(y - y_\oplus) \left[1 - \frac{(R_\odot - R_\oplus)^2}{D_\odot^2} \right]^{1/2} = (R_\odot - R_\oplus)(x - x_\oplus)$$

The line L_u crosses the artificial satellite's orbit at (x_s, y_s) so

$$y_s = y_\oplus + \frac{(x_s - x_\oplus)(R_\odot - R_\oplus)}{D_\odot [1 - (R_\odot - R_\oplus)^2/D_\odot^2]^{1/2}} = +(R_s^2 - x_s^2)^{1/2}$$

There is a lower line analogous to L_u (not shown in Figure 16) for which a minus sign would appear on the far right-hand side of the formula immediately above. This fixes x_s (<0). In fact, after a few pages of algebra

$$x_s = -\frac{R_\oplus(R_\odot - R_\oplus)}{D_\odot} - (R_s^2 - R_\oplus^2)^{1/2} \left[1 - \frac{(R_\odot - R_\oplus)^2}{D_\odot^2} \right]^{1/2}$$

$$= x_\oplus - \left(\frac{R_s^2}{R_\oplus^2} - 1 \right)^{1/2} y_\oplus$$

I could find no simpler form for y than $(R_s^2 - x_s^2)^{1/2}$.

To recap the results for the umbral shadow cone,

$$x_\oplus = -\frac{R_\oplus(R_\odot - R_\oplus)}{D_\odot}, \qquad y_\oplus = \frac{R_\oplus[D_\odot^2 - (R_\odot - R_\oplus)^2]^{1/2}}{D_\odot}$$

$$x_\odot = D_\odot - \frac{R_\odot(R_\odot - R_\oplus)}{D_\odot}, \qquad y_\odot = \frac{R_\odot[D_\odot^2 - (R_\odot - R_\oplus)^2]^{1/2}}{D_\odot}$$

$$= D_\odot + \frac{R_\odot x_\oplus}{R_\oplus}, \qquad\qquad = \frac{R_\odot y_\oplus}{R_\oplus}$$

$$x_s = x_\oplus - \frac{(R_s^2 - R_\oplus^2)^{1/2} y_\oplus}{R_\oplus}, \qquad y_s = (R_s^2 - x_s^2)^{1/2}$$

and

$$\tan e_u = |y_s / x_s|$$

In an exactly analogous fashion one can show that for the intersections of the penumbral line (L_p)

$$x_\oplus = \frac{R_\oplus(R_\odot + R_\oplus)}{D_\odot}, \qquad y_\oplus = \frac{R_\oplus[D_\odot^2 - (R_\odot + R_\oplus)^2]^{1/2}}{D_\odot}$$

$$x_\odot = D_\odot - \frac{R_\odot(R_\odot + R_\oplus)}{D_\odot}, \qquad y_\odot = -\frac{R_\odot[D_\odot^2 - (R_\odot + R_\oplus)^2]^{1/2}}{D_\odot}$$

$$= D_\odot - \frac{R_\odot x_\oplus}{R_\oplus}, \qquad\qquad = -\frac{R_\odot y_\oplus}{R_\oplus}$$

$$x_s = x_\oplus - \frac{(R_s^2 - R_\oplus^2)^{1/2}}{R_\oplus} y_\oplus, \qquad y_s = (R_s^2 - x_s^2)^{1/2}$$

$$\tan e_p = |y_s / x_s|$$

It is left to the reader to derive from these expressions the results usually found in the astronomy texts:

$$e_p \simeq \frac{R_\oplus}{R_s} + \frac{R_\oplus}{D_\odot} + \frac{R_\odot}{D_\odot}$$

$$e_u \simeq \frac{R_\oplus}{R_s} + \frac{R_\oplus}{D_\odot} - \frac{R_\odot}{D_\odot}$$

Among the other effects left out of the above analysis is the 1–2% enlargement of the Earth's shadow due to refraction effects [see *Sky and Telescope* **57**, 12 (1979)]. Meeus has treated the ellipticity of the shadow (1969).

Direct Visibility, Rise–Set

An artificial satellite is potentially visible (at whatever wavelength band) only if it is up, that is, altitude > 0. For an observer located at $\boldsymbol{\rho} = \rho(\cos \phi' \cos \lambda,$

$\cos \phi' \sin \lambda$, $\sin \phi'$), with sidereal time τ this constraint is

$$\mathbf{k} \cdot R[R_3(\tau) \cdot S \cdot \mathbf{q} - \boldsymbol{\rho}] \geq 0 \qquad (5.1)$$

where S was defined in Eq. 2.12a in terms of the Euler angles of the orientation of the orbit and R is the matrix given by

$$R = R_3\left(\frac{\pi}{2}\right) R_2\left(\frac{\pi}{2} - \phi'\right) R_3(\lambda)$$

The unit vector $\mathbf{k} = (0, 0, 1)$ as usual. Using Eq. 2.13b for $\mathbf{r} = S\mathbf{q}$, Eq. 5.1 reduces to

$$f(\tau) \cos E + g(\tau) \sin E - h(\tau) \geq 0 \qquad (5.2)$$

where

$$f(\tau) = a[S_{11} \cos \phi' \cos(\lambda + \tau) + S_{21} \cos \phi' \sin(\lambda + \tau) + S_{31} \sin \phi']$$

$$g(\tau) = a(1 - e^2)^{1/2}[S_{12} \cos \phi' \cos(\lambda + \tau) + S_{22} \cos \phi' \sin(\lambda + \tau) + S_{32} \sin \phi']$$

$$h(\tau) = \rho + ef(\tau)$$

with $S = (S_{ij})$. Equation 5.2 is an (implicit) quadratic equation in (say) $\cos E$. The solution is (Kaula 1966)

$$\sin E = \frac{gh \mp f(f^2 + g^2 - h^2)^{1/2}}{f^2 + g^2}$$

$$\cos E = \frac{fh \pm g(f^2 + g^2 - h^2)^{1/2}}{f^2 + g^2}$$

and, obviously, one requires that $f^2 + g^2 \geq h^2$. Kaula (1966) also presents entrance (exit) into (out of) the Earth's shadow cone in a similar fashion. Rise and set times are determined by the two instances per orbit when the left-hand side of Eq. 5.1 or 5.2 is zero.

Kepler's Laws.   This section is a small diversion one might want to skip on a first reading. First Kepler's three laws of planetary motion will be presented and then how, for instance, he might have derived the third from observation. So, following Moulton (1902),

1.   The radius vector of each planet with respect to the Sun sweeps out equal areas in equal times.
2.   The orbit of each planet is an ellipse with the Sun at one focus.
3.   The squares of the (sidereal) periods of the planetary revolutions are proportional to the cubes of their semi-major axes.

It is known that these relationships are only (excellent) approximations. For instance, in the case of the Earth it is the Earth–Moon barycentric radius vector that satisfies $r^2\dot\phi = $ const, not that of the Earth with respect to the Sun (more properly the solar system barycenter). Likewise, no planet (moonless or not) revolves about the Sun in a perfect ellipse as the other planets are continually perturbing its motion.

How could Kepler have derived, for instance, his third law from observations? After all, the simplest quantity to calculate from the observations (for a superior planet, one that is further from the Sun than the Earth is) is the *synodic period S*. The synodic period is the time interval between similar configurations such as opposition. When a superior planet is at opposition, one can draw a straight line through the planet, the Earth, and the Sun. (This ignores the differing inclinations of the planetary orbits. A more precise definition would require equality of the planet and Earth heliocentric longitudes or right ascensions.) A moment after opposition the Earth and the planet have moved, with the Earth moving more rapidly. Another way to say this is $n_\oplus > n$ where n is the mean motion of the superior planet in question.

Suppose that the first opposition occurred at $t = t_0$, the next at $t = t_1$, and so forth. The synodic periods (differing slightly due to planetary perturbations or observational errors) are $S_k = t_{k+1} - t_k$. Now for each synodic period

$$\int_{t_k}^{t_{k+1}} (n_\oplus - n)\, dt = 2\pi$$

Adding the first K of these equalities leads to

$$\int_{t_0}^{t_K} (n_\oplus - n)\, dt = 2\pi K \tag{5.3}$$

Let the two sidereal periods be $P_\oplus$ and P. Then the total time interval $t_K - t_0$ can be written as some whole number plus a fraction of each of these periods, say as

$$\Delta t = t_K - t_0 = \begin{cases} (N_\oplus + f_\oplus)P_\oplus \\ (N + f)P \end{cases}$$

where the N's are integers and the f's are proper fractions. Now we show that $N_\oplus - N = K$.

To see this, expression 5.3 is calculated piece by piece. For some proper fractions $g_\oplus$ and g the integrals can be written as

$$\int_{t_0}^{t_K} n_\oplus\, dt = 2\pi(N_\oplus + g_\oplus)$$

$$\int_{t_0}^{t_K} n\, dt = 2\pi(N + g)$$

Substituting into Eq. 5.3 yields

$$N_\oplus - N - K = g - g_\oplus$$

But since $g, g_\oplus \in (0, 1)$, it follows that $|g - g_\oplus| < 1$. However, the left-hand side is an integer, whence $g = g_\oplus$ and

$$N_\oplus - N = K$$

Remember that both $N_\oplus$ and K are known (because we counted). Thus, if f is known, P can be obtained. An asymptotic estimate (which we will improve on momentarily) is

$$P = \frac{\Delta t}{N+f} = \frac{\Delta t}{N+\frac{1}{2}} \cdot \frac{1+1/(2N)}{1+f/N} + \frac{\Delta t}{N+\frac{1}{2}}\left[1+\frac{\frac{1}{2}-f}{N}\right]$$

To improve on this, the mean synodic period S is defined as $\Delta t / K$. Then

$$P = \frac{KS}{N+f}$$

but because $N_\oplus - N = K$, it can recast as

$$P = \frac{KS}{N_\oplus - K + f}$$

By dealing with the reciprocal of this, it follows that

$$\frac{1}{P} = \frac{1}{P_\oplus} - \frac{1}{S} + \frac{f - f_\oplus}{P_\oplus N_\oplus}$$

which shows that the value of $1/P$ can be arbitrarily refined as $N_\oplus \to \infty$. Thus, for a superior planet

$$\frac{1}{P} = \frac{1}{P_\oplus} - \frac{1}{S}$$

but for an inferior planet the problem is slightly different and one has

$$\frac{1}{P} = \frac{1}{P_\oplus} + \frac{1}{S}$$

Searches

What is known as the search with discrete effort for a fixed target is formulated below. Asteroid searches are used as a model to illustrate the mathematical formalism. Relatively little simplification of the physics or astronomy is necessary to do this. Next the results of optimal search theory are stated and the search problem is solved. Then an optimal search for a minor planet is explicitly constructed and exhibited (Taff 1981b).

My search team looks for asteroids on the celestial sphere. In the largest sense this forms the two-dimensional *search space* of the problem. In practice we delineate a limited area of the celestial sphere (say above altitude 30° or along the ecliptic) that we actually search in. This search space is denoted *J*.

We search using a telescope with a finite field of view. In practice we always look at the entire field, never a fraction of a field or more than one field at a time. Hence the search space J is a discrete set of fields of view. These fields are numbered by the index $j = 1, 2, \ldots$, with $\max(j) < \infty$ (because the celestial sphere only encompasses 4π steradians).

Before the minor planet is found, we can assign an a priori *target distribution* on the search space J, $p: J \rightarrow [0, 1]$ (the notation means that p is a function defined over the set J which maps elements of J into the domain zero to unity inclusive). The target distribution specifies the a priori probability of finding an asteroid in field of view $j \in J$ before we start the search. For main belt asteroids a reasonable model for p is p is uniform over all geocentric ecliptic longitudes and over the geocentric ecliptic latitude range $<10°$ (or $5°$ or $20°$). For Earth-approaching minor planets p is uniform over the topocentric celestial sphere both because of parallax effects and the broad range of these minor planets' inclinations. Note that

$$\sum_{j \in J} p(j) \le 1$$

for the asteroid might not be in the search space J at all.

When we examine a field of view for an asteroid we expend a certain amount of effort trying to detect it. We may look at the same field of view several times. To account for this, a *cost function* is defined as $c: J \times \{0, 1, 2, \ldots\} \rightarrow [0, \infty]$ which measures the cost of performing k searches in the jth field of view, $c(j, k)$. Clearly $c(j, 0) = 0 \forall j \in J$. I could measure cost by the time spent examining a field of view plus the time spent in moving to the next field of view (this makes c nonlocal). Operationally most searches spend the same time in each field of view (more or less). Also, because $[\text{area } J]^{1/2}/(\text{slew speed}) \ll (\text{time spent examining the field of view})$, the nonlocal element of c is both unimportant and varies little. Thus, I shall measure cost by time and, in the case of the asteroid search, specialize to the case when the incremental cost of the kth examination of field of view j, namely,

$$\gamma(j, k) = c(j, k) - c(j, k-1)$$

is a constant independent of both j and k.

When we examine a field of view of the search space looking for our quarry, there is a conditional probability of detecting the target on or before the kth look in that field of view given that it is present. This function, for field of view j, is denoted by $b(j, k): J \times \{0, 1, 2, \ldots\} \rightarrow [0, 1]$. Naturally, $b(j, 0) = 0 \forall j \in J$. From the *detection function* b I can construct the probability of failing to detect the asteroid on the first $k-1$ examinations of field of view number j and succeeding on the kth one (given that the minor planet is there). That is,

$$\beta(j, k) = b(j, k) - b(j, k-1)$$

There is a lot of physics and mathematics subsumed in the detection function. Clearly it depends on the asteroid's brightness, the resolution element

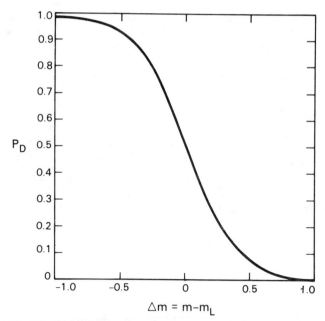

Figure 17. A heuristic probability of detection curve. The limiting magnitude (m_L) is at the 50% point (by definition).

size of the detector(s), the false-alarm probability one is willing to accept, how tired one is, and so on. Because the zodiacal region is unchanging, because atmospheric extinction can be computed, and because the location of the Moon is known, and so on, this is a computable function. Operationally, for a fixed set of external parameters, detection probability has the shape shown in Fig. 17, where m_L is the quoted (50% probability of detection) limiting magnitude. The form shown in the diagram will be used to compute the optimal search plans below.

Finally I need to define a *search plan*. A discrete search plan is a sequence $\xi = (\xi_1, \xi_2, \xi_3, \ldots)$, which tells one to first look in field of view number ξ_1, if the asteroid is found then terminate the search; if it isn't found there, one looks next in field of view number ξ_2, and so forth. Another way to describe this is by a function that conveys the *allocation of effort* devoted to each field of view. To this end, we define $f: J \to [0, \infty]$, which tells one the number of inspections in field of view j.

Above I referred to searches for a fixed target. Clearly, the minor planets that we are trying to find are moving. The mathematical formulation of this approximation is $[\text{area}(J)/\text{search rate}] \cdot$ asteroidal angular speed $\ll$ field of view. Take area$(J) = 1000$ square degrees, search rate $= 100$ square degrees/hr, asteroid geocentric angular speed $= 0\overset{\circ}{.}25$/day, field of view $= 2°$. These are typical numbers for us and they result in $0.10 \ll 2$.

The total cost of carrying out the search plan ξ with allocation f is denoted by $C[f]$,

$$C[f] = \sum_{j \in J} c(j, f(j))$$

The total number of scrutinizations involved is $\sum_{j \in J} f(j)$. Similarly, the total probability of target detection with this allocation of effort is $P[f]$,

$$P[f] = \sum_{j \in J} p(j) b(j, f(j))$$

There are four types of searches one might define as optimal. One might be interested in maximizing the total probability of detection for a given number of inspections (say K). If the incremental cost function γ is a constant, then (after a suitable renormalization) one is demanding that $P[f]$ be a maximum for $C[f] < K$. Such a search is termed *totally optimal*. If one further demanded this for all $K = 1, 2, 3, \ldots$, then the search would be called *uniformly optimal*. A third type of search plan that might be considered maximizes the probability of detection with respect to the incremental cost and does so at every stage of the search. Mathematically one finds the value of j that maximizes $p(j)\beta(j, k)/\gamma(j, k)$ at each k. These searches are called *locally optimal*. Lastly one might entertain a search plan that minimizes the total expected cost (i.e., was *the fastest*) to find the target.

The essential assumptions necessary to cast the asteroid search into the simplest form of the mathematical superstructure that Stone (1975) outlines are (1) that the minor planet is fixed in the search space, (2) that the search space is discrete, (3) that the allocation of effort is discrete, and (4) that γ is bounded away from zero and $p(j)\beta(j, k)/\gamma(j, k)$ is a decreasing function of j. I do not believe that the physics or astronomy is strained by these strictures. In fact, (5) $\gamma = \mathrm{const}$ is not unreasonable. The important point is that under these five limitations the totally optimal search plan, the uniformly optimal search plan, the locally optimal search plan, and the fastest searches are all *identical*. Not only that, it can be explicitly exhibited. See Stone's text for the rigorous mathematical statements of the relevant theorems and their proofs.

A bit more mathematics is needed before one can see the solution to the problem. The search plan $\xi = (\xi_1, \xi_2, \xi_3, \ldots)$ is a sequence of values $\xi_i \in J$ for $i = 1, 2, 3, \ldots$, which specifies that the ith examination be in field of view number ξ_i if the previous $i - 1$ inspections failed to detect the asteroid in fields of view $\xi_1, \xi_2, \ldots, \xi_{i-1}$. Let the set of all such search plans be denoted by Ξ. Introduce the probability $P[n, \xi]$ (and the cost $C[n, \xi]$) of detecting the minor planet on or before the nth examination while performing search plan $\xi \in \Xi$ (of the first n examinations). Finally, let $r(j, n, \xi)$ be the number of scrutinizations out of the first n that are placed in the jth field of view while following search plan ξ. A uniformly optimal search plan [for $\gamma(j, k) \neq 0$] $\xi^* \in \Xi$ is one such that

$$P[n, \xi^*] = \max\{P[n, \xi]: \xi \in \Xi\}, \qquad n = 1, 2, \ldots, K$$

A locally optimal search plan ξ^* is one such that ξ_1 is determined by ($\gamma \neq 0$ necessarily)

$$\frac{p(\xi_1)\beta(\xi_1, 1)}{\gamma(\xi_1, 1)} = \max_{j \in J} \frac{p(j)\beta(j, 1)}{\gamma(j, 1)}$$

and having determined the fields of view for the first $n-1$ looks $(\xi_1, \xi_2, \ldots, \xi_{n-1})$, the field of view for the nth look is determined from

$$\frac{p(i)\beta(i, r(i, n-1, \xi)+1)}{\gamma(i, r(i, n-1, \xi)+1)} = \max_{j \in J} \frac{p(j)\beta(j, r(j, n-1, \xi)+1)}{\gamma(j, r(j, n-1, \xi)+1)}$$

with $\xi_n^* = i$. Now define $k_n = r(\xi_n, n, \xi^*)$. The notation means that the nth examination of search plan ξ is placed in field of view ξ_n and that it is the k_nth time that this field of view has been searched. The total cost to find the asteroid can be expressed in a variety of ways if the limit as $n \to \infty$ of $P[n, \xi]$ is unity:

$$\mu(\xi) = \sum_{n=1}^{\infty} C[n, \xi](P[n, \xi] - P[n-1, \xi])$$

$$= \sum_{n=1}^{\infty} \sum_{m=1}^{n} \gamma(\xi_m, k_m) p(\xi_n) \beta(\xi_n, k_n)$$

$$= \sum_{m=1}^{\infty} \sum_{n=m}^{\infty} \gamma(\xi_m, k_m) p(\xi_n) \beta(\xi_n, k_n)$$

$$= \sum_{m=1}^{\infty} \gamma(\xi_m, k_m)(1 - P[m-1, \xi])$$

since $P[0, \xi] = 0$. If $\gamma(j, k) = 1$, then this reduces to

$$\mu(\xi) = \sum_{n=0}^{\infty} (1 - P[n, \xi])$$

Under the assumptions outlined above, if q_j is the probability of detecting the asteroid after a single look at field of view j (given that it is there), then, as each examination is an independent event,

$$\beta(j, k) = q_j(1 - q_j)^{k-1} \qquad \text{for } j \in J, k = 1, 2, \ldots$$

We normalize such that $\gamma(j, k) = 1 \forall j \in J, k = 1, 2, \ldots$, and have an allocation $f(j)$ such that the total cost (e.g., number of inspections) is K,

$$\sum_{j \in J} f(j) = K$$

The total probability of detection is

$$P[f] = \sum_{j \in J} p(j)b(j, f(j)) = \sum_{j \in J} p(j)[1 - (1 - q_j)^{f(j)}]$$

One makes the nth search in field of view $i \in J$ such that

$$p(i)q_i(1-q_i)^{r(i,n-1,\xi)} = \max_{j \in J} p(j)q_j(1-q_j)^{r(j,n-1,\xi)} \qquad (5.4)$$

then $\xi = \xi^*$ and is optimal (in all four senses); Chew (1967). Since J is finite, the existence of an i satisfying the above is guaranteed. When the uniformity of the target distribution p is exploited, the result in Eq. 5.4 is even simpler:

$$q_i(1-q_i)^{r(i,n-1,\xi)} = \max_{j \in J} q_j(1-q_j)^{r(j,n-1,\xi)} \qquad (5.5)$$

I have already argued that the a priori target distribution $p(j)$ can be approximated by a defective uniform distribution over the search space. In fact, for an Earth-approaching asteroid search $\sum_{j \in J} p(j) = \text{area}(J)/4\pi$. I have also argued that the incremental cost function is homogeneous over J and independent of the number of looks, $\gamma(j, k) = 1$ (in appropriate units). The probability of detection on a single look (given that the asteroid is there) is q_j. This depends on the brightness of the asteroid and the night sky background. Three effects tend to make minor planets fainter: atmospheric extinction, loss owing to increasing phase angle (Sun–asteroid–observer angle with vertex at the asteroid), and increasing distance (heliocentric or geocentric).

The extinction is modeled as usual,

$$\varepsilon = \varepsilon_z \sec z$$

where z is the topocentric zenith distance and ε_z is the extinction per unit air mass. For the phase function in magnitudes z the Gehrels and Tedesco (1979) results are used:

$$B(1, 0) = B(1, \theta) + 0.538 - 0.134|\theta|^{0.714} - 7\Theta \qquad \text{for } |\theta| \leqslant 7°$$

$$B(1, 0) = B(1, \theta) - \theta\Theta \qquad \text{for } |\theta| > 7°$$

where $B(1, 0)$ is the absolute B magnitude and $B(1, \theta)$ is the apparent magnitude at phase angle θ. The parameter of the linear part of the phase function in magnitudes $\Theta = 0^m039/\text{deg}$.

For asteroids very much brighter than the limiting magnitude of the search $(m - m_L < -1^m)$ the probability of detection is essentially unity. For asteroids very much fainter than the limiting magnitude $(m - m_L > 1^m)$ the probability of detection is essentially zero. See Fig. 17. Hence, the most interesting range from the point of view of planning a search is the regime $|m - m_L| < \frac{1}{2}^m$. The calculations shown here are for midnight on a winter solstice night, $\varepsilon_z = 0^m13/\text{air mass}$, and a $B(1, 0) = m_L - 1$ and $m_L - \frac{1}{2}$. The search space J is the $20° \times 2^h$ (declination $\times$ right ascension) area on the celestial sphere centered at opposition. (The latitude of the observatory is $33°49'$.) Each cell of the search space is a square, two degrees per side so that there are 150 cells in the search space. We look first in the cell with the highest probability of detection and choose subsequent cells based on Eq. 5.5. A simple, repetitive enumeration allows us to choose the cell order. Figure 18 shows the order of

NORTH

145	136	121	100 187	81 154	72 166	60 194	53	54	59 195	71 167	82 155	99 186	122	135
139	126	104 193	80 153	63 185	46	38	29	30	37	45	64 184	79 152	103 192	125
137	118	96 177	70 170	48	27	20	13	14	19	28	47	69 171	95 176	117
131	112	86 161	62 190	40	21	10	3	4	9	22	39	61 191	85 160	111
132	110	84 158	58 198	35	17	8	1	2	7	18	36	57 199	83 159	109
138	116	90 169	66 182	43	25	12	×5	6	11	26	44	65 183	89 168	115
140	124	102 188	76 156	50	34	24	15	16	23	33	49	75 157	101 189	123
146	134	114	92 173	74 164	52	42	31	32	41	51	73 165	91 172	113	133
149	144	127	108	94 175	77 151	68 180	55 201	56 200	67 181	78 150	93 174	107	128	143
202	148	142	130	120	105 196	98 179	87 162	88 163	97 178	106 197	119	129	141	147

SOUTH

WEST   EAST

RIGHT ASCENSION

(DECLINATION on vertical axis)

Figure 18. A portion of an optimal search plan, in $2° \times 2°$ fields of view, for midnight on a winter solstice night. The apparent magnitude of the asteroid at its brightest is $m_L - 1^m$.

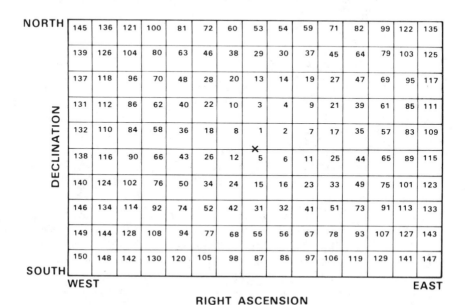

Figure 19. Same as for Fig. 18 but the minor planet is $0^m.5$ brighter.

Figure 20. Same as for Fig. 19 but on a summer solstice night.

NORTH · SOUTH · EAST · WEST · RIGHT ASCENSION · DECLINATION

155

search until each box has been examined once (the geocentric distance of the asteroid is 0.5 A.U.). Figure 19 is the search plan for the same geometrical circumstances but $0^{m}.5$ brighter.

As a contrast Fig. 20 shows the same problem solved for midnight on a summer solstice night, $m = m_L - \frac{1}{2}$. I argue that this is a nonintuitive search plan. A further sophistication of these algorithms includes eclipses, avoidance of the Moon, and so on.

ORBIT-RELATED TOPICS

A fairly common thing to do with artificial satellites is to launch them. The typical deep-space launch consists of several phases. First there is a direct ascent via rocket to the parking orbit. After some number of revolutions in the parking orbit a booster fires. This takes the payload via the transfer orbit out to deep space. If a circular orbit is desired, another booster firing is required. The more rocket firings necessary, the larger the initial vehicle is and the heavier the weight of the initial complement of fuel. The larger the initial quantity of fuel, the larger the ascent stage must be. All of this costs money, so optimizing all of these rocket firings and maneuvers is desirable. This leads to such topics as, for example, minimum energy transfers. All of the above subjects will be touched on below. Finally some interplanetary considerations are dealt with.

Rockets

Rockets are a simple classical system that vividly points out Newton's pre-science in writing $\mathbf{F} = d\mathbf{p}/dt$ instead of $\mathbf{F} = m\mathbf{a}$. To derive the equation of motion of a rocket, let its instantaneous mass be m and assume that it is moving with velocity $\mathbf{v}$ relative to an inertial system. If the exhaust is being expelled at the rate dm/dt with velocity $\mathbf{v} + \mathbf{u}$, then

$$\frac{d(m\mathbf{v})}{dt} - (\mathbf{v} + \mathbf{u})\frac{dm}{dt} = \mathbf{F}$$

where $\mathbf{F}$ is the total external force (air drag, gravity, etc.). Simplifying,

$$m\frac{d\mathbf{v}}{dt} = \mathbf{u}\frac{dm}{dt} + \mathbf{F}$$

Let $\mathbf{F}$ be a constant gravitational force $-mg(0, 0, 1)$, let the rocket be going upward ($\dot{v} > 0$), and the exhaust downward:

$$m\frac{dv}{dt} = -u\frac{dm}{dt} - mg$$

The thrust developed by the rocket is $u\dot{m}$. Sometimes u is written as the product of g and the specific impulse I. Assuming that u, $\dot{m}$, and g are all constant, then one can integrate the equation of motion to obtain the change in speed Δv gained in a time Δt,

$$\Delta v = gI \ln\left(\frac{m_i}{m_f}\right) - g\,\Delta t$$

Here m_i is the initial mass of the rocket before the burn of duration Δt and m_f is the final mass of the rocket after this expenditure of fuel.

Rockets are usually built in stages rather than developing one tremendously large vehicle. Quantities of interest are the optimum stage mass ratios for a given orbital injection speed. This problem can be solved if we remember that each stage has an initial (ground) weight w_i and a final (ground) weight $w_f = w_i -$ weight of fuel $-$ weight of empty stage (i.e., its structural weight). Let the latter two be w and W. Then, adding numerical subscripts for the N stages,

$$\frac{w_{i_1}}{w_p} = \left(\frac{w_{i_1}}{w_{i_1} - w_1 - W_1}\right)\left(\frac{w_{i_2}}{w_{i_2} - w_2 - W_2}\right)\cdots\left(\frac{w_{i_N}}{w_p}\right)$$

where $w_p =$ final weight of the last stage $=$ payload. Define

$$\rho_n = \frac{w_{i_n}}{w_{i_n} - w_i}, \qquad \sigma_n = \frac{W_i}{w_i + W_i}$$

then, if $w_{i_1} = w_T$ is the total weight at liftoff,

$$\frac{w_T}{w_p} = \prod_{n=1}^{N} \frac{\rho_n(1 - \sigma_n)}{1 - \rho_n\sigma_n}$$

Now, from the solution of the rocket equation of motion, the injection speed is simply

$$V = \sum_{n=1}^{N} gI_n \ln \rho_n - gT$$

where T is the total duration of all of the separate burns. We want to minimize the mass ratio w_T/w_p for a given V. To do so, we introduce the Lagrange multiplier λ and form [it's easier to deal with the logarithm of the mass ratio and the logarithm is a monotonic function of its (real) argument], because additive constants are irrelevant,

$$\ln\left(\frac{w_T}{w_p}\right) + \lambda\left(V - \sum_{n=1}^{N} gI_n \ln \rho_n\right)$$

This is a function of the $\{\rho_n\}$, which is a minimum if

$$\rho_n = \frac{\lambda gI_n - 1}{\lambda gI_n\sigma_n}$$

where λ is determined from the implicit relationship

$$V = \sum_{n=1}^{N} gI_n \ln \rho_n - gT$$

Orbit Transfers

Once in orbit one might want to change the orbit. This could consist of in-plane changes, say to a different orbit or by rendering the current one circular. The goal might be a specific target (say the Moon) rather than an orbit of a particular size or shape. Again optimization is an economic necessity. It might also be desirable to change the inclination, which is much more expensive to do and difficult to analyze (in general).

The Hohmann Transfer

Hohmann (1925) found the minimum energy coplanar transfer orbit between two circular orbits. Suppose the rocket is in a circular orbit of radius a and one wants to send it to a larger circular orbit of radius A. To do this, the booster is fired while in the smaller orbit applying thrust only tangentially. We approximate this as an impulsive (i.e., very short time) thrust. Since the rocket's kinetic energy relative to the primary is increased by the application of a positive tangential thrust, the rocket follows a new elliptical transfer orbit (see Fig. 21). At the apogee ($=A$, perigee $=a$) of this elliptical transfer orbit another impulsive, positive, tangential thrust is applied. This serves to change the elliptical orbit with apogee A into a circular one of radius A. To reverse the effects, the procedure is reversed.

In order to calculate the speed and energy changes due to this transfer, we will exploit the conservation of energy equation, Eq. 2.3a, in the form

$$\frac{\mathbf{v} \cdot \mathbf{v}}{2} - \frac{\mu}{r} = -\frac{\mu}{2\alpha}$$

for velocity $\mathbf{v}$, semi-major axis α, $\mu = GM_\oplus$. In the first (circular) orbit

$$\frac{v^2}{2} - \frac{\mu}{a} = -\frac{\mu}{2a}$$

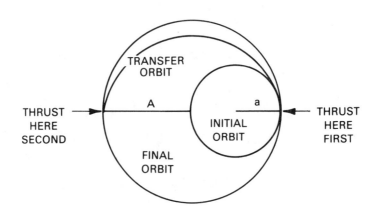

Figure 21. Geometry of a simple Hohmann transfer orbit.

or

$$v = \left(\frac{\mu}{a}\right)^{1/2}$$

Clearly the semi-major axis of the transfer orbit is $\alpha = (a+A)/2$ so that after the first impulse the rocket's speed ν (remember that it is an impulsive change so the rocket is still at $r=a$) is

$$\nu^2 = \mu\left(\frac{2}{a} - \frac{2}{a+A}\right) = \frac{2\mu}{a}\left(\frac{A}{a+A}\right)$$

The difference between ν and v is the amount of speed that must be added,

$$\Delta v = \nu - v = \left[\frac{2\mu}{a}\left(\frac{A}{a+A}\right)\right]^{1/2} - \left(\frac{\mu}{a}\right)^{1/2}$$

When the rocket reaches apogee on the elliptical transfer orbit, its distance is $r = A+a$ and we must apply an impulse appropriate to a final speed of $V = (\mu/A)^{1/2}$. Using the conservation of energy equation twice, we calculate

$$\Delta V = \left(\frac{\mu}{A}\right)^{1/2} - \left[\frac{2\mu}{A}\left(\frac{a}{a+A}\right)\right]^{1/2}$$

and, therefore, the total speed change is $\Delta v + \Delta V$ and the total energy change is proportional to $(\Delta v + \Delta V)^2$.

More can be learned about the Hohmann transfer if we parameterize the problem by the ratio of orbital radii

$$R = \frac{A}{a}$$

and treat the dimensionless ratio $(\Delta v + \Delta V)/v$. Thus, we investigate

$$S_H = \left(1 - \frac{1}{R}\right)\left(\frac{2R}{1+R}\right)^{1/2} + 1/\sqrt{R} - 1$$

Clearly $S_H(R=1) = 0$ and $S_H(R=\infty) = \sqrt{2}-1$. Less clear is that S_H has a single maximum for $R \in [1, \infty]$. To find it, we calculate dS_H/dR and set it equal to zero. After factoring out common (and nonzero) terms, one derives

$$R^3 - 15R^2 - 9R - 1 = 0$$

This equation has a real root near $R = 15.5817187$, which represents a maximum for $S_H(\approx 0.5363)$. Let the actual value of R that maximizes S_H be denoted by R_m. Since once $R > R_m$ the necessary relative speed expenditure diminishes, ascension to very high orbits is possible; the limiting value was already obtained above $S_H(R=\infty) = \sqrt{2}-1$. Consider escape to infinity from $r=a$ and then returning from infinity to $r=A$. The total speed change is $(\sqrt{2}-1)v + (\sqrt{2}-1)V$ so $S' = (\sqrt{2}-1)(1+1/\sqrt{R})$. If $S_H = S'$, then R is a solution of

$$R^3 - (7+4\sqrt{2})R^2 + (3+4\sqrt{2})R - 1 = 0$$

or $R \simeq 11.9387655$. Let the exact value of this intersection point be labeled R_i. Then for orbital size ratios $R \in [1, R_i]$ the ordinary Hohmann transfer is best. For ratios $R \in [R_i, R_m]$ a Hohmann transfer to and from infinity is possible. For ratios $R > R_m$ it turns out that the bielliptic transfer is optimum.

Bielliptic Transfer

There is a three-impulse set of maneuvers that will result in going from an orbit of radius a to another coplanar one of radius A. The starting point is as in a Hohmann transfer except that the initial impulsive velocity change results in an elliptic transfer orbit with a semi-major axis larger than A. At apogee in this orbit another positive impulse is applied, putting the rocket into a second elliptic transfer orbit that has a perigee distance equal to A. When at perigee a *negative* impulse is applied, this results in a circular orbit of radius A. This mode of transfer was proposed by Hoelker and Silber (1959) and is reversible.

With a and A as above and α the semi-major axis of the second ellipse (see Fig. 22) one can write the total dimensionless speed transfer ($R = A/a$ again and $\rho = 2\alpha/a - R$) as

$$S_B = \left(\frac{2\rho}{1+\rho}\right)^{1/2} + \left(\frac{2\rho/R}{\rho+R}\right)^{1/2} + \left(\frac{2R/\rho}{\rho+R}\right)^{1/2} - 1 - 1/\sqrt{R} - \left(\frac{2/\rho}{1+\rho}\right)^{1/2}$$

Now we fix R and consider the extrema of S_B with respect to ρ. They occur for $\partial S_B/\partial \rho = 0$ or

$$\rho = -\frac{3R+1}{3(R+1) \pm 2[R(3R-2)]^{1/2}}$$

But $\rho > 0$ so the lower sign must be taken. Since we have portrayed the instance

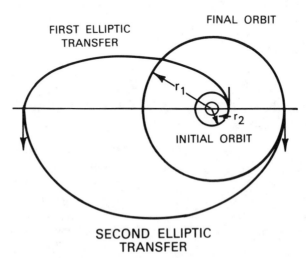

Figure 22.   Geometry for a bielliptic transfer orbit.

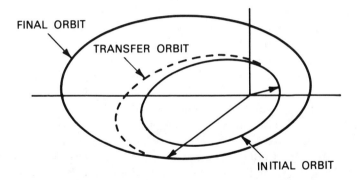

Figure 23.   Geometry for a cotangential transfer orbit.

where $\alpha > A > a$, it follows that $\rho > R > 1$ or

$$(3R+1)(R^3-15R^2-9R-1)<0$$

We have met this cubic before and the interesting root is R_m.

Finally the issue of what type of transfer is optimum for $R > R_i$ can be addressed. If $R = 9$, then the ρ value for which $\partial S_B/\partial\rho = 0$ is infinite. Hence the bielliptic transfer should not be used for $R < 9$ (where we have already shown that the Hohmann transfer is optimal). It turns out [see Escobal (1968) for the details of the analysis] that for $R < R_i$ the Hohmann transfer is best, for $R > R_m$ the bielliptic transfer is best, but for $R \in (R_i, R_m)$ one must do a more complicated test.

Other Types of Transfer

Another special case easily analyzed is the coplanar, cotangential, coaxial elliptic transfer shown in Fig. 23. This is a two-impulse transfer utilizing an elliptical transfer orbit. There is a more general set of coplanar, cotangential transfers between elliptic orbits that can be analyzed relatively simply. See Escobal (1968) for a discussion. The last interesting case is that of a plane change. The relationship between a simple plane change by an amount θ and the impulsive change in the velocity $\Delta\mathbf{v}$ is

$$|\Delta\mathbf{v}| = 2|\mathbf{v}| \sin\left(\frac{\theta}{2}\right)$$

Just as semi-major axis and eccentricity changes are initiated at a perifocus, plane changes frequently occur during nodal passage. Note that to change the plane by 60° requires an energy expenditure equivalent to the launch itself.

The Patched Conic Approximation

So far orbital transfers about the primary have been considered. Now we turn to orbit transfers that involve a change of primary. As an illustration of the process consider a trajectory from the Earth (or its vicinity) to the Moon (or its vicinity). Very near the Earth the effects of the Moon are small and one can regard the rocket to be on an unperturbed orbit about the Earth. If we interchange the Moon for the Earth in the previous sentence, then it is still approximately true. Consequently we are led to the idea of treating the path between the two as a discontinuous one—two-body motion dominated by the Earth followed by two-body motion dominated by the Moon. Thus the phrase "patched conic." The location of the boundary point must be a function of the Earth–Moon mass ratio ($\sim 81:1$) and the Earth–Moon distance (d). Exactly where was deduced by Battin (1964) with the result that the patch point is $D = d(m_{\mathbb{C}}/m_{\oplus})^{2/5}$ from the center of the Moon. Moreover, although the patched conic approximation is reasonable for Earth-to-Moon trajectories, it is not good the other way around.

Let the rocket start at $t = t_0$ with geocentric values of distance, speed, flight path angles, and phase angle at departure (these last two are engineering terms) of r_0, v_0, ϕ_0, and γ_0. See Fig. 24. This set of variables turns out to be

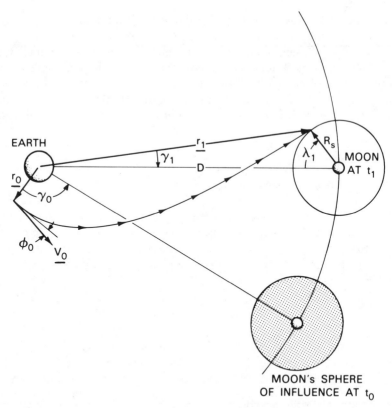

Figure 24. Geometry for the patched conic approximation. (Used by permission of Dover Publications, Inc., New York, from R. R. Bate, D. D. Mueller, and J. F. White, *Fundamentals of Astrodynamics*, © 1971.)

inconvenient because one has to iterate through the solution in order to obtain the correct launch time (remember that the flight time is unknown and therefore the amount of lunar motion during the flight is unknown). Instead of γ_0 we use the selenocentric longitude at the distance $D(\approx d/6)$. The initial energy is $\mathcal{E} = v_0^2/2 - GM_\oplus/r_0$ and the initial angular momentum is $L = r_0 v_0 \cos \phi_0$. The geocentric distance at the lunar sphere of influence crossing is $r_1 = (d^2 + D^2 - 2\,dD \cos \lambda_1)^{1/2}$. At this point the speed $v_1 = [2(\mathcal{E} + GM/r_1)]^{1/2}$ and the flight path angle is $\phi_1 = \cos^{-1}(L/r_1 v_1)$. Finally, $r_1 \sin \gamma_1 = D \sin \lambda_1$.

In order to calculate the time of flight $t_1 - t_0$, we need to invert Kepler's equation,

$$t_1 - t_0 = \left(\frac{a^3}{GM_\oplus}\right)^{1/2} [(E_1 - e \sin E_1) - (E_0 - e \sin E_0)]$$

The eccentric anomalies E_0, E_1 are related to the corresponding true anomalies v_0, v_1. These in turn are given by $r = a(1 - e^2)/(1 + e \cos v)$. The semi-major axis is $a = -GM_\oplus/2\mathcal{E}$ and the eccentricity $e = [1 - L^2/GM_\oplus a)]^{1/2}$. Now we can calculate the location of the Moon; it has moved through a geocentric arc of $\omega_{\mathbb{C}}(t_1 - t_0)$ where $\omega_{\mathbb{C}}$ is the geocentric angular speed of the Moon ($\approx 0\rlap{.}{''}546/\text{sec}$). Finally $\gamma_0 = v_1 - v_0 - \gamma_1 - \omega_{\mathbb{C}}(t_1 - t_0)$.

Having gotten the rocket to the patch point, we need to bring it to the Moon; to do this, the appropriate selenocentric conic must be constructed. Using uppercase letters to denote selenocentric values, we have $R_1 = D$ and $V_1 = [v_1^2 + d^2\omega_{\mathbb{C}}^2 - 2v_1\,d\omega_{\mathbb{C}}\cos(\phi_1 - \gamma_1)]^{1/2}$. The orientation of the selenocentric velocity vector is given by Θ_1,

$$V_1 \sin \Theta_1 = d\omega_{\mathbb{C}} \cos \lambda_1 - v_1 \cos(\lambda_1 + \gamma_1 - \phi_1)$$

When Θ_1 vanishes, the rocket will hit the Moon.

Generalized Rise and Set

A problem of practical importance is the direct visibility of two objects in orbit. Satellite-to-satellite, satellite-to-moon/planet, and so on, have become considerations of weight for communications, scientific, and military applications. In the next few pages the major elements of computations of this nature will be outlined.

Satellite-to-Satellite Visibility

Suppose two satellites are orbiting the Earth. When is one visible to the other? If their geocentric locations are $\mathbf{r}_1$ and $\mathbf{r}_2$, then they will be visible when the length of the normal from the Earth's center to $\mathbf{r}_1 - \mathbf{r}_2$ exceeds the Earth's radius. See Fig. 25. Suppose that this normal vector is called $\mathbf{N}$. Call the remaining part of the satellite–Earth center–normal intersection triangles $\mathbf{s}_1$ and $\mathbf{s}_2$ as shown in Fig. 25. Then

$$\mathbf{N} + \mathbf{s}_1 = \mathbf{r}_1, \qquad \mathbf{N} + \mathbf{s}_2 = \mathbf{r}_2$$

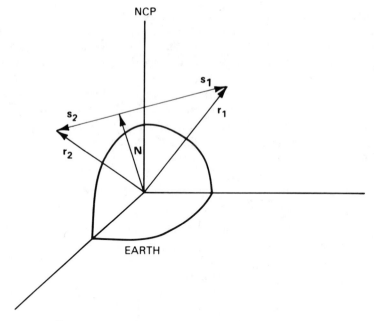

Figure 25. Rise-set geometry for two satellites at $\mathbf{r}_1$ and $\mathbf{r}_2$.

At the instant of visibility $\mathbf{N} \cdot \mathbf{s}_1 = \mathbf{N} \cdot \mathbf{s}_2 = 0$ so the total chord length $|\mathbf{r}_1 - \mathbf{r}_2|$ is

$$|\mathbf{r}_1 - \mathbf{r}_2| = s_1 + s_2 = (r_1^2 - N^2)^{1/2} + (r_2^2 - N^2)^{1/2}$$
$$= (r_1^2 + r_2^2 - 2\mathbf{r}_1 \cdot \mathbf{r}_2)^{1/2}$$

Square twice and reduce. One then obtains the rise–set function or visibility function V,

$$V = (\mathbf{r}_1 \cdot \mathbf{r}_2)^2 - r_1^2 r_2^2 - 2\mathbf{r}_1 \cdot \mathbf{r}_2 N^2 + (r_1^2 + r_2^2) N^2$$

If $V < 0$, there is visibility (consider opposition when $\mathbf{r}_1 \cdot \mathbf{r}_2 = -r_1 r_2$), whereas if $V > 0$, the Earth intervenes. Actually, for unimpeded communication, N should exceed the Earth's radius by the thickness of the atmosphere.

Interplanetary Visibility

This section follows the work of Escobal and Affatali (1961). Let the heliocentric location of the Earth be $\mathbf{r}_\oplus$, that of the satellite be $\mathbf{r}$, the topocentric location of the satellite be $\mathbf{R}$, and the geocentric location of the Earth-bound observer be $\boldsymbol{\rho}$. Then the closure of the Sun–Earth–observer–satellite quadrilateral requires that

$$\mathbf{r}_\oplus - \mathbf{r} + \mathbf{R} + \boldsymbol{\rho} = 0$$

By solving for $\mathbf{R}$ and squaring, we obtain the magnitude of the topocentric distance,

$$R^2 = r^2 + \rho^2 + r_\oplus^2 - 2\mathbf{r} \cdot \boldsymbol{\rho} - 2\mathbf{r} \cdot \mathbf{r}_\oplus + 2\boldsymbol{\rho} \cdot \mathbf{r}_\oplus$$

To be visible, one demands that

$$\boldsymbol{\rho} \cdot \mathbf{R} = R\rho \sin a$$

represent a positive altitude. This is equivalent to the visibility function V being negative (as above),

$$V = \boldsymbol{\rho} \cdot (\mathbf{r}_\oplus + \boldsymbol{\rho} - \mathbf{r}) + \rho(r^2 + \rho^2 + r_\oplus^2 - 2\boldsymbol{\rho} \cdot \mathbf{r}$$
$$- 2\mathbf{r} \cdot \mathbf{r}_\oplus + 2\boldsymbol{\rho} \cdot \mathbf{r}_\oplus)^{1/2} \sin a$$

The visibility function varies most rapidly with the eccentric anomaly of the spacecraft and approximate solutions can be found (for a given day) by regarding the Earth as fixed. Under this approximation V can be written as a quartic equation in the cosine (or sine) of the sidereal time.

Galactic Visibility

In 1973 Klebesadel, Strong, and Olson were allowed to reveal the detection of bursts of extra–solar system γ rays (at ~ 1 MeV). The radiation had been detected by some of the *Vela* satellites ($e \approx 0$, $i \approx 0$, $n \approx \frac{1}{4}$ rev/day) placed into orbit by the United States to detect nuclear weapons tests. As the astronomical importance of these events became clear, more information was released, new instruments were put on board spacecraft designed for other purposes, and new spacecraft were launched. A recent paper on this topic mentions the *Vela 5A*, *5B*, *6A*, and *6B* satellites, the *Orbiting Geophysical Observatory 3* and *5* satellites, the *International Monitoring Platform 6* and *7* satellites, the *Orbiting Solar Observatory 6*, *7*, and *8* satellites, the two *Solrad* satellites *11A* and *11B*, *Helios 2*, the *Pioneer Venus Orbiter*, the Russian *Prognoz 6* and *7*, *SIGNE 3*, the *International Sun–Earth Explorer 3*, and finally the Franco–Soviet pair *Venera 11* and *12*. [As most of these references carry 5–20 names and the astrophysics of the bursts is not pertinent herein, we simply list the principal references: *Ap. J.* **188**, L1 (1974); *Ap. J.* **189**, L9 (1974); *Ap. J.* **229**, L47 (1979); *Sci.* **205**, 119 (1979); *Ap. J.* **232**, L1 (1979); *Sov. Astr. Lett.* **5**, 314 (1979); *Ap. J.* **237**, L1 (1980); *Ap. J.* **237**, L7 (1980); *Ap. J.* **246**, L133 (1981); *Ap. J.* **255**, L45 (1982); and *Ap. J.* **259**, L51 (1982).] What is of a bit more interest is how one uses the available information (i.e., satellite geocentric location at the time of detection) to deduce a position for the source.

In this particular application there are large sources of error due to timing errors on board the spacecraft, the triggering threshold of the detector as a function of the burst's shape, the differing sensitivities of the various detectors to different energy bands, the poor angular resolution of the detectors, and the possible nonsimultaneous rise of the burst over its spectrum. This is not the place to delve into any of this. Rather, we want to abstract the essence of the triangulation problem and briefly consider the nature of the necessary astrometric corrections.

The spacecraft are distributed across the solar system, some of them several astronomical units away from the Earth. Clearly the longer the baseline of the array, the more accurate the eventual source position will be. Since the sources appear to be galactic in nature (say 1 pc or further) as seen within the solar system, the burst of radiation can be viewed as an incoming plane wave. As the wave crosses the location of each detector, it triggers it, and the time of this event is sent back to the Earth. In order to perform a meaningful analysis of these timings, one must know the locations of each detecting satellite in the same coordinate system (typically 1950.0 solar system barycentric) and know how to correct each of the individual spacecraft's clocks to a common time system (say UTC). In principle, if Chapters Three and Four have been read, these reductions present no especial difficulty. There are, however, both special relativistic and general relativistic clock corrections that need to be included. To lowest order these corrections are independent of each other and mimic the usual aberration formulas [for discussion see Bisnovaty-Kogan, Estulin, Havenson, Kurt, Mersov, and Novikov (1981)].

After all of this preliminary analysis what can you tell about the source location in this interplanetary surveying mode? If there were detections by only two spacecraft, then the result would be a small circle of uncertainty on the sky. Both the center and radius of this circle fall out of the analysis. Two sets of two detections yield two small circles that presumably intersect twice. The ambiguity cannot be resolved. Similarly a set of detections from a triplet of satellites yields two sets of circles whose intersections represent the reflection of the source's location about the plane determined by the three satellites (e.g., are diametrically opposite each other on the sky). Once there are four or more detections, the problem becomes redundant and the source's position could be pinned down very well except for the poor resolution of these detectors and the aforementioned sources of error.

PROBLEMS

1. Satellites are launched along great circle routes eastward so as to gain the Earth's rotational speed. Calculate the gain from launch at a place with latitude θ and compare it to the speed in a 1000 km elevation orbit. As a corollary prove that $|i| \leq \theta$ for a direct orbit. Deduce the corresponding limits for a retrograde orbit and the launch azimuths.

2. Define the synodic period of an artificial satellite with respect to a fixed geographical location as the time interval between crossings of the celestial meridian. Prove that for a circular equatorial orbit the synodic period S is related to the sidereal period P by

$$S = \frac{2\pi P}{2\pi - \omega_\oplus P}$$

Further show that the time interval, in hours, for two zenith transits at a latitude θ of an inclined orbit (inclination $= i$, $e = 0$ still), is

$$\frac{24(1 - \sin^2 \theta \csc^2 i)}{(2\pi/P) - (2\pi/S)(1 - \sin^2 \theta \csc^2 i)^{1/2}} + \frac{2\pi \sin^2 \theta \csc^2 i}{(2\pi P) - (2\pi/S) \cos \theta \sec i}$$

3. Design an artificial satellite orbit that could provide a continuous communications link between Moscow and Siberia for at least 8 hr. Can it be done for 12 hr?

4. Verify the astronomical results for e_p and e_u from the exact relationships.

5. Compute the size of the Earth's shadow at geosynchronous distance.

6. Formulate an along-orbit search for an artificial satellite that is expected to be no more than ± 20 min off. How does it differ from the all-sky search developed in the text?

7. Two artificial satellites are both orbiting the Earth in non-coplanar circular orbits of different size. What sequence of orbital plane and semi-major axis changes most efficiently places the lower one in the higher one's orbit. Only a plane change or a semi-major axis change is allowed per rocket firing.

8. Suppose that the Earth is spherical, nonrotating, and without an atmosphere. Launch a rocket from a point P with speed v making an angle θ with the upward vertical. Suppose that it returns to the Earth at Q an angular distance $\omega < \pi$ away. Show that the semi-major axis a is given by

$$a = \frac{[1 - e \cos(\omega/2)]R_\oplus}{1 - e^2}$$

where e is the eccentricity of the orbit. Further deduce that

$$\frac{v^2 R_\oplus}{GM_\oplus} = 1 + e\left(\frac{e - \cos(\omega/2)}{1 - e \cos(\omega/2)}\right)$$

and that

$$\tan \theta = \frac{[1 - e \cos(\omega/2)]}{e \sin(\omega/2)}$$

Further deduce that v^2 has a minimum for a given value of ω with two different values of the eccentricity.

9. Suppose you wanted to launch a rocket from the Earth and land it on the Moon. Should you approach the Moon from in front or behind? Why?

10. Determine a "free-return" lunar trajectory via the patched conic approximation (i.e., no firing near the Moon is necessary to return to the Earth).

11. Show that the minimum energy ballistic orbit whose total span in true anomaly is 2ν has

$$a = \frac{R_\oplus}{2}(1 + \sin \nu), \qquad e = \frac{\cos \nu}{1 + \sin \nu}$$

and that the maximum elevation is given by

$$\left(\frac{R_\oplus}{2}\right)\left[\sqrt{2}\sin\left(\nu + \frac{\pi}{4}\right) - 1\right]$$

12. Determine the time of flight for a Hohmann transfer.

13. A rocket is traveling in a circular orbit. An impulsive thrust is applied normal to the orbit such that the final velocity vector makes an angle θ outward from the tangent to the initial orbit. Show that the area enclosed from the point of application to the new apogee is given by

$$A = \frac{a^2}{2}(1 - e^2)^{1/2}\left[e(1 - e^2)^{1/2} + \sin^{-1}(e) + \frac{\pi}{2}\right]$$

where a is the new semi-major axis and e is the new eccentricity. Determine a and e in terms of θ, the original orbital radius R, and the speed increment Δv.

14. Show that for coplanar, coaxial, elliptical transfers the minimum energy transfers are when the impulses are applied at perigee of the inner orbit and at apogee of the outer one (for a two-impulsive thrust maneuver).

chapter six

Other Soluble Problems

Exactly soluble problems are precious. They yield total knowledge, allow approximations from a position of strength, and sometimes yield computationally useful frameworks for practical problems. Within the realm of celestial mechanics there are several exactly soluble problems. One of them is central to all celestial mechanics texts—the two-body problem for spherically symmetric bodies. When one of the bodies becomes oblate, the general solution is available only if the second (spherically symmetric) object stays in the plane of the oblate object's equator. In order to explicitly exhibit the solution in terms of known transcendental functions, even further restrictions are necessary—in particular to allow only the ρ_{20} term in Eqs 1.22 to be nonzero. (The usual abbreviation for this quantity is J_2.) These two restrictions yield an analytically tractable problem in that the orbit and the time dependence of the location may be explicitly displayed. Even with these two restrictions the problem is still important within the solar system—most natural satellites, be they moons or rings, lie in the equatorial plane of their (oblate) primaries and many artificial satellites of the Earth lie in the plane of the celestial equator. Later, in Chapters Nine and Ten, the general oblate primary problem will be approximately solved via Lagrange's planetary equations. In this chapter some of these results will be derived from the exact solution. Moreover, other approximation techniques will be exploited to investigate the orbit and the time dependence.

It turns out that a third-order polynomial in the distance is critical to determining the nature of the solution. When $J_2 = 0$ the cubic has zero for one of its roots and the solution for the motion is accomplished in terms of trigonometric functions. For $J_2 \neq 0$ the full cubic must be considered and the solution involves elliptic functions. Below we formulate the problem, develop the central role of this cubic, and analyze its properties. The following subsections discuss the orbit, the time dependence of the longitude, and the time dependence of the distance. As appropriate, approximations are used to rederive the classical results referred to above. Finally, because J_2 is small in all practical applications, the notation of the Keplerian two-body problem has been maintained. Of course, for $J_2 \neq 0$, a is not the semi-major axis nor is e

the eccentricity. However, as $J_2 \to 0$ the quantities denoted in this discussion by $a, e, \ldots$ do approach their Keplerian values.

Another exactly soluble problem in celestial mechanics is that of two fixed centers. This was first discussed by Euler in 1760. As the name implies two (spherically symmetric) finite bodies are fixed in inertial space and one solves for the motion of a particle in their combined gravitational field. Requiring the three bodies to be coplanar simplifies the analysis somewhat. As a start to a discussion of third-body perturbations by the Moon or by the Sun on an artificial satellite, the utility of the exact problem is manifest. A less obvious but similar case is the perturbation of a minor planet (revolving about the Sun) by Jupiter. It turns out that the exact solution to the problem of two fixed centers is so complicated that it is useless in practice. One can still exploit approximations to it in order to anticipate results derived in Chapters Nine and Ten. Hence, this problem is also discussed in this chapter.

Of more general interest are the five exactly soluble cases of the restricted three-body problem. Now we relax the unphysical assumptions that the two massive bodies are fixed in inertial space. Instead, we allow them to revolve about each other in circular orbits (hence the adjective restricted). Again we insist on coplanarity of all three objects for all times. These solutions were discovered by Lagrange in 1772. Two of them are stable (we will also discuss linear stability analyses herein) and used in the solar system by the Trojan asteroids. The three-body problem is a rich subject that we shall barely touch on. Fortunately Szebehely (1967) has written a very nice introduction to the general subject. We will also look at the Kirkwood gaps and Hill's problem before concluding this part of the discussion. Finally we will outline a proof of Bertrand's (1873) theorem and solve the simple harmonic oscillator problem.

OBLATE SPHEROIDS

Formulation

Physics and Geometry

We choose as the origin the center of mass of the oblate primary. The equatorial plane of the primary coincides with the plane $z = 0$. Since the primary is a spheroid, there is no preferred origin of longitudes. Cartesian coordinates x, y, and z are related to spherical coordinates r, θ, ϕ (distance, latitude, and longitude) by

$$x = r \cos \theta \cos \phi, \qquad y = r \cos \theta \sin \phi, \qquad z = r \sin \theta$$

This coordinate system is taken to be an inertial one. The gravitational potential of the oblate primary, whose total mass is M, is assumed to be

$$U(\mathbf{r}) = -\frac{GM}{r}\left[1 + \left(\frac{J_2}{2r^2}\right)(1 - 3\sin^2\theta)\right] \tag{6.1a}$$

where G is the universal constant of gravitation and $r = |\mathbf{r}|$. The physical interpretation of J_2 is that it is the difference between the polar and the equatorial moments of inertia per unit mass. $J_2 \geq 0$ herein (if J_2 were negative, then we would be describing prolate spheroids). Note too that the dimensions of J_2 are those of a length squared.

The equations of motion of a particle are

$$\ddot{\mathbf{r}} = -\nabla U(\mathbf{r}) \tag{6.1b}$$

where

$$\ddot{\mathbf{r}} = [\ddot{r} - r\dot{\phi}^2\cos^2\theta - r\dot{\theta}^2, \; -r\ddot{\theta} - 2\dot{r}\dot{\theta} - r\dot{\phi}^2\sin\theta\cos\theta, \tag{6.1c}$$
$$(r\ddot{\phi} + 2\dot{r}\dot{\phi})\cos\theta - 2r\dot{\phi}\dot{\theta}\sin\theta]$$

and

$$\nabla U(\mathbf{r}) = \left(\frac{\partial U}{\partial r}, \; \frac{-1}{r}\frac{\partial U}{\partial\theta}, \; \frac{\sec\theta}{r}\frac{\partial U}{\partial\phi}\right)$$

Now $\partial U/\partial\theta = 0$ at $\theta = 0$ so that if at some arbitrary instant of time θ and $\dot{\theta}$ vanish, then so does $\ddot{\theta}$ as well as all of the higher time derivatives of θ. Therefore, with the initial conditions $\theta(T) = \dot{\theta}(T) = 0$, $\theta(t) = 0 \forall t$. With this simplification the equations of motion 6.1 reduce to (since $\partial U/\partial\phi = 0$)

$$\ddot{r} - r\dot{\phi}^2 = -\frac{\partial U}{\partial r}\bigg|_{\theta=0} \tag{6.1d}$$

$$r\ddot{\phi} + 2\dot{r}\dot{\phi} = 0 \quad \text{or} \quad \frac{d}{dt}(r^2\dot{\phi}) = 0$$

As in the Keplerian two-body problem define the constant L by

$$L = r^2\dot{\phi}$$

and rewrite the radial equation of motion as

$$\ddot{r} - \frac{L^2}{r^3} = -\frac{\partial U}{\partial r}\bigg|_{\theta=0}$$

This is directly integrable to

$$\frac{\dot{r}^2}{2} = \mathscr{E} - U|_{\theta=0} - \frac{L^2}{2r^2}$$

where $\mathscr{E}$ is a constant of integration easily identified as the total energy per unit mass.

Instead of dealing with $\mathscr{E}$ and L, we choose a different parameterization. Define $(\mu = GM)$ a and e via

$$\mathscr{E} = -\frac{\mu}{2a}, \qquad L^2 = \mu a(1 - e^2)$$

Then, since $U|_{\theta - 0} = -(\mu/r)[1 + J_2/2r^2]$, the conservation of energy equation may be written as

$$\dot{r}^2 = -\frac{\mu f(r)}{ar^3} \tag{6.2}$$

with the cubic $f(r)$ defined by

$$f(r) \equiv r^3 - 2ar^2 + a^2(1 - e^2)r - aJ_2$$

Note that for bound motion $\mathscr{E} < 0$ so $a > 0$. We can also adjust the handedness of the coordinate system such that $L \geqslant 0$, therefore $1 \geqslant e^2$. This is all one can deduce, in general, about the ranges of a and e. However, since the physically meaningful constants of integration are $\mathscr{E}$ and L, the parameterization in terms of a, e is valid only if the Jacobian $\partial(\mathscr{E}, L)/\partial(a, e) \neq 0$. This implies $a < \infty$ and $e \neq 0$.

The Cubic $f(r)$

One can write $f(r)$ in a variety of forms,

$$\begin{aligned}
f(r) &= r^3 - 2ar^2 + a^2(1 - e^2)r - aJ_2 \\
&= r[r^2 - 2ar + a^2(1 - e^2)] - aJ_2 \\
&= r[r - a(1 + e)][r - a(1 - e)] - aJ_2 \\
&= r(r - r_+)(r - r_-) - aJ_2 \\
&= (r - R_a)(r - R_p)(r - R_0)
\end{aligned} \tag{6.3}$$

where $r_\pm = a(1 \pm e)$ and R_a, R_p, R_0 are the three roots of f. Were $J_2 = 0$, then the three roots of f would be r_+, r_-, and 0. Since if J_2 vanishes a and e are the semi-major axis and eccentricity, it follows that when $J_2 = 0$, r_+ and r_- are the maximum and minimum distances of the particle from the primary. The labeling of the three roots of f when $J_2 \neq 0$ is suggestive of this, and we assume that $R_a \geqslant R_p \geqslant R_0$. First we show that f has three real roots for small J_2.

The discriminant of the cubic $z^3 + a_1 z^2 + a_2 z + a_0$ is given by $Q^3 + R^2$ where

$$Q = \frac{a_1}{3} - \left(\frac{a_2}{3}\right)^2, \qquad R = \frac{a_1 a_2 - 3a_0}{6} - \left(\frac{a_2}{3}\right)^3$$

Here $a_2 = -2a$, $a_1 = a^2(1 - e^2)$, and $a_0 = -aJ_2$ so the discriminant is equal to

$$Q^3 + R^2 = \frac{-a^6 e^2(1 - e^2)^2}{27} + \frac{a^4 J_2(-1 + 9e^2)}{27} + \frac{a^2 J_2^2}{4} \tag{6.4}$$

For small enough J_2 (>0) this is negative. When the discriminant of a cubic is negative, then the cubic has three real, unequal roots. If J_2 increases from 0 until $Q^3 + R^2 = 0$, then the cubic has three real roots and at least two of them are equal. This corresponds to the case of a circular orbit and will be discussed separately below.

The special case of $e = 0$, $J_2 \neq 0$, implies that f has the form $f(r) = r(r-a)^2 - aJ_2$. In this case one would expect f to have three real roots near a, a, and 0. Directly from the analytical solution of a cubic for small J_2, one may show that the roots are approximately equal to $a + \sqrt{J_2}$, $a - \sqrt{J_2}$, and 0 in this case. In the second order of approximation one finds (from Newton's method) that the roots are approximately equal to $a \pm \sqrt{J_2} - J_2/(2a)$ and J_2/a. The special case of $J_2 = 0$, $e \neq 0$, has for roots $r_\pm$, 0. Further specializing to $e = 0$ finds that the roots are a, a, and 0 again.

Returning to the general problem assume that J_2 is small enough that $Q^3 + R^2 < 0$. One may obtain the values of R_a, R_p, and R_0 directly from the analytical solution for a cubic, but this yields nothing transparent. Instead, we will compute approximate values for the three quantities by exploiting the fact that as $J_2 \to 0$, R_a, R_p, and R_0 must approach r_+, r_-, and 0. Using Newton's method we find that the first-order approximations to the roots of f are

$$R_a \simeq r_+ + \frac{J_2}{2er_+} > r_+$$

$$R_p \simeq r_- - \frac{J_2}{2er_-} < r_- \tag{6.5}$$

$$R_0 \simeq 0 + \frac{J_2}{p} > 0$$

where $p = a(1 - e^2)$ as usual. Continuing the iteration process of Newton's method we calculate, to the second order in J_2 (but *not* to the second order in the iteration scheme),

$$R_a \simeq r_+ + \frac{J_2}{2er_+} - \frac{J_2^2(1 + 3e)}{(2er_+)^3}$$

$$R_p \simeq r_- - \frac{J_2}{2er_-} + \frac{J_2^2(1 - 3e)}{(2er_-)^3} \tag{6.6}$$

$$R_0 \simeq 0 + \frac{J_2}{p} + \frac{2J_2^2}{p^3}$$

Note that the $J_2 \neq 0$, $e = 0$, results cannot be obtained from these formulas merely by setting $e = 0$. From Eqs. 6.5 and 6.6 we conjecture that

$$R_a \geqslant r_+ \geqslant r_- \geqslant R_p \geqslant R_0 \geqslant 0$$

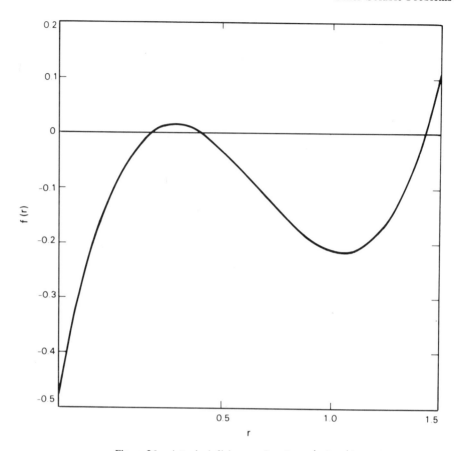

Figure 26. A typical $f(r)$ curve ($a = 1$, $e = \frac{1}{3}$, $J_2 = \frac{1}{10}$).

To further elucidate the nature of $f(r)$, compute df/dr from Eq. 6.3 and then solve for its roots. The results are that df/dr vanishes if $r = R_\pm$: $R_\pm = (a/3)[2 \pm (1 + 3e^2)^{1/2}]$. From the values of d^2f/dr^2 at these two points we deduce that at R_+, f has a local minimum whereas at R_-, f has a local maximum. Since $f(\pm\infty) = \pm\infty$ and $f(0) = -J_2 < 0$, I further conjecture that

$$R_a \geqslant r_+ \geqslant R_+ \geqslant r_- \geqslant R_p \geqslant R_- \geqslant R_0 \geqslant 0$$

Figure 26 shows $f(r)$ for $a = 1$, $e = \frac{1}{3}$, and $J_2 = \frac{1}{10}$. The numerical values of R_0, R_-, R_p, r_-, R_+, r_+, and R_a are 0.176200992, 0.281766487, 0.398063916, 0.666666667, 1.051566846, 1.333333333, and 1.425735091. Note that in order for the extrema of f to be real, one must have $e^2 \geqslant -\frac{1}{3}$.

The results of this analysis, remembering Eq. 6.2, are that the motion is confined to that range of r such that $f(r) \leqslant 0$, that is, $r \in [R_p, R_a]$. When $r = R_p$ or R_a, the radial velocity vanishes and these represent the turning points of the motion. Lastly, from the different forms of f in Eq. 6.3, one can deduce that

$$R_a + R_p + R_0 = 2a = r_+ + r_-$$

$$R_a R_p + R_a R_0 + R_p R_0 = a^2(1 - e^2) = ap = r_+ r_-$$

$$R_a R_p R_0 = aJ_2$$

Having gleaned the maximum amount of information concerning f and dr/dt, we now turn to a discussion of the orbit, $r(\phi)$.

The Orbit

The Exact Solution

From the definition of L as $r^2\dot\phi$ and the formula for $\dot r$ in Eq. 6.2 one may derive that $r' = dr/d\phi$ is related to $\dot r = dr/dt$ via

$$\dot r = \frac{Lr'}{r^2}$$

Therefore, the equation of the orbit is $[p = a(1 - e^2)$ still$]$

$$(r')^2 = -\frac{rf(r)}{ap} \tag{6.7}$$

Orient the xy axes (arbitrary because the primary is a spheroid) such that at $t = T$, $r = R_p$ and $\phi = 0$. Suppose that the earliest time for which $r = R_a$ is $t = T + t_a$ where $\phi = \phi_a$. Let the next time that r equals R_p be $t = T + t_p$ where $\phi = \phi_p$. Note that since the differential equation of the orbit is even in ϕ, one can build up the orbit by reflecting about $\phi = 0$ either one of these segments. Because we know that $R_p \leq r \leq R_a$ and that $\dot r^2$ and $(r')^2$ are nonnegative, it follows that $\dot r$ and r' are greater than 0 for $r \in (R_p, R_a)$, $t - T \in (0, t_a)$, and that $\dot r$ and r' are less than 0 for $t - T \in (t_a, t_p)$. Both $\dot r$ and r' vanish if $r = R_a$ or R_p. Hence, the solution of Eq. 6.7 is

$$\int_{R_p}^{r} \frac{ds}{[-sf(s)]^{1/2}} = \frac{+1}{(ap)^{1/2}} \int_0^{\phi} d\psi, \qquad \phi \in [0, \phi_a], \, t - T \in [0, t_a]$$

$$\int_{R_a}^{r} \frac{ds}{[-sf(s)]^{1/2}} = \frac{-1}{(ap)^{1/2}} \int_{\phi_a}^{\phi} d\psi, \qquad \phi \in [\phi_a, \phi_p], \, t - T \in [t_a, t_p]$$

Define the angles λ and Λ by

$$\sin^2 \lambda = \frac{R_a - R_0}{R_a - R_p} \cdot \frac{r - R_p}{r - R_0}, \qquad \sin^2 \Lambda = \frac{R_p}{R_a - R_p} \cdot \frac{R_a - r}{r}$$

and the modulus $k(>0)$ by

$$k^2 = \frac{R_a - R_p}{R_a - R_0} \cdot \frac{R_0}{R_p}, \qquad k^2 \in [0, 1]$$

Then the results of the above two integrations are

$$\left[\frac{(R_a - R_0)R_p}{ap}\right]^{1/2} \phi = 2F(\lambda, k)$$

$$\left[\frac{(R_a - R_0)R_p}{ap}\right]^{1/2} (\phi - \phi_a) = 2F(\Lambda, k)$$

where F is the incomplete elliptic integral of the first kind.
Introduce $\gamma(>0)$ by (note as $J_2 \to 0$, $k^2 \to 0$, $\gamma^2 \to 1$)

$$\gamma^2 = \frac{(R_a - R_0)R_p}{ap} = \frac{(R_a - R_0)R_p}{(R_a + R_0)R_p + R_a R_0}$$

Then the two forms above may be more compactly written as

$$\gamma\phi = 2F(\lambda, k), \qquad \phi \in [0, \phi_a]$$
$$\gamma(\phi - \phi_a) = 2F(\Lambda, k), \qquad \phi \in [\phi_a, \phi_p]$$

(6.8)

Elliptic Functions

Elliptic integrals arise in much work in higher analysis. Basically, every single-valued analytic function that has an algebraic addition theorem is an elliptic function (or a limiting case of one). In particular if $R(z)$ is a quartic in z, then

$$\int \frac{f(z)}{\sqrt{R(z)}} dz$$

is an elliptic integral when f is a rational function of z. It turns out that this integral can assume only one of (or a linear combination of) three distinct forms. These are

$$\int_0^\psi \frac{dz}{[(1-z^2)(1-k^2z^2)]^{1/2}} = \int_0^\theta \frac{d\phi}{(1-k^2\sin^2\phi)^{1/2}} = F(\theta, k)$$

$$\int_0^\psi \left(\frac{1-k^2z^2}{1-z^2}\right)^{1/2} dz = \int_0^\theta (1-k^2\sin^2\phi)^{1/2} d\phi = E(\theta, k)$$

$$\int_0^\psi \frac{dz}{(1-nz^2)[(1-z^2)(1-k^2z^2)]^{1/2}} = \int_0^\theta \frac{d\phi}{(1-n\sin^2\phi)(1-k^2\sin^2\phi)^{1/2}}$$
$$= \Pi(\theta, n, k)$$

where $\psi = \sin\phi$. These are called Legendre's normal forms for the *incomplete elliptic integrals* of the first, (F), second (E), and third (Π) kinds. The quantity $k \in [0, 1]$ (usually) is called the modulus. $(1-k^2)^{1/2} \equiv k'$ is known as the complementary modulus. The quantity n is known as the parameter. When the argument $\psi = 1$ or $\theta = \pi/2$ these integrals are said to be *complete*. The notation used for the complete elliptic integrals of the first, second, and third kinds is

$$K(k) = F\left(\frac{\pi}{2}, k\right)$$

$$E(k) = E\left(\frac{\pi}{2}, k\right)$$

$$\Pi(n, k) = \Pi\left(\frac{\pi}{2}, n, k\right)$$

The reader should have no trouble seeing what happens if $k = 0$ or $k = 1$. Hence, it should be no surprise that these are really *inverse* functions. In further analogy with the trigonometric functions one defines the sine amplitude function via

$$\text{sn}^{-1}(\sin\theta, k) = F(\theta, k)$$

Similarly there is a cosine amplitude function $\text{cn}(\psi, k)$, which can be defined by

$$\text{sn}^2(\psi, k) + \text{cn}^2(\psi, k) = 1$$

There is also a tangent-like function $\text{tn}(\psi, k) = \text{sn}(\psi, k)/\text{cn}(\psi, k)$. More important is the third Jacobian elliptic function $\text{dn}(\psi, k)$,

$$k^2\, \text{sn}^2(\psi, k) + \text{dn}^2(\psi, k) = 1$$

These functions are *doubly* periodic functions. The periods of the sine amplitude function are $4K$ and $2iK'$; of the cosine amplitude $4K$ and $2K + 2iK'$; of the third Jacobian elliptic function $2K$ and $4iK'$; the quantity $K' = K(k')$. Moreover, because they are doubly periodic, elliptic functions are a much richer subject than are trigonometric or hyperbolic functions.

It can now be shown that $\phi_p = 2\phi_a$. If $r = R_a$, then $\lambda = \pi/2$ so $\gamma\phi_a = 2K(k)$ where K is the complete elliptic integral of the first kind. Also, if $r = R_p$, then $\Lambda = \pi/2$ so $\gamma(\phi_p - \phi_a) = 2K(k)$ too or $\phi_p = 2\phi_a = 4K(k)/\gamma$.

Equation 6.8 gives $\phi(r)$ explicitly. The standard discussions of the Keplerian two-body problem deal with the inverse of this, $r(\phi)$. As

$$\text{sn}^{-1}(\sin\psi, \kappa) = F(\psi, \kappa)$$

where $\text{sn}(\psi, \kappa)$ is the sine amplitude function of argument ψ, modulus κ, from Eq. 6.8, with $u = \gamma\phi/2$,

$$r(\phi) = \frac{R_p[1 - k^2\, \text{sn}^2(u, k)]}{1 - m\, \text{sn}^2(u, k)} = \frac{R_p\, \text{dn}^2(u, k)}{1 - m\, \text{sn}^2(u, k)} \tag{6.9}$$

where $\text{dn}(\psi, \kappa)$ is the delta amplitude function of argument ψ, modulus κ, and m is an abbreviation for

$$m = \frac{R_a - R_p}{R_a - R_0} = (R_p/R_0)k^2, \qquad 1 \geq m \geq k^2$$

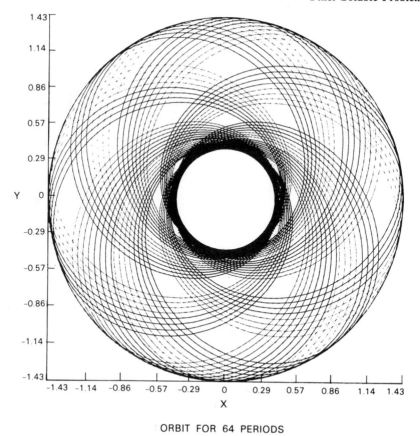

ORBIT FOR 64 PERIODS

Figure 27. Sixty-four orbits ($a = 1$, $e = \frac{1}{3}$, $J_2 = \frac{1}{10}$) with an advance of periapse almost equal to 180° (hence the double-lobed structure).

Equation 6.9 was derived from the upper form in Eq. 6.8. Had the lower form and the addition theorem for $\mathrm{sn}(\psi, \kappa)$ been used, then the identical result would have been obtained. This demonstrates the analytical equivalence of the two and that Eq. 6.9 holds over the entire orbit. It is clear that the orbit is periodic because the elliptic functions are periodic; $\mathrm{sn}(\psi, \kappa)$ and $\mathrm{dn}(\psi, \kappa)$ share $4K(\kappa)$ as a period. The orbit does not repeat, however, as can be clearly seen in Fig. 27. This is for $a = 1$, $e = \frac{1}{3}$, and $J_2 = \frac{1}{10}$ and shows $r(\phi)$ for 64 complete revolutions. The original plot was in three colors, which can be seen in the gradations of grey. The values of k, m, γ, and ϕ_a in this case are 0.603365954, 0.822443482, 0.748043275, and 268°.59733.

The Advance of the Periapse

A well-known result from perturbation theory, via Lagrange's planetary equations, is that there is a secular advance of the argument of periapse. From

Eq. 6.9 one can see that $r(0) = r(\phi_p) = r(2\phi_a)$ so that this advance is evenly distributed over each half of the orbit. Its amount is $\Delta\omega = \phi_p - 2\pi = 4K(k)/\gamma - 2\pi$. To first order in J_2

$$k^2 = \frac{2eJ_2}{p^2}, \quad \gamma^2 = 1 - \frac{J_2(3-e)}{p^2} \tag{6.10}$$

and as $k^2 \to 0$, $K(k) \to (\pi/2)(1+k^2/4)$. Therefore, one calculates that

$$\Delta\omega = \frac{3\pi J_2}{p^2}$$

confirming the first-order perturbation theory value.

An Approximate Orbit.   As $\kappa^2 \to 0$

$$sn(\psi, \kappa) \to \sin\psi - \left(\frac{\kappa}{2}\right)^2 (\psi - \sin\psi\,\cos\psi)\cos\psi$$

$$dn(\psi, \kappa) \to 1 - 2\left(\frac{\kappa}{2}\right)^2 \sin^2\psi$$

Utilizing these approximations (and those in Eq. 6.10) in Eq. 6.9, we derive, after some algebra, the limiting expression for $r(\phi)$,

$$r(\phi) \to \frac{p}{1+e\cos\phi}$$
$$\times \left[1 - \left\{J_2 \frac{[\cos\phi + 3e + 3e^2(\cos\phi + \phi\sin\phi) + e^3(1+\sin^2\phi)]}{2ep^2(1+e\cos\phi)}\right\}\right] \tag{6.11}$$

The first term is just the Keplerian orbit. Also note that, to first order in J_2, $r(0) = r(2\pi) = R_p$ and $r(\pi) = R_a$. We can recover the value of $\Delta\omega$ by remembering that at an apse $r' = 0$. Computing this from Eq. 6.11 and assuming that $\phi_a = \pi + \Delta\omega/2$ or $\phi_p = 2\pi + \Delta\omega$, where $\Delta\omega$ is of order J_2, we immediately recover the above value for $\Delta\omega$.

   An Alternative Derivation.   The approximate form of $r(\phi)$ in Eq. 6.11 can be derived directly from the differential equation 6.7. Set

$$r(\phi) = r_0(\phi) + r_1(\phi)$$

where $r_0(\phi)$ is the Keplerian solution $p/(1+e\cos\phi)$ and r_1 is of order J_2. Using this in Eq. 6.7 and linearizing in r_1 yields

$$r_1' + fr_1 = \frac{J_2 g}{a^2}$$

with f and g given by

$$f(\phi) = \frac{r_0[2r_0^2 - 3ar_0 + a^2(1-e^2)]}{apr_0'} \quad , \quad g(\phi) = \frac{a^2r_0}{2pr_0'}$$

The general solution for $r_1(\phi)$ is

$$r_1(\phi) = C \exp\left[-\int^{\phi} f(\phi') \, d\phi'\right] + \left(\frac{J_2}{a^2}\right) \exp\left[-\int^{\phi} f(\phi') \, d\phi'\right] \int^{\phi} g(\phi')$$

$$\times \exp\left[\int^{\phi'} f(\phi'') \, d\phi''\right] d\phi'$$

where C is the arbitrary constant of integration. After integration,

$$r_1(\phi) = \frac{C \sin \phi}{(1 + e \cos \phi)^2} - \frac{J_2[\cos \phi + 3e + 3e^2(\cos \phi + \phi \sin \phi) + e^3(1 + \sin^2 \phi)]}{2ep(1 + e \cos \phi)^2}$$

$C = 0$ because it is not needed to reproduce $r(0) = r(2\pi) = R_p$ or $r(\pi) = R_a$ to first order in J_2. The result is now identical to that obtained above from the expansion of the exact solution, Eq. 6.11.

A Precessing Ellipse.   Define a pseudo semi-major axis and eccentricity α and ε, via

$$2\alpha = R_a + R_p$$

$$2\alpha\varepsilon = R_a - R_p$$

To first order in J_2 they are given by

$$\alpha = a - \frac{J_2}{2p}$$

$$\varepsilon = e + \frac{J_2(1+e^2)}{2aep}$$

The function

$$r(\phi) = \frac{\alpha(1-\varepsilon^2)}{1 + e \cos \gamma\phi}$$

represents a precessing ellipse of semi-major axis α and eccentricity ε. The argument of periapse advances by $2\pi/\gamma - 2\pi$, which falls short of the actual value by $\pi k^2/2\gamma$ as $J_2 \to 0$. Hence, although this form is intuitively attractive and correctly indicates the limits of the motion, it cannot adequately replace Eq. 6.11 as an approximation to the orbit—even in lowest order. In particular, it reproduces all of Eq. 6.11 except for the e^3 term, which is now $e^3(1 - \phi \sin \phi)$ instead of $e^3(1 - \sin^2 \phi)$.

Circular Orbits

If the orbit is a circle, then r is a constant. From Eq. 6.9 this implies that $dn^2(u, k)$ is proportional to $1 - m \, sn^2(u, k) \forall u$. As $dn^2(\psi, \kappa) = 1 - \kappa^2 \, sn^2(\psi, \kappa)$, this is possible if and only if the proportionality constant is unity and $m = k^2$. For m to be equal to k^2, then (1) $k^2 = 0$, $m = 0$, and $R_a = R_p$ or (2) $k^2 = 1$, $m = 1$, and $R_p = R_0$. In either case $f(r)$ has a double root. It is a simple matter to show from Eq. 6.2 that if $f(r) = (r - R_1)^2 (r - R_2)$ and one applies a small perturbation to the orbit $r = R_1$, then for this orbit to be stable it must be that $R_1 > R_2$. As unstable orbits are of no interest in this discussion, it follows that the interesting case is (1) with $k^2 = m = 0$ and $r = R_a = R_p > 0$.

From the analytical solution of a cubic equation it is known that a cubic has a double root when its discriminant vanishes. From Eq. 6.4 one can see that forcing $Q^3 + R^3$ to vanish implies a relationship between J_2, a, and e. Since J_2 is a given of the problem and circularity is a statement concerning angular momentum rather than energy, it is really a relationship for $e(J_2, a)$. As $Q^3 + R^2 = 0$ is a cubic in e^2 but a quadratic in J_2, we will deal with $J_2(a, e)$. A vanishing discriminant implies that

$$\frac{27 J_2}{2a^2} = 1 - 9e^2 \pm (1 + 3e^2)^{3/2}$$

Since J_2 is real and nonnegative, it must be that $e^2 \geq -\frac{1}{3}$ (cf. above). Because L is real and positive, it must also be true that $1 \geq e^2$. If we take the upper sign, then J_2 is always ≥ 0 for $e^2 \in [-\frac{1}{3}, 1]$, whereas if we take the lower sign, then J_2 is ≥ 0 only for $e^2 \in [-\frac{1}{3}, 0]$.

If we actually solve for the roots of $f(r) = 0$ under the constraint $Q^3 + R^2 = 0$, we find that they are $(a/3)[2 \pm 2(1 + 3e^2)^{1/2}]$ and $(a/3)[2 \mp (1 + 3e^2)^{1/2}]$ (twice) with the signs in the same order as above [i.e., in the $J_2(a, e)$ quadratic]. The stability analysis demonstrated that the physically interesting solution demanded that the double root be the larger. This occurs for the lower sign and

$$R_0 = \left(\frac{2a}{3}\right)[1 - (1 + 3e^2)^{1/2}]$$

$$R_a = R_p = \left(\frac{a}{3}\right)[2 + (1 + 3e^2)^{1/2}] \quad \text{with } e^2 \in [-\tfrac{1}{3}, 0]$$

Note that if $e^2 = -1/3$, then f has a triple root at $2a/3$ and $J_2 = 8a^2/27$. Figure 28 shows f for $e^2 = -1/9$, $a = 1$, and $J_2 = 0.107827329$. Note also that if $e^2 = 0$, $J_2 = 0$, then we return to the familiar meaning of a circular orbit as one with zero eccentricity. Remember, though, that if $J_2 = 0$ then a circular orbit is the one with maximum angular momentum. With $J_2 \neq 0$, again the angular momentum is larger for a circular orbit (because $e^2 \leq 0$) than it is for a noncircular one ($e^2 \geq 0$). The triple root represents an unstable orbit but has the maximum angular momentum.

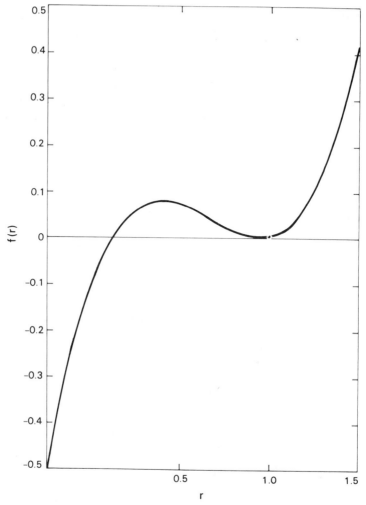

Figure 28. An $f(r)$ curve with a double root ($a = 1$, $e^2 = -\frac{1}{9}$, $J_2 = 0.107827329$).

A last note: Once J_2 is large enough, $Q^3 + R^2 > 0$, $e^2 \in [-\frac{1}{3}, 1]$, and f has a single real root and a pair of complex conjugate roots. The single real root represents a circular orbit. This happens for $a^2/10 < J_2 < a^2/2$. As $J_2 \to \infty$, the radius of this unstable orbit $\to (aJ_2)^{1/3} + 2a/3$.

The Time Dependence of ϕ and r

The Exact Solution for ϕ and the Period

By combining $L = r^2\dot{\phi}$ and Eq. 6.9,

$$\dot{\phi} = \frac{(L/R_p^2)[1 - m \, \mathrm{sn}^2(u, k)]}{\mathrm{dn}^4(u, k)}$$

where $u = \gamma\phi/2$. The initial condition is $\phi = 0$ when $t = T$. If the incomplete elliptic integrals of the second and third kinds, $E(\psi, \kappa)$ and $\Pi(\psi, n, \kappa)$ are introduced, then the above can be integrated to obtain

$$
\begin{aligned}
\frac{2\pi(t-T)}{P} = \Big\{ & [R_0^2 - R_a R_p + R_0(R_a + R_p)]F(\beta, k) \\
& + R_p(R_a - R_0)E(\beta, k) + (R_p - R_0)(R_a + R_p + R_0)\Pi(\beta, m, k) \\
& - \frac{R_p(R_a - R_p)\sin\beta\cos\beta(1 - k^2\sin^2\beta)^{1/2}}{1 - m\sin^2\beta} \Big\} \Big/ [a^2 R_p(R_a - R_0)]^{1/2}
\end{aligned}
$$

Here $P^2 \equiv 4\pi^2 a^3/\mu$ (i.e., $P =$ the period were $J_2 = 0$). The angle β is related to ϕ via $\sin\beta = \mathrm{sn}(u, k)$ or $\beta = \mathrm{am}(u, k)$ where $\mathrm{am}(\psi, k)$ is the amplitude function of argument ψ, modulus κ.

Now if $t = T + t_a$, then $\phi = \phi_a$ and $u = \gamma\phi_a/2 = K(k)$. Note that $\mathrm{sn}[K(k), k] = 1$ so $\beta = \pi/2$. As the incomplete elliptic integrals $F(\psi, \kappa)$, $E(\psi, \kappa)$, and $\Pi(\psi, n, \kappa)$ become the corresponding complete elliptic integrals $K(\kappa)$, $E(\kappa)$, and $\Pi(n, \kappa)$ when their argument is $\pi/2$,

$$
\begin{aligned}
\frac{2\pi t_a}{P} = \{ & [R_0^2 - R_a R_p + R_0(R_a + R_p)]K(k) + R_p(R_a - R_0)E(k) \\
& + (R_p - R_0)(R_a + R_p + R_0)\Pi(m, k)\}/[a^2 R_p(R_a - R_0)]^{1/2} \quad (6.12)
\end{aligned}
$$

Similarly, if $t = T + t_p$, then $\phi = \phi_p$, $u = \gamma\phi_p/2 = 2K(k)$. But $\mathrm{sn}[2K(k), k] = 0$ so $\beta = \pi$. Because all of the incomplete elliptic integrals obey the addition theorem $I(\pi, \kappa) = 2I(\pi/2, \kappa)$, it follows immediately that $t_p = 2t_a$. Therefore, not only is the orbit periodic but the time development of ϕ is periodic too. Furthermore, the period is t_p and it requires as much time for ϕ to increase from 0 to ϕ_a as it does for ϕ to increase from ϕ_a to ϕ_p. For $a = 1$, $e = \frac{1}{3}$, and $J_2 = \frac{1}{10}$, $t_p/P = 0.976979$.

Our next quest is the value of t_p as $J_2 \to 0$. From Eq. 6.12 and the fact that as $\kappa^2 \to 0$.

$$
K(\kappa) \to \frac{\pi}{2}\left(1 + \frac{\kappa^2}{4}\right)
$$

$$
E(\kappa) \to \frac{\pi}{2}\left(1 - \frac{\kappa^2}{4}\right)
$$

$$
\Pi(n, \kappa) \to \frac{\pi}{2}(1 - n)^{-1/2}\left\{1 + \left(\frac{\kappa^2}{2}\right)[1 + (1 - n)^{-1/2}]^{-1}\right\}
$$

We can show, laboriously, that

$$
t_p = 2t_a \to P + \text{terms of order } J_2^2
$$

Since $P^2 = 4\pi^2 a^3/\mu$, it follows that there is no secular change in a, because of J_2, to first order in J_2. This is a standard result of perturbation theory. The same statement is true of the eccentricity and can be derived herein from the constancy of L (by definition) and the absence of a first-order term in P.

The Exact Solution for r

We can write Eq. 6.2 as

$$\dot{r}^2 = -\left(\frac{2\pi a}{P}\right)^2\left[\frac{rf(r)}{r^4}\right]$$

Remembering the discussion above for the limits of the orbit, one should realize that the solution to this differential equation is

$$\int_{R_p}^{r}\frac{s^2\,ds}{[-sf(s)]^{1/2}}=\frac{2\pi a}{P}\int_T^t d\tau,\qquad t-T\in[0,t_a]$$

$$\int_{R_a}^{r}\frac{s^2\,ds}{[-sf(s)]^{1/2}}=\frac{-2\pi a}{P}\int_{T+t_a}^t d\tau,\qquad t-T\in[t_a,t_p]$$

In order to actually integrate this, the following result is needed: If $P(u)$ is the quartic polynomial $P=a_0u^4+a_1u^3+a_2u^2+a_3u+a_4$, $a_0\neq 0$ and k any integer greater than 1, then

$$2(k-1)a_0\int^v\frac{u^k\,du}{[P(u)]^{1/2}}=2v^{k-3}P^{1/2}(v)+\sum_{j=1}^4(2+j-2k)a_j\int^v\frac{u^{k-j}\,du}{[P(u)]^{1/2}}$$

Here $P(s)=-sf(s)$, $a_0=-1$, $a_1=2a$, $a_2=-a^2(1-e^2)$, $a_3=aJ_2$, $a_4=0$, and $k=2$. Consequently,

$$\frac{2\pi(t-T)}{P}=\left\{2[(R_p-R_0)\Pi(\lambda,m,k)+R_0F(\lambda,k)]\right.$$

$$\left.+\left(\frac{R_a}{a}\right)[(R_p-R_0)\Pi(\lambda,k^2,k)-R_pF(\lambda,k)]\right\}\Big/[R_p(R_a-R_0)]^{1/2}$$

$$-\left(\frac{-F(r)}{a^2r}\right)^{1/2}$$

for $t-T\in[0,t_a]$ and

$$\frac{2\pi(t-T-t_a)}{P}=\left\{2R_a\Pi(\Lambda,m',k)-\left(\frac{J_2}{k^2R_a}\right)\right.$$

$$\times[(k^2-m')F(\Lambda,k)$$

$$\left.+m'E(\Lambda,k)]\right\}\Big/[R_p(R_a-R_0)]^{1/2}+\left(\frac{-F(r)}{a^2r}\right)^{1/2}$$

for $t-T\in[t_a,t_p]$. Here $m'=-(m-k^2)/(1-m)=1-R_a/R_p$. One could show that $2t_a=t_p$ again and demonstrate the absence of a first-order term in t_p/P as $J_2\rightarrow 0$. As in the $J_2=0$ problem these two expressions are not identical; it is $r(t)$, $\phi(t)$, and $r(\phi)$ that hold over both parts of the orbit, not their inverses.

Second-Order Solutions

The practical problem for artificial satellites orbiting the Earth is the solution for the orbit including the J_2^2 and J_4 terms because, for the Earth, $J_2^2 \approx |J_4|$, $|J_3|$. The potential now has the form (note that the J_3 term produces no force if $\theta = 0$)

$$U(\mathbf{r}) = -\left(\frac{GM}{r}\right)\left[1 + \left(\frac{J_2}{2r^2}\right)(1 - 3\sin^2\theta) + \left(\frac{J_3}{2r^3}\right)(3 - 5\sin^2\theta)\sin\theta\right.$$
$$\left. - \left(\frac{J_4}{8r^4}\right)(3 - 30\sin^2\theta + 35\sin^4\theta)\right]$$

The full equations of motion, 6.1b,c,d, have already been given. Anthony and Perko (1961) have obtained the solution to this problem in a straightforward fashion. Their method is common, will be illustrated several times in this book (including the polar orbit problem discussed below), and leads to many long equations. We will not reproduce their work here. Brouwer (1946) also looked at this problem and solved it in a rotating reference frame. He takes a circular orbit as the zeroth-order solution, transforms into the corotating reference system, and then solves the problem. This is reminiscent of Hill's lunar theory and offers additional insights. Finally Anthony and Fosdick (1961) solve the problem to second order for *inclined* orbits.

Second-Order Effects

An equatorial section of the Earth does not behave as a circular section in its gravitational effects. That is, in Eq. 1.22a, ρ_{22}, ρ_{2-2} are not both zero. The effects of equatorial ellipticity are more easily studied via conventional perturbation theory; see Chapter 10.

A Polar Orbit

A problem of interest (both theoretically and practically—dozens of artificial satellites are in close, nearly polar orbits) is a polar orbit under the potential 6.1a. The explicit form of Eqs. 6.1 is

$$\ddot{r} - r\dot{\theta}^2 - r\dot{\phi}^2\cos^2\theta = -\frac{GM}{r^2}\left[1 + \frac{3J_2(1 - 3\sin^2\theta)}{2r^2}\right]$$

$$\frac{1}{r}\frac{d}{dt}(r^2\dot{\theta}) + r\dot{\phi}^2\sin\theta\cos\theta = -\left(\frac{3GMJ_2}{r^4}\right)\sin\theta\cos\theta$$

$$\frac{1}{r\cos\theta}\frac{d}{dt}(r^2\dot{\phi}\cos^2\theta) = 0$$

Above we found the solution for $\theta = 0 \forall t$. Now we look at an approximate solution that has $\phi = \text{const} \forall t$. This will allow me to illustrate a useful device of Lindstedt's (1882) on eliminating secular terms.

We specialize to $\phi = 0$ (a rotation of the coordinate system about the oblate spheroid's axis of symmetry can insure this) and initial conditions in the equatorial plane with the particle moving horizontally,

$$r(0) = r_0, \qquad \dot{r}(0) = 0$$

$$\theta(0) = 0, \qquad \dot{\theta}(0) = \frac{v_0}{r_0}$$

(6.13)

The simplified equations of motion are

$$\ddot{r} - r\dot{\theta}^2 = -\left(\frac{GM}{r^2}\right)\left[1 + \frac{3J_2(1 - 3\sin^2\theta)}{2r^2}\right]$$

$$\frac{d}{dt}(r^2\dot{\theta}) = -\left(\frac{3GMJ_2}{r^3}\right)\sin\theta\cos\theta$$

(6.14)

A classic change of variables in the two-body problem is $u = 1/r$ (try it in Eq. 2.3a). In addition, introduce $U = du/d\theta$ and $h = r^2\dot{\theta}$ and change independent variables from θ to t (try this too for the two-body problem). The coupled set 6.14 becomes

$$h^2\frac{dU}{d\theta} + hU\frac{dh}{d\theta} + h^2u = GM\left[1 + \frac{3J_2u^2(1 - 3\sin^2\theta)}{2}\right]$$

$$h\frac{dh}{d\theta} = -3GMJ_2u\sin\theta\cos\theta$$

$$U = \frac{du}{d\theta}$$

To obtain a first-order solution in terms of J_2, u, U, and h are expanded in a power series in J_2. In addition, to remove any possible secular terms, the independent variables are simultaneously changed from θ to χ, namely (remember that J_2 has dimensions, the choice of numerical factors simplifies the final formulas, and R is the equatorial radius of the oblate spheroid),

$$u = u_0(\chi) + \left(\frac{3J_2}{2R^2}\right)u_1(\chi) + \cdots$$

$$U = U_0(\chi) + \left(\frac{3J_2}{2R^2}\right)U_1(\chi) + \cdots$$

$$h = h_0(\chi) + \left(\frac{3J_2}{2R^2}\right)h_1(\chi) + \cdots$$

$$\theta = \chi\left[1 + \left(\frac{3J_2}{2R^2}\right)\theta_1 + \cdots\right]$$

where θ_1 is a constant. Performing the above substitution and isolating powers of J_2 results in (prime implies $d/d\chi$)

$$h_0' = 0$$

$$u_0' = U_0$$

$$h_0' U_0 + h_0(U_0' + u_0) = \frac{GM}{h_0}$$

$$h_1' h_0 = -GMR^2 u_0 \sin 2\chi$$

$$u_1' = U_1 + U_0 \theta_1$$

$$h_0(U_1' + u_1 + u_0\theta_1) + h_1(U_0' + u_0) + h_0' U_1 + h_1' U_0$$

$$= \left(\frac{GM}{h_0}\right)\left[-\frac{h_1}{h_0} + \theta_1 + u_0^2 R^2(1 - 3\sin^2\chi)\right]$$

The initial conditions 6.13 imply that

$$u_0 = \frac{1}{r_0}, \qquad h_0 = r_0 v_0$$

$$u_1 = u_1' = u_0' = U_0 = U_1 = h_1 = 0$$

We introduce $\Gamma = r_0^{1/2} v_0 / \sqrt{GM}$ and an arbitrary constant of integration e (the pseudo eccentricity). Then, upon integration of the new set of equations,

$$h_0 = r_0 v_0$$

$$u_0 = \frac{1 + e \cos \chi}{r_0 \Gamma^2}$$

$$h_1 = -\frac{r_0 v_0}{\Gamma^4}\left(\frac{R}{r_0}\right)^2 \frac{(3 + 4e) - 3e \cos \chi - 3\cos 2\chi - e \cos 3\chi}{6}$$

Using these results the u_1' differential equation takes the form

$$u_1'' + u_1 = \alpha + \beta$$

where

$$\alpha = -\left(\frac{e}{r_0 \Gamma^2}\right)\left[2\theta_1 + \left(\frac{R}{r_0 \Gamma^2}\right)^2\right]\cos \chi$$

$$r_0 \beta = \left(\frac{R}{r_0 \Gamma^3}\right)^2\left[\left(\frac{1}{2} + \frac{4e}{3} - \frac{e^2}{8}\right) + \frac{1 + e^2}{2}\cos 2\chi + \frac{5}{3}e \cos 3\chi + \frac{5}{8}e^2 \cos 4\chi\right]$$

Now θ_1 is a constant as yet undetermined. The choice

$$\theta_1 = -\frac{(R/r_0 \Gamma^2)^2}{2}$$

makes α vanish and simplifies the differential equation for u_1. We make this choice now (and this is Lindstedt's point) so that

$$\theta = \left[1 - \left(\frac{3J_2}{4\Gamma^4 r_0^2}\right)\right]\chi$$

and

$$r_0 u_1 = \left(\frac{R}{r_0\Gamma^3}\right)^2 \left[\left(\frac{1}{2} + \frac{4e}{3} - \frac{e^2}{8}\right) + \left(-\frac{1}{3} - \frac{9e}{8} + \frac{e^2}{3}\right)\cos\chi\right.$$
$$\left. - \frac{1+e^2}{6}\cos 2\chi - \frac{5e}{24}\cos 3\chi - \frac{e^2}{24}\cos 4\chi\right]$$

Note that since r, h, and the tangential speed v are periodic functions of χ (with period 2π), they are periodic functions of the latitude θ but with period $2\pi[1 - 3J_2 R^2/(4\Gamma^4 r_0^2)]$. Hence, the line of apsides regresses. The problems at the end of this chapter extend the analysis to arbitrary initial latitudes, nearly circular orbits, and so on. See also Anthony and Fosdick (1961).

The final results for r, θ, the speed v, and the quantity $h = r^2\dot\theta$ are

$$\theta = \left[1 - \frac{3J_2}{4\Gamma^4 r_0^2}\right]\chi$$

$$\frac{r}{r_0} = \frac{1+e}{1+e\cos\chi}\left(1 - \frac{J_2/(4r_0)^2}{(1+e)(1+e\cos\chi)^2}f(\chi)\right)$$

$$v = h\left[u^2 + \left(\frac{du}{d\theta}\right)^2\right]^{1/2}$$

$$= \frac{(1+2e\cos\chi + e^2)^{1/2}}{1+e}v_0\left(1 + \frac{J_2}{32r_0^2}\frac{g(\chi)}{(1+e)^2(1+2e\cos\chi + e^2)}\right)$$

$$\frac{h}{r_0 v_0} = 1 - \frac{J_2}{6\Gamma^4 r_0^2}[3 + 4e - 3e\cos\chi - 3\cos 2\chi - e\cos 3\chi]$$

wherein the functions f and g are of the form of truncated Fourier cosine series,

$$f(\chi) = 12 + 32e - 3e^2 - (8 + 27e - 8e^2)\cos\chi - 4(1 + e^2)\cos 2\chi$$
$$- 5e\cos 3\chi - e^2\cos 4\chi$$
$$g(\chi) = 16e(1 - 3e - e^2) - 2(8 + 21e - 8e^2 - 3e^3)\cos\chi + 16(1 + e^2)\cos 2\chi$$
$$+ e(26 + 7e^2)\cos 3\chi + 16e^2\cos 4\chi + 3e^2\cos 5\chi$$

Hamilton–Jacobi Theory

It would appear from this text that there is no perspective on classical mechanics other than that of Newton's three laws of motion. This is not the case and there are several other formulations of the foundations of mechanics. These both extend Newtonian mechanics by illuminating different aspects of the subject and sometimes greatly simplify the solution of certain types of problems. One such alternative is that of Hamilton–Jacobi theory (whose fundamental equation provides a bridge from classical mechanics to quantum mechanics). In the artificial satellite problem it does one very important thing—it provides that most general potential for which the nonspherically symmetric two-body problem may be solved. In particular, by studying the separability of the Hamilton–Jacobi equation in spherical coordinates (it is a partial differential equation) Garfinkel (1964) was able to prove that the potential

$$V = -\left(\frac{GM}{r}\right)\left\{1 - \left(\frac{3J_2}{2r}\right)\left[\alpha(\sin^2\theta - \beta) + \gamma r + \frac{\delta}{r}\right]\right\}$$

is the most general function $V(r, \theta)$ such that (1) it preserves the gross features of the principal part of the oblateness potential U,

$$U(r, \theta) = -\left(\frac{GM}{r}\right)\left[1 + \frac{J_2 P_2(\sin\theta)}{r^2}\right]$$

(2) the Hamilton–Jacobi equation separates, (3) it leads to a solution of the Hamilton–Jacobi equation in terms of (no worse than) elliptic functions, and (4) there exists a choice of the parameters α, β, γ, and δ such that all of the first-order secular effects of J_2 can be incorporated.

Obviously $V \neq U$; indeed the leading J_2 dependence is wrong ($1/r^2$ vs. $1/r^3$; remember that the $1/r^2$ term in U was eliminated by choosing the center of mass of the oblate primary as the origin of the reference frame). We know, however, that the main problem of artificial satellite theory is insoluble in spherical coordinates. Another interesting potential was invented by Aksnes (1965)

$$V_A = -\left(\frac{GM}{r}\right)\left[1 - \frac{J_2 P_2(\sin\theta)}{pr^2}\right]$$

where $p = a(1 - e^2)$. The time average, over the unperturbed ellipse, of $V_A - U$ is zero. Further work along these lines can be found in Sterne (1957, 1958), Garfinkel (1958, 1959), Garfinkel and Aksnes (1970), and Aksnes (1970).

TWO FIXED CENTERS

Imagine two finite, spherically symmetric objects with masses M and M'. Suppose them to be held, in an inertial coordinate system, on the x axis, say at $(a, 0, 0)$ and $(-a, 0, 0)$. The potential of these objects at any point $\mathbf{r}$ in space will be

$$U(\mathbf{r}) = -\frac{GM}{|\mathbf{r} - a\mathbf{i}|} - \frac{GM'}{|\mathbf{r} + a\mathbf{i}|} \tag{6.15}$$

where $\mathbf{i}$ is the unit vector in the positive x direction. We now release a third body of infinitesimal mass m. It moves under the combined influence of these two bodies while not influencing M or M'. This is the physical statement of the problem of two fixed centers. An immediate specialization is to the case where all three objects remain coplanar, say in the plane $z = 0$.

The applicability of this to the perturbations of an artificial satellite by the Moon or the Sun should be obvious. The relevant approximation is that $P/P_{\text{(}}$ or $P/P_\odot \ll 1$ where P is the period of revolution of the artificial satellite about the Earth and $P_{\text{(}}$, $P_\odot$ are the same quantities for the Moon and Sun. As most artificial satellites have $P \leqslant 1$ day and most have $P \leqslant 6$ hr, the approximation is a very good one. The next step would be to exploit the fact that $\varepsilon = (GM/4a^2)/(GM_\oplus/D^2) \ll 1$ where D is the Earth-satellite distance, M is $M_{\text{(}}$ or $M_\odot$, and $2a$ is the Earth-Moon or Earth-Sun distance. For a geosynchronous satellite $\varepsilon_{\text{(}} = 0.000148$, $\varepsilon_\odot = 0.0261$. For a 2-hr-period satellite $\varepsilon_{\text{(}} = 0.000954$, $\varepsilon_\odot = 0.00000541$.

Unfortunately, the analytical solution of the planar two-fixed-center problem is so obtuse, with its own naturally occurring distance scale (i.e., a), that it appears to be useless as a computational aid. Here computational means in either the analytical or numerical sense. Considering how long the solution has been known and the total absence of any presentation such as the above for the $J_2 \neq 0$ problem, we will not spend too much time on it. The analytical first-order treatment in Chapter Nine is of some interest.

The Planar Solution

The natural coordinate system for the potential 6.15 is the elliptic one ξ, η. These coordinates are defined in terms of the rectangular coordinates x, y via

$$x = a \cosh \xi \cos \eta, \qquad y = a \sinh \xi \sin \eta$$

The curves $\xi = \text{const}$ are ellipses, those for $\eta = \text{const}$ are hyperbolas. The potential $(z = 0)$ in these coordinates takes the simple form

$$U = -\frac{GM/a}{\cosh \xi - \cos \eta} - \frac{GM'/a}{\cosh \xi + \cos \eta}$$

The equations of motion $\ddot{\mathbf{r}} = -\nabla U$ can be obtained by a straightforward coordinate transformation.* One gets for ξ

$$a^2 \frac{d}{dt}[(\cosh^2 \xi - \cos^2 \eta)\dot{\xi}] - a^2(\dot{\xi}^2 + \dot{\eta}^2) \cosh \xi \sinh \xi = -\frac{\partial U}{\partial \xi}$$

or

$$a^2 \frac{d}{dt}[(\cosh^2 \xi - \cos^2 \eta)^2 \dot{\xi}^2] - 2a^2 \dot{\xi}(\dot{\xi}^2 + \dot{\eta}^2)(\cosh^2 \xi - \cos^2 \eta) \cosh \xi \sinh \xi$$

$$= -2\dot{\xi}(\cosh^2 \xi - \cos^2 \eta)\frac{\partial U}{\partial \xi}$$

---

* This is inelegant and inefficient. The best way to do this is via Lagrange's equations.

Using the equation of energy conservation, $K + U = \mathscr{E}$, where K is the kinetic energy,

$$a^2 \frac{d}{dt}[(\cosh^2 \xi - \cos^2 \eta)^2 \dot{\xi}^2] = -2\dot{\xi}(\cosh^2 \xi - \cos^2 \eta)\frac{\partial U}{\partial \xi}$$

$$+ 2(\mathscr{E} - U)\dot{\xi}\frac{\partial}{\partial \xi}(\cosh^2 \xi - \cos^2 \eta)$$

$$= 2\dot{\xi}\frac{\partial}{\partial \xi}[(\mathscr{E} - U)(\cosh^2 \xi - \cos^2 \eta)]$$

$$= 2\dot{\xi}\frac{\partial}{\partial \xi}\left[\mathscr{E}(\cosh^2 \xi - \cos^2 \eta)\right.$$

$$+ \frac{GM}{a}(\cosh \xi + \cos \eta)$$

$$\left. + \frac{GM'}{a}(\cosh \xi - \cos \eta)\right]$$

$$= 2\frac{d}{dt}\left[\mathscr{E}\cosh^2 \xi + \frac{G(M + M')}{a}\cosh \xi\right]$$

because

$$K = \tfrac{1}{2}\dot{x}^2 + \tfrac{1}{2}\dot{y}^2$$

$$= \tfrac{1}{2}a^2(\dot{\xi}^2 + \dot{\eta}^2)(\cosh^2 \xi - \cos^2 \eta)$$

Integrating,

$$\frac{a^2}{2}(\cosh^2 \xi - \cos^2 \eta)^2 \dot{\xi}^2 = \mathscr{E}\cosh^2 \xi + \frac{G(M + M')}{a}\cosh \xi - \varepsilon$$

where ε is the constant of integration.

Subtracting this from the equation of energy conservation, which can be written as

$$\frac{a^2}{2}(\dot{\xi}^2 + \dot{\eta}^2)(\cosh^2 \xi - \cos^2 \eta)^2 = \mathscr{E}(\cosh^2 \xi - \cos^2 \eta) + \frac{GM}{a}(\cosh \xi + \cos \eta)$$

$$+ \frac{GM'}{a}(\cosh \xi - \cos \eta)$$

we deduce that

$$\frac{a^2}{2}(\cosh^2 \xi - \cos^2 \eta)^2 \dot{\eta}^2 = -\mathscr{E}\cos^2 \eta - \frac{G(M' - M)}{a}\cos \eta + \varepsilon$$

Finally, eliminating dt between these two equations, we have the separated, first-order differential equation

$$\frac{(d\xi)^2}{\mathscr{E}\cosh^2\xi+[G(M+M')/a]\cosh\xi-\varepsilon}$$

$$=\frac{(d\eta)^2}{-\mathscr{E}\cos^2\eta-[G(M'-M)/a]\cos\eta+\varepsilon}$$

Each side can now be integrated in terms of elliptic functions. See Langebartel (1981) for an example of using this as an intermediate orbit.

THE THREE-BODY PROBLEM

The problem of two fixed centers is unrealistic for two reasons—the attracting bodies are fixed and the third body exerts no gravitational force on them. If we relax the first constraint so that the two-force-exerting bodies move along their normal Keplerian orbits (conic sections about their common center of mass), then we have the three-body problem. This serves as a much more realistic model for the motion of an artificial satellite within lunar space, or for the motion of a planetary moon in the planet–Sun system, or for the motion of an asteroid in the Jupiter–Sun system. The three-body problem is the simplest physically meaningful problem of classical celestial mechanics that is not soluble in closed form [the simplest physically meaningful problem of modern celestial mechanics is known as the main problem of artificial satellite theory (i.e., $J_2\neq0$ only) and it too is not soluble in closed form]. There is a rich literature on the three-body problem, an excellent introductory text by Szebehely (1967), much numerical work, analytical results concerning periodic solutions, and so forth. Moreover, just as there is a special case of the main problem of artificial satellite theory that is completely soluble in closed form, there are two special cases of the (planar, restricted) three-body problem that are completely soluble in closed form. In one set of cases (three physical solutions) the three bodies are collinear; in the other set of cases (two physical solutions) the three bodies are at the vertices of an equilateral triangle. The latter solutions are especially important philosophically (Lagrange found them in 1772). That nature actually uses them in the Jupiter–Sun system for minor planets was not known until Max Wolf found 588 Achilles on February 22, 1906. Celestial mechanics has *predicted* very few things and this is one of them.

The general problem of three bodies is usually confined to the restricted problem of three bodies by forcing coplanarity of the three bodies and circular orbits on the massive objects. Including non-coplanar motions, allowing for elliptical orbits for the primaries, allowing one of the primaries to be oblate, and so on, represent various extensions of the problem. Given Szebehely's book and a nice introduction to the subject in Moulton's (1902) text, we point out the highlights and use the subject as a vehicle to introduce a more general form of stability analysis than heretofore exhibited. We will follow Szebehely's notation in order to ease the reader's transition to that reference. Additional references can be found in Szebehely (1973).

Formulation

Two spherically symmetric objects revolve about their center of mass in circular orbits as a result of their mutual gravitational attraction. A third body, which always lies in the orbital plane of the other two, feels the gravitational attraction of the other two but does not influence their motion. Let the masses of the primaries be m_1 and m_2. If their circular orbits have radii b and a about their common center of mass (see Fig. 29), then ($k^2 = G$)

$$\frac{k^2 m_1 m_2}{l^2} = m_2 a n^2 = m_1 b n^2$$

where $l = a + b$ and n is their mean motion or angular speed of revolution. Accordingly,

$$k^2 m_1 = a n^2 l^2, \qquad k^2 m_2 = b n^2 l^2$$

and Kepler's third law takes the form

$$k^2(m_1 + m_2) = n^2 l^3$$

Furthermore, if $M = m_1 + m_2$, then

$$a = \frac{m_1 l}{M}, \qquad b = \frac{m_2 l}{M}$$

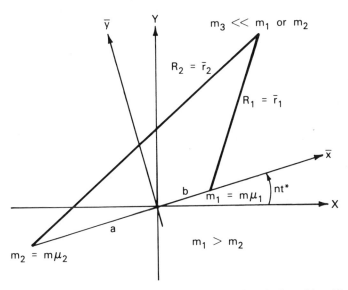

Figure 29. The coordinate system for the restricted, planar, three-body problem. The infinitesimal mass is m_3 and at $t = 0$ m_1 is at the right.

Let $\mathbf{R} = (X, Y)$ be the location of m_3 in the inertial reference frame and let t^* denote the time. Two reference frames and both dimensional and non-dimensional units will be needed, hence symbols are at a premium. The equation of motion for m_3 are

$$\ddot{\mathbf{R}} = \nabla F \tag{6.16}$$

where F is the negative of the potential (the disturbing function of celestial mechanics, see Chapters Nine and Ten). Here

$$F = \frac{k^2 m_1}{R_1} + \frac{k^2 m_2}{R_2}$$

$$R_i^2 = (X - X_i)^2 + (Y - Y_i)^2, \qquad i = 1, 2$$

and m_1 is at (X_1, Y_1), m_2 is at (X_2, Y_2).

Let us choose the origin of t^* such that both m_1 and m_2 are on the X axis at that instant. Then

$$X_1 = b \cos nt^*, \qquad Y_1 = b \sin nt^*$$

$$X_2 = -a \cos nt^*, \qquad Y_2 = -a \sin nt^*$$

The fact that the locations of m_1 and m_2 are time dependent makes this problem nasty to handle in general. The explicit form of Eq. 6.16 is

$$\frac{d^2 X}{dt^{*2}} = -k^2 \left[m_1 \frac{(X - b \cos nt^*)}{R_1^3} + m_2 \frac{(X + a \cos nt^*)}{R_2^3} \right]$$

$$\frac{d^2 Y}{dt^{*2}} = -k^2 \left[m_1 \frac{(Y - b \sin nt^*)}{R_1^3} + m_2 \frac{(Y + a \sin nt^*)}{R_2^3} \right]$$

The Rotating Coordinate System

It would facilitate matters if the explicit time dependence above were eliminated. Such a transformation can be performed by moving into a noninertial coordinate system, namely one rotating about the origin with period $P = 2\pi/n$. So, if

$$\mathbf{R} = R_3(-nt^*)\bar{\mathbf{r}}$$

then

$$R_1 = [(\bar{x} - b)^2 + \bar{y}^2]^{1/2}, \qquad R_2 = [(\bar{x} + a)^2 + \bar{y}^2]^{1/2}$$

and Eq. 6.16 is transformed into

$$\frac{d^2 \bar{x}}{dt^{*2}} - 2n \frac{d\bar{y}}{dt^*} - n^2 \bar{x} = -k^2 \left[m_1 \frac{(\bar{x} - b)}{R_1^3} + m_2 \frac{(\bar{x} + a)}{R_2^3} \right]$$

$$\frac{d^2 \bar{y}}{dt^{*2}} + 2n \frac{d\bar{x}}{dt^*} - n^2 \bar{y} = -k^2 \left[\frac{m_1 \bar{y}}{R_1^3} + \frac{m_2 \bar{y}}{R_2^3} \right] \tag{6.17}$$

Note, however, that neither R_1 nor R_2 are explicitly time dependent because of the motion of m_1 or m_2, for in the rotating system they are fixed.

Jacobi's Integral. We multiply the top of Eqs. 6.17 by $d\bar{x}/dt^*$ and the bottom by $d\bar{y}/dt^*$, and add. With little difficulty, even in this cumbersome notation, one should arrive at, after an integration with respect to t^*,

$$\left(\frac{d\bar{x}}{dt^*}\right)^2 + \left(\frac{d\bar{y}}{dt^*}\right)^2 = 2F^* - C^* \tag{6.18}$$

where

$$F^* = \frac{n^2}{2}(\bar{x}^2 + \bar{y}^2) + k^2\left(\frac{m_1}{R_1} + \frac{m_2}{R_2}\right)$$

(Szebehely prefers to write R_1 as $\bar{r}_1$ and R_2 as $\bar{r}_2$ in the rotating coordinate system in order to emphasize that their time dependence is solely due to the movement of m_3.)

The quantity C^* is a constant of integration. The relationship Eq. 6.18 was first found by Jacobi (1836) and is known as Jacobi's integral—the constant is called Jacobi's constant. This relationship is extremely useful in characterizing possible motions, for the speed (a scalar and therefore an invariant under rotations or translations) must be nonnegative.

The Dimensionless Coordinate System

We can simplify the above discussion by introducing dimensionless quantities. In particular, we set

$$x = \frac{\bar{x}}{l}, \quad y = \frac{\bar{y}}{l}, \quad t = nt^*$$

$$r_1 = \frac{R_1}{l}, \quad r_2 = \frac{R_2}{l}, \quad \Omega = \frac{F^*}{l^2 n^2} + \frac{\mu_1 \mu_2}{2}$$

where $\mu_{1,2} = m_{1,2}/M$. Then the equations of motion are

$$\ddot{x} - 2\dot{y} = \frac{\partial \Omega}{\partial x}, \quad \ddot{y} + 2\dot{x} = \frac{\partial \Omega}{\partial y}$$

$$\Omega = \mu_1\left(\frac{r_1^2}{2} + \frac{1}{r_1}\right) + \mu_2\left(\frac{r_2^2}{2} + \frac{1}{r_2}\right) \tag{6.19}$$

$$= \frac{\mu_1 r_1^2 + \mu_2 r_2^2}{2} + \frac{\mu_1}{r_1} + \frac{\mu_2}{r_2}$$

and Jacobi's integral $(C = C^* + \mu_1 \mu_2)$ is

$$\dot{x}^2 + \dot{y}^2 = 2\Omega - C \tag{6.20}$$

Note that $\mu_1 + \mu_2 = 1$ so that one of them can be eliminated. This is done by placing the larger of m_1, m_2 to the right at $t = 0$ in Fig. 29. Hence $0 \leq \mu \leq \frac{1}{2}$.

Thus, finally,

$$\Omega = \frac{(1-\mu)r_1^2 + \mu r_2^2}{2} + \frac{1-\mu}{r_1} + \frac{\mu}{r_2}$$

$$r_1^2 = (x-\mu)^2 + y^2, \qquad r_2^2 = (x+1-\mu)^2 + y^2 \tag{6.21}$$

The Equilibrium Solutions

"Equilibrium" solutions in the rotating coordinate system refer to those solutions of Eqs. 6.19 and 6.21—of which Eq. 6.20 is a first integral reducing the planar, restricted three-body problem to a third-order system—for which $\mathbf{r} = (x, y)$ is a constant for all times. From the form of Eq. 6.19 this obviously implies that $\nabla \Omega = 0$ at such locations. These equations are precisely

$$x - \frac{(1-\mu)(x-\mu)}{r_1^3} - \mu \frac{(x+1-\mu)}{r_2^3} = 0$$

$$y \left(1 - \frac{1-\mu}{r_1^3} - \frac{\mu}{r_2^3} \right) = 0 \tag{6.22}$$

When $y = 0$, the top line of Eq. 6.22 is a cubic in x; hence there are three collinear equilibrium solutions. If $y \neq 0$, then in order to satisfy the lower line of Eq. 6.22, it must be that $r_1 = r_2 = 1$. Coincidentally, this satisfies the upper line too. Thus, there are a total of five equilibrium solutions, three collinear ones and two equilateral triangle ones. These are usually labeled L_1-L_5. The coordinates of L_4 are $x = \mu - \frac{1}{2}$, $y = +\sqrt{3}/2$ and those of L_5 are $x = \mu - \frac{1}{2}$, $y = -\sqrt{3}/2$. The equation for the x values of L_1, L_2, and L_3 is a quintic. L_1 lies to the left of the less massive primary, L_2 lies in between the two primaries, and L_3 lies to the right of the more massive primary. The orientation is still that of Fig. 29. One can show that

$$\mu - 2 \leqslant x_1 \leqslant \mu - 1$$

$$\mu - 1 \leqslant x_2 \leqslant \mu$$

$$\mu \leqslant x_3 \leqslant \mu + 1$$

Stability Analysis

There are many concepts of stability in dynamics (and mathematics). One kind has already been explored, wherein if a particle remained "close" to where it would have been, then the new motion would be stable. This concept can be formalized. Since the dimensionality of the problem is not relevant to the essentials of the formulation, we do so in one dimension.

Consider a particle on the x axis in the potential field $U(x)$. The equation of motion is

$$\ddot{x} = -\frac{dU}{dx}$$

Suppose this equation has a solution $x = a$. Is it stable? In order to answer this question, we need to define stability, and this is done in a semi-intuitive fashion. If the difference, over long times, between the new motion and where the particle would have been had it not been disturbed is small, then the original motion (or lack thereof) is stable.

In order to settle the question, the *approximate* solution of the equation of motion subject to a new initial condition will be investigated. Instead of $x = a$, $\dot{x} = 0$, $x = a + \delta x$, $\delta \dot{x} \neq 0$, will be used. Moreover, since $|\delta x|$ is small (in some ill-defined sense), we can expand U in a Taylor series about $x = a$,

$$U(a + \delta x) \simeq U(a) + U'(a)\, \delta x + \tfrac{1}{2} U''(a)(\delta x)^2 + \cdots$$

and as a consequence

$$\frac{dU(a + \delta x)}{dx} = U'(a) + U''(a)\, \delta x + \cdots$$

However $U'(a) = 0$ and since $\dot{x} = 0$,

$$\delta \ddot{x} = U''(a)\, \delta x + \cdots$$

This is the equation of motion for a harmonic oscillator of frequency ω, $\omega^2 = -U''(a)$. If ω is real, then δx is a sinusoidal function of t and the motion is stable. If ω is imaginary, then δx is an exponential function of t and, no matter how small its initial value is, the motion with the new initial condition will (eventually) depart from it by an arbitrarily large amount. Such motion is unstable. Should ω vanish, then we need to keep higher-order terms in the Taylor series expansion of U in order to decide the issue.

In several dimensions the analysis is similar. We need to examine the eigenvalues of the Hessian matrix (the matrix of all second partial derivatives) of the potential U. If the eigenvalue problem is written as

$$|U^2 - \lambda I| = 0$$

(where U^2 is the Hessian matrix of U, λ is an eigenvalue, and I is an appropriately dimensioned unit matrix), then the motion is said to be stable if all of the eigenvalues are real and unstable if at least one is imaginary. Note that this linearized stability analysis cannot be used to truly predict long-term motion—the neglected terms in U may come to dominate.

Stability of L_1-L_5. To discuss the stability of the equilibrium points, set $x = a + \xi$, $y = b + \eta$ in the equations of motion 6.19 and then only keep first-order terms in ξ and η. The result is ($\Omega_{xy} = \Omega_{yx}$)

$$\ddot{\xi} - 2\dot{\eta} = \Omega_{xx}(a, b)\xi + \Omega_{xy}(a, b)\eta$$

$$\ddot{\eta} + 2\dot{\xi} = \Omega_{xy}(a, b)\xi + \Omega_{yy}(a, b)\eta$$

where the notation $\Omega_{xy}(a, b)$ means $\partial^2 \Omega(x, y)/\partial x\, \partial y$ evaluated at $x = a$, $y = b$.

The eigenvalue problem is

$$\begin{vmatrix} \Omega_{xx} - \lambda^2 & \Omega_{xy} + 2\lambda \\ \Omega_{xy} - 2\lambda & \Omega_{yy} - \lambda^2 \end{vmatrix} = 0$$

or

$$\lambda^4 + (4 - \Omega_{xx} - \Omega_{yy})\lambda^2 + \Omega_{xx}\Omega_{yy} - (\Omega_{xy})^2 = 0$$

If we in turn set $(a, b) = (x, y)$ corresponding to $L_1, L_2, \ldots, L_5$, evaluate the derivatives of Ω there, and solve the above quartic, then we can directly determine the absence or presence of stability. The algebra is not enlightening. It turns out that all of the collinear points are unstable. Even in the rotating reference frame our physical intuition should have predicted this. About the equilateral triangle points (L_4 and L_5) the motion is stable for $\mu < [1 - \sqrt{69/9}]/2 = 0.038521$, unstable for μ larger than this value. When μ is equal to this special value, there are secular terms (e.g., terms proportional to t) in the solution. Therefore, these solutions are also unstable.

Other Aspects

We have exhibited five exact solutions to the equations of motion of the planar, restricted three-body problem. No other solutions are known in general (there are other special values of mass ratios and distance ratios that allow exact solutions but these are sterile). The motion of the infinitesimal particle cannot be found, but the existence of Jacobi's integral Eq. 6.20 restricts the possible realms of motion. The reason is clear—it must be that $\dot{x}^2 + \dot{y}^2 \geqslant 0$ or $2\Omega \geqslant C$. Jacobi's constant C is a function of the initial conditions. Depending on the value of C, different areas of the xy plane become accessible to the particle. Forbidden and allowed regions are separated by the curves $2\Omega = C$, the curves of zero velocity. We shall take a quick look at the different types of zero velocity curves and then briefly explore more complex three-body problems.

Zero-Velocity Curves

Consider separately retreat to infinity ($r_1, r_2 \to \infty$) or close approach ($r_1 \to 0$ or $r_2 \to 0$). In either case we can see from Eq. 6.21 that $\Omega \to \infty$. Thus, as $C \to \infty$, the zero velocity curves have three branches. We first discuss the tightly bound orbits and then the large orbits about both.

If $r_1 \to 0$, then the infinitesimal particle is closely bound to the more massive primary. As $r_1 \to 0$, $r_2 \to 1$ and $C \to 2(1 - \mu)/r_1$, so that the curves of zero velocity approach circles. On the other hand, if $r_2 \to 0$, then $r_1 \to 1$ and the zero-velocity curves are nearly circular around the less massive primary ($C \to 2\mu/r_2$). In either case the small body is tightly bound to one of the primaries and can only orbit near it.

As both r_1 and $r_2 \to \infty$, their difference approaches 0, so we can call either of them r. Now $C \to r^2$ and the Jacobi integral is simply $\dot{x}^2 + \dot{y}^2 = x^2 + y^2 - C$. Once again the curves of zero velocity are nearly circular. See Fig. 30. As r

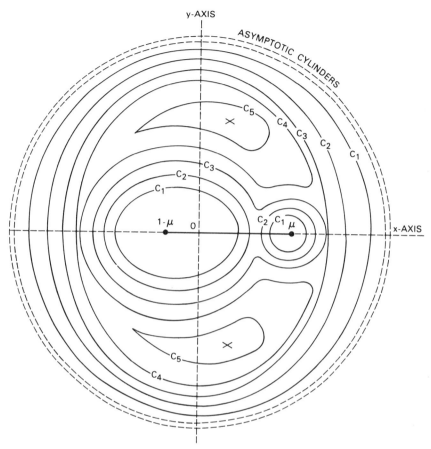

Figure 30. Curves of zero velocity for various values of the Jacobi constant, $C_1 > C_2 > \cdots > C_5$. The equilateral Lagrangian points are marked by the crosses. (Used by permission of Dover Publications, Inc., New York, from F. R. Moulton *An Introduction to Celestial Mechanics*, © 1914, 1942, 1970.)

decreases in distance from infinity, the value of C decreases. The curves of zero velocity become noticeably distorted circles and eventually branch about the equilateral triangle points. The interpretation of closed zero-velocity curves about L_4 and L_5 is that motion bound to these places is possible—that is, they are stable equilibrium solutions. Note that there are no such branches about the collinear solutions L_1, L_2, and L_3. The curves of zero velocity become increasingly complex for smaller values of $C(C_n > C_{n+1}$ in Fig. 30), allowing various horseshoe- and tadpole-shaped orbital regions. Eventually C increases again as $r_1 \to 0$ or $r_2 \to 0$.

Generalizations

One can generalize the planar, restricted, three-body problem in several ways. One can allow the primaries to revolve about their common center of mass in ellipses instead of in circles. Once one does this their angular motion becomes nonuniform and there no longer exists a unique coordinate system (obtained by a rotation) which freezes the primaries on a line. Moreover, their separation changes so that the analogs of r_1 and r_2 are now explicitly time dependent. Little theoretical work appears to have been done on this generalization.

A simpler generalization is to allow the infinitesimal body full three-dimensional motion. This preserves the special coordinate system obtained by a constant angular speed rotation. It also preserves the special solutions found above ($z = 0$ for all of them). The extra dimension complicates the analysis but does not render it beyond our capabilities. One reason is that to a certain extent the motion separates; this is exactly true for the linearized stability analysis whence these results are unchanged.

Hill's Problem. Hill examined a very special three-body problem in an attempt to gain insight into the Earth–Moon–Sun system. In particular, he further specialized the planar, restricted three-body problem by assuming that the equatorial horizontal parallax of the Sun vanished (i.e., the Sun is infinitely distant). Because Hill's problem is of some interest and its direct formulation illustrates several techniques of classical planetary theory, it will be presented in some detail. In an inertial reference frame the equations of motion of the Moon and of the Earth are

$$\ddot{\mathbf{r}}_\zeta = -\frac{GM_\oplus}{|\mathbf{r}_{\oplus\zeta}|^3}(\mathbf{r}_\zeta - \mathbf{r}_\oplus) - \frac{GM_\odot}{|\mathbf{r}_{\zeta\odot}|^3}(\mathbf{r}_\zeta - \mathbf{r}_\odot)$$

$$\ddot{\mathbf{r}}_\oplus = -\frac{GM_\odot}{|\mathbf{r}_{\oplus\odot}|^3}(\mathbf{r}_\oplus - \mathbf{r}_\odot) - \frac{GM_\zeta}{|\mathbf{r}_{\oplus\zeta}|^3}(\mathbf{r}_\oplus - \mathbf{r}_\zeta)$$

Here $\mathbf{r}_\oplus$, $\mathbf{r}_\zeta$, and $\mathbf{r}_\odot$ are the location vectors of the Earth, Moon, and Sun in the inertial system. Choose the Earth as the origin of a new coordinate system and define $\boldsymbol{\rho}_\zeta$ and $\boldsymbol{\rho}_\odot$ by

$$\boldsymbol{\rho}_\zeta = \mathbf{r}_\zeta - \mathbf{r}_\oplus, \qquad \boldsymbol{\rho}_\odot = \mathbf{r}_\odot - \mathbf{r}_\oplus$$

The equation of motion of the Moon relative to the Earth is then found to be

$$\ddot{\boldsymbol{\rho}}_\zeta + \frac{G(M_\oplus + M_\zeta)}{|\boldsymbol{\rho}_\zeta|^3}\boldsymbol{\rho}_\zeta = -GM_\odot\left(\frac{\boldsymbol{\rho}_\zeta - \boldsymbol{\rho}_\odot}{|\boldsymbol{\rho}_{\zeta\odot}|^3} + \frac{\boldsymbol{\rho}_\odot}{|\boldsymbol{\rho}_\odot|^3}\right) \tag{6.23}$$

If in the Earth-centered coordinate system $\boldsymbol{\rho}_\zeta = (X, Y, Z)$ and $\boldsymbol{\rho}_\odot = (X', Y', Z')$, then the x-component of Eq. 6.23 is

$$\ddot{X} + \frac{\kappa}{\rho_\zeta^3}X = -GM_\odot\left(\frac{X - X'}{\rho_{\zeta\odot}^3} + \frac{X'}{\rho_\odot^3}\right) \tag{6.24}$$

where $\kappa = G(M_\oplus + M_\zeta)$. The right-hand side of Eq. 6.24 may be written as the derivative of the disturbing function R,

$$\frac{\partial R}{\partial X} = \frac{\partial}{\partial X}\left[GM_\odot\left(\frac{1}{\rho_{\zeta\odot}} - \frac{\boldsymbol{\rho}_\zeta \cdot \boldsymbol{\rho}_\odot}{\rho_\odot^3}\right)\right]$$

The equations of motion of the Moon relative to the Earth, therefore, are

$$\ddot{\boldsymbol{\rho}}_{\mathbb{C}} + \frac{\kappa \boldsymbol{\rho}_{\mathbb{C}}}{|\boldsymbol{\rho}_{\mathbb{C}}|^3} = \nabla_{\rho_{\mathbb{C}}} R$$

Now $\rho_{\mathbb{C}\odot}^2 = \rho_{\mathbb{C}}^2 + \rho_{\odot}^2 - 2\rho_{\mathbb{C}}\rho_{\odot}\cos S$ where S is the angle at the Earth in the Moon–Earth–Sun triangle. Hence,

$$R = \frac{GM_{\odot}}{\rho_{\odot}}\left(\frac{\rho_{\odot}}{\rho_{\mathbb{C}\odot}} - \frac{\rho_{\mathbb{C}}}{\rho_{\odot}}\cos S\right)$$

The first term in the parentheses can be expanded in a series of Legendre polynomials as

$$\frac{\rho_{\odot}}{\rho_{\mathbb{C}\odot}} = \left[1 - 2\left(\frac{\rho_{\mathbb{C}}}{\rho_{\odot}}\right)\cos S + \left(\frac{\rho_{\mathbb{C}}}{\rho_{\odot}}\right)^2\right]^{-1/2}$$

$$= P_0 + \left(\frac{\rho_{\mathbb{C}}}{\rho_{\odot}}\right)P_1 + \left(\frac{\rho_{\mathbb{C}}}{\rho_{\odot}}\right)^2 P_2 + \cdots$$

Note that $\rho_{\mathbb{C}}/\rho_{\odot} \sim 3 \times 10^{-3}$. Observe too that

$$GM_{\odot}\left(1 + \frac{M_{\oplus} + M_{\mathbb{C}}}{M_{\odot}}\right) = n_{\odot}^2 a_{\odot}^3$$

or

$$GM_{\odot} \cong n_{\odot}^2 a_{\odot}^3$$

since $(M_{\oplus} + M_{\mathbb{C}})/M_{\odot} \sim 3 \times 10^{-6}$. Here $n_{\odot}$ is the solar mean motion and $a_{\odot}$ is the mean solar distance. The new form of the disturbing function becomes

$$R = \frac{a_{\odot}^3 n_{\odot}^2}{\rho_{\odot}}\left[1 + \left(\frac{\rho_{\mathbb{C}}}{\rho_{\odot}}\right)^2 P_2 + \left(\frac{\rho_{\mathbb{C}}}{\rho_{\odot}}\right)^3 P_3 + \left(\frac{\rho_{\mathbb{C}}}{\rho_{\odot}}\right)^4 P_4 + \cdots\right]$$

where the first term will be dropped since it does not contribute to the derivatives of R. Multiplying and dividing the members of R with the appropriate powers of the mean lunar distance $a_{\mathbb{C}}$, we can rearrange it in the form

$$R = n_{\odot}^2 a_{\mathbb{C}}^2\left[\left(\frac{\rho_{\mathbb{C}}}{a_{\mathbb{C}}}\right)^2\left(\frac{a_{\odot}}{\rho_{\odot}}\right)^3 P_2 + \left(\frac{a_{\mathbb{C}}}{a_{\odot}}\right)\left(\frac{\rho_{\mathbb{C}}}{\rho_{\odot}}\right)^3\left(\frac{a_{\odot}}{\rho_{\odot}}\right)^4 P_3 + \cdots\right]$$

If, in addition, the eccentricity of the orbit of the Sun, $e_{\odot} = 0.0168$, is neglected, then $a_{\odot}/\rho_{\odot} = 1$ everywhere in the last expression for R. Furthermore, if $a_{\mathbb{C}}/a_{\odot} = 0.0025$ is neglected, then the disturbing function reduces to

$$R = n_{\odot}^2 \rho_{\mathbb{C}}^2\left(-\tfrac{1}{2} + \tfrac{3}{2}\cos^2 S\right) \tag{6.25}$$

The quantity $a_{\mathbb{C}}/a_{\odot}$ represents the ratio between the sines of the solar and lunar parallaxes; therefore, neglecting this quantity in the expansion of the disturbing function corresponds to one of Hill's simplifying assumptions. Furthermore, if the lunar inclination is also neglected (this makes the problem planar), then we have the following equations of motion of the Moon in an Earth-centered coordinate system:

$$\ddot{X} + \frac{\kappa}{\rho_{\mathbb{C}}^3} X = \frac{\partial R}{\partial X}, \qquad \ddot{Y} + \frac{\kappa}{\rho_{\mathbb{C}}^3} Y = \frac{\partial R}{\partial Y}$$

The disturbing function R is given by Eq. 6.25.

Now we select a coordinate system that is centered on the Earth and rotates in the (fixed) plane of the ecliptic with angular speed $n_\odot$. The axis $\tilde{x}$ is directed constantly toward the Sun. The transformation between the previous system XY and the new $\tilde{x}\tilde{y}$ is given by

$$\boldsymbol{\rho}_{\mathbb{C}} = R_3(-n_\odot t^*)\tilde{\mathbf{r}}$$

and t^* is the (dimensional) time. The two equations of motion in the Earth-centered rotating coordinate system become

$$\ddot{\tilde{x}} - 2n_\odot \dot{\tilde{y}} = \frac{\partial R^*}{\partial \tilde{x}}, \qquad \ddot{\tilde{y}} + 2n_\odot \dot{\tilde{x}} = \frac{\partial R^*}{\partial \tilde{y}} \tag{6.26}$$

because the form of the disturbing function is now

$$R^* = \frac{n_\odot^2}{2}(\tilde{x}^2 + \tilde{y}^2) + \frac{\kappa}{\rho_{\mathbb{C}}} + n_\odot^2 \rho_{\mathbb{C}}^2 \left(-\frac{1}{2} + \frac{3}{2}\cos^2 S \right)$$

or

$$R^* = \frac{\kappa}{\rho_{\mathbb{C}}} + \frac{3}{2} n_\odot^2 \tilde{x}^2 \tag{6.27}$$

Hill's problem is described by Eq. 6.26 with Eq. 6.27 defining R^*. The dots signify derivatives with respect to t^*. Other derivations of Hill's problem are possible and there is a rich literature concerning its solution.

KAM Theory

The three-body problem has been further illuminated in the last decade by the work of Kolmogorov, Arnold, and Moser. Their work can be found in Arnold (1978), Moser (1970), and Siegel and Moser (1971). These researchers use the techniques of differential topology to formulate classical mechanics and then attempt to find solutions, study issues of stability, and so on. The mathematical formalism necessary to discuss these problems is daunting. So far the practical results are not worth mastery of the material. It also appears that perturbations pertinent to artificial satellite mechanics cannot be incorporated into this representation of Newton's laws. More important is that a fresh new point of view has been introduced that has the potential of advancing our understanding and knowledge of some of the fundamental problems of celestial mechanics.

The Kirkwood Gaps

Kirkwood (1867) first pointed out that the minor planets tend not to have mean motions that are simple rational multiples of Jupiter's. Figure 31 illustrates this for the first 2829 numbered asteroids (from Taff 1973) and those class 1 PLS (Palomar–Leiden Survey; van Houten et al., 1970) that are not already included in the numbered group. The gaps at mean motion ratios of 2:1, 9:4, 7:3, 5:2, 3:1, and 4:1 are crystal clear. Those at 8:3 and 10:3 are less so. Similar phenomena are now apparent in the Saturnian ring system. Either we are seeing numerology of the most perverse kind or something simple (?) is at work. Recent reviews (Scholl 1979; Froeschlé and Scholl 1979) indicate that the situation is still unsettled.

To begin to study this subject we will solve, in rectangular coordinates, the planar restricted three-body problem to second order in perturbation theory. Picard's method of successive substitutions will be used with a circular orbit

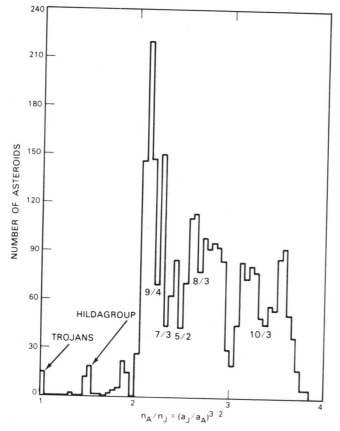

Figure 31. The distribution of minor planet mean motions in units of Jupiter's. The narrow regions deficient in asteroids are known as the Kirkwood gaps.

as the starting point (and the point of this exercise is mainly to illustrate it). Both the coordinates and velocity components can be obtained exactly and analytically in terms of a single quadrature. Although secular terms do appear in the second-order terms, there is no specialization at the gaps, indeed nowhere except at $1:1$ (where the analysis presented here is invalid). If, to first order (i.e., $\cos i = 1$, $\sin i = i$), one includes inclination effects, then the formal solution is identical. The inclusion of eccentricity effects for the asteroids, which may be critical to an understanding of the Kirkwood gaps, appears to be a task too formidable for the present technique.

Formulation

The assumptions are (1) that the Sun, of mass $M_\odot$, is at rest at the origin of an inertial reference frame; (2) that Jupiter, of mass $M(\varepsilon = M/M_\odot)$ revolves about the Sun in a circular orbit of radius A, $\Omega^2 \equiv GM_\odot/A^3$; and (3) that the asteroid, of mass $m \ll M$, would revolve about the Sun in a circular orbit of radius a, $\omega^2 \equiv GM_\odot/a^3$, in Jupiter's orbital plane, were $M = 0$. Uppercase symbols will be used for Jovian parameters and the corresponding lowercase ones for the minor planet's. In the inertial frame

$$\mathbf{R}(t) = A(\cos \Omega t, \sin \Omega t, 0)$$

$$\mathbf{r}(t) = a[\cos(\omega t + \phi), \sin(\omega t + \phi), 0]$$

where ϕ is an arbitrary phase at $t = 0$. The asteroid's equations of motion are

$$\ddot{\mathbf{r}} = -\frac{GM_\odot \mathbf{r}}{|\mathbf{r}|^3} - \frac{GM(\mathbf{r} - \mathbf{R})}{|\mathbf{r} - \mathbf{R}|^3}$$

If $\mathbf{r} = (\xi, \eta, \zeta)$ in the inertial frame and $\mathbf{r} = (x, y, z)$ in the rotating frame, where their interrelationship is given by

$$\xi = x \cos \Omega t - y \sin \Omega t$$

$$\eta = x \sin \Omega t + y \cos \Omega t$$

$$\zeta = z$$

then the asteroid's new equations of motion are (in the rotating frame)

$$\dot{v} - 2\Omega u = \frac{\partial U}{\partial x}$$

$$\dot{u} + 2\Omega v = \frac{\partial U}{\partial y} \tag{6.28}$$

$$\dot{w} = \frac{\partial U}{\partial z}$$

Here $\dot{\mathbf{r}} = (u, v, w)$ and $U = \Omega^2(x^2 + y^2)/2 + GM_\odot/r + GM/|\mathbf{r} - \mathbf{R}|$. From the form of Eq. 6.28 one is immediately led to Jacobi's integral and the result that there is no first-order secular perturbation in a. If $M = 0$, then

$\mathbf{r} = a[\cos \theta(t), \sin \theta(t), 0]$ in the rotating frame with

$$\theta(t) \equiv (\omega - \Omega)t + \phi$$

We will also need

$$\alpha = \frac{a^2 + A^2}{2aA} > 1 \qquad \text{for } a \neq A$$

$$\kappa^2 = \frac{2}{\alpha + 1} = \frac{4aA}{(a+A)^2} \qquad \kappa > 0 \text{ (and } \kappa < 1 \text{ for } a \neq A\text{)}$$

$$\psi(t) = \sin^{-1}\left[\frac{1}{\kappa^2} \frac{1 - \cos \theta(t)}{\alpha - \cos \theta(t)}\right]^{1/2}$$

$$\Delta F(t) = F[\psi(t), \kappa] - F[\psi(0), \kappa]$$

$$\Delta E(t) = E[\psi(t), \kappa] - E[\psi(0), \kappa]$$

where F and E are the usual elliptic integrals of the first and second kind. Lastly we abbreviate $\varepsilon A^2 \Omega^2 / (\omega - \Omega)$ as δ, the small parameter of the problem.

Elementary Solution

The zeroth-order solution is ($z = \dot{z} = \ddot{z} = 0$ now)

$$x_0(t) = a \cos \theta(t), \qquad u_0(t) = -a(\omega - \Omega) \sin \theta(t)$$

$$y_0(t) = a \sin \theta(t), \qquad v_0(t) = a(\omega - \Omega) \cos \theta(t)$$

The first-order solution, via Picard's method, is

$$x_1(t) = x_0(t), \qquad y_1(t) = y_0(t)$$

$$u_1(t) = u_0(t) + \delta\left[\frac{\Delta F(t)}{A+a} + \frac{\Delta E(t)}{A-a}\right]$$

$$v_1(t) = v_0(t) = \frac{\delta}{(2aA)^{1/2}}\left(\frac{1}{[\alpha - \cos \theta(t)]^{1/2}} - \frac{1}{[\alpha - \cos \theta(0)]^{1/2}}\right)$$

The change of variable $\chi = \alpha - \cos \theta$ was used to derive these. The second-order solution is

$$x_2(t) = x_1(t) + \delta \int_0^t \left[\frac{\Delta F(\tau)}{A+a} + \frac{\Delta E(\tau)}{A-a}\right] d\tau$$

$$y_2(t) = y_1(t) + \delta\left[\frac{2}{\omega - \Omega} \frac{\Delta F(t)}{A+a} - \frac{t/(2aA)^{1/2}}{[\alpha - \cos \theta(0)]^{1/2}}\right]$$

$$u_2(t) = u_1(t) + 2\delta\left[\frac{2\Omega}{\omega - \Omega} \frac{\Delta F(t)}{A+a} - \frac{\Omega t/(2aA)^{1/2}}{[\alpha - \cos \theta(0)]^{1/2}}\right]$$

$$v_2(t) = v_1(t) - 2\Omega \delta \int_0^t \left[\frac{\Delta F(\tau)}{A+a} + \frac{\Delta E(\tau)}{A-a}\right] d\tau$$

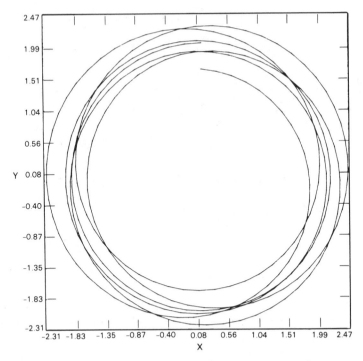

Figure 32. A perturbed minor planet orbit for $\mu = 10^{-3}$ and angular speed ratio of 2. The duration is six unperturbed periods.

The presence of the secular terms in x and y (and therefore of terms such as $t \cos \Omega t$ and $t \sin \Omega t$ in ξ and η) are clear. It is also clear that, for example, $\omega = 2\Omega$ produces nothing special. Moreover, the presence of the elliptic functions in r_2 and $\dot{r}_2$ prohibits the reduction of r_3 or $\dot{r}_3$ to simple quadratures.

The Orbits.   Figures 32 and 33 show the orbits in the inertial frame for $\omega/\Omega = 2$ and 4 with $\phi = 0$. The 2:1 resonance orbits are for $\varepsilon = 0.001$ and the 4:1 are for $\varepsilon = 0.01$. It appears that the path of the asteroid merely fills an annulus in the Jovian orbital plane without developing any unusual features. Of course, these solutions are only valid for times $\sim 2\pi\delta/\omega$. The plots are for six unperturbed periods.

THE SIMPLE HARMONIC OSCILLATOR

Bertrand's Theorem

According to Plummer (1918), in 1873 Bertrand proved that the inverse square law force and the direct linear force law are the only two central force laws, derivable from a potential function, which yield bounded closed orbits. On

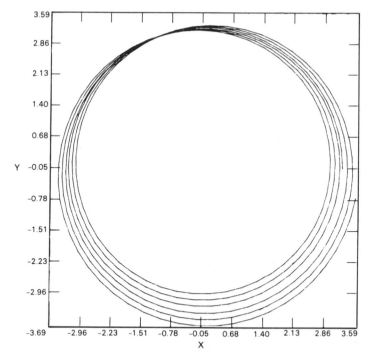

Figure 33. Same as Fig. 32 but $\mu = 10^{-2}$ and angular speed ratio of 4.

pages 2–9 of his book Plummer presents a proof of this result. Goldstein's second edition of *Classical Mechanics* (1980) reproduces Plummer's result in modern language. Arnold (1978), without ever mentioning Bertrand, also provides the outline of a proof. Although I will not fill in all of the analytical steps for the reader, immediately below I flesh out Arnold's logic. The importance of the result is that it provides a deep connection between the two-body problem and the three-dimensional, isotropic, simple harmonic oscillator problem.

The logic of Arnold's argument can be simply outlined. If all bounded orbits are closed, then since a circular orbit is always possible for a central force, all nearly circular bounded orbits should be closed too. Moreover, this (i.e., being closed) must be true independent of the radius of the circular orbit. The implications of these statements are that the potential energy is either a power law $U(r) = Ar^p$ ($p \geq -2$, $p \neq 0$) or a logarithmic function $U(r) = B \ln r$. By considering the constraints of "closedness" on the apsidal angle, Arnold then shows that the logarithmic case cannot be closed while there are two possibilities for the power law case. These two deal with positive power laws ($p > 0$) and the meaning of bounded and with negative power laws that yield infinitely tight binding. After an examination of these two limiting cases, it turns out that only $p = +2$ and $p = -1$ are left. Q.E.D.

The conservation of energy equation for a central force, after the usual reduction to the equivalent one-body problem, may be expressed as

$$\frac{m\dot{r}^2}{2} + \frac{L^2}{2mr^2} + U = \mathscr{E}$$

Hence the "pseudo-potential energy" $V = U + L^2/2mr^2$. From this we can deduce that

$$\int dt = \pm \int \frac{dr}{[2(\mathscr{E} - V)/m]^{1/2}}$$

and from $L = mr^2\dot{\phi}$, ϕ being the azimuthal angle, it follows that

$$\phi = \int \frac{(L/r^2)\, dr}{[2m(\mathscr{E} - V)]^{1/2}}$$

Orbits exist such that $\mathscr{E} \geq V(r)$ for a fixed value of L. This inequality gives a single (or several, depending on V) region(s) with

$$0 \leq r_{min} \leq r_{max} \leq \infty$$

Clearly, if $0 \leq r_{min} \leq r_{max} < \infty$, then the motion is bounded ($r_{max} < \infty$) and also takes place in the annulus with inner radius r_{min} and outer radius r_{max}. As ϕ increases, r varies between these two limits. The angular extent between successive apsides (extrema of r) is called the apsidal angle. It is given by the expression

$$\Phi = \int_{r_{min}}^{r_{max}} \frac{(L/r^2)\, dr}{\{2m[\mathscr{E} - V(r)]\}^{1/2}} \tag{6.29}$$

A closed orbit is one such that Φ is commensurate with 2π. Finally, in the special case when $\mathscr{E}$ is equal to a minimum of V, $r_{min} = r_{max} = R$ and the orbit is a circle of radius R. Now that the stage has been set, the argument can be filled in.

First we calculate the value of Φ for a nearly circular orbit. Since the orbit is nearly circular, $\mathscr{E}$ must be slightly larger than the minimum value of V. Near $r = R$ the pseudopotential V can be written as [since it is a minimum, $V'(R) = 0$, $V''(R) > 0$]

$$V(r) = V(R) + \frac{(r-R)^2}{2} V''(R)$$

The extremes of r are found by solving $\mathscr{E} - V(r) = 0$. They are simply

$$R \pm \left\{ \frac{2[\mathscr{E} - V(R)]}{V''(R)} \right\}^{1/2}$$

with r_{max} given by the plus sign and r_{min} given by the minus sign. We use the definition in Eq. 6.29 of Φ plus the change of variables

$$x = (r - R)\left[\frac{V''(R)}{2[\mathscr{E} - V(R)]}\right]^{1/2}$$

to rewrite the expression for Φ in the form

$$\Phi = \frac{L}{[V''(R)m]^{1/2}} \int_{-1}^{1} \frac{dx}{(1-x^2)^{1/2}}\left[R + x\left\{\frac{2[\mathscr{E} - V(R)]}{V''(R)}\right\}^{1/2}\right]^{-2}$$

Hence

$$\Phi_{\text{circ}} \simeq \frac{\pi L}{R^2[V''(R)m]^{1/2}}$$

or, in terms of U,

$$\Phi_{\text{circ}} \simeq \frac{\pi[U'/(3U' + RU'')]^{1/2}}{\sqrt{m}}$$

We want this result to be independent of the value of R; that is,

$$\frac{U'}{3U' + RU''} = \text{const}$$

We choose the form of the constant to be $1/(2+p)$ and integrate. One deduces directly that

$$U = Ar^p \quad (p \geq -2, p \neq 0) \qquad \text{or} \qquad U = B \ln r$$

within an arbitrary, unimportant, additive constant. From the formula for Φ_{circ} we now accordingly find that $(m = 1 \text{ now}; \text{ its value is irrelevant})$

$$\Phi_{\text{circ}} = \frac{\pi}{(2+p)^{1/2}} \qquad (p = 0 \text{ is the logarithmic case})$$

This completes the first part of the result.

If $p = 0$ (the logarithmic case), then $\Phi_{\text{circ}} = \pi/\sqrt{2}$, which is not a rational multiple of 2π. Hence, this functional form can be dismissed from further consideration. If $p > 0$, then as $r \to \infty$, $U(r) \to \infty$. Therefore, to remain bounded, $\mathscr{E} \to \infty$ too. The general limit as $\mathscr{E} \to \infty$ of Φ is $\pi/2$. To see this, we make the change of $u = L/r$ in Eq. 6.29. Then

$$\Phi = \int_{u_{\min}}^{u_{\max}} \frac{dx}{\{2m[\mathscr{E} - W(x)]\}^{1/2}}$$

where $W(x) = U(L/x) + x^2/2m$. Further let $y = x/u_{\max}$ so that

$$\Phi = \int_{y_{\min}}^{1} \frac{dy}{\{2m[w(1) - w(y)]\}^{1/2}}$$

where $w(y) = y^2/2m + U(L/yu_{\max})/u_{\max}^2$. Clearly, as $\mathscr{E} \to \infty$, $u_{\max} \to \infty$, $y_{\min} \to 0$

and the last term in w is negligible. The $\pi/2$ result follows after a trivial integration. Now if $\pi/(2+p)^{1/2} = \pi/2$ for $p > 0$, $p = 2$ and we have the isotropic simple harmonic oscillator as a possibility.

The last piece involves negative values for p. There is the possibility that $\mathcal{E} \to -\infty$ as $r \to 0$. First we show that in general as $\mathcal{E} \to -\infty$, $\Phi = \pi/(2-q)$ for power law potential energies $U = -kr^{-q}$, $q \in (0, 2)$. The proof follows the type of analysis presented immediately above and leads to

$$\Phi = \int_0^1 \frac{dx}{(x^q - x^2)^{1/2}} = \frac{\pi}{2-q}$$

(This is one of several misprints in this section of Arnold's book.) Then we replace $-q$ and p and demand equivalence with the circular orbit result, or

$$\frac{\pi}{2+p} = \frac{\pi}{(2+p)^{1/2}}$$

Only $p = 1$ satisfies this constraint—the inverse square law.

The Isotropic, Three-Dimensional, Simple Harmonic Oscillator

To ensure that the terminology is understood the potential energy is $U = kr^2/2$. The force constant is k and it is the same in all directions (isotropic). If the force constant were different in different directions, then $U = \mathbf{r} \cdot \mathbf{k} \cdot \mathbf{r}/2$ where $\mathbf{k}$ is now a diagonal tensor,

$$\mathbf{k} = \begin{pmatrix} k_x & 0 & 0 \\ 0 & k_y & 0 \\ 0 & 0 & k_z \end{pmatrix}$$

This would be an anisotropic, three-dimensional, simple harmonic oscillator. Since $r^2 = x^2 + y^2 + z^2$, this explains the three-dimensional aspect. The simple harmonic part is due to the fact that U is quadratic in x, y, z. The resulting motion is then sinusoidal with a period independent of the amplitude.

The force is $\mathbf{F} = -\nabla U = -k\mathbf{r} = m\ddot{\mathbf{r}}$. Energy conservation is obvious and we set

$$\mathcal{E} = U + \frac{m}{2}\dot{\mathbf{r}} \cdot \dot{\mathbf{r}} \geq 0 \tag{6.30}$$

Furthermore, since $\mathbf{F}$ is central $\mathbf{L} = \mathbf{r} \times m\dot{\mathbf{r}}$ is a constant vector. Therefore, the orbit lies in a plane given by $\mathbf{r} \cdot \mathbf{L} = 0$ and the problem can be reduced to a two-dimensional one. We set

$$\mathbf{L} = L(\sin i \sin \Omega, -\sin i \cos \Omega, \cos i)$$

where i (the inclination) and Ω (the longitude of the ascending node) specify the orientation of $\mathbf{L}$ and L is its magnitude. Finally polar coordinates r, ϕ are introduced in the orbital plane so that the equations of motion are

$$m(\ddot{r} - r\dot{\phi}^2) = f(r) = -\frac{dU}{dr} = -kr$$

$$\frac{m}{r}\frac{d}{dt}(r^2\dot{\phi}) = 0$$

The Orbit

As usual we introduce Binet's transformation $u = 1/r$ and switch from time derivatives to derivatives with respect to ϕ via (prime $= d/d\phi$)

$$\dot{r} = r'\dot{\phi} = -\frac{u'\dot{\phi}}{u^2} = -\frac{Lu'}{m}$$

Accordingly, the equations of motion are

$$u'' + u = \frac{-mf(u^{-1})}{L^2 u^2} = \frac{mk}{L^2 u^3} \tag{6.31}$$

The mean motion n is defined by $n^2 = k/m$. After a single integration of Eq. 6.31 we have the relationship

$$(u')^2 = \text{const} - \left(u^2 + \frac{m^2 n^2}{L^2 u^2}\right)$$

Comparison with the conservation of energy equation 6.30 in the u form shows that the constant is $2\mathscr{E}m/L^2$. Thence,

$$(u')^2 = \frac{(u_-^2 - u^2)(u^2 - u_+^2)}{u^2}$$

where

$$u_+^2 = \frac{1}{a^2}, \qquad u_-^2 = \frac{1}{b^2} = \frac{1}{a^2(1 - e^2)}$$

and

$$r_{\max}^2 = a^2 = \frac{L^2/m}{\mathscr{E} - (\mathscr{E}^2 - L^2 n^2)^{1/2}} = \frac{\mathscr{E} + (\mathscr{E}^2 - L^2 n^2)^{1/2}}{mn^2}$$

$$r_{\min}^2 = b^2 = a^2(1 - e^2) = \frac{L^2/m}{\mathscr{E} + (\mathscr{E}^2 - L^2 n^2)^{1/2}} = \frac{\mathscr{E} - (\mathscr{E}^2 - L^2 n^2)^{1/2}}{mn^2}$$

At $r = a$ or $r = b$ both $\dot{r} = 0$ and $r' = 0$; hence their identification with the extrema of r.

Suppose that the motion starts at $\phi = \phi_0$ where $r = a$. Then immediately thereafter $r' < 0$, so

$$\int_{1/a}^{u} \frac{x\,dx}{[(b^{-2} - x^2)(x^2 - a^{-2})]^{1/2}} = + \int_{\phi_0}^{\phi} d\psi$$

or

$$r^2 = \frac{a^2(1-e^2)}{1-e^2+e^2\sin^2(\phi-\phi_0)}$$

Without loss of generality let ϕ_0 be zero and let another orientation angle, ω, define the line of the major axis ("the argument of periapse"). Then

$$r^2 = \frac{a^2(1-e^2)}{1-e^2+e^2\sin^2\phi}$$

is the equation of the orbit. If $\xi = r\cos\phi$, $\eta = r\sin\phi$ are rectangular coordinates in the orbital plane, then this is equivalent to

$$\frac{\xi^2}{a^2} + \frac{\eta^2}{b^2} = 1$$

This is the equation of an ellipse, center at the origin, major axis a, minor axis b (and therefore eccentricity e), whose major axis (the line of apsides) lies on $\eta = 0$.

The Time Dependence

We return to the conservation of energy equation 6.30 and solve for $\dot{r}^2$,

$$\dot{r}^2 = \frac{n^2}{r^2}(a^2-r^2)(r^2-b^2)$$

Suppose that $r = a$, $\phi = 0$ corresponds to $t = T$ ("the time of periapse passage"). Since $a \geq r$, $\dot{r} < 0$ for $t > T$, then

$$\int_a^r \frac{x\,dx}{[(a^2-x^2)(x^2-b^2)]^{1/2}} = -n\int_T^t d\tau$$

or

$$r^2 = a^2\cos^2 M + b^2\sin^2 M \qquad (6.32)$$

where the mean anomaly $M = n(t-T)$. Note that ϕ is the true anomaly as well as the "eccentric anomaly." However $r^2 = \xi^2 + \eta^2$; therefore

$$\xi = a\cos M, \qquad \eta = b\sin M$$

Note too that $L = mr^2\dot{\phi} = 2m\,dA/dt$ where A is the areal rate. Whence,

$$L = 2m\left(\frac{\pi ab}{P}\right) = mnab$$

where the period is $P = 2\pi/n$.

To determine the time dependence of ϕ, we use $L = mr^2\dot{\phi}$ and the expression 6.32 for $r^2(t)$, namely,

$$\int_0^\phi d\psi = \frac{L}{m}\int_T^t \frac{d\tau}{a^2\cos^2\mu + b^2\sin^2\mu}, \qquad \mu = n(\tau-T)$$

The result is

$$\tan \phi = (1 - e^2)^{1/2} \tan M, \qquad M = n(t - T)$$

This is "Kepler's equation."

To summarize, the three-dimensional motion is given by

$$\mathbf{r} = S\mathbf{q} = rS \begin{pmatrix} \sin \phi \\ \cos \phi \\ 0 \end{pmatrix} = aS \begin{pmatrix} \cos M \\ (1 - e^2)^{1/2} \sin M \\ 0 \end{pmatrix} \tag{6.33}$$

where the unitary rotation matrix $S = R_3(-\Omega) R_1(-i) R_3(-\omega)$. The R matrices are the usual elementary rotation matrices. The angular momentum has already been given; the energy $\mathcal{E}$ is

$$\mathcal{E} = \frac{mn^2 a^2}{2}(2 - e^2)$$

PROBLEMS

1. Consider the nearly circular polar orbit about an oblate planet. Show that to first order in both e and J_2 the solution reduces to

$$\theta = \left(1 - \frac{3J_2}{4r_0^2}\right)\chi$$

$$h = h_0 \left[1 - \left(\frac{3J_2}{4r_0^2}\right)(1 - \cos 2\chi)\right]$$

$$r_0 u = 1 - e(1 - \cos \chi) + \left(\frac{J_2}{4r_0^2}\right)(3 - 2\cos \chi - \cos 2\chi)$$

$$\frac{r}{r_0} = 1 + e(1 - \cos \chi) - \left(\frac{J_2}{4r_0^2}\right)(3 - 2\cos \chi - \cos 2\chi)$$

$$\frac{v}{v_0} = 1 - e(1 - \cos \chi) + \left(\frac{J_2}{2r_0^2}\right)(-\cos \chi + \cos 2\chi)$$

2. Again consider the slightly oblate spherical problem. Suppose that the solution, to first order in J_2, has been obtained separately for the equatorial plane and for a polar orbit. Call these $R(0)$ and $R(\pi/2)$. Explain why the general first-order solution is given by

$$R(0) \cos^2 i + R\left(\frac{\pi}{2}\right) \sin^2 i$$

for arbitrary inclination i. Further show that the relationship between the central angle (no longer θ) and χ is given by

$$\text{Central angle} = \left[1 + \frac{3J_2}{2r_0^2\Gamma^4}\left(1 - \frac{3}{2}\sin^2 i\right)\right]\chi$$

3. Show that even if the full potential is

$$U = -\frac{GM}{r}\left[1 - \sum_{n=2}^{\infty} J_n P_n(\sin\theta)\right]$$

that circular orbits at constant latitude are possible but that they are not in the equatorial plane. In fact for a circular orbit of radius a and small $|J_n|$ the declination is given by the solution of ($|\delta| \ll \pi/2$) (Blitzer 1962)

$$\delta\left[1 - \sum_{n=2}^{\infty} \frac{n+1}{a^n} J_n P_n(\delta)\right] + \sum_{n=2}^{\infty} \frac{J_n P'_n(\delta)}{a^n} = 0$$

or

$$\delta = \frac{\sum_{n\,\text{odd}} A_n J_n/a^n}{1 - \sum_{n\,\text{even}} B_n J_n/a^n}$$

where

$$A_n = \frac{(-1)^{n+1/2}(n+1)!}{2^n[(n+1)/2]![(n-1)/2]!}$$

$$B_n = \frac{(-1)^{n/2}(n+1)(n+1)!}{2^n[(n/2)!]^2}$$

Whence

$$\delta = \frac{3J_3}{2a^2} - \frac{15J_5}{8a^5} + \frac{35J_7}{16a^7}$$

4. Show explicitly that, including the J_2 term, a circular equatorial orbit is stable to small radial oscillations. Further deduce that the radius for an unstable orbit is ~150 miles (about the Earth).

5. Describe the connection between the problem of two fixed centers and the special oblate spherical potential

$$U = -\frac{GM}{r}\left[1 + \sum_{n=1}^{\infty} \frac{(-J_2)^n}{r^{2n}} P_{2n}(\sin\theta)\right]$$

6. Planetary rings are composed of aggregations of individual particles each of which orbit the oblate primary in a nearly elliptic orbit. It appears that the particles in a single ring have a common apsidal line. Show that this line rotates at a rate

$$\frac{d\omega}{dt} = \left(\frac{GM}{a^3}\right)^{1/2}\left[\frac{3J_2}{2p^2} - \frac{15J_4}{4p^4}\right]$$

where $p = a(1 - e^2)$ is the semi-latus rectum of the ellipse.

7. Show that in the restricted three-body problem the total energy is not conserved.

8. Three-point masses m_1, m_2, and m_3 are located at the corners of an equilateral triangle with sides of length L. The only forces acting on the particles are those of their mutual gravitational attraction. (1) Show that the net force acting on any one of the particles is directed toward the center of mass of the system. (2) Show that a uniform rotational motion of the system as a whole about its center of mass can be found such that the particles move without changing their positions relative to one another. Find the required angular speed.

9. Consider the planar restricted three-body problem wherein the larger primary is an oblate spheroid whose equatorial plane lies in the plane of motion. Discuss the nature of the "triangular" equilibrium solutions and their stability (Subbarao and Sharina 1975).

10. From Jacobi's integral derive Tisserand's criteria for the identity of comets in the Sun–Jupiter–comet system, namely that

$$C = \frac{1}{a} + 2[a(1-e^2)]^{1/2} \cos i$$

is an approximate constant of the motion [where a, e, and i are the (ecliptic) orbital elements of the comet].

chapter seven

Laplacian-Type Initial Orbit Determination

This chapter deals with initial orbit determination utilizing passively acquired angles-only data—a classic problem in astronomy. So well developed is the literature on the subject that the educated reader might wonder what might be added. My rejoinder is the following questions: "Have you ever really computed an orbit given only three sets of optical observations when it was of importance to do so?" and "Assuming that your answer was yes, did it work?" I would hazard a guess that less than one-tenth of 1% of the readers of this book can answer yes to the first question. Further I will speculate that the answer to the second question has always been no. Moreover, with the advent of ballistic missiles and artificial satellites and the contemporaneous development and refinement of radars, many of the classical astronomical techniques were reformulated for these new problems with their new observables. The use of laser radars, beacon tracking, electro-optical cameras, and so forth, has increased the meaningful data acquisition rate on near-Earth objects both because of their mode of operation and because near-Earth objects move so much faster than do the natural objects of the solar system. As an example, a typical minor planet has a geocentric angular speed of $0°\!.25/$day, a typical high-altitude satellite has a topocentric angular speed of $15''/$sec ($=360°/$day), whereas a low-altitude satellite might have a topocentric angular speed in excess of $10°/$min ($=14{,}400°/$day). And yet only one initial orbit determination technique breaks with those of the past (see Taff and Hall 1977, 1980, and later in this chapter).

EDITORIAL

Astronomical Initial Orbit Determination

Consider the instances in astronomy when one computes orbits. One might do it for a binary star system (this topic is discussed in Chapter Thirteen) or

216

for a moon of another planet. Obviously in these two cases there is no danger of losing the object because of an inaccurate set of orbital elements. Nor are there any impediments to the acquisition of arbitrarily large amounts of data. Planets do go into conjunction and most stars do set between observing seasons, but the planets always return to opposition and the stars always rise again. Thus, except for curiosity or the challenge of being able to deduce an orbit from minimal information, there is no compelling observational need to use only three sets of observations nor is there any compelling need to compute an accurate orbit with alacrity for these types of objects.

There are three other instances in astronomy where initial orbit determination is practiced—planets, comets, and meteors. There could be a problem for these objects before the advent of photography, but not after. The reasons are several: Photographic plates cover large areas of the sky, obviating the need for precision pointing; photographic plates yield a permanent record and old plates can always be searched for prediscovery images thereby yielding more than the minimum number of observations; and photographic plates were developed at about the same time as rapid worldwide communication so that local events lost their ability to adversely influence the making of an observation. Now we return to the prephotographic era.

Meteors are so common, the sporadic rate is $\sim 10/\text{hr}$, that one never contemplated computing orbits for any but a few. Since a parallax was determined for a meteor only in 1798, and scientific interest in the subject was not stimulated until the great Leonid shower of November 1833 (when the radiant was discovered), I'll take the position that they have played a very minor role in initial orbit determination. They continue to do so even after the realization that meteor showers are highly correlated with extinct comets.

Comets are another case in three regards. Most of them that are visible to the naked eye are not periodic (Halley's is a notable exception). Hence, recovery was never in question. Moreover, comets have comae and tails and are difficult to lose or lose sight of. Finally Olbers (in 1797) solved the initial orbit problem for *parabolic* comet orbits. This is but a sidelight on the general problem.

This leaves us planets. Planets were important (at least the modern major ones and the first few minor ones). Remember, however, that Uranus (discovered by Herschel in 1781) was visible to the naked eye and 19 (some sources indicate 20) prediscovery observations were quickly found. Also with a sidereal period of 83.7 yr one was not in much danger of losing Uranus. Since both Neptune and Pluto were discovered on the basis of *predictions*, one clearly needed no initial orbit determination method (and they have even longer sidereal periods). This leaves us minor planets.

Let us review the solar systems circa 1800. We knew of the planets of antiquity—Mercury, Venus, Earth, Mars, Jupiter, and Saturn. (The seven classical planets of antiquity were the above minus Earth, plus the Moon and the Sun.) By and large these lay in the plane of the ecliptic revolving about the Sun in nearly circular orbits (see Table 2). Moreover, the semi-major axes

Table 2. Planetary Data

Planet	Semi-major Axis (A.U.)	Bode's Law Value (A.U.)	Eccentricity	Inclination
Mercury	0.387	0.4	0.206	$7°00$
Venus	0.723	0.7	0.007	3.39
Earth	1.000	1.0	0.017	0
Mars	1.524	1.6	0.093	1.85
(Ceres)	2.767	2.8	0.076	10.62
Jupiter	5.204	5.2	0.049	1.31
Saturn	9.580	10.0	0.051	2.49
Uranus	19.141	19.6	0.046	0.77

of the planets satisfied a progression discovered by Titius and recently published (in 1772) by Bode. As yet, however, there was no object for the fifth place. Presciently Laplace (in 1780) published his technique of initial orbit determination, for the following year Herschel discovered the first new planet in the history of the world. It fits the above scheme very well. This plus earlier (i.e., the moons of Jupiter, etc.) discoveries show the scientific community that the solar system *does* contain additional bodies. Believing in the numerology of the Titius–Bode relationship, a search is organized for the missing planet. On the first night of the nineteenth century Piazzi discovered Ceres. He kept the discovery to himself for 3 weeks; illness then forced him from the telescope after 41 nights (*not* 3!) of observing. (Ceres beats the slow mails of the winter of 1801 to conjunction.) Is this object the missing planet and how shall we find it after conjunction?

The Historical Myth—Ceres

I regard this as the first time an orbit was really needed. The stories in modern-day astronomy books recount that Gauss heard of the difficulty in September, invented his method of initial orbit determination in October, predicted (in November) a position for Ceres and von Zach (a member of the original search team; Piazzi was a potential member and Bode organized it) found it within $0°5$ of that position on January 1, 1802 (it was cloudy in December*). What better triumph can one ask of science? How far off could the prediction have been before we stopped trumpeting this epic? How did Gauss *really* do it? Before giving my answers to the last two questions, I need to complete the story. Olbers too found Ceres (on January 2) and then Pallas in the Spring. In 1804 Juno was discovered and in 1807 Olbers found Vesta. The next minor planet was discovered in 1845. Photographic searches for asteroids were started in 1891.

---

* Some accounts have one clear night, December 7, during which von Zach "glimpsed" it. If so, this position obviously allows a differential correction of large weight.

The above is the historical myth concerning the reacquisition of Ceres. Now to my questions posed above and then to some speculations. I would guess that the errors could have been as large as 5° before the power of the tale pales. With perfect hindsight, and both theoretical and observational expertise in what I'm about to propose, I'd have done it as follows: The solar system lies in a plane and this new object is discovered near it. Hence the inclination is zero and the longitude of the ascending node is superfluous. All orbits are circular* so I'll assume zero for the eccentricity and the argument of perihelion is meaningless. I believe in Bode's law so I know the semi-major axis. Moreover, the assumed semi-major axis correctly reproduces the *observed* angular speed. This leaves a single orbital element, the mean longitude (say) to fix. Finally, had I attempted all of this for the just-discovered Uranus, it would have worked (*see* Table 2). I would have tried it for Ceres, and since I am temporarily Gauss, the inventor of least squares, I would have found an intelligent way to use Piazzi's 41 nights of data and perform a simple differential correction of the orbit.

We do not know how Gauss actually computed the orbit of Ceres or any of the other big four minor planets. We do know that however he did it; Gauss *did not* use the method published in his *Theoria Motus* in 1809. I quote at length, and in context, from the Preface to that work, first concerning Uranus:

> As soon as it was ascertained that the motion of the new planet, discovered in 1781, could not be reconciled with the parabolic hypothesis, astronomers undertook to adapt a circular orbit to it, which is a matter of simple and very easy calculation. By a happy accident the orbit of this planet had but a small eccentricity, in consequences of which the elements resulting from the circular hypothesis sufficed at least for an approximation of which could be based the determination of the elliptic elements. There was a concurrence of several other very favorable circumstances. For, the slow motion of the planet, and the very small inclination of the orbit to the plane of the ecliptic, not only rendered the calculations much more simple, and allowed the use of special methods not suited to other cases; but they removed the apprehension, lest the planet, lost in the rays of the sun, should subsequently elude the search of observers, (an apprehension which some astronomers might have felt, especially if its light had been less brilliant); so that the more accurate determination of the orbit might be safely deferred, until a selection could be made from observations more frequent and more remote, such as seemed best fitted for the end in view.

The next paragraph of the Preface discusses the general problem:

> Thus, in every case in which it was necessary to deduce the orbits of heavenly bodies from observations, there existed advantages not to be despised, suggesting, or at any rate permitting, the application of special methods; of which advantages the chief one was, that by means of hypothetical assumptions an approximate knowledge of some elements could be obtained before the computation of the

---

* Note that of the inner planets Mercury has both the largest eccentricity and the highest inclination. Even its values are not huge.

elliptic elements was commenced. Notwithstanding this, it seems somewhat strange that the general problem. . . .

To determine the orbit of a heavenly body, without any hypothetical assumption, from observations not embracing a great period of time, and not allowing a selection with a view to the application of special methods, was almost wholly neglected up to the beginning of the present century; or, at least, not treated by any one in a manner worthy of its importance; since it assuredly commended itself to mathematics by its difficulty and elegance, even if its great utility in practice were not apparent. An opinion had universally prevailed that a complete determination from observations embracing a short interval of time was impossible,—an ill-founded opinion,—for it is now clearly shown that the orbit of a heavenly body may be determined quite nearly from good observations embracing only a few days; and this without any hypothetical assumption.

The last statement is false. Finally, Gauss on Gauss and Ceres:

Some ideas occurred to me in the month of September in the year 1801, engaged at the time on a very different subject, which seemed to point to the solution of the great problem of which I have spoken. Under such circumstances we not unfrequently, for fear of being too much led away by an attractive investigation, suffer the associations of ideas, which, more attentively considered, might have proved most fruitful in results, to be lost from neglect. And the same fate might have befallen these conceptions, had they not happily occurred at the most propitious moment for their preservation and encouragement that could have been selected. For just about this time the report of the new planet, discovered on the first day of January of that year with the telescope at Palermo, was the subject of universal conversation; and soon afterwards the observations made by that distinguished astronomer PIAZZI from the above date to the eleventh of February were published. Nowhere in the annals of astronomy do we meet with so great an opportunity, and a greater one could hardly be imagined, for showing most strikingly, the value of this problem, than in this crisis and urgent necessity, when all hopes of discovering in the heavens this planetary atom, among innumerable small stars after the lapse of nearly a year, rested solely upon a sufficiently approximate knowledge of its orbit to be based upon these very few observations. Could I ever have found a more seasonable opportunity to test the practical value of my conceptions, that now in employing them for the determination of the orbit of the planet Ceres, which during these forty-one days had described a geocentric arc of only three degrees, and after the lapse of a year must be looked for in a region of the heavens very remote from that in which it was last seen? This first application of the method was made in the month of October, 1801, and the first clear night, when the planet was sought for* as directed by the numbers deduced from it, restored the fugitive to observation. Three other new planets, subsequently discovered, furnished new opportunities for examining and verifying the efficiency and generality of the method.

Several astronomers wished me to publish the methods employed in these calculations immediately after the second discovery of Ceres; but many things—

* By de Zach, December 7, 1801.

other occupations, the desire of treating the subject more fully at some subsequent period, and, especially, the hope that a further prosecution of this investigation would raise various parts of the solution to a greater degree of generality, simplicity, and elegance,—prevented my complying at the time with these friendly solicitations. I was not disappointed in this expectation, and have no cause to regret the delay. For, the methods first employed have undergone so many and such great changes, that scarcely any trace of resemblance remains between the method in which the orbit of Ceres was first computed, and the form given in this work.

Modern Reality—Chiron

It would also be instructive to review the published history of the slow-moving object discovered by C. T. Kowal in 1977. It is now known as minor planet 2060 Chiron. Its preliminary designation was 1977UB.

The first International Astronomical Union circular containing information about "Slow-moving Object Kowal" was number 3129 (dated November 4, 1977). It reported two accurate positions by Kowal (separated by 25 hr) and one approximate position by T. Gehrels from a photographic plate taken a week (October 11) earlier. The motion was very slow and retrograde, at least one-third of that of a main belt asteroid. Presumably it was the strange motion that kept the Minor Planet Center from publishing (a potentially embarrassing) orbital element set. Four days later IAU circular 3130 reported two accurate positions from Gehrels (replacing his preliminary one) and two more positions from Mt. Palomar acquired on November third and fourth. An orbital element set accompanied this and it was labeled "extremely indeterminate." The important parameters are the eccentricity $e = 0.031$ and the period $P = 66.1$ yr. We are advised that this orbit was "selected so as to minimize the aphelion distance." Since e is essentially zero, this means minimizing the period.

Seven days later IAU circular 3134 reported another pair of obervations (November 9 and 10). It also contained the comment "that a near-circular orbit solution (cf. IAUC 3130) is still viable, but an ellipse of very high eccentricity is not." Very high is not defined (0.9 or 0.5?). Additional observations from mid-November are reported in circulars 3140 and 3143. Finally, by the end of November, Circular 3145 reported two observations from 1969 based on the work of J. G. Williams. A new orbital element set was also included, $e = 0.37860$, $P = 50.70$ yr. Not very circular. Within another week prediscovery images from 1952 and the early 1940s are reported (circular 3147). What is not reported there (but is in Kowal et al., 1979) is the error in the predicted position—1°.1 for the 1969 observations based on the original orbital element set, 0°.25 for the 1952 observations based on the improved element set, and 0°.5 for the 1941 points. Williams had at least 15 positions to use to deduce the orbit that allowed him to find the 1969 positions. Also not mentioned is that finding the short, faint, trail of Chiron is as much luck as celestial mechanics—the trail was marked on the 1941 plate *in* 1941 but subsequently ignored.

Finally, by mid-December, an observation from 1895 has been reported and yet another orbit produced by the process of differential correction. Now $e = 0.378623$ and $P = 50.68$ yr. The small inclination had changed by 33% from its original value of $i = 5°2$ (to $6°9229$) but the effect and importance of this are small. After the publication of this circular (3151) additional observations from the 1940s and 1976 appear (3156, 3215).

In my opinion it is clear that the deduction of a reasonable orbit for 2060 Chiron depended much more on modern communications and large-scale ($6° \times 6°$) photographic plates than it did on Carl Fredrich Gauss.

Modern Initial Orbit Determination

Modern initial orbit determination is concerned with rockets and artificial satellites. Although optical observations (both passive and active) have been made and continue to be made on these objects, radar, *sans doute*, is the premier observing technique. Radars give distance (*radio detecting and ranging*) and distance determination is the essence of initial orbital determination by optical means.

In the only instance of which I am aware, when Gauss's method [really it was Gibbs's (1888) refinement but that is a detail] was used on high-altitude artificial satellites, it failed (see Taff 1979b). The causes of the failure were twofold. The permissible range of validity of the method was exceeded and this is clearly no fault of the method. The other reason is simply that the method does not work.

Outline of the Remainder.   The next section reviews, in a more rigorous fashion than usual, the fundamentals of Laplacian initial orbit determination. Most solar system problems either have the Sun or the Earth as a force center and both of these cases are covered. We also present an elucidation of the essential difference between the Laplacian and Gaussian forms of initial orbit determination.

Next new tests of the high-quality, data-rich, Laplacian method are discussed. It has been applied to an asteroid discovered by this author (1982HS). This is in a very rare high-inclination, high-eccentricity orbit. It was also applied to the original Earth-approaching minor planet (1862 Apollo) and another, new unusual minor planet (1982SA).*

The last section discusses the problem of orbit determination for artificial satellites using angles and angular rate data. One can approach this problem traditionally or, uniquely, *explicitly* solve the problem analytically. The section concludes with a modern look at the determination of $\mathbf{l}$ and $\mathbf{\dot{l}}$.

Other Variants.   My literature search reveals two other types of variants on data-rich, angles-only, initial orbit determination. One is a pair of papers

---

* A summary of many artificial satellite tests of the Gauss–Gibbs and Laplace–Taff methods are in the next chapter.

by Kiselev and Bykov (1974, 1976). They prefer to analyze the problem in terms of the concepts of analytic geometry (i.e., tangents, normals, curvature, the binomial, and the torsion). Herget (1948) too elaborates somewhat on this point of view (pp. 37-39). Beyond the geometrical perspective I see no utility in these methods. A paper by Neutsch (1981) exploits the f and g *functions* (not series!) in a statistical initial orbit determination technique. I cannot tell if this is practical. Finally, the methods for angles and angular velocity presented below have been extended to include multiple observations and perturbing forces. However they require very good angular velocity data to be successful.

LAPLACE'S METHOD

In this section Laplace's method from a heliocentric point of view for minor planets and from a geocentric point of view for artificial satellites will be presented. The last section stresses the essential mathematical differences between the Laplacian and Gaussian techniques.

Artificial Satellites

Let the observer's geodetic datum be (H, λ, ϕ')—his height above the reference ellipsoid, his geocentric longitude, and his geocentric latitude. From this data one can compute the observer's geocentric distance ρ and the local sidereal time τ corresponding to the local mean solar time t (which is simply related to Universal Time; Ephemeris Time buffs will have to wait their turn). Thus, the geocentric location of the observer $\boldsymbol{\rho}$ is known and the derivatives of $\boldsymbol{\rho}$ with respect to t can be calculated.

Consider now some artificial satellite whose geocentric location is $\mathbf{r}$ in the equatorial coordinate system. Its topocentric location $\mathbf{R}$ is related to $\mathbf{r}$ and $\boldsymbol{\rho}$ via

$$\mathbf{r} = \mathbf{R} - \boldsymbol{\rho} \tag{7.1}$$

The direction to the satellite can be measured. To stress this, $\mathbf{R}$ is written as

$$\mathbf{R} = R\mathbf{l} \tag{7.2}$$

where $\mathbf{l}$ is the unit vector of topocentric direction cosines. Substitute Eq. 7.2 into Eq. 7.1 and differentiate twice with respect to t,

$$\mathbf{r} = R\mathbf{l} + \boldsymbol{\rho}$$
$$\dot{\mathbf{r}} = \dot{R}\mathbf{l} + R\dot{\mathbf{l}} + \dot{\boldsymbol{\rho}} \tag{7.3}$$
$$\ddot{\mathbf{r}} = \ddot{R}\mathbf{l} + 2\dot{R}\dot{\mathbf{l}} + R\ddot{\mathbf{l}} + \ddot{\boldsymbol{\rho}}$$

Since the satellite does orbit the Earth,

$$\ddot{\mathbf{r}} = \frac{-GM_{\oplus}\mathbf{r}}{r^3} \tag{7.4}$$

where $r = |\mathbf{r}|$. Replace the left-hand side of the last line of Eq. 7.3 with its equivalent in Eq. 7.4 to find that

$$\frac{-GM_\oplus \mathbf{r}}{r^3} = \ddot{R}\mathbf{l} + 2\dot{R}\dot{\mathbf{l}} + R\ddot{\mathbf{l}} + \ddot{\boldsymbol{\rho}} \tag{7.5}$$

We can isolate R by finding a vector perpendicular to both $\mathbf{l}$ and $\dot{\mathbf{l}}$. Clearly $\mathbf{l} \times \dot{\mathbf{l}}$ will do and

$$R = \frac{-1}{(\mathbf{l} \times \dot{\mathbf{l}}) \cdot \ddot{\mathbf{l}}} \left[\frac{GM_\oplus}{r^3} (\mathbf{l} \times \dot{\mathbf{l}}) \cdot \mathbf{r} + (\mathbf{l} \times \dot{\mathbf{l}}) \cdot \ddot{\boldsymbol{\rho}} \right] \tag{7.6}$$

Next take the scalar product of the uppermost line of Eq. 7.3 with itself to deduce

$$r^2 = R^2 + 2R\mathbf{l} \cdot \boldsymbol{\rho} + \rho^2 \tag{7.7}$$

Finally, replace $\mathbf{r}$ on the right-hand side of Eq. 7.6 with its value from Eq. 7.1, namely,

$$R = \frac{-1}{(\mathbf{l} \times \dot{\mathbf{l}}) \cdot \ddot{\mathbf{l}}} \left[\frac{GM_\oplus}{r^3} (\mathbf{l} \times \dot{\mathbf{l}}) \cdot \boldsymbol{\rho} + (\mathbf{l} \times \dot{\mathbf{l}}) \cdot \ddot{\boldsymbol{\rho}} \right] \tag{7.8}$$

Equations 7.7 and 7.8 are two-coupled, nonlinear equations in the two unknowns r and R. They are equivalent to a single eighth-order polynomial in r.

If we step back from the algebra and look at the terms appearing in these two equations, they are of three types—quantities that are known or can be computed $(\boldsymbol{\rho}, \dot{\boldsymbol{\rho}}, \ddot{\boldsymbol{\rho}})$, quantities that can be measured $(\mathbf{l})$, and quantities for which a method of calculating is needed ($\dot{\mathbf{l}}$ and $\ddot{\mathbf{l}}$). The formal solution to the initial orbit determination process requires $\mathbf{r}$ and $\dot{\mathbf{r}}$ (or R and $\dot{R}$). Thus we need an equation for $\dot{R}$. Clearly we may obtain one from Eq. 7.5, for upon scalar multiplication by $\mathbf{l} \times \ddot{\mathbf{l}}$,

$$2\dot{R} = \frac{-1}{(\mathbf{l} \times \ddot{\mathbf{l}}) \cdot \dot{\mathbf{l}}} \left[\frac{GM_\oplus}{r^3} (\mathbf{l} \times \ddot{\mathbf{l}}) \cdot \mathbf{r} + (\mathbf{l} \times \ddot{\mathbf{l}}) \cdot \ddot{\boldsymbol{\rho}} \right]$$

or after replacement of $\mathbf{r}$ by $R\mathbf{l} + \boldsymbol{\rho}$

$$2\dot{R} = \frac{-1}{(\mathbf{l} \times \ddot{\mathbf{l}}) \cdot \dot{\mathbf{l}}} \left[\frac{GM_\oplus}{r^3} (\mathbf{l} \times \ddot{\mathbf{l}}) \cdot \boldsymbol{\rho} + (\mathbf{l} \times \ddot{\mathbf{l}}) \cdot \ddot{\boldsymbol{\rho}} \right] \tag{7.9}$$

The set of equations 7.7–7.9 represents a complete solution to the problem without any assumptions concerning the nature of the orbit. The reader should note that these results are completely rigorous. Methods to obtain approximations to the topocentric angular velocity $\dot{\mathbf{l}}$ and the topocentric angular acceleration $\ddot{\mathbf{l}}$ will be discussed below.

Asteroids

Let

   $\mathbf{r}$ = heliocentric equatorial location of the asteroid

   $\mathbf{R}_\odot$ = geocentric equatorial location of the Sun = $(X_\odot, Y_\odot, Z_\odot)$

   $\boldsymbol{\rho}$ = geocentric location of the observer = $\rho\mathbf{l}(\tau, \phi') = \rho(\xi, \eta, \zeta)$

   $\mathbf{R}$ = topocentric equatorial location of the asteroid = $R\mathbf{l}(A, \Delta) = R(\lambda, \mu, \nu)$

where the vector of direction cosines is given by

$$\mathbf{l}(\alpha, \delta) = (\cos \delta \cos \alpha, \cos \delta \sin \alpha, \sin \delta)$$

From Fig. 34 one can see that

$$\mathbf{r} = -\mathbf{R}_\odot + \boldsymbol{\rho} + \mathbf{R} \tag{7.10}$$

The minor planet orbits the Sun. Ignoring the mass of the asteroid compared to that of the Sun ($= M_\odot$) and planetary perturbations,

$$\ddot{\mathbf{r}} = \frac{-GM_\odot\mathbf{r}}{r^3} \tag{7.11}$$

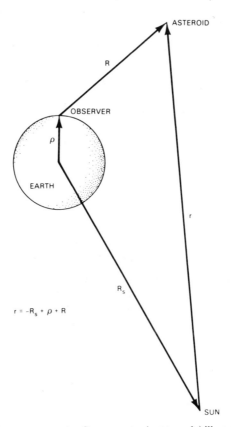

Figure 34. Asteroid–observer–geocenter–Sun geometry (not to scale) illustrating diurnal parallax.

Substituting Eq. 7.10 into Eq. 7.11 results in

$$\ddot{\mathbf{R}} + \ddot{\boldsymbol{\rho}} - \ddot{\mathbf{R}}_\odot = \frac{-GM_\odot(\mathbf{R} + \boldsymbol{\rho} - \mathbf{R}_\odot)}{r^3} \tag{7.12}$$

But the Earth revolves about the Sun (approximately; neglecting the mass of the Moon, the geocentric distance of the Earth–Moon barycenter, and planetary perturbations again) so

$$\ddot{\mathbf{R}}_\odot = \frac{-GM_\odot \mathbf{R}_\odot}{R_\odot^3} \tag{7.13}$$

Thus, after using Eq. 7.13 in Eq. 7.12,

$$\ddot{\mathbf{R}} + \frac{GM_\odot \mathbf{R}}{r^3} = -GM_\odot \mathbf{R}_\odot \left(\frac{1}{R_\odot^3} - \frac{1}{r^3} \right) - \ddot{\boldsymbol{\rho}} - \frac{GM_\odot \boldsymbol{\rho}}{r^3} \tag{7.14}$$

Consider (say) the x component of Eq. 7.14. As $d^2(R\lambda)/dt^2 = \ddot{R}\lambda + 2\dot{R}\dot{\lambda} + R\ddot{\lambda}$, and ρ and ϕ' are constants, one deduces that

$$\lambda \ddot{R} + \dot{\lambda}(2\dot{R}) + \left(\ddot{\lambda} + \frac{GM_\odot \lambda}{r^3} \right) R = -GM_\odot X_\odot \left(\frac{1}{R_\odot^3} - \frac{1}{r^3} \right) + \rho\xi \left(\ddot{\tau}^2 - \frac{GM_\odot}{r^3} \right) \tag{7.15}$$

With the y and z equations corresponding to Eq. 7.15, they can be regarded as a system of three linear, inhomogeneous equations in the three unknowns $\ddot{R}$, $2\dot{R}$, and R. The determinant of the system, D, is given by

$$D = \begin{vmatrix} \lambda & \dot{\lambda} & \ddot{\lambda} + \dfrac{GM_\odot \lambda}{r^3} \\[2mm] \mu & \dot{\mu} & \ddot{\mu} + \dfrac{GM_\odot \mu}{r^3} \\[2mm] \nu & \dot{\nu} & \ddot{\nu} + \dfrac{GM_\odot \nu}{r^3} \end{vmatrix} = \begin{vmatrix} \lambda & \dot{\lambda} & \ddot{\lambda} \\[2mm] \mu & \dot{\mu} & \ddot{\mu} \\[2mm] \nu & \dot{\nu} & \ddot{\nu} \end{vmatrix}$$

The solution for R is

$$R = \frac{D_1'}{D} \tag{7.16}$$

with D_1' given by

$$D_1' = \begin{vmatrix} \lambda & \dot{\lambda} & -GM_\odot X_\odot \left(\dfrac{1}{R_\odot^3} - \dfrac{1}{r^3} \right) + \rho\xi \left(\ddot{\tau}^2 - \dfrac{GM_\odot}{r^3} \right) \\[3mm] \mu & \dot{\mu} & -GM_\odot Y_\odot \left(\dfrac{1}{R_\odot^3} - \dfrac{1}{r^3} \right) + \rho\eta \left(\ddot{\tau}^2 - \dfrac{GM_\odot}{r^3} \right) \\[3mm] \nu & \dot{\nu} & -GM_\odot Z_\odot \left(\dfrac{1}{R_\odot^3} - \dfrac{1}{r^3} \right) + \rho\zeta \left(0 - \dfrac{GM_\odot}{r^3} \right) \end{vmatrix}$$

If $\rho = 0$ then D_1' simplifies to D_1:

$$D_1' = -GM_\odot \left(\frac{1}{R_\odot^3} - \frac{1}{r^3} \right) \begin{vmatrix} \lambda & \dot{\lambda} & X_\odot \\ \mu & \dot{\mu} & Y_\odot \\ \nu & \dot{\nu} & Z_\odot \end{vmatrix} \equiv D_1$$

Note that if one knows the topocentric direction cosines [i.e., (λ, μ, ν)], the topocentric angular velocity [i.e., $(\dot{\lambda}, \dot{\mu}, \dot{\nu})$], and the topocentric angular acceleration [i.e., $(\ddot{\lambda}, \ddot{\mu}, \ddot{\nu})$], then the formal solution for R in Eq. 7.16 is really an expression for $R(r)$ since the observer's location, velocity, and acceleration as well as the solar location, velocity, and acceleration are known. Hence, to solve the problem, one more relationship between R and r is needed. To obtain this relationship the basic geometrical relationship of Eq. 7.10 is squared. The result is

$$r^2 = R^2 + 2\rho R \cos \mathcal{Z} - 2RR_\odot \cos \psi + \rho^2 + R_\odot^2 - 2\boldsymbol{\rho} \cdot \mathbf{R}_\odot \qquad (7.17)$$

where $\mathcal{Z}$ is the topocentric zenith distance of the minor planet $[\boldsymbol{\rho} \cdot \mathbf{R} = \rho R \cos \mathcal{Z}]$ and $\cos \psi = (\lambda X_\odot + \mu Y_\odot + \nu Z_\odot)/R_\odot$. This completes the solution of the problem. We have two equations in two unknowns. One equation expresses the physics, the other the geometry. If ρ is set equal to zero, then Eqs. 7.16 and 7.17 reduce to

$$R = \left(\frac{D_1}{D} \right) \left[\frac{1}{R_\odot^3} - \frac{1}{r^3} \right]$$

$$r^2 = R^2 - 2RR_\odot \cos \psi + R_\odot^2 \qquad (7.18)$$

The form of these is classic in angles-only initial orbit determination as are the facts that the coupled system is equivalent to a single polynomial equation of the eighth degree,

$$s^8 - (a^2 - 2a \cos \psi + 1)s^6 + 2a(a - \cos \psi)s^3 - a^2 = 0 \qquad (7.19)$$

(where $s = r/R_\odot$ and $a = D_1/DR_\odot^4$) and that $s = 1$, $R = 0$ is a solution. This represents the Earth. Note that the explicit inclusion of the diurnal parallax removes this degeneracy—a useful point for Earth-approaching asteroids.

Let us summarize: If one knows one's location, velocity, and acceleration relative to the center of the Earth (as one can), and if one knows the Earth's (equivalently the Sun's) location, velocity, and acceleration (again, as one can), and if one knows the minor planet's position, angular velocity, and angular acceleration (see below), then one can compute the topocentric distance of the asteroid and its topocentric radial velocity. The first statement follows since R is a solution of the coupled $R(r)$ set (Eq. 7.18). One finds $\dot{R}$ by returning to Eq. 7.15 and solving for $2\dot{R}$, namely,

$$2\dot{R} = \frac{D_2'}{D} \qquad (7.20)$$

where

$$D'_2 = \begin{vmatrix} \lambda & -GM_\odot X_\odot \left(\dfrac{1}{R_\odot^3} - \dfrac{1}{r^3} \right) + \rho\xi \left(\dot{\tau}^2 - \dfrac{GM_\odot}{r^3} \right) & \ddot{\lambda} \\[2ex] \mu & -GM_\odot Y_\odot \left(\dfrac{1}{R_\odot^3} - \dfrac{1}{r^3} \right) + \rho\eta \left(\dot{\tau}^2 - \dfrac{GM_\odot}{r^3} \right) & \ddot{\mu} \\[2ex] \nu & -GM_\odot Z_\odot \left(\dfrac{1}{R_\odot^3} - \dfrac{1}{r^3} \right) + \rho\zeta \left(0 - \dfrac{GM_\odot}{r^3} \right) & \ddot{\nu} \end{vmatrix}$$

Note that

$$D'_2(\rho = 0) = -GM_\odot \left(\frac{1}{R_\odot^3} - \frac{1}{r^3} \right) \begin{vmatrix} \lambda & X_\odot & \ddot{\lambda} \\ \mu & Y_\odot & \ddot{\mu} \\ \nu & Z_\odot & \ddot{\nu} \end{vmatrix} \equiv D_2$$

Clearly, once a complete topocentric specification of location and velocity is available (coupled with the preserved ancillary information ρ, $\dot{\rho}$, $\mathbf{R}_\odot$, and $\dot{\mathbf{R}}_\odot$), one can produce a corresponding heliocentric set. Going from the data $\mathbf{r}$ and $\dot{\mathbf{r}}$ at some time $t = t_0$ to the orbital element set is a straightforward algebra problem.

The essential points are that (1) there is no appreciable neglect of physics and (2) the geometry and physics are enforced at a single instant of time. In practice one relaxes the proviso that no mathematical approximations are made by numerically differentiating both $\mathbf{R}_\odot$ and $\mathbf{l}(A, \Delta) = (\lambda, \mu, \nu)$ to obtain $\ddot{\mathbf{R}}_\odot$, $\dot{\mathbf{l}}$, and $\ddot{\mathbf{l}}$. Note that this enforces the geometrical constraint at more than one point while still enforcing the physics at a single instant of time. Note too that although numerical differentiation may well be inaccurate, it is neither theoretically impossible, nor forbidden, nor can the time span of the observations preclude orbit determination.

How does one obtain $\dot{\mathbf{l}}$ and $\ddot{\mathbf{l}}$? At a minimum three values of $\mathbf{l}$ are needed to determine $\ddot{\mathbf{l}}$. Using only three sets amounts to a trick as far as I'm concerned. Furthermore, on slow-moving objects such as asteroids ($\sim$0°.5–2°/day for the fastest ones) numerical differentiation of three observations is criminal. (Clearly, with observational data of low accuracy acquired on slowly moving objects, the inherently unstable process of central difference approximations for second derivatives will produce a value for $\ddot{\mathbf{l}}$ of little utility or resemblance to reality. Thus, in the past, in practice, Laplace's method failed.) On the other hand, smoothing a dozen or two observations over a night and then analytically differentiating might work.

The Gaussian Difference

What Gauss did that was different was to exploit the fact that the orbit lies in a plane. Then for three points at t_1, t_2, and t_3 one can write

$$a\mathbf{r}_1 + b\mathbf{r}_2 + c\mathbf{r}_3 = 0$$

since the 3 three-dimensional location vectors are linearly dependent. From the fact that the vector cross product yields an area and the known properties of central force motion, one can derive formulas for a, b, and c. The essential step that makes the computation feasible is the use of power series (vs. numerical differentiation) to express r_1 and r_3 in terms of r_2. These series are called the f and g series. As discussed at length in Chapter Eight, the radius of convergence of these series rapidly approaches zero as the eccentricity of the orbit approaches unity. Therefore, one may be *a priori* forbidden to use the Gaussian technique and not know it. Thus, the central element of contrariness between the methods of Laplace and Gauss lies in the nature of an approximation. Laplace's method forces one to make a *numerical* approximation in order to calculate the angular velocity and the angular acceleration. Note that there is no inherent restriction to only three observations. Gauss's method forces one to make an *analytical* approximation. Therein the use of only three sets of angular measurements is absolute. Note too that although a particular numerical differentiation may be more or less accurate [and can be arbitrarily refined *without* expanding the observational time span (but by increasing the data acquisition rate)], Gauss's approximation may fail catastrophically without notice or means of redress *ex post facto*.

NUMERICAL EXAMPLES

A logical thing to do after the explanation of a computational procedure is to illustrate its use. However, when the numerical aspects of the calculation are sufficiently complex that a computer is utilized to perform them, then the mode of presentation of such an example becomes problematical. Less than full disclosure is clearly necessary but not always desirable. The course adopted below is to provide a full textual explanation of the computations, full numerical starting information, the critical intermediary value(s), and full numerical ending information. The observational data is real, acquired at M.I.T.'s artificial satellite observatory (Example 3.7).

Since the observational data consists of topocentric right ascension and declination (plus their associated time), and the method requires not only position but angular velocity and angular acceleration as well, some means of numerically differentiating the data is necessary. The background for this is briefly discussed below. The smoothing done is simple, unweighted, least squares fits to polynomials—each coordinate independently. Following the order of presentation for the theory above, numerical examples of artificial satellite orbit determination are discussed first, then those for asteroids. The details of going from r and $\dot{r}$ or r_1 and r_2 to $a = (a, t, \Omega, e, \omega, M_0)$ are found in Chapter Eight.

Data Smoothing

The fundamental decisions to be made concern the time interval over which the smoothing is to be performed and the form of the smoothing function. In both the fast-moving minor planet and the artificial satellite cases I rely on Weierstrass's approximation theorem and use a polynomial form. The degree of the polynomial is not so quickly decided upon and will be discussed below. In addition, whereas one would clearly smooth at most over a few hours for an artificial satellite, one might do so for as long as a week (or be forced to do so for such a duration) for an asteroid. Hence we first turn to the relative advantages of night-to-night preliminary smoothing for a fast-moving minor planet versus a simultaneous fit of several nights' observations. While doing this, we must keep in mind that what is wanted are the best estimators for R and $\dot{R}$. Considering the transcendental nature of the dependence of R and $\dot{R}$ on the observations (see Eqs. 7.16 and 7.20), we will settle for the best estimators of A, $\dot{A}$, and $\ddot{A}$ and for Δ, $\dot{\Delta}$, and $\ddot{\Delta}$ (if possible to do so simultaneously).

Asteroids

At first glance the principal advantage of a nightly reduction of the observations is that a low-order polynomial will do. On the other hand, since the total observing time is not a priori set and there may be an order of magnitude difference in the angular speeds of the asteroids observed, a single polynomial may well not fit all cases of interest. The primary drawback of night-to-night reductions is the lack of a rigorous, clearly beneficial method of combination to yield position, angular velocity, and angular acceleration at a simultaneous epoch.

Let us first try to decide on the appropriate order of the smoothing polynomial for an individual night's observations. A *fortiori* nothing less than a linear model will do—even if one has a "single" observation.* Since the general premise of this analysis is that of an abundance of data, the question of the highest-order polynomial to be used must be addressed. The analysis of the physics and geometry presented above is relatively complete. Therefore, an excessively high-order polynomial is not necessary to absorb unmodeled effects. Thus, considering the quarry, the inclusion of fifth-order terms seems superfluous, quadratic minimally sufficient.

Another digression. [I am aware of the fact that I have just argued myself into the position of (theoretically) being able to determine an orbit based on an extremely short time span.] All that Laplace's method requires are direction cosines, their first and their second derivatives. Once the smoothing polynomial

---

* I must now digress for one never acquires a single observation of an artificial satellite or of an asteroid in a search mode. Consider a photographic search first. One discriminates the object *because* of its motion—a trail is left on the photographic plate (exposed with the telescope in sidereal drive so that the stars are held fixed) marking the passage of the object. Hence, even if this is the only record, one can (and does in extreme cases) independently measure the endpoints of the trail to deduce two positions. Streak formation by electronic means is analogous as is forming a broken image by chopping with a rotating shutter. Hence, there are always at least two observations.

is second order, these are all computable. However, a much better description of what one knows are estimators for the direction cosines, their rates of change, and their second derivatives. One also has estimates of the variances and covariances of these quantities, which should preclude any premature attempt at initial orbital determination.

Ideally one would prefer to use some objective criteria for ascertaining the correct number and type of terms to include in the interpolating polynomial. Tests of significance, based on the F test, can be constructed but lack a rigorous logical basis. Hence I would rely on experience and judgment to determine the correct degree of the interpolating polynomial and eschew apparently formal procedures.

Finally each night is likely to be different. Some will be completely cloudy, some cloudy in only the first half, some clear, and so on. Therefore, although a quartic or a quintic fit might be appropriate for the perigee passage of an Earth-approaching asteroid observed near the winter solstice, a quadratic fit the next evening (which is almost completely clouded out) would suffice. One cannot afford to lose this element of flexibility. So, after observing the minor planet for some number of nights we have, for each (at least partially clear) night, a position, probably a good angular velocity, and likely an estimate for the angular acceleration. In addition, the results of a given night's fit may not be simultaneously epoched (to obtain the minimum variance estimator, see below). From this inhomogenous and incomplete set of intermediate reductions we must now deduce the values for A, $\dot{A}$, and $\ddot{A}$ and for Δ, $\dot{\Delta}$, and $\ddot{\Delta}$ to begin the calculations of Laplace's method.

How? How indeed. This author knows of no theoretically sound method of combination that will unambiguously produce, in some well-defined and meaningful statistical sense, "best" estimates for a position, angular velocity, and angular acceleration (all at one epoch). We can concoct a large variety of apparently reasonable procedures to do this. None of these appears to be more attractive than an overall fit of all of the data *ab initio*.

Artificial Satellites

In general the observing span should be as short as possible commensurate with obtaining good estimates for the desired quantities. The order of the polynomial necessary to do this depends on the topocentric appearance of the orbit—contrast two circular orbits with the same period but one with an inclination of zero and the other with an inclination of $>30°$. The variety of combinations of orbital element sets and geographical circumstances are too numerous for any but the most general rules of thumb. It also becomes imperative that not only is the set of observations large, but dense—performing an observation must be rapid enough that high-order polynomial fits are necessary because of the duration of the observing span. In any case, and the one to be dealt with in practice, at least a quadratic form is necessary and a quintic would usually be excessive. Finally, one can easily visualize situations wherein the order of the right ascension (or azimuth) polynomial is different from that of the declination (or altitude) polynomial. Experience and wisdom are needed in general in the artificial satellite case.

Artificial Satellite Examples

Three examples of the computation of artificial satellite orbital elements by my revision of Laplace's method are presented below. The first case is of a Russian Cosmos satellite rocket body on its way to apogee in the northern hemisphere. The second case is a similar object (i.e., another Russian Cosmos rocket body) already at apogee in the southern hemisphere. The third case is also a Russian satellite, but this time soon after equator crossing and almost doubling its topocentric distance during the time span of observation. The format of the examples will be identical. First a list of the observational data (A, Δ, t) will be given. Then columns for R, $\dot{A}$, and $\dot{\Delta}$ (computed from the orbital element set) will be provided for reference. The next set of numbers will correspond to the second-, third-, fourth-, and possibly fifth-order polynomial fits of the right ascensions and declinations (separately). The results of these least squares fits are an average time of observation $\langle t \rangle$, corresponding values of $\langle \alpha \rangle$, $\langle \delta \rangle$, $\langle \dot{\alpha} \rangle$, $\langle \dot{\delta} \rangle$, $\langle \ddot{\alpha} \rangle$, $\langle \ddot{\delta} \rangle$, and the residuals from the fits (ε_α and ε_δ). In all cases the units are seconds of arc ("/sec or "/sec²). R is in kilometers. From a fit to the right ascension data and one to the declinations an element set is computed and compared to the actual one. The 0.5-, 1-, and 2-hr prediction errors in the positions due to the generated initial element sets are also given. The numerical procedures used are exactly those embodied in Eqs. 7.7–7.9, so there is little narrative discussion within the examples themselves.

Example 7.1. These are 11 observations each separated by 4 min from day 35 of 1982 (times are UTC).

t	A	Δ	R (km)	$\dot{A}$	$\dot{\Delta}$
$15^h 24^m 0^s$	$09^h 20^m 01\overset{s}{.}6$	$+44° 28' 03''$	17063.2	$48\overset{''}{.}6$/sec	$47\overset{''}{.}6$/sec
15 28 0	09 33 13.4	47 29 19	17529.1	50.4	43.1
15 32 0	09 46 54.6	50 13 00	18009.2	52.3	38.8
15 36 0	10 01 07.1	52 40 05	18500.1	54.3	34.8
15 40 0	10 15 52.2	54 51 30	18998.6	56.3	31.0
15 44 0	10 31 09.9	56 48 07	19502.0	58.3	27.4
15 48 0	10 46 58.6	58 30 44	20008.1	60.2	24.0
15 52 0	11 03 15.6	60 00 07	20515.0	61.9	20.8
15 56 0	11 19 56.8	61 16 59	21021.1	63.2	17.7
16 00 0	11 36 56.2	62 22 05	21525.1	64.2	14.9
16 04 0	11 54 07.2	63 16 05	22026.0	64.6	12.2

$\langle t \rangle = 15^h 44^m 0^s$ and the result of the quadratic through quintic least squares fits are as follows:

Fit Order	Quadratic	Cubic	Quartic	Quintic
$\langle\alpha\rangle$	$10^h31^m13^s95$	Same	$10^h31^m09^s89$	Same
$\langle\delta\rangle$	$+56°48'14\rlap{.}''7$	Same	$+56°48'07\rlap{.}''0$	Same
$\langle\dot\alpha\rangle$	$57\rlap{.}''948$/sec	$58\rlap{.}''351$/sec	Same	$58\rlap{.}''346$/sec
$\langle\dot\delta\rangle$	$27\rlap{.}''960$/sec	$27\rlap{.}''367$/sec	Same	$27\rlap{.}''371$/sec
$\langle\ddot\alpha\rangle$	$0\rlap{.}''007385$/sec^2	Same	$0\rlap{.}''008118$/sec^2	Same
$\langle\ddot\delta\rangle$	$-0\rlap{.}''014671$/sec^2	Same	$-0\rlap{.}''014579$/sec^2	Same
ε_α	$138\rlap{.}''6$	$51\rlap{.}''7$	$1\rlap{.}''21$	$0\rlap{.}''75$
ε_δ	$189\rlap{.}''8$	$6\rlap{.}''5$	$0\rlap{.}''87$	$0\rlap{.}''19$

The mathematics of equally weighted least squares with equally spaced data has some degenerate redundancies as the order of the fit increases. Thus, when reading across the table, an entry such as "Same" indicates that the quantity to the left is to be repeated.

The element set from the quintic fits at $35^d15^h43^m59^s93$ (planetary aberration!) is displayed alongside the actual one at $37^d02^h40^m17^s49$.

	5, 5	True
i	$64°28$	$64°29$
Ω	$121°56$	$121°41$
e	0.6506	0.6519
ω	$318°94$	$319°41$
M_0	$40°52$	$6°99$
n	2.0315 rev/day	1.9980 rev/day

Finally the predicted positional errors 0.5, 1, and 2 hr into the future were $0\rlap{.}'6$, $0\rlap{.}'6$, and $3\rlap{.}'2$.

Example 7.2. In this case there are 11 observations separated by 2 min each on day 32 of 1982.

t	A	Δ	R (km)	$\dot A$	$\dot\Delta$
$04^h00^m0^s$	$05^h52^m52^s8$	$-52°45'\ 18''$	42398.4	$3\rlap{.}''6$/sec	$-8\rlap{.}''1$/sec
4 02 0	05 53 21.6	$-53\ 01\ 32$	42386.6	3.6	-8.1
4 04 0	05 53 50.8	$-53\ 17\ 48$	42373.0	3.7	-8.1
4 06 0	05 54 20.5	$-53\ 34\ 04$	42357.6	3.7	-8.1
4 08 0	05 54 50.5	$-53\ 50\ 21$	42340.5	3.8	-8.1
4 10 0	05 55 20.9	$-54\ 06\ 40$	42321.5	3.8	-8.2
4 12 0	05 55 51.8	$-54\ 22\ 59$	42300.7	3.9	-8.2
4 14 0	05 56 23.1	$-54\ 39\ 19$	42278.2	4.0	-8.2
4 16 0	05 56 54.8	$-54\ 55\ 41$	42253.8	4.0	-8.2
4 18 0	05 57 27.1	$-55\ 12\ 04$	42227.6	4.1	-8.2
4 20 0	05 57 29.8	$-55\ 28\ 28$	42199.6	4.1	-8.2

$\langle t \rangle = 04^h 10^m 0^s$ and the results of the quadratic through quintic least squares fits are as follows:

Fit Order	Quadratic	Cubic	Quartic	Quintic
$\langle \alpha \rangle$	$05^h 55^m 20^s\!.92$	Same	$05^h 55^m 20^s\!.93$	Same
$\langle \delta \rangle$	$-54°06'39''\!.6$	Same	$-54°06'39''\!.6$	Same
$\langle \dot{\alpha} \rangle$	$3''\!.836/\text{sec}$	$3''\!.831/\text{sec}$	Same	$3''\!.830/\text{sec}$
$\langle \dot{\delta} \rangle$	$-8''\!.158/\text{sec}$	$-8''\!.157/\text{sec}$	Same	$-8''\!.156/\text{sec}$
$\langle \ddot{\alpha} \rangle$	$0''\!.000448/\text{sec}$	Same	$0''\!.000435/\text{sec}^2$	Same
$\langle \ddot{\delta} \rangle$	$-0''\!.000074/\text{sec}^2$	Same	$-0''\!.000071/\text{sec}^2$	Same
ε_α	$0''\!.87$	$0''\!.42$	$0''\!.37$	$0''\!.34$
ε_δ	$0''\!.32$	$0''\!.25$	$0''\!.24$	$0''\!.23$

Since adding a free parameter to adjust in a least squares analysis always reduces the residuals, it is not clear to me that anything beyond the cubic fit is necessary. The element set at $32^d 04^h 09^m 59^s\!.86$ is displayed below next to the real one at epoch $34^d 20^h 53^m 09^s\!.05$.

	3, 3	True
i	$71°\!.71$	$71°\!.28$
Ω	$248°\!.49$	$248°\!.07$
e	0.5973	0.6434
ω	$39°\!.44$	$41°\!.67$
M_0	$215°\!.22$	$352°\!.85$
n	1.7701 rev/day	2.0065 rev/day

The predicted positional errors 0.5, 1, and 2 hr into the future were $0''\!.4$, $2''\!.1$, and $15''\!.6$.

Example 7.3. In this last artificial satellite case there are 13 observations separated by 2 min each early on day 304 of 1981.

t	A	Δ	R	$\dot{A}$	$\dot{\Delta}$
$0^h 24^m 0^s$	$02^h 36^m 32^s\!.7$	$+34°56'52''$	10629.6	$66''\!.7/\text{sec}$	$73''\!.4/\text{sec}$
0 26 0	02 45 23.6	37 14 03	11183.9	66.0	64.0
0 28 0	02 54 08.6	39 13 48	11737.6	65.2	55.9
0 30 0	03 02 46.5	40 58 34	12288.9	64.3	49.0
0 32 0	03 11 16.4	42 30 26	12836.5	63.2	43.0
0 34 0	03 19 37.6	43 51 08	13379.3	62.1	37.8
0 36 0	03 27 49.4	45 02 06	13916.7	60.9	33.3
0 38 0	03 35 51.2	46 04 36	14448.1	59.6	29.3
0 40 0	03 43 42.7	46 59 38	14973.2	58.3	25.8
0 42 0	03 51 23.5	47 48 07	15491.6	56.9	22.7
0 44 0	03 58 53.3	48 30 49	16003.4	55.5	20.0
0 46 0	04 06 12.0	49 08 22	16508.2	54.1	17.6
0 48 0	04 13 19.5	49 41 21	17006.2	52.7	15.4

$\langle t \rangle = 0^h 36^m 0^s$. Quadratic, cubic, and quartic fits give the following results:

Fit Order	Quadratic	Cubic	Quartic
$\langle \alpha \rangle$	$03^h 27^m 48^s\!.74$	Same	$03^h 27^m 49^s\!.38$
$\langle \delta \rangle$	$+45°03'17''\!.0$	Same	$+45°02'05''\!.5$
$\langle \dot{\alpha} \rangle$	$60''\!.604/\sec$	$60''\!.866/\sec$	Same
$\langle \dot{\delta} \rangle$	$35''\!.734/\sec$	$33''\!.211/\sec$	Same
$\langle \ddot{\alpha} \rangle$	$-0''\!.010035/\sec^2$	Same	$-0''\!.010362/\sec^2$
$\langle \ddot{\delta} \rangle$	$-0''\!.037677/\sec^2$	Same	$-0''\!.035242/\sec^2$
ε_α	$50''\!.8$	$8''\!.29$	$0''\!.55$
ε_δ	$485''\!.8$	$62''\!.0$	$7''\!.12$

There was no further appreciable reduction of the residuals in the quintic fit. Another good sign of the stability of the least squares process is the uniformity of the $\ddot{\alpha}$ and $\ddot{\delta}$ values. Finally, on occasion, it may make sense to (say) use the results of a cubic fit for the right ascension with (say) the quartic fit for the declination. In this case the 4, 5 combination may be arguably better than the 4, 4 orbital element set shown below.

	4, 4	True
i	$62°\!.85$	$62°\!.87$
Ω	$333°\!.25$	$333°\!.13$
e	0.7390	0.7397
ω	$316°\!.16$	$316°\!.13$
M_0	$22°\!.40$	$4°\!.81$
n	1.9952 rev/day	1.9926 rev/day

The epoch for the true set is $305^d 0^h 06^m 09^s\!.57$ whereas that of the 4, 4 set is $304^d 0^h 35^m 59^s\!.99$. The three prediction errors were $0''\!.8$, $1''\!.3$, and $1''\!.8$ at 0.5, 1, and 2 hr.

Discussion of this process in more detail as well as comparisons with the Gauss–Gibbs techniques are given in Chapter Eight.

Minor Planet Examples

Why does one *need* an orbital element set for an asteroid today? To recover it at the next opposition (~ 1.3 yr = 1 synodic period). Not to recover it tomorrow, not even to recover it at the next dark of the Moon. Therefore, it is difficult to get excited about constructing orbital element sets for main belt minor planets. For an Earth-approaching asteroid one does need an element

set for a subsequent month. Both 1982HS and 1982SA were discovered by my
search program. Unbeknown to me 1982SA was also discovered prior (by two
days) by E. Shoemaker and E. F. Helin. The data discussed below for these
objects and 1862 Apollo (recovered by us accidentally) are real observations
acquired by us. Their precision is 2-3″. The data utilized for the artificial
satellite tests above were pseudo-observations good to 1″.

1982HS and 1982SA

Both 1982HS and 1982SA are inner main belt, high-inclination, high-eccentricity, minor planets. Since they are minor planets, and orbit the Sun, one
needs a heliocentric initial orbit generator (Fig. 34). One also needs to take
into account the fact that the observer is on the surface of the Earth. We did
this above, where the fundamental equations of the problem are given—the
coupled pair of Eqs. 7.16 and 7.17. In practice, this set is solved as follows:
From the observed angular speed one can tell that the minor planet is not
close, hence the diurnal parallax correction can be momentarily ignored. We
therefore solve the simpler system (Eq. 7.18), which results in Eq. 7.19.
However, because $s = 1$ is an exact root of that eighth-order polynomial, we
actually use Eq. 7.19 divided by $s - 1$; namely,

$$s^7 + s^6 - a(a - 2 \cos \psi)(s^5 + s^4 + s^3) + a^2(s^2 + s + 1) = 0$$

This also protects against unusual geometrical circumstances. Having a guess
for $r = sR$ one now can correct the original observations for diurnal parallax,
redo the least squares fit, and again solve the above seventh-order equation
for s. Note that since the topocentric observations have been adjusted for
diurnal parallax, this is now a rigorous procedure. We cycle through this
sequence until convergence is achieved. Typically this requires at most
three iterations. If this process does not work initially, or fails to converge,
then recourse to the rigorous set of formulas, Eqs. 7.16 and 7.17, is always
available.

The software is set up to perform quadratic, cubic, and quartic fits of both
the right ascension and declination separately. The residuals are exhibited in
an effort to discern which order is appropriate for which coordinate (as above
for the artificial satellites). For these asteroids fits for ecliptic longitude and
ecliptic latitude have also been considered. Obviously no advantage will be
gained for Earth-approaching asteroids by such a change of coordinate
system.

Example 7.4. The data is given below in the standard Minor Planet Center
format for 1982SA; year, month, decimal day (all UTC to the nearest half
second) followed by right ascension and declination relative to the mean
equator and equinox of 1950.0.

t	A	Δ
1982 09 22$\overset{d}{.}$38078	23^h 51^m 02$\overset{s}{.}$04	+08° 22′ 20″.1
1982 09 22.39328	23 51 00.34	+08 22 32.4
1982 09 23.20965	23 49 28.12	+08 33 32.8
1982 09 23.22277	23 49 26.51	+08 33 36.3
1982 09 23.23686	23 49 24.62	+08 33 44.6
1982 09 23.28413	23 49 19.06	+08 34 25.0
1982 09 24.20074	23 47 36.98	+08 46 31.8
1982 09 24.20360	23 47 34.14	+08 46 32.8
1982 09 24.23294	23 47 30.75	+08 47 00.5
1982 09 24.27779	23 47 25.48	+08 47 37.5
1982 09 24.30990	23 47 21.68	+08 48 01.5
1982 09 24.32741	23 47 19.41	+08 48 14.5
1982 09 25.17829	23 45 43.15	+08 59 17.5
1982 09 25.20297	23 45 40.21	+08 59 36.8
1982 09 25.21793	23 45 38.47	+08 59 50.5
1982 09 25.24294	23 45 35.51	+09 00 11.8

Of course the A and Δ values are not corrected for planetary aberration or diurnal parallax. In a format similar to the artificial satellite cases the results of the quadratic, cubic, and quartic fits are given below with $\langle t \rangle = 267^d 0^h 10^m 58\overset{s}{.}69$.

Fit Order	Quadratic	Cubic	Quartic
$\langle \alpha \rangle$	23^{h}47^{m}56$\overset{s}{.}$83	23^{h}47^{m}56$\overset{s}{.}$79	23^{h}47^{m}57$\overset{s}{.}$68
$\langle \delta \rangle$	+08°44′01″.4	+08°44′01″.5	+08°43′59″.8
$\langle \dot{\alpha} \cos \delta \rangle$	−1692″.03/day	−1695″.84/day	−1732″.96/day
$\langle \dot{\delta} \rangle$	790″.42/day	790″.85/day	795″.50/day
$\langle \ddot{\alpha} \cos \delta \rangle$	−10″.81/day^2	−3″.35/day^2	−137″.11/day^2
$\langle \ddot{\delta} \rangle$	−10″.94/day^2	−11″.19/day^2	5″.54/day^2
$\varepsilon_\alpha \cos \delta$	9″.63	9″.52	9″.03
ε_δ	2″.54	2″.53	2″.51

Clearly only a quadratic fit is necessary, but in order to illustrate the sensitivity of the method to the fit order, the four combinations $(\alpha, \delta) = (2, 2)$, $(2, 3)$, $(3, 2)$, and $(3, 3)$ were utilized in constructing orbital element sets. The next new Moon was on October 17, 1982 $= 290^d 1982$ and we computed positions for 0^h UTC on that date. The four element sets, an earlier Minor Planet Center (MPC) element set, predictions from them and the initial and final values of R from the seventh-order equation are all given immediately below.

α, δ Fit order	2, 2	2, 3	3, 2	3, 3	MPC
a (A.U.)	1.91	1.88	1.86	1.82	1.85
e	0.14	0.13	0.14	0.14	0.10
ω (°)	61.0	60.9	65.1	65.6	27.8
i (°)	21.0	20.1	19.5	18.5	20.0
Ω (°)	350.6	350.3	350.1	349.7	350.1
M_0 (°)	321.4	321.2	318.5	318.1	346.2
R (A.U.)					
Final	1.72	1.69	1.68	1.65	—
Initial	1.51	1.49	1.60	1.59	—
α	23^h11^m7	23^h11^m9	23^h11^m8	23^h11^m9	23^h10^m9
δ	$+13°10'$	$+13°09'$	$+13°10'$	$+13°09'$	$+12°45'$

We recovered the asteroid on day 288 near $\alpha = 23^h12^m5$, $\delta = +12°34'$. Lastly the geocentric coordinates of the Sun at $\langle t \rangle$ were

$$\mathbf{R}_\odot = (-1.00320630, -0.00291902, -0.00125620) \text{ A.U.}$$

$$\dot{\mathbf{R}}_\odot = (3.419111 \times 10^{-4}, -1.573003 \times 10^{-2}, -6.821144 \times 10^{-3}) \text{ A.U./day}$$

Example 7.5. For 1982HS, in the same format as above, the observational data is as follows:

t	A	Δ
1982 04 28.20491	$14^h27^m33^s35$	$-07°52'29''2$
1982 04 28.21557	14 27 32.34	$-07\ 52\ 41.8$
1982 04 28.22978	14 27 30.54	$-07\ 52\ 53.4$
1982 04 28.24988	14 27 28.48	$-07\ 53\ 14.7$
1982 04 28.27527	14 27 25.47	$-07\ 53\ 38.9$
1982 04 29.14892	14 25 53.49	$-08\ 08\ 20.0$
1982 04 29.18805	14 25 49.49	$-08\ 09\ 00.9$
1982 04 29.21636	14 25 46.49	$-08\ 09\ 29.5$
1982 04 29.25215	14 25 42.62	$-08\ 10\ 02.7$
1982 04 29.27658	14 25 40.02	$-08\ 10\ 30.7$
1982 04 29.31022	14 25 35.98	$-08\ 11\ 00.6$
1982 05 04.14304	14 16 53.16	$-09\ 35\ 14.1$
1982 05 04.20412	14 16 46.46	$-09\ 36\ 22.6$
1982 05 04.24907	14 16 41.42	$-09\ 37\ 08.1$
1982 05 04.29774	14 16 35.82	$-09\ 38\ 02.3$

The results of the fits were, at $\langle t \rangle = 120^d 05^h 32^m 19\overset{s}{.}22$,

Fit Order	Quadratic	Cubic	Quartic
$\langle \alpha \rangle$	$14^h 23^m 58\overset{s}{.}26$	$14^h 23^m 58\overset{s}{.}64$	$14^h 23^m 58\overset{s}{.}03$
$\langle \delta \rangle$	$-08°26'42''.7$	$-08°26'40''.8$	$-08°26'30''.5$
$\langle \dot{\alpha} \cos \delta \rangle$	$-1587''.21/\text{day}$	$-1580''.13/\text{day}$	$-1589''.10/\text{day}$
$\langle \dot{\delta} \rangle$	$-1026''.21/\text{day}$	$-1023''.52/\text{day}$	$-1013''.43/\text{day}$
$\langle \ddot{\alpha} \cos \delta \rangle$	$-12''.50/\text{day}^2$	$-11''.20/\text{day}^2$	$-7''.34/\text{day}^2$
$\langle \ddot{\delta} \rangle$	$-12''.40/\text{day}^2$	$-12''.09/\text{day}^2$	$-16''.26/\text{day}^2$
$\langle \varepsilon_\alpha \cos \delta \rangle$	$2''.36$	$2''.31$	$2''.30$
$\langle \varepsilon_\delta \rangle$	$1''.35$	$1''.33$	$1''.29$

In this instance

$$\mathbf{R}_\odot = (0.781606685, 0.58302948, 0.252795595) \text{ A.U.}$$

$$\dot{\mathbf{R}}_\odot = (-1.057823 \times 10^{-2}, 1.229954 \times 10^{-2}, 5.333857 \times 10^{-3}) \text{ A.U./day}$$

and the next new Moon was on May 23 (day 143). Predictions were again performed for midnight.

α, δ Fit Order	2, 2	2, 3	3, 2	3, 3	MPC
a (A.U.)	1.90	1.97	2.19	2.32	2.47
e	0.19	0.21	0.27	0.29	0.33
ω (°)	221.4	222.2	227.6	227.9	229.5
i (°)	18.5	19.8	23.3	25.2	26.4
Ω (°)	44.7	44.4	43.8	43.5	43.0
M_0 (°)	326.9	327.4	328.4	330.1	329.5
R (A.U.)					
Final	1.62	1.66	1.75	1.80	—
Initial	1.71	1.79	1.60	1.66	—
α	$01^h 48^m.4$	$01^h 48^m.2$	$01^h 47^m.1$	$01^h 46^m.9$	$01^h 42^m.6$
δ	$-15°41'$	$-15°39'$	$-15°43'$	$-15°41'$	$-15°16'$

Apollo

Example 7.6. Apollo 1862 is the prototypical Earth-approaching asteroid. This author's search team accidentally recovered it on April 21, 1982. A total of nine observations were secured that night, three on the next.

t	A	Δ
1982 04 27$^\text{d}$27392	14$^\text{h}$07$^\text{m}$00$^\text{s}$77	$-13°\,16'\,30''9$
1982 04 27.27868	14 06 59.47	$-13\,16\,39.2$
1982 04 27.28755	14 06 57.17	$-13\,16\,45.9$
1982 04 27.29945	14 06 53.48	$-13\,16\,52.3$
1982 04 27.31551	14 06 49.28	$-13\,16\,59.7$
1982 04 27.33134	14 06 44.43	$-13\,17\,16.0$
1982 04 27.33359	14 06 43.95	$-13\,17\,17.5$
1982 04 27.35243	14 06 38.48	$-13\,17\,29.4$
1982 04 27.36315	14 06 35.68	$-13\,17\,40.0$
1982 04 28.15672	14 02 59.58	$-13\,27\,53.4$
1982 04 28.23203	14 02 36.52	$-13\,28\,52.5$
1982 04 28.27278	14 02 23.83	$-13\,29\,23.6$

The table below shows the results from solving the full coupled set (Eqs. 7.16 and 7.17) for the topocentric distance. Also shown is the radial velocity result from Eq. 7.20. In addition the results of the cubic least squares fits for the right ascension, the declination, and their rates are listed. The second column gives these same quantities as derived by cubic interpolation within the 2-day tabulations of the 1982 Ephemerides of Minor Planets (EMP). The good agreement is clear for all but $\dot{R}$—the quantity most sensitive to angular acceleration.

	Observational/Computed	EMP
r	1.2316 A.U.	1.1779 A.U.
α	14$^\text{h}$05$^\text{m}$47$^\text{s}$25	14$^\text{h}$05$^\text{m}$49$^\text{s}$60
δ	$-13°19'52''2$	$-13°19'13''4$
R	0.2251 A.U.	0.1711 A.U.
$\dot{r}$		-9.5154×10^{-3} A.U./day
$\dot{\alpha}$	$-265^\text{s}96$/day	$-269^\text{s}77$/day
$\dot{\delta}$	$-763''39$/day	$-751''29$/day
$\dot{R}$	-1.9220×10^{-2} A.U./day	-9.0496×10^{-3} A.U./day

The orbital element set deduced from these values is an unreasonable one and we do not know why. The values used for $\mathbf{R}_\odot$ and $\dot{\mathbf{R}}_\odot$ at $\langle t \rangle = 117^\text{d}12^\text{h}59^\text{m}39^\text{s}44$ are

$$\mathbf{R}_\odot = (0.809229347,\ 0.549347203,\ 0.238188842)\ \text{A.U.}$$

$$\dot{\mathbf{R}}_\odot = (-9.959679\times10^{-3},\ 1.274498\times10^{-2},\ 5.526976\times10^{-3})\ \text{A.U./day}$$

THE USE OF ANGLES AND ANGULAR RATES

Lincoln Laboratory (M.I.T.) has developed the prototype for the Ground-based Electro-Optical Deep Space Surveillance (GEODSS) network. The purpose of these observatories is the acquisition and tracking of artificial satellites. This system will supplant the Baker–Nunn camera network. Included in the GEODSS's telescope system is the (theoretical) capability to deduce instantaneous topocentric angular rates. The (current M.I.T.) accuracy is $\approx 0\rlap{.}''01/\text{sec}$ for topocentric angular speeds less than $\approx 500''/\text{sec}$. We were therefore led to consider a new initial orbit problem: The calculation of orbital elements from observations of angles and angular rates. In this section a complete, exact,[*] analytical solution to the problem using only two sets of observational data is presented (a data set consists of a topocentric unit vector, its derivative, and the time). Approximate solutions to the same problem can also be constructed using modifications of the classical methods of Lagrange and of Laplace (see Taff and Hall 1977). In the last section the simultaneous deduction of $\mathbf{l}$ and $\dot{\mathbf{l}}$ is considered.

The fundamental geometrical relationship is

$$\mathbf{r} = \mathbf{R} + \boldsymbol{\rho} = R\mathbf{l} + \boldsymbol{\rho} \tag{7.1, 7.2}$$

Two successive differentiations yield

$$\dot{\mathbf{r}} = \dot{R}\mathbf{l} + R\dot{\mathbf{l}} + \dot{\boldsymbol{\rho}}$$
$$\ddot{\mathbf{r}} = \ddot{R}\mathbf{l} + 2\dot{R}\dot{\mathbf{l}} + R\ddot{\mathbf{l}} + \ddot{\boldsymbol{\rho}} \tag{7.3}$$

where the dots denote differentiation with respect to the time. Finally, the two-body equations of motion are

$$\ddot{\mathbf{r}} = -\frac{\mu \mathbf{r}}{|\mathbf{r}|^3} \tag{7.4}$$

where $\mu = GM_{\oplus}$.

The Exact Method

The equations of motion have four integrals given by

$$\mathbf{L} = \mathbf{r} \times \dot{\mathbf{r}} = \text{const vector}$$
$$\mathcal{E} = \frac{\dot{\mathbf{r}} \cdot \dot{\mathbf{r}}}{2} - \frac{\mu}{|\mathbf{r}|} = \text{const scalar} \tag{7.21}$$

Equations 7.21 express the conservation of the total angular momentum and the conservation of the total energy (both per unit mass). In particular, since these quantities *are* constants of motion, the left-hand sides of Eqs. 7.21 can

---

[*] *Exact* is used in the sense that no approximations have been made in the physics or in the geometry.

be evaluated at the two observation times, t_1 and t_2, and then set equal to each other. In terms of the topocentric variables the result is

$$\mathbf{L}(\mathbf{R}_1, \dot{\mathbf{R}}_1, \boldsymbol{\rho}_1, \dot{\boldsymbol{\rho}}_1) = \mathbf{L}(\mathbf{R}_2, \dot{\mathbf{R}}_2, \boldsymbol{\rho}_2, \dot{\boldsymbol{\rho}}_2)$$
$$\mathscr{E}(\mathbf{R}_1, \dot{\mathbf{R}}_1, \boldsymbol{\rho}_1, \dot{\boldsymbol{\rho}}_1) = \mathscr{E}(\mathbf{R}_2, \dot{\mathbf{R}}_2, \boldsymbol{\rho}_2, \dot{\boldsymbol{\rho}}_2) \tag{7.22}$$

where

$$\mathbf{L}(\mathbf{R}, \dot{\mathbf{R}}, \boldsymbol{\rho}, \dot{\boldsymbol{\rho}}) = \mathbf{r} \times \dot{\mathbf{r}} = R^2(\mathbf{l} \times \dot{\mathbf{l}}) + R(\mathbf{l} \times \dot{\boldsymbol{\rho}} - \dot{\mathbf{l}} \times \boldsymbol{\rho})$$
$$+ \dot{R}(\boldsymbol{\rho} \times \mathbf{l}) + \boldsymbol{\rho} \times \dot{\boldsymbol{\rho}}$$

$$\mathscr{E}(\mathbf{R}, \dot{\mathbf{R}}, \boldsymbol{\rho}, \dot{\boldsymbol{\rho}}) = \frac{\dot{\mathbf{r}} \cdot \dot{\mathbf{r}}}{2} - \frac{\mu}{|\mathbf{r}|}$$
$$= \frac{[\dot{R}^2 + R^2(\dot{\mathbf{l}} \cdot \dot{\mathbf{l}}) + \dot{\boldsymbol{\rho}} \cdot \dot{\boldsymbol{\rho}}]}{2} + \dot{R}(\mathbf{l} \cdot \dot{\boldsymbol{\rho}}) + R(\dot{\mathbf{l}} \cdot \dot{\boldsymbol{\rho}}) - \frac{\mu}{|\mathbf{R} + \boldsymbol{\rho}|}$$

These relations represent four independent, nonlinear, equations relating the 12 variables $\mathbf{R}_1, \dot{\mathbf{R}}_1, \mathbf{R}_2, \dot{\mathbf{R}}_2$. In the situation considered here eight of these quantities $(\mathbf{l}_1, \mathbf{l}_2, \dot{\mathbf{l}}_1, \dot{\mathbf{l}}_2)$ have been measured. Therefore, the problem is well posed. The system 7.22 may be algebraically reduced to (at worst) one equation of the 24th order in one unknown. However, a multidimensional Newton-Raphson technique was used to obtain the results below. The procedure used to obtain the initial estimates for R and $\dot{R}$ (from $\mathbf{l}$ and $\dot{\mathbf{l}}$) for the Newton-Raphson iteration will also be discussed.

Upon completion of the solution we know topocentric position and velocity at the same time. We can obtain geocentric position and velocity directly from Eqs. 7.1 and 7.3. An alternative to solving Eq. 7.1 is to obtain r by squaring Eq. 7.1; namely,

$$r^2 = R^2 + \rho^2 + 2R\rho(C + S)$$
$$C = \cos \phi' \cos \Delta \cos H, \qquad S = \sin \phi' \sin \Delta, \qquad H = \tau - A$$

Geocentric right ascension and declination are then accurately calculated from

$$\sin(\alpha - A) = \left(\frac{\rho}{r}\right) \cos \phi' \sec \delta \sin H$$
$$\sin(\delta - \Delta) = \left(\frac{\rho}{r}\right) \sin \phi' \csc \gamma \sin(\gamma - \Delta) \tag{3.22}$$

where

$$\tan \gamma = \tan \phi' \cos\left(\frac{\alpha - A}{2}\right) \sec\left(H - \frac{\alpha - A}{2}\right) \tag{3.23}$$

Numerical Examples

To demonstrate the method tests were performed on six different artificial satellites. Their orbits represent a cross section of the types encountered in practice. The sample includes a low-altitude satellite, a high-eccentricity synchronous satellite, a low-eccentricity synchronous satellite, a high-eccentricity trans-synchronous satellite, a low-eccentricity trans-synchronous satellite, and a high-eccentricity 12-hr-period satellite. Their (*NORAD*) Space Defense Center identification numbers, class name, eccentricities, and periods are listed in the first column of Table 3.

Table 3. Percent Error in the Topocentric Distance

Satellite	Total Time Span (min.)	Percent Error in R^a	
		Approximate Method	Exact
SDC 16	5	−17.46	+0.24
Vanguard RB	10	−21.39	+1.13
$e = 0.206$	20	−26.71	+0.64
$P = 2^h30$	40	−18.08	+6.80
SDC 748	5	−11.02	−0.12
Elektron	10	−11.11	−0.07
$e = 0.781$	20	−11.37	−0.06
$P = 22^h6$	40	−12.13	−0.07
SDC 2608	5	−10.97	+1.96
ATS	10	−11.00	−0.36
$e = 0.00053$	20	−11.04	−0.03
$P = 23^h9$	40	−11.09	+0.19
SDC 2805	5	+1.03	−0.88
Cosmos	10	+1.09	−0.76
$e = 0.789$	20	+0.99	−0.36
$P = 19^h6$	40	+0.54	−0.60
SDC 3955	5	−12.88	+10.79
Vela	10	−12.88	
$e = 0.0344$	20	−12.88	+1.67
$P = 112^h$	40	−12.90	+13.86
SDC 5367	5	+5.06	−0.00
Molniya	10	+5.21	
$e = 0.734$	20	+5.49	+0.09
$P = 12^h0$	40	+6.02	+0.09

[a] No entry indicates convergence to an incorrect root.

Since the determination of an orbit ultimately rests on the precision with which the topocentric distance (R) is known, Table 3 lists the percentage error in R for each of four time spans. The table includes the results for the initial estimation procedure (see immediately below) and the exact method. Perusal of Table 3 shows that the estimation procedure (due to the neglect of the radial velocity) does not work well for the low-altitude satellite. However, the $\approx 10\%$ accuracy obtained for the other objects is gratifying. Considering the very small ratio (time span/orbital period) it is surprising that R is usually determined to better than 0.5%. The exact method is inaccurate for this (particular) Vela satellite because of the poor accuracy in the angular rates coupled with the fact that, relative to our observatory, the satellite's motion was direct, stationary, and then retrograde during the observational time span. This method is very sensitive to errors in the value of $\mathbf{i}$.

Estimating the Distance. A first-order guess for r can be obtained by assuming that the orbit is circular and that the topocentric angular speed $(\Omega; \Omega^2 = \dot{A}^2 \cos^2 \Delta + \dot{\Delta}^2 = \mathbf{i} \cdot \mathbf{i})$ is equal to the geocentric angular speed $(\omega; \omega^2 = \dot{\alpha}^2 \cos^2 \delta + \dot{\delta}^2)$, namely,

$$r_1 = \left(\frac{\mu}{\Omega^2}\right)^{1/3}$$

A second-order estimate can be obtained from r_1 by correcting, on the average, for a nonzero eccentricity e. Depending on the information available, this is expressed by

$$r_2 = F(e)r_1 \qquad e \text{ roughly known} \tag{7.23}$$

or

$$r_2 = \langle F(e)\rangle r_1 \qquad e \text{ unknown} \tag{7.24}$$

$F(e)$ and its average value $\langle F(e)\rangle$ are discussed below. When $r_2 \geq 3\rho$, it is worthwhile to improve upon the approximation $\omega = \Omega$ by using

$$\omega^2 \approx \Omega^2 + \left(\frac{\rho}{r_2}\right)[-\Omega^2(C+S) + \dot{r}(\dot{A}C + \dot{A}S \cos \phi')]$$

where S and C were defined earlier. One then computes r_3 from

$$r_3 = \left(\frac{\mu}{\omega^2}\right)^{1/3}$$

and r_4 from either Eq. 7.23 or 7.24 with r_3 replacing r_1.

The eccentricity correction factor, $F(e)$, is calculated by using the exact relationship

$$r^2\omega = [\mu a(1-e^2)]^{1/2}$$

which expresses the fact that $\mathbf{L} \cdot \mathbf{L} = \text{const.}$ The unknown semi-major axis a is eliminated via $r = a(1 - e \cos E)$ where E is the eccentric anomaly. The result is

$$\frac{r}{r_c} = \left(\frac{1-e^2}{1-e \cos E}\right)^{1/3}, \qquad r_c = \left(\frac{\mu}{\omega^2}\right)^{1/3}$$

$F(e)$ is defined as the average, over an orbital period P of r/r_c, namely,

$$F(e) = \left(\frac{1}{P}\right) \int_0^P \left(\frac{r}{r_c}\right) dt = \frac{P_{2/3}(\varepsilon)}{\varepsilon^{4/3}}, \quad \varepsilon = \frac{1}{(1-e^2)^{1/2}}$$

$P_\lambda(u)$ is the Legendre polynomial of order λ. Table 4 lists $F(e)$ for $e = 0(0.05)1$.

Table 4. Eccentricity Correction Factor

e	$F(e)$	e	$F(e)$
0	1	0.55	0.870938
0.05	0.999027	0.60	0.843115
0.10	0.996101	0.65	0.811248
0.15	0.991197	0.70	0.774575
0.20	0.984275	0.75	0.731972
0.25	0.975276	0.80	0.681663
0.30	0.964119	0.85	0.620560
0.35	0.950697	0.90	0.542382
0.40	0.934873	0.95	0.429811
0.45	0.916469	1	0
0.50	0.895257		

Since knowledge of the satellite class may yield an accurate (± 0.1) estimate for e before the orbit is computed, Eq. 7.23 may be appropriate. In the absence of any prior knowledge the probability distribution of eccentricities may be assumed to be uniform (a poor representation of the actual case but unimportant here), whence

$$\langle F(e) \rangle = \int_0^1 F(e)\, de = 0.8263$$

and Eq. 7.24 is used instead. The latter procedure was followed in all of the numerical cases presented earlier.

From the estimate for r an estimate for R is obtained. Finally, the approximations $\dot{R}_1 = \dot{R}_2 = (R_2 - R_1)/(t_2 - t_1)$ are then used to begin the Newton–Raphson iteration.

The Direct Reduction of the Observations

Motivation

If one uses a solid-state device camera (CID or CCD) or an electron beam tube with digitized output to observe artificial satellites, then a whole new field of optical observing and data reduction is thrown open. With such a device connected to a telescope moving at the sidereal rates the time history of the motion of a satellite can be obtained. That is, one can know the coordinates of the light deposited on the camera target and *when* that light was deposited. This is possible with a photographic plate, but difficult to do accurately with good time resolution. Within this knowledge is a set of complete information concerning the satellite's geocentric location and velocity. Hence, an analysis of the image's location as a function of time can yield a full set of initial conditions for orbital analysis. Moreover, there should be an extremely high internal precision in these positions as they all suffer from nearly the same systematic errors. The random and systematic errors associated with the telescope pointing enter weakly or not at all. Therefore, one has the promise of more information than just the object's position and of extremely high precision in this additional data.

The analysis of the imaged motion is straightforward. It is given in the next few sections. In practice the distance and radial velocity can only be determined if the motion on the camera target deviates from a straight line and this won't normally occur. However, the angular velocity, should be precisely ascertained and this information should then be used in initial orbit construction as well as in the differential correction of orbits (see Chapter Eleven).

The Geocentric Geometry. A celestial object's location is given by $\mathbf{r} = r\mathbf{l}$, $\mathbf{l} = (\cos \delta \cos \alpha, \cos \delta \sin \alpha, \sin \delta)$. As usual the geocentric distance is r, the geocentric declination is δ, and the geocentric right ascension is α. Over very short times, very much less than an orbital period ($<0.3\%$ in practice), we can regard the motion as uniform and rectilinear in the geocentric coordinate system. Because this is different from uniform "rectilinear" motion on the celestial sphere (i.e., along a great circle), there is a coupling between the quotient of radial velocity ($\dot{r}$) and the distance (r) and the object's position (i.e., $\mathbf{l}$). To see how this comes about, consider the orthonormal basis given by $\mathbf{l}, \mathbf{a}$, and $\mathbf{d}$; $\mathbf{a} = (-\sin \alpha, \cos \alpha, 0)$, $\mathbf{d} = (-\sin \delta \cos \alpha, -\sin \delta \sin \alpha, \cos \delta)$. I can write

$$\dot{r} = \mathbf{l} \cdot \dot{\mathbf{r}}, \qquad \boldsymbol{\omega} = \dot{\alpha} \cos \delta \, \mathbf{a} + \dot{\delta} \mathbf{d}$$

In terms of the constant unit vector $\mathbf{n} = (0, 0, 1)$, which points toward the North Celestial Pole,

$$\dot{r} = \mathbf{l} \cdot \dot{\mathbf{r}}$$

$$\dot{\alpha} \cos \delta = \mathbf{a} \cdot \boldsymbol{\omega} = \frac{\mathbf{n} \cdot (\mathbf{l} \times \dot{\mathbf{r}})}{r \cos \delta} \tag{7.25}$$

$$\dot{\delta} = \mathbf{d} \cdot \boldsymbol{\omega} = \frac{\mathbf{n} \cdot \dot{\mathbf{r}} \sec \delta - \dot{r} \tan \delta}{r}$$

By differentiating Eq. 7.25 and enforcing the constraint $\ddot{\mathbf{r}} = \mathbf{0}$, we obtain

$$\ddot{r} = r\omega^2$$

$$\ddot{\alpha} = 2\dot{\alpha}\left(\dot{\delta} \tan \delta - \frac{\dot{r}}{r} \right)$$

$$\ddot{\delta} = \dot{\alpha}^2 \sin \delta \cos \delta - 2\dot{\delta}\left(\frac{\dot{r}}{r} \right)$$

Thus, even though the space motion is along a geodesic, the position of the object as a function of time changes because of its purely radial motion. This term is known as the foreshortening term in astrometry. (It can be measured for some stars.) To convince the reader that $\dot{r}/r$ is not numerically much smaller than $\dot{\alpha}$ or $\dot{\delta}$ for artificial satellites, consider a satellite with argument of perigee $= 270°$, inclination $= 60°$, eccentricity $= 1/\sqrt{2}$. These numbers are typical of the Molniya type of satellite of which, with their associated rocket bodies, there are hundreds. For such a satellite observed at the equator $\dot{\alpha} = n\sqrt{2}$, $\dot{\delta} = n\sqrt{6}$, $\dot{r}/r = 2n$ where n is the mean motion (~ 2 rev/day). Clearly, the three quantities are comparable and, since $\delta = 0$, the foreshortening terms drive the acceleration of the position.

To derive the rigorous results, the equations of motion are used:

$$\mathbf{r}(t) = \mathbf{r}(t_0) + \dot{\mathbf{r}}(t_0)(t - t_0), \qquad \dot{\mathbf{r}}(t) = \dot{\mathbf{r}}(t_0) \qquad (7.26)$$

By manipulating the component forms of Eq. 7.26, one can obtain [$d = r(t)/r(t_0)$, $\mu_0 = \dot{\alpha}(t_0)$, $\mu_0' = \dot{\delta}(t_0)$, $v_0 = \dot{r}(t_0)/r(t_0)$, $T = t - t_0$, $\delta_0 = \delta(t_0)$, $\alpha = \alpha(t)$, $\omega_0^2 = \mu_0^2 \cos^2 \delta_0 + (\mu_0')^2$, etc.]

$$d^2 = 1 + 2v_0 T + (v_0^2 + \omega_0^2) T^2$$

$$\tan(\alpha - \alpha_0) = \frac{\mu_0 T}{1 + (v_0 - \mu_0' \tan \delta_0) T} \qquad (7.27)$$

$$\tan\left[\frac{\delta - \delta_0}{2}\right] = \frac{\mu_0' T + (1 - d + v_0 T) \tan \delta_0}{d + \{[1 + (v_0 - \mu_0' \tan \delta_0) T]^2 + \mu_0^2 T^2\}^{1/2}}$$

$$\frac{\dot{r}}{r} = \frac{\dot{d}}{d} = [v_0 + (v_0^2 + \omega_0^2) T] d^2$$

$$\dot{\alpha} = \frac{\mu_0}{[1 + (v_0 - \mu_0' \tan \delta_0) T]^2 + \mu_0^2 T^2} \qquad (7.28)$$

$$\dot{\delta} = \frac{\mu_0'(1 + v_0 T) - \omega_0^2 T \tan \delta_0}{d^2 \{[1 + (v_0 - \mu_0' \tan \delta_0) T]^2 + \mu_0^2 T^2\}^{1/2}}$$

These equations were first derived by Eichhorn and Rust (1970).

Equations 7.27 and 7.28 give a complete description of the motion in the form required. The position at any time depends on the five initial conditions α_0, δ_0, v_0, μ_0, and μ_0'. To recover the sixth quantity, one needs to refer the motion to the observer's location instead of the center of the Earth. Before this is done, one could put the physics back into the problem. The main purpose in considering uniform rectilinear motion is to bring to the reader's attention the geometrical effect of the foreshortening terms. Equation 7.26 could now be replaced by the formulas of two-body Keplerian motion with concomitant changes in Eqs. 7.27 and 7.28. The development of the next two sections is independent of the physics, or lack thereof, in Eqs. 7.27 and 7.28. It should also be clear that, since $\dot{r}/r$ enters first into $\ddot{\alpha}$ and $\ddot{\delta}$, it can only be determined if the imaged path has measurable curvature. The radius of curvature is essentially infinite in all realistic scenarios. Hence, as a practical matter, the physics does not matter.

Diurnal Parallax

Except for time variables, the corresponding uppercase letter is used to denote a topocentric quantity. Thus, $A = A(t) =$ the topocentric right ascension at time t, $M_0' = \dot{\Delta}_0 = \dot{\Delta}(t_0) =$ the proper motion in the topocentric declination at time t_0, $D = R/R_0$, and so on. Let $\boldsymbol{\rho} = \rho(\cos\phi'\cos\tau, \cos\phi'\sin\tau, \sin\phi')$ be the observer's geocentric location. Here ρ is the observer's geocentric distance, ϕ' the geocentric latitude, and τ the local mean sidereal time. The translation to the observer is accomplished by

$$\mathbf{r} = \mathbf{R} + \boldsymbol{\rho}$$

If we write the right-hand side of this in component for M and use Eq. 7.26 similarly expressed, then we can derive results analogous to those in Eqs. 7.27 and 7.28. They are, for the location,

$$D^2 + 2DS\cos\mathcal{Z} = 1 + 2V_0T + (V_0^2 + \Omega_0^2)T^2 + 2S\cos\mathcal{Z}_0$$

$$+ 2ST\left(V\cos\mathcal{Z}_0 + \frac{d\cos\mathcal{Z}_0}{dt}\right) + (S\dot{\tau}T\cos\phi')^2$$

$$- 2S\dot{\tau}T^2\cos\phi'\cos\Delta_0\sin H_0(V_0 - M_0'\tan\Delta_0 - M_0\cot H_0)$$

$$\tan(A - A_0) = \frac{M_0T + S_1}{1 + (V_0 - M_0'\tan\Delta_0)T + S_2} \tag{7.29}$$

$$\tan\left(\frac{\Delta - \Delta_0}{2}\right) = \frac{M_0'T + (1 - D + V_0T)\tan\Delta_0}{D + C}$$

The missing initial condition is $S = \rho/R_0$. The other quantities are defined by ($\mathcal{Z} =$ topocentric pseudo-zenith distance, $H =$ topocentric hour angle)

$$\cos\mathcal{Z} = \sin\phi'\sin\Delta + \cos\phi'\cos\Delta\cos H, \qquad H = \tau - A$$

$$C^2 = [1 + (V_0 - M_0'\tan\Delta_0)T]^2 + M_0^2T^2 + 2\{S_1M_0T + S_2[1 + (V_0 - M_0'\tan\Delta_0)T]\}$$

$$+ (S\cos\phi'\sec\Delta_0)^2\{(\dot{\tau}T)^2 + 2[1 - \cos(\tau - \tau_0) - \dot{\tau}T\sin(\tau - \tau_0)]\}$$

$$S_1 = S\cos\phi'\sec\Delta_0[\sin H_0 + \dot{\tau}T\cos H_0 - \sin(\tau - A_0)] \tag{7.30}$$

$$S_2 = S\cos\phi'\sec\Delta_0[\cos H_0 - \dot{\tau}T\sin H_0 - \cos(\tau - A_0)]$$

Necessarily, for example, as $S \to 0$, $D \to d$, and Eq. 7.29 reduces to Eq. 7.27. The analog of Eq. 7.28 can be obtained from Eq. 7.29 by differentiation with respect to the time. Equations 7.29 and 7.30 are rigorous. Equations 7.27–7.30 completely specify the topocentric location of the object in terms of the initial conditions $S = \rho/r_0$, α_0, δ_0, $v_0 = \dot{r}_0/r_0$, $\mu_0 = \dot{\alpha}_0$, and $\mu_0' = \dot{\delta}_0$. When the telescope-camera combination images the motion, with the telescope in sidereal drive, on a photographic plate or camera target, the resulting streak contains, albeit implicitly, a complete description of the motion. The last step then is to project a portion of the celestial sphere onto a plane.

Standard Coordinates

If the camera is equivalent to a pinhole camera, then the projection of the imaged portion of the celestial sphere onto the plane of the camera target (or photographic plate) will be as in a gnomonic projection. The resulting-coordinate system in this plane is a rectangular one known as the *standard coordinate* system. To relate standard coordinates, symbolized by (ξ, η), to right ascension and declination, we choose the unit of length for the standard coordinates to be the telescope's focal length and we need to know the coordinates where a plane, parallel to the plane of the camera target, touches the celestial sphere. This is known as the *tangential point.* One end of the optical axis of the telescope pierces the celestial sphere at this point, the other end intersects the center of the plate. The positive η axis points toward the North Celestial Pole. The positive ξ axis points toward the east point.

If (α^*, δ^*) are the coordinates of the tangential point, then the relationship between ξ, η, and a corresponding α, δ is given by

$$\xi = \frac{\sec \delta^* \sin(\alpha - \alpha^*)}{\tan \delta^* \tan \delta + \cos(\alpha - \alpha^*)}$$

$$\eta = \frac{\tan \delta - \tan \delta^* \cos(\alpha - \alpha^*)}{\tan \delta^* \tan \delta + \cos(\alpha - \alpha^*)}$$

(7.31)

The inverse of these relationships is given by

$$\cot \delta \sin(\alpha - \alpha^*) = \frac{\xi}{\sin \delta^* + \eta \cos \delta^*}$$

$$\cot \delta \cos(\alpha - \alpha^*) = \frac{\cos \delta^* - \eta \sin \delta^*}{\sin \delta^* + \eta \cos \delta^*}$$

The formula

$$\sin \delta = \frac{\sin \delta^* + \eta \cos \delta^*}{(1 + \xi^2 + \eta^2)^{1/2}}$$

is useful near the equator.

The problem is now completely solved. Because one obviously observes topocentrically, we replace α, δ in Eqs. 7.31 by A, Δ computed from Eqs. 7.29 and 7.30. The geocentric coordinates appearing in Eqs. 7.29 and 7.30 are given as a function of time by Eqs. 7.27. Because we measured the properties of the streak imaged on the plate, we have an implicit problem for the determination of the location and velocity of the moving object. Once this problem has been solved, the orbital elements can be computed. It is also clear from Eqs. 7.31

that it is simplest if the position of the telescope is not changed during the exposure of the plate.

A matrix form is ($f = $ focal length; this equation is not restricted to $f = 1$ as the above are)

$$l(\alpha, \delta) = \frac{R_3(-\alpha^* - \pi/2)R_1(\delta^* - \pi/2)}{(\xi^2 + \eta^2 + f^2)^{1/2}}(\xi, \eta, f)$$

Approximate Formula

For small fields of view Eqs. 7.31 may be replaced, approximately, by

$$\xi = (\alpha - \alpha^*)\cos \delta^* - (\alpha - \alpha^*)(\delta - \delta^*)\sin \delta^*$$

$$\mu = (\delta - \delta^*) + \frac{(\alpha - \alpha^*)^2}{4}\sin 2\delta$$

(7.32)

Similarly, through second order in the field of view, Eqs. 7.29 and 7.30 take the form

$$D = 1 + V_0 T + \frac{(\Omega_0^2 + S\dot{\tau}^2 \cos \phi' \cos \Delta_0 \cos M_0)T^2}{2}$$

$$A = A_0 + M_0 T$$

$$+ \frac{(2M_0 M_0' \tan \Delta_0 - 2V_0 M_0 + S\dot{\tau}^2 \cos \phi' \sec \Delta_0 \sin M_0)T^2}{2}$$

(7.33)

$$\Delta = \Delta_0 + M_0' T$$

$$- \frac{(M^2 \sin \Delta_0 \cos \Delta_0 + 2V_0 M_0' + S\dot{\tau}^2 \cos \phi' \sin \Delta_0 \cos M_0)T^2}{2}$$

Now α and δ in Eqs. 7.32 are to be interpreted as the observed (i.e., topocentric) right ascension and declination. Thus, one replaces α in Eq. 7.32 with A from Eq. 7.33 and similarly for Δ. $E = A_0 - \alpha^*$, $E' = \Delta_0 - \delta^*$, $M_0 T$, $M_0' T, \ldots$, are the small parameters of the problem (approximately equal to the field of view). Formally one has the measured parameters (ξ, η) expressed as quadratic functions of T and the six unknowns A_0, Δ_0 (or E, E'); M_0, M_0'; V_0 and $S = \rho/R_0$. To first order in the field of view,

$$\xi = E \cos \delta^* + M_0 T \cos \delta^*$$

$$\mu = E' + M_0' T$$

(7.34)

showing that the angular velocity is available for determination as long as there is a streak at all.

The demonstration that the radius of curvature $\to 0$ as $S \to 0$ requires the keeping of all second-order terms in expressions such as Eqs. 7.34. Enjoy.

PROBLEMS

1. The problem is the following: Given a number of observations of a quantity $x(t)$ and their associated times, say $\{x_n, t_n\}$ where $x_n = x(t_n)$. Let the weight of observation number n be w_n. Find, using polynomial forms to model $x(t)$, the best estimates for $\dot{x}$, dx/dt, and d^2x/dt^2 at some epoch t'. Ideally one would solve the minimization problem via a maximum likelihood technique, form estimators for the parameters of the polynomial, deduce estimators and weights for x, $\dot{x}$, and $\ddot{x}$, and then require that these quantities be evaluated at a time ($=t'$) such that their variances are absolute minima. This may or may not yield a coeval set for x, $\dot{x}$, and $\ddot{x}$. Note that the search for the minimum variance estimators requires functional differentiation with respect to the probability distribution of performing an observation. In general this problem is intractable and we make two reasonable simplifying assumptions to speed the analysis (rather what is left of it). The first assumption is that the observations are executed, in time, symmetrically about $\langle t \rangle$,

$$\langle t \rangle = \frac{\sum_n w_n t_n}{\sum_m w_m}$$

Hence, for all odd k,

$$\sum_n w_n \tau_n^k = 0, \qquad \tau_n = t_n - \langle t \rangle$$

The second, much more restrictive, assumption is that the observations are equally spaced with interval T,

$$\tau_n = nT, \qquad n = -N, -N+1, \ldots, N$$

and have equal weights $w_n = w \; \forall n \in [-N, N]$. These two assumptions are unnecessary when treating a zeroth-order, first-degree polynomial form for $x(t)$. Of course, such forms do not yield an interesting estimator for $\ddot{x}$.

Derive the formulas S_k for the sums

$$S_k = \sum_{n=-N}^{+N} w_n \tau_n^k = \begin{cases} (2N+1)w & k = 0 \\ 0 & k \text{ odd} \\ 2wT^k \sum_{n=1}^{N} n^k & k \text{ even} \end{cases}$$

In particular, deduce that

$$S_0 = (2N+1)w, \qquad S_2 = \tfrac{1}{6}[2wT^2 N(N+1)(2N+1)]$$

$$S_4 = \tfrac{1}{30}[2wT^4 N(N+1)(2N+1)(3N^2+3N-1)]$$

$$S_6 = \tfrac{1}{42}[2wT^6 N(N+1)(2N+1)(3N^4+6N^3-3N+1)]$$

Also show that as $N \to \infty$

$$\sum_{n=-N}^{+N} w_n \tau_n^k \to \frac{w}{T} \int_{-NT}^{NT} u^k \, du = \begin{cases} \dfrac{2w}{T} \dfrac{(NT)^{k+1}}{k+1} & k \text{ even} \\[2mm] 0 & k \text{ odd} \end{cases}$$

Note too that for large N it matters little whether there were $2N+1$ or $2N$ observations.

2. The model considered here is $x(t) = a$ and the data is $\{x_n, t_n\}$, $n = -N, -N+1, \ldots, N$ where $x_n = x(t_n)$. The nth datum has a weight w_n. Minimize the sum of the square of the residuals

$$R = \sum_{n=-N}^{+N} w_n (x_n - a)^2$$

with respect to the parameter a. Show that this leads to the normal equations

$$MA = D$$

where $M = S_0$ is the matrix of the normal equations, $A = a$ is the vector of polynomial parameters, and D is the vector of observations $D = \sum w_n x_n$. Invert M to find the estimators for the elements of A,

$$\tilde{a} = \frac{\sum_n w_n x_n}{\sum_m w_m}$$

and along the diagonal of M^{-1} the estimators for the variances of the elements of A,

$$\text{var}(\tilde{a}) = \frac{1}{\sum_n w_n}$$

The off-diagonal elements of M^{-1} provide estimators for the covariances of the elements of A ($=0$ here).

Now use this model to propagate x in time and obtain an estimator for $x(t)$ at any time $\tilde{x}(t)$. Then calculate an estimate for the variance of $\tilde{x}(t)$ via (in this simple case)

$$\text{var}[\tilde{x}(t)] = \left[\frac{\partial \tilde{x}(t)}{\partial \tilde{a}}\right]^2 \text{var}(\tilde{a}) = \frac{1}{\sum_n w_n}$$

Justify all steps.

3. The model considered here is $x(t) = a + b(t - t_0)$ where t_0 is an arbitrary epoch. Let $\tau = t - t_0$ and again form the weighted sum of the residuals

$$R = \sum_n w_n (x_n - a - b\tau_n)^2$$

Next minimize R with respect to a and b. Show that this leads to $MA = D$

where

$$M = \begin{pmatrix} S_0 & S_1 \\ S_1 & S_2 \end{pmatrix}, \qquad A = \begin{pmatrix} a \\ b \end{pmatrix}, \qquad D = \begin{pmatrix} \sum w_n x_n \\ \sum w_n x_n \tau_n \end{pmatrix}$$

Now deduce that

$$M^{-1} = |M|^{-1} \begin{pmatrix} S_2 & -S_1 \\ -S_1 & S_0 \end{pmatrix}$$

with $|M| = \det(M) = S_0 S_2 - S_1^2$. Again $A = M^{-1} D$ and

$$\mathrm{var}[\tilde{x}(t)] = \left(\frac{\partial \tilde{x}}{\partial \tilde{a}}\right)^2 \mathrm{var}(\tilde{a}) + \left(\frac{\partial \tilde{x}}{\partial \tilde{b}}\right)^2 \mathrm{var}(\tilde{b}) + 2\left(\frac{\partial \tilde{x}}{\partial \tilde{a}}\right)\left(\frac{\partial \tilde{x}}{\partial \tilde{b}}\right) \mathrm{cov}(\tilde{a}, \tilde{b})$$

$$= \frac{\tau^2 S_0 - 2\tau S_1 + S_2}{|M|}$$

Find the value of $\tau = t - t_0$ such that $\mathrm{var}[\tilde{x}(t)]$ is a minimum. Show that it is

$$\tau = \tau_e \equiv \frac{S_1}{S_0} = \langle t \rangle - t_0$$

Show further that $\partial^2 \, \mathrm{var}[\tilde{x}(t)]/\partial t^2 = 2S_0/|M| \geq 0$ so it is a minimum. (That $|M| \geq 0$ follows from an application of the Cauchy inequality.) The value of $\mathrm{var}[\tilde{x}(t)]$ at $t = \langle t \rangle$ is $1/S_0$, just as in the previous case. The variance of $\dot{\tilde{x}}(t)$ is equal to $\mathrm{var}(\tilde{b}) = S_0/|M|$. Now choose $t_0 = \langle t \rangle$; then $S_1 = 0$ and $\mathrm{cov}(\tilde{a}, \tilde{b}) = 0$. Since $|M| = S_0 S_2$, deduce that the general expression for $\mathrm{var}[\tilde{x}(t)]$ is

$$\mathrm{var}[\tilde{x}(t)] = \frac{1}{S_0} + \frac{\tau^2}{S_2}$$

which clearly shows that $\tau = 0$ ($t = \langle t \rangle$) is that instant when $\mathrm{var}[\tilde{x}(t)]$ is a minimum. Further note that min $\mathrm{var}[\tilde{x}(t)]$ is independent of the distribution in time of the observations. For $\dot{\tilde{x}}$ the situation is different, $\mathrm{var}[\dot{\tilde{x}}(t)] = 1/S_2$ and one desires the largest possible spread of observing instants to minimize this quantity.

4. The model considered here is $x(t) = a + b\tau + c\tau^2$, $\tau = t - \langle t \rangle$. Deduce the normal equations $MA = D$,

$$M = \begin{pmatrix} S_0 & S_1 & S_2 \\ S_1 & S_2 & S_3 \\ S_2 & S_3 & S_4 \end{pmatrix}, \qquad A = \begin{pmatrix} a \\ b \\ c \end{pmatrix}, \qquad D = \begin{pmatrix} \sum w_n x_n \\ \sum w_n x_n \tau_n \\ \sum w_n x_n \tau_n^2 \end{pmatrix}$$

and show that

$$M^{-1} = |M|^{-1} \begin{pmatrix} S_2 S_4 - S_3^2 & S_2 S_3 - S_1 S_4 & S_1 S_3 - S_2^2 \\ S_2 S_3 - S_1 S_4 & S_0 S_4 - S_2^2 & S_1 S_2 - S_0 S_3 \\ S_1 S_3 - S_2^2 & S_1 S_2 - S_0 S_3 & S_0 S_2 - S_1^2 \end{pmatrix}$$

where

$$|M| = S_0 S_2 S_4 + 2 S_1 S_2 S_3 - S_2^3 - S_0 S_3^2 - S_4 S_1^2.$$

Prove that

$$|M| \operatorname{var}[\tilde{x}(t)] = S_2 S_4 - S_3^2 + 2(S_2 S_3 - S_1 S_4)\tau$$
$$+ (S_0 S_4 + 2 S_1 S_3 - 3 S_2^2)\tau^2$$
$$+ 2(S_1 S_2 - S_0 S_3)\tau^3 + (S_0 S_2 - S_1^2)\tau^4$$

Introduce the assumption of symmetry $(S_1 = S_3 = 0)$ and prove that $\partial \operatorname{var}[\tilde{x}(t)]/\partial t = 0$ if $\tau = 0$ or $\tau = \tau_\pm$ where

$$\tau_\pm \equiv \pm\left(\frac{3 S_2^2 - S_0 S_4}{2 S_0 S_2}\right)^{1/2}$$

Further calculate that

$$\frac{|M|}{(3 S_2^2 - S_0 S_4)} \frac{\partial^2 \operatorname{var}[\tilde{x}(t)]}{\partial t^2} = \begin{cases} -2 & \tau = 0 \\ +4 & \tau = \tau_\pm \end{cases}$$

Now turn to a general analysis of the variance of $\tilde{x}(t) = \tilde{b} + 2\tilde{c}\tau$. Start with a demonstration of

$$|M| \operatorname{var}[\tilde{x}(t)] = S_0 S_4 - S_2^2 + 4(S_1 S_2 - S_0 S_3)\tau + 4(S_0 S_2 - S_1^2)\tau^2$$

and then deduce that $\partial \operatorname{var}[\tilde{x}(t)]/\partial t$ vanishes if

$$\tau = \dot{\tau}_e = \frac{S_0 S_3 - S_1 S_2}{2(S_0 S_2 - S_1^2)}$$

Furthermore, at this value of τ, prove that $|M|\partial^2 \operatorname{var}[\tilde{x}(t)]/\partial t^2 = 8(S_0 S_2 - S_1^2)$ so that this represents a minimum (the proof of the positiveness of the second derivative of $\operatorname{var}[\tilde{x}(t)]$ with respect to t follows from a double application of the Cauchy inequality). Introduce the symmetry assumption $(S_1 = S_3 = 0)$; this value of τ is just 0 and min $\operatorname{var}[\tilde{x}(t)] = 1/S_2$ as in the linear model. Prove that general $\ddot{x}(t) = 2\tilde{c}$ so $\operatorname{var}[\ddot{x}(t)] = 4(S_0 S_2 - S_1^2)/|M| \geq 0$.

5. Introduce the uniform time separation assumption of the observations into the results of Problem 4. Then prove that as $N \to \infty$

$$\tau_\pm^2 \to \frac{(2N^2 + 2N + 1)T^2}{10}$$

$$\operatorname{var}[\tilde{x}(0)] \to \frac{9(1 - 1/2N)}{8Nw}$$

$$\min \operatorname{var}[\tilde{x}(t)] \to \frac{3(1 - 3/2N)}{8wN^3 T^2}$$

$$\min \text{var}[\tilde{x}(t)] \rightarrow \frac{45(1-5/2N)}{2wN^5T^4}$$

and the reduction in the variance of $\tilde{x}(t)$ from $\tau = 0$ to $\tau = \tau_\pm$ is one-fifth of its value.

6. Measurements on photographic plates of double-star images are used to deduce the distance ρ between the stars and the position angle θ of the companion (fainter star) relative to the primary (brighter star). Position angle is measured from the North through the East. Show that, if (ξ_p, η_p) and (ξ_s, η_s) are the standard coordinates of the primary and secondary,

$$\theta = \tan^{-1}\left(\frac{\xi_P - \xi_S}{\eta_P - \eta_S}\right) + \frac{180}{\pi}\xi_S \tan \delta^*$$

$$\rho^2 = (\xi_P - \xi_S)^2 + (\eta_P - \eta_S)^2$$

for tangential point (α^*, δ^*).

7. Deduce that the rigorous formulas for ρ and θ in terms of the standard coordinates are

$$\tan \theta = \frac{[(\xi_P - \xi_S)\cos \delta^* + (\xi_S \eta_P - \xi_P \eta_S)\sin \delta^*](1 + \xi_P^2 + \eta_P^2)^{1/2}}{[(\eta_P - \eta_S) + \xi_S(\xi_S \eta_P - \xi_P \eta_S)]\cos \delta^*}$$

$$-[\xi_S(\xi_P - \xi_S) + \eta_S(\eta_P - \eta_S)]\sin \delta^*$$

$$\sin^2 \rho = \frac{(\xi_P - \xi_S)^2 + (\eta_P - \eta_S)^2 + (\xi_P \eta_S - \xi_S \eta_P)^2}{(1 + \xi_P^2 + \eta_P^2)(1 + \xi_S^2 + \eta_S^2)}$$

(Eichhorn 1981).

chapter eight

Gaussian-Type Initial Orbit Determination

REALITY

If you are ever in the business of true, honest to goodness orbit determination using angles-only data then first use the statistical Laplacian method outlined in the previous chapter. However, if the problem you face is more specialized, say orbit determination for main belt asteroids, near-stationary artificial satellites, or some other specialized class of objects, then use a specially adopted differential correction process. An example of this type of analysis is discussed in Chapter Eleven. As a last resort use a Gaussian technique. For as the dean of orbit computers, Paul Herget, has written (1965):

> It would be a constructive achievement to dispel the myth that "a preliminary orbit can be computed from three observations." This dictum is based upon a deceptively simple deduction: the problem requires six arbitrary constants for the solution of three simultaneous, second-order, differential equations, and each observation provides two observed coordinates, α and δ. Several factors mitigate the success of such simple logic. The theoretical aspects of the subject have been fully discussed by the author (Herget 1939, 1948). The practical aspects include the inevitable errors of observations, sometimes inconsistent observations, and the indeterminateness that is due to a short observed arc and/or to the nearly great-circle motion of the object. For all these reasons, we wish to have at hand an electronic computer program which is able to handle as many observations as may be available, which indicates the degree of indeterminateness of the resulting solution, and which readily permits the operator to impose his judgment upon the progress of the computations.

Herget (1948) does *not* fully discuss the theoretical aspects of the subject. The essential missing material is contained in this chapter. Gauss himself wrote much like Herget did. First, on specialized classes of objects (comets and

256

planets in this case; Gauss 1809):

> As soon as it was ascertained that the motion of the new planet, discovered in 1781 [Uranus], could not be reconciled with the parabolic hypothesis, astronomers undertook to adapt a circular orbit to it, which is a matter of simple and very easy calculation. By a happy accident the orbit of this planet had but a small eccentricity, in consequence of which the elements resulting from the circular hypothesis sufficed at least for an approximation on which could be based the determination of the elliptic elements. There was a concurrence of several other very favorable circumstances. For, the slow motion of the planet, and the very small inclination of the orbit to the plane of the ecliptic, not only rendered the calculations much more simple, and allowed the use of special methods not suited to other cases; but they removed the apprehension, lest the planet, lost in the rays of the sun, should subsequently elude the search of observers, (an apprehension which some astronomers might have felt, especially if its light has been less brilliant); so that the more accurate determination of the orbit might be safely deferred, until a selection could be made from observations more frequent and more remote, such as seemed best fitted for the end of view.

> Thus, in every case in which it was necessary to deduce the orbits of heavenly bodies from observations, there existed advantages not to be despised, suggesting, or at any rate permitting, the application of special methods; of which advantages the chief one was, that by means of hypothetical assumptions an approximate knowledge of some elements could be obtained before the computation of the elliptic elements was commenced.

That last phrase is crucial to clever initial orbit determination. Later, from the beginning of Book II:

> Seven elements are required for the complete determination of the motion of a heavenly body in its orbit, the number of which, however, may be diminished by one, if the mass of the heavenly body is either known or neglected; neglecting the mass can scarcely be avoided in the determination of an orbit wholly unknown, where all the quantities of the order of the perturbations must be omitted, until the masses on which they depend become otherwise known. Wherefore, in the present inquiry, the mass of the body being neglected, we reduce the number of the elements to six, and, therefore, it is evident, that as many quantities depending on the elements, but independent of each other, are required for the determination of the unknown orbit. These quantities are necessarily the places of the heavenly body observed from the earth; since each one of which furnishes two data, that is, the longitude and latitude, or the right ascension and declination, it will certainly be the most simple to adopt three geometric places which will, in general, be sufficient for determining the six unknown elements. This problem is to be regarded as the most important in this work, and, for this reason, will be treated with the greatest care in this section.

> But in the special case, in which the plane of the orbit coincides with the ecliptic, and thus both the heliocentric and geocentric latitudes, from their nature, vanish, the three vanishing geocentric latitudes cannot any longer be considered

as three data independent of each other: then, therefore, this problem would remain indeterminate, and the three geocentric places might be satisfied by an infinite number of orbits. Accordingly, in such a case, four geocentric longitudes must, necessarily, be given, in order that the four remaining unknown elements (the inclination of the orbit and the longitude of the node being omitted) may be determined. But although, from an indiscernible principle, it is not to be expected that such a case would ever actually present itself in nature, nevertheless, it is easily imagined that the problem, which, in orbit exactly coinciding with the plane of the ecliptic, is absolutely indeterminate, must, on account of the limited accuracy of the observations, remain nearly indeterminate in orbits very little inclined to the ecliptic, where the very slightest errors of the observations are sufficient altogether to confound the determination of the unknown quantities. Whereof, in order to examine this case, it will be necessary to select six data: for which purpose we will show in section second, how to determine an unknown orbit from four observations, of which two are complete, but the other two incomplete, the latitudes or declinations being deficient.

Finally, as all our observations, on account of the imperfection of the instruments and of the senses, are only approximations to the truth, an orbit based only on the six absolutely necessary data may be still liable to considerable errors. In order to diminish these as much as possible, and thus to reach the greatest precision attainable, no other method will be given except to accumulate the greatest number of the most perfect observations, and to adjust the elements, not so as to satisfy this or that set of observations with absolute exactness, but so as to agree with all in the best possible manner. For which purpose, we will show in the third section how, according to the principles of the calculus of probabilities, such an agreement may be obtained, as will be, if in no one place perfect, yet in all the places the strictest possible.

The determination of orbits in this manner, therefore, so far as the heavenly bodies move in them according to the laws of Kepler, will be carried to the highest degree of perfection that is desired. Then it will be proper to undertake the final correction, in which the perturbations that the other planets cause in the motion, will be taken account of: we will indicate briefly in the fourth section, how these may be taken account of, so far at least, as it shall appear consistent with our plan.

If one is using complete radar data, the method outlined immediately below should be used. As a last resort, or if your data is incomplete, use Gauss's method as refined by Gibbs. A full discussion of the Gauss–Gibbs method occupies most of this chapter. It is clear that its importance has been overblown and that it suffers from a fatal flaw that Laplace's method (or variants thereof for radar applications) does not have. Extensive numerical and experimental experience with it on hundreds of high-altitude artificial satellites and a few interesting (e.g., non–main belt) asteroids shows that it leaves much to be desired. There is no need to utilize minimum data methods. The middle part of the chapter discusses extensive numerical tests of three different, angles-only, initial orbit determination methods. The last part of this chapter is devoted to the construction of the orbital element set itself.

A COMPLETE RADAR DATA METHOD

The best radars provide excellent values for position, topocentric distance, and topocentric radial velocity. For instance the Millstone Hill Radar has precisions of $0°.005$ in each angle, 30 m in distance, and 30 mm/sec in radial velocity on a high ($\sim 6:1$) signal-to-noise deep-space artificial satellite. Almost all low-altitude artificial satellites are in circular orbits. Hence, for them, much poorer radar data, even without radial velocity information, provides enough information to yield an orbit. The reasons are several: (1) The orbit is circular by design and necessity (as will be discussed in more detail in the next two chapters) so that there are only four orbital elements to fix. (2) Two of these remaining orbital elements are the inclination and the longitude of the ascending node. These angles specify the direction of the angular momentum vector and can always be accurately determined on a minimum of data. Why this is so I don't know, that it is so in practice is abundantly clear for all types of satellites and asteroids. (3) Since one knows that the orbit is circular and the topocentric location (and not just the topocentric *position*), the semi-major axis is immediately known. Thus, in this case, the hardest quantity to determine is the time of perigee passage or (better) the mean longitude at epoch. Therefore, low-altitude artificial satellite initial orbit determination based on radar data is essentially trivial.

For high-altitude satellites observed completely by a radar (that is $\mathbf{R}$ and $\dot{R}$ are determined simultaneously) the following minimal method is analytically very simple. Once again we utilize the four constants of the motion given by the energy and angular momentum integrals (per unit mass). The argument is the same as the one presented in Chapter Seven and we again write

$$\mathscr{E}(\mathbf{r}, \dot{\mathbf{r}}) = \frac{\dot{\mathbf{r}} \cdot \dot{\mathbf{r}}}{2} - \frac{\mu}{|\mathbf{r}|} = \text{const scalar}$$

$$\mathbf{L}(\mathbf{r}, \dot{\mathbf{r}}) = \mathbf{r} \times \dot{\mathbf{r}} = \text{const vector}$$

This is expressed topocentrically ($\mathbf{r} = \mathbf{R} + \boldsymbol{\rho} = R\mathbf{l} + \boldsymbol{\rho}$) as

$$\mathscr{E} = \frac{\dot{R}^2 + R^2(\dot{\mathbf{l}} \cdot \dot{\mathbf{l}}) + \dot{\boldsymbol{\rho}} \cdot \dot{\boldsymbol{\rho}}}{2} + \dot{R}(\mathbf{l} \cdot \dot{\boldsymbol{\rho}}) + R(\dot{\mathbf{l}} \cdot \dot{\boldsymbol{\rho}}) - \frac{\mu}{(R^2 + 2\mathbf{R} \cdot \boldsymbol{\rho} + \rho^2)^{1/2}}$$

$$\mathbf{L} = R^2(\mathbf{l} \times \dot{\mathbf{l}}) + R(\mathbf{l} \times \dot{\boldsymbol{\rho}} - \dot{\mathbf{l}} \times \boldsymbol{\rho}) - \dot{R}(\mathbf{l} \times \boldsymbol{\rho}) + \boldsymbol{\rho} \times \dot{\boldsymbol{\rho}}$$

$$(8.1)$$

The assumption is that we have complete topocentric radar information (e.g., R, $\dot{R}$, and $\mathbf{l}$ but not $\dot{\mathbf{l}}$) at two times, say t_1 and t_2. The equations

$$\mathscr{E}_1 = \mathscr{E}_2, \qquad \mathbf{L}_1 = \mathbf{L}_2 \tag{8.2}$$

represent four equations in the four unknowns $\dot{\mathbf{l}}_1, \dot{\mathbf{l}}_2$. Note that the angular momentum is a linear function of $\dot{\mathbf{l}}$. Hence the angular momentum part of Eqs. 8.2 can be regarded as three linear, inhomogeneous equations in three unknowns, say $\dot{A}_1$, $\dot{A}_2$, and $\dot{\Delta}_1$. The explicit algebraic solution of this triplet allows one to write each of these three unknowns parametrically in terms of known quantities (ρ and $\dot{\rho}$), measured quantities ($\mathbf{R}$ and $\dot{R}$), and the fourth element of $\mathbf{l}_1$ and $\mathbf{l}_2$, Δ_2 here. When these formulas are utilized in the conservation of energy equation, which is only quadratic in the topocentric angular velocity, see Eq. 8.1, initial orbit determination has been reduced to the solution of a single quadratic equation. Simpler than this it is not likely to get. Note too that this is exact, analytic, and explicit.

This technique has not been received with open arms by the radar community even though it was published long ago (Taff and Hall 1977). The reason is simple—the radar community uses variants of the Gauss–Gibbs methods. These variants frequently include the distance as given (after all it is a radar that they are using) and then immediately relax the (poor) initially determined orbit via a differential correction process (see Chapter Eleven). Because the radar has maintained a continuous track on the missile, rocket, or artificial satellite the entire time, acquiring additional data at its pulse repetition frequency, almost any method will work to start the process. Hence there is no observational pressure to start better.

GAUSS'S METHOD

So what did Gauss do? Gauss realized that the conservation of angular momentum in the two-body problem implied that the motion must occur in a plane. Because space is three dimensional, this implies that three location vectors, say $\mathbf{r}_1$, $\mathbf{r}_2$ and $\mathbf{r}_3$ at times t_1, t_2 and t_3, are linearly dependent. The mathematical expression of this is

$$c_1\mathbf{r}_1 + c_2\mathbf{r}_2 + c_3\mathbf{r}_3 = \mathbf{0} \tag{8.3}$$

where the c_j, $j = 1, 2, 3$ are constants not all zero. This is the first step in the Gaussian method. The second step consists of finding an approximate way to express $\mathbf{r}_1$ and $\mathbf{r}_3$ in terms of quantities at time t_2. This problem can be approached in a general fashion.

It has been stressed (see Chapter One) that the initial conditions for Newton's equations of motion 1.7 are the location and velocity vectors evaluated at some time, say $t = t_0$. In particular, one might suspect that the solution to the two-body problem could be written as

$$\mathbf{r}(t) = f(\mathbf{r}_0, \dot{\mathbf{r}}_0, t - t_0)\mathbf{r}_0 + g(\mathbf{r}_0, \dot{\mathbf{r}}_0, t - t_0)\dot{\mathbf{r}}_0 \tag{8.4}$$

where $\mathbf{r}_0 = \mathbf{r}(t = t_0)$, and so on. Thus, f and g satisfy the differential equations (assuming that $\ddot{\mathbf{r}} = -\mu\mathbf{r}/r^3$)

$$\ddot{f} = -\frac{\mu f}{r^3}, \qquad \ddot{g} = -\frac{\mu g}{r^3}$$

with $r^2 = f^2 r_0^2 + 2fg\mathbf{r}_0 \cdot \dot{\mathbf{r}}_0 + g^2 \dot{r}_0^2$. Note too that

$$f\dot{g} - \dot{f}g = 1$$

follows from the conservation of the angular momentum. We do not have to solve the differential equations for f and g; all we need do is equate Eq. 8.4 to the explicit parametric solution of the two-body problem given in Eq. 2.13. With $L^2 = \mu a(1-e^2)$ the result is

$$Lf = rr_0\dot{r}_0 \cos(v-v_0) - r\dot{r}_0 \sin(v-v_0)$$
$$Lg = rr_0 \sin(v-v_0) \tag{8.5}$$

where v is the true anomaly at time $t[v_0 = v(t=t_0)]$ and r is the distance at time $t[r = a(1-e^2)/(1+e\cos v)]$. Equations 8.5 represent an exact solution to the problem of relating $\mathbf{r}$ at time t to $\mathbf{r}$ and $\dot{\mathbf{r}}$ at some other time t_0. Clearly to be useful one must *know* the orbital element set, and just as clearly, at the onset of initial orbit determination, one has no such knowledge.

Gauss may have known of Eqs. 8.5 (or their variants when expressed in terms of the eccentric anomaly). In any case he arrived at his approximations to the f and g series (and hence values for the c_j in Eq. 8.3) by geometric arguments. As usually expounded in the celestial mechanics texts this is phrased in terms of approximating the sectorial areas of an ellipse by the inscribed triangle. Geometrically this follows because the solution of Eq. 8.3 for the c_j can be written as

$$\frac{\mathbf{r}_3 \times \mathbf{r}_1}{c_2} = \frac{\mathbf{r}_1 \times \mathbf{r}_2}{c_3} = \frac{\mathbf{r}_2 \times \mathbf{r}_3}{c_1} \tag{8.6}$$

coupled with the interpretation of the norm of the vector cross product as an area. In fact, see Fig. 35, c_1/c_2 is the ratio of the area of the triangle OP_2P_3

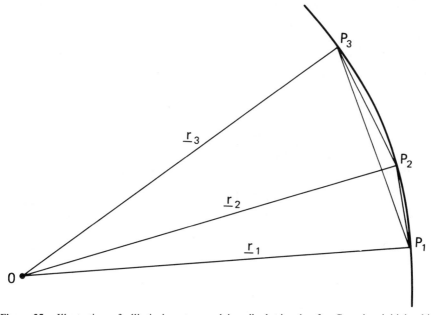

Figure 35. Illustration of elliptical sectors and inscribed triangles for Gaussian initial orbit determination.

to triangle OP_1P_3 whereas c_3/c_2 is the ratio of the area of the triangle OP_1P_2 to triangle OP_1P_3. The equal areas swept out in equal times consequence of the central nature of the gravitational force is now invoked, together with Kepler's equation to replace areas of triangles with elliptical sector areas and thence time differences. Although this picture of the Gaussian process is interesting, it is also irrelevant. What is needed, however motivated or deduced, is an approximate expression for $\mathbf{r}_1$ and $\mathbf{r}_3$ in terms of quantities at $t = t_2$. Since these expressions are power series, the very power series obtained by expanding the f and g functions in Eqs. 8.5, a more direct approach will be taken here.

The first few terms of the f and g series are ($T = t - t_0$)

$$f = 1 - \tfrac{1}{2}h_0 T^2 + \tfrac{1}{2}h_0 p_0 T^3 + \tfrac{1}{24}[h_0(3q_0 - 15p_0^2 + h_0)T^4]$$
$$+ \tfrac{1}{8}[h_0 p_0(7p_0^2 - 3q_0 - h_0)T^5] - \tfrac{1}{720}[h_0(945p_0^4 - 210h_0 p_0^2$$
$$- 630p_0^2 q_0 + 45q_0^2 + 24h_0 q_0 + h_0^2)T^6] + \cdots \qquad (8.7)$$
$$g = T - \tfrac{1}{6}h_0 T^3 + \tfrac{1}{4}h_0 p_0 T^4 + \tfrac{1}{120}[h_0(9q_0 + h_0 - 45p_0^2)T^5]$$
$$+ \tfrac{1}{24}[h_0 p_0(14p_0^2 - 6q_0 - h_0)T^6] + \cdots$$

where the quantities h, p, and q are defined by

$$h = \frac{\mu}{r^3}, \qquad p = \frac{\mathbf{r} \cdot \dot{\mathbf{r}}}{r^2}, \qquad q = \frac{\dot{\mathbf{r}} \cdot \dot{\mathbf{r}}}{r^2} - h \qquad (8.8)$$

It is important to notice that the actual dependence of f and g on $\mathbf{r}_0$ and $\dot{\mathbf{r}}_0$ is really on $|\mathbf{r}_0|$, $|\dot{\mathbf{r}}_0|$, and $|\mathbf{r}_0 \cdot \dot{\mathbf{r}}_0|$. Now we restrict ourselves to the first two terms in each series. With the choice of t_2 as the middle observation define ($k^2 = \mu$)

$$T_1 = k(t_3 - t_2), \qquad T_2 = k(t_3 - t_1), \qquad T_3 = k(t_2 - t_1)$$

Accordingly, in Eq. 8.4,

$$\mathbf{r}_1 = f_1 \mathbf{r}_2 + g_1 \dot{\mathbf{r}}_2, \qquad \mathbf{r}_3 = f_3 \mathbf{r}_2 + g_3 \dot{\mathbf{r}}_2$$

where, from Eq. 8.7 and the definitions just introduced,

$$f_1 = 1 - \tfrac{1}{2}h_2 T_3^2, \qquad g_1 = -T_3(1 - \tfrac{1}{6}h_2 T_3^2)$$
$$f_3 = 1 - \tfrac{1}{2}h_2 T_1^2, \qquad g_3 = +T_1(1 - \tfrac{1}{6}h_2 T_1^2)$$

Finally, from Eq. 8.6, the ratios of the c_j are found to be

$$\frac{c_1}{c_2} = \left(\frac{T_1}{T_2}\right)\left[1 + \frac{h_2(T_2^2 - T_1^2)}{6}\right]$$
$$\frac{c_3}{c_2} = \left(\frac{T_3}{T_2}\right)\left[1 + \frac{h_2(T_2^2 - T_3^2)}{6}\right]$$

Thus the constants in Eq. 8.3 are known in terms of the observed times and the unknown distance r_2. Of course, because only the first pair of terms in the f and g series has been kept, and especially no radial velocity terms, the values obtained for the c_j will be very poor ones when the orbit is eccentric. Should the f and g series not converge (unbeknown to us), then the values for the c_j will be meaningless.

The next step in the procedure is to rewrite Eq. 8.3 or Eq. 8.6 in the topocentric form utilizing $\mathbf{r} = R\mathbf{l} + \boldsymbol{\rho}$. Once this is done, and Gauss's expressions for the c_j inserted, one is left with an equation of the form

$$\rho_2 = A + B/R_2^3$$

A and B depend only on the observations $\mathbf{l}_j$, the times t_j $(j = 1, 2, 3)$, and the physical constants (e.g., $\mu = GM$) of the problem. This relationship, coupled with the basic geometrical relationship (obtained by squaring $\mathbf{r} = R\mathbf{l} + \boldsymbol{\rho}$ at $t = t_2$), closes the system. One typically completes the process of initial orbit determination by computing R_1 and R_3 and thence $\mathbf{r}_1$ and $\mathbf{r}_3$. Now one has six initial conditions from which to compute an orbital element set.

Gibbs's Refinement

What did Gibbs (1888) do? Gibbs realized that one could include higher-order terms in the f and g series rather than merely the quadratic ones. Presumably, because the radial velocity could now be taken into account, this would extend the realm of applicability of the technique to more highly eccentric orbits. This supposition is false.

In particular, Gibbs realized that not just quadratic but quartic terms could be included. To see how, we write

$$\mathbf{r}_j = \mathbf{A} + \mathbf{B}t_j + \mathbf{C}t_j^2 + \mathbf{D}t_j^3 + \mathbf{E}t_j^4, \qquad \text{for } j = 1, 2, 3 \tag{8.9}$$

The vectors $\mathbf{A}, \mathbf{B}, \ldots, \mathbf{E}$ are constants. Now, with the physics of the situation enforced,

$$\ddot{\mathbf{r}} = -\frac{\mu \mathbf{r}}{r^3}, \qquad \text{at } t = t_j \tag{8.10}$$

Equations 8.9 and 8.10 represent six linear inhomogeneous equations in the five unknowns $\mathbf{A}$–$\mathbf{E}$ for Eq. 8.10 is equivalent to, after a double differentiation of Eq. 8.9,

$$2\mathbf{C} + 6\mathbf{D}t_j + 12\mathbf{E}t_j^2 = -\frac{\mu \mathbf{r}_j}{r_j^3} \tag{8.11}$$

There can be a consistent solution to such an overdetermined system if and only if the augmented matrix of the system has a vanishing determinant (see, e.g., Murdoch 1957). This constraint on Eqs. 8.9 and 8.11 at the three times $t_1, t_2 = 0$ and t_3 may be expressed as

$$t_3\left(1 + \frac{t_1^2 - t_1 t_3 - t_3^2}{12 r_1^3}\right)\mathbf{r}_1 - (t_3 - t_1)\left(1 - \frac{t_1^2 - 3 t_1 t_3 + t_3^2}{12 r_2^3}\right)\mathbf{r}_2 - t_1\left(1 - \frac{t_1^2 + t_1 t_3 - t_3^2}{12 r_3^3}\right)\mathbf{r}_3 = \mathbf{0}$$

If we now go back to the uppercase t notation for the coefficients c_j, then

$$c_1 = -T_1\left(1 + \frac{\Gamma_1}{r_1^3}\right), \qquad \Gamma_1 = \frac{T_1 T_3 + T_2(T_3 - T_1)}{12}$$

$$c_2 = -T_2\left(1 - \frac{\Gamma_2}{r_2^3}\right), \qquad \Gamma_2 = \frac{T_1 T_3 + T_2^2}{12}$$

$$c_3 = -T_3\left(1 + \frac{\Gamma_3}{r_3^3}\right), \qquad \Gamma_3 = \frac{T_1 T_3 - T_2(T_3 - T_1)}{12}$$

The remaining steps of the analysis are as outlined above. This extension beyond quadratic terms (and the limitation to no more than quartic) follows from the fact that the evaluation of the constants $A-E$ in Eq. 8.7 is not at issue when one has three sets of angles-only data. Rather, what is desired is an expression of the linear dependence of three three-dimensional vectors. Because such a relationship implies that six of the elements of these vectors are independent, one must set up six equations among the unknowns, thus up to and including quartic terms but no more. Herget (1948) recounts yet another improvement of a similar type developed by Koziel.

The f and g Series

The Radius of Convergence

First, I show why one might suspect that there is a problem with the f and g series for large eccentricities. Then I provide several rigorous proofs for the following theorem.

Theorem: The f and g series defined by Eq. 8.4 have the following radius of convergence in $T = t - t_0$: if $e = 0$, then the radius of convergence is infinite; if e is unity, then the radius of convergence is $(8Q^3/9\mu)^{1/2}$ where $\mu = GM$ and Q is the distance from the focus of the parabola to its directrix; and if $e \in (0, 1)$, then the radius of convergence is given by $P \imath/(2\pi)$ where P is the period and

$$\imath = (M_0^2 + \{\ln[1 + (1 - e^2)^{1/2}] - \ln e - (1 - e^2)^{1/2}\}^2)^{1/2} \qquad (8.12)$$

Here M_0 is the value of the mean anomaly corresponding to $t = t_0$, $M_0 \in [-\pi, \pi]$. Below I discuss the implications of this for orbit determination using the Gaussian method.

The coefficients of $T = t - t_0$ in Eqs. 8.7 are algebraic combinations of the auxiliary variables $h, p,$ and q (see Eq. 8.8). In order to get a quantitative feeling for the relative size of these three quantities, their average over an entire orbit will be computed*:

$$\langle h \rangle = \frac{\mu}{a^3(1 - e^2)^{3/2}}, \qquad \langle q \rangle = (1 - e^2)\langle h \rangle$$

$$\langle p \rangle = 0, \qquad \langle p^2 \rangle = \frac{e^2 \langle h \rangle}{2}$$

* For any quantity u, $\langle u \rangle = \int_0^P u \, dt / P$.

Hence the coefficient of T^k in f is on the order, on the average (given a very heuristic averaging procedure), of $h^{k/2}$. For g the coefficient of T^k is on the order of $h^{k-1/2}$. A simple proof by induction coupled with exact relationship,

$$\frac{dh}{dt} = -3hp$$

shows that this is true for all values of k. Hence, for the series to converge, one would need (by Cauchy's root test)

$$T\langle h\rangle^{1/2} < 1 \qquad \text{or} \qquad \frac{T}{P} < \frac{(1-e^2)^{3/4}}{2\pi} \tag{8.13}$$

Clearly as $e \to 1$, $T/P \to 0$.

The above can be easily criticized, especially for the use of averaging and the fashion in which the limit, as the eccentricity approaches unity, was taken. The proper limit of the formulas of elliptical motion involves both letting $e \to 1$ and $a \to \infty$ such that $Q = a(1-e)$ is finite. Hence, an asymptotic expansion in $1-e$ that starts from the parabolic formulas is much more appropriate. This will be investigated below. More generally, it appears that the convergence of the f and g series depends on the object's distance (since $h = \mu/r^3$); that is, Eq. 8.13 without averaging implies that

$$r > (T^2\mu)^{1/3}$$

is necessary for convergence. This leads to the question "What fraction of an orbit is the distance greater than some preselected lower bound?" The answer is left to the reader.

If the motion is assumed to be circular, then $p = q = 0$, $h = \mu/a^3 = n^2$, and, by induction on the coefficients of T^k in the f and g series, one finds that

$$f = \cos[v(T) - v_0]$$

$$g = \frac{1}{n}\sin[v(T) - v_0]$$

But if $e = 0$, $v = E = M$, and since the sine and cosine are analytic for all real values of their argument, the power series expansions of f and g in terms of $t - t_0$ converge everywhere. This proves the first part of the theorem.

We now consider the case $e = 1$. The analog of Kepler's equation for a parabola is

$$\tan\frac{v}{2} + \frac{1}{3}\tan^3\left(\frac{v}{2}\right) = \left(\frac{\mu}{2Q^3}\right)^{1/2}(t - t_0)$$

where $v(t_0) = 0$. The explicit solution for $\tan(v/2)$ is

$$\tan\tfrac{1}{2}v = [cT + (1 + c^2T^2)^{1/2}]^{1/3} + [cT - (1 + c^2T^2)^{1/2}]^{1/3}$$

with

$$c^2 = \frac{9\mu}{8Q^3}, \qquad c > 0, \ T = t - t_0$$

From this result one can compute $\sin v$ and $\cos v$ in terms of T. If this is substituted into Eqs. 8.4 and the binomials expanded, we will have the appropriate power series. They will converge when the binomial series converges, for example, for $c|T| \leq 1$. Now we can address the earlier self-criticism concerning the limit as $e \to 1$. Since

$$Q = \lim_{\substack{a \to \infty \\ e \to 1}} a(1-e), \qquad Q \text{ finite}$$

it will be asymptotically correct to replace Q by $a(1-e)$ in c. Thus, as $e \to 1$, we can expect convergence for the f and g series when

$$\frac{3\pi|T|}{P} < \sqrt{2}(1-e)^{3/2} \qquad (8.14)$$

Although of a different functional form than the result in Eq. 8.13, the problem with $e \to 1$ does not go away.*

Having provided this much motivation for the existence of a problem, we now turn to solving it rigorously. It is simpler to work with the eccentric anomaly than the mean anomaly so we rewrite the f and g functions as

$$f = \left(\frac{a}{Lr_0}\right)(r_0\dot{r}_0 \sin v_0 + L \cos v_0)(\cos E - e)$$

$$+\frac{a(1-e^2)^{1/2}}{r_0 L}(L \sin v_0 - r_0\dot{r}_0 \cos v_0) \sin E$$

$$g = -\left(\frac{ar_0}{L}\right)\sin v_0(\cos E - e)$$

$$+\left[\frac{ar_0(1-e^2)^{1/2}}{L}\right]\cos v_0 \sin E$$

We need not only expand the trigonometric functions of the eccentric anomaly in power series (which converge for all real values of E) but also expand E in a convergent power series in M, substitute this, term by term, in the series for $\cos E$ and $\sin E$, rearrange terms to obtain a power series for f and g in M, and then show convergence. The essential point of this chapter is that Gauss *did not* solve the general initial orbit determination problem. He only solved the *nearly circular* initial orbit determination problem unless the eccentricity, the semi-major axis, and the time of periapse passage (or half of the orbital element set) are known in advance.

Taff's Proof

Taff's proof is long-winded but involves simpler mathematics than does Moulton's. Hence, it will be presented first. From Kepler's equation, that is

$$E - M = e \sin E$$

* Equation 8.14 is the correct limit of Eq. 8.12 when $M_0 = 0$ and $e \to 1^-$.

and as discussed in Chapter Two, $e \sin E$ can be expanded in a Fourier sine series,

$$e \sin E = \sum_{k=1}^{\infty} a_k \sin kM$$

with

$$a_k = \frac{2}{\pi} \int_0^{\pi} e \sin E \sin kM \, dM$$

If one integrates by parts and then replaces $e \sin E$ by $E - M$, one deduces that

$$\frac{\pi k a_k}{2} = \int_0^{\pi} \cos kM \, dE - \int_0^{\pi} \cos kM \, dM$$

The last integral vanishes and if Kepler's equation is used again, the expression for a_k becomes

$$a_k = \left(\frac{2}{\pi k}\right) \int_0^{\pi} \cos[k(E - e \sin E)] \, dE$$

By Eq. 2.20, the integral expression for a Bessel function,

$$a_k = \frac{2}{\pi} J_k(ke)$$

completing the derivation of Eq. 2.9,

$$E = M + \sum_{\nu=1}^{\infty} \left(\frac{2}{\nu}\right) J_\nu(\nu e) \sin(\nu M) \tag{2.9}$$

This series converges for $e \in [0, 1]$. However, a Taylor's series representation for $E(M)$ is needed. To compute the derivatives $d^p E / dM^p$, we differentiate Eq. 2.9 p times term by term and then sum. The sum will be $d^p E / dM^p$ if the series obtained by $(p-1)$-fold differentiation term by term converges and if the series obtained by p-fold differentiation term by term converges uniformly. Thus, we need to show that

$$\sum_{\nu=1}^{\infty} \nu^{p-1} J_\nu(\nu e) \binom{\sin}{\cos}(\nu M) \tag{8.15}$$

converges uniformly $\forall p \geq 1$ and $e \in (0, 1)$.

By Weierstrass's comparison test, since for all real M, $|\sin \nu M| \leq 1$, $|\cos \nu M| \leq 1$, if

$$\sum_{\nu=1}^{\infty} |\nu^{p-1} J_\nu(\nu e)|, \qquad p \geq 1, \qquad e \in (0, 1) \tag{8.16}$$

converges, then the series 8.15 converges uniformly. But $J_\nu(\nu e)$ for $e \in (0, 1)$ is a positive decreasing function of ν (Watson 1922, Section 8.5). It therefore follows from D'Alembert's ratio test that the series 8.16 converges whence the series 8.15 all converge uniformly for $p \geqslant 1$, $e \in (0, 1)$. Therefore,

$$
\frac{d^p E}{dM^p} =
\begin{cases}
\delta_{p_1} + 2(-1)^{(p-1)/2} \sum_{\nu=1}^{\infty} \nu^{p-1} J_\nu(\nu e) \cos(\nu M), & p \text{ odd} \\[2ex]
2(-1)^{p/2} \sum_{\nu=1}^{\infty} \nu^{p-1} J_\nu(\nu e) \sin(\nu M), & p \text{ even}
\end{cases}
$$

Since all of the derivatives of E with respect to M exist, E is an analytic function of M. It is, therefore, expressible in a Taylor's series and the Taylor series, in a sufficiently small neighborhood, does converge to E. At $M = 0^*$, one has

$$
\left. \frac{d^p E}{dM^p} \right|_{M=0} =
\begin{cases}
\delta_{p_1} + 2(-1)^{(p-1)/2} \sum_{\nu=1}^{\infty} \nu^{p-1} J_\nu(\nu e), & p \text{ odd} \\[2ex]
0, & p \text{ even}
\end{cases}
$$

Thus, since $\sum_{n=1}^{\infty} J_n(ne) = e/2(1-e)$,

$$
E = \frac{M}{1-e} + \sum_{k=1}^{\infty} \frac{S_{2k+1} M^{2k+1}}{\Gamma(2k+2)}
$$

where $\Gamma(z)$ is the gamma function and

$$
S_{2k+1} = 2(-1)^k \sum_{\nu=1}^{\infty} \gamma_{2k}(\nu, e), \qquad \gamma_{2k}(\nu, e) \equiv \nu^{2k} J_\nu(\nu e)
$$

The radius of convergence of the $E(M)$ power series, $\imath$, can be obtained by Cauchy's root test:

$$
\frac{1}{\imath} = \limsup_{k \to \infty} \left| \frac{S_{2k+1}}{\Gamma(2k+2)} \right|^{1/(2k+1)}
$$

Since for $e \in (0, 1)$ $J_\nu(\nu e)$ is a positive decreasing function of ν, and for $k \geqslant 1$, $\nu \geqslant 1$, ν^{2k} is a positive increasing function of ν, it follows that $\gamma_{2k}(\nu, e)$ has a single maximum as a function of ν. Call this value of ν, N (not necessarily an integer). Let $[N]$ be the greatest integer less than or equal to N. Then, following the logic used to derive the integral test for infinite series convergence,

$$
\gamma_{2k}(m-1, e) < \int_{m-1}^{m} \gamma_{2k}(\nu, e)\, d\nu < \gamma_{2k}(m, e), \qquad \forall m \leqslant [N]
$$

So, since $J_0(0) = 1$,

$$
\sum_{m=0}^{[N]-1} \gamma_{2k}(m, e) < \int_0^{[N]} \gamma_{2k}(\nu, e)\, d\nu < \sum_{m=0}^{[N]} \gamma_{2k}(m, e)
$$

---

* The expansion about $M_0 = 0$ is equivalent to expanding about the instant of periapse passage. For any other value of M_0 we would find that the f and g series in $M - M_0$ has a radius of convergence $\imath$ given by Eq. 8.12 with $M_0 = 0$. Hence, once the $M_0 = 0$ result is established, the $M_0 \neq 0$ result follows immediately.

Also,

$$\gamma_{2k}([N]+m, e) > \int_{[N]+m}^{[N]+m+1} \gamma_{2k}(\nu, e) \, d\nu > \gamma_{2k}([N]+m+1, e), \qquad \forall m \geq 1$$

so

$$\sum_{m=[N]+1}^{\infty} \gamma_{2k}(m, e) > \int_{[N]+1}^{\infty} \gamma_{2k}(\nu, e) \, d\nu > \sum_{m=2+[N]}^{\infty} \gamma_{2k}(m, e)$$

Combining the inequalities yields

$$\int_0^{\infty} \gamma_{2k}(\nu, e) \, d\nu - \int_{[N]}^{[N]+1} \gamma_{2k}(\nu, e) \, d\nu + \gamma_{2k}([N], e) + \gamma_{2k}([N]+1, e)$$

$$> \sum_{m=0}^{\infty} \gamma_{2k}(m, e)$$

However,

$$\gamma_{2k}(N, e) > \int_{[N]}^{N} \gamma_{2k}(\nu, e) \, d\nu > \gamma_{2k}([N], e)$$

and

$$\gamma_{2k}(N, e) > \int_{N}^{[N]+1} \gamma_{2k}(\nu, e) \, d\nu > \gamma_{2k}([N]+1, e)$$

so

$$\int_{[N]}^{[N]+1} \gamma_{2k}(\nu, e) \, d\nu > \gamma_{2k}([N], e) + \gamma_{2k}([N]+1, e)$$

Therefore

$$I_{2k} = \int_0^{\infty} \gamma_{2k}(\nu, e) \, d\nu > \sum_{m=0}^{\infty} \gamma_{2k}(m, e) = \frac{|S_{2k+1}|}{2}$$

We now have, for all $k \geq 1$, $e \in (0, 1)$ an upper bound for $|S_{2k+1}|$. The next step is the evaluation of the integral I_{2k}.

We can write (Watson 1922, Section 8.5)

$$J_{\nu}(\nu e) = \frac{1}{\pi} \int_0^{\pi} \exp[-\nu F(\theta, e)] \, d\theta$$

where

$$F(\theta, x) = \ln[\theta + (\theta^2 - x^2 \sin^2 \theta)^{1/2}] - \ln(x \sin \theta) - (\theta^2 - x^2 \sin^2 \theta)^{1/2} \cot \theta \tag{8.17}$$

and we can prove the sequence of inequalities

$$F(\theta, x) \geq F(0, x) \geq F(0, 1) = 0 \tag{8.18}$$

Replace $J_\nu(\nu e)$ in the integral for I_{2k} with this expression, interchange the order of integration (this requires the uniform convergence of the improper integral I_{2k}; we shall explicitly show this below), and make the change of variable $\mu = \nu F(\theta, x)$ in the ν integral. Then

$$2I_{2k} = \frac{2}{\pi} \int_0^\pi \frac{d\theta}{F^{2k+1}(\theta, e)} \int_0^\infty \mu^{2k} \exp(-\mu)\, d\mu > |S_{2k+1}|$$

The μ integral is $\Gamma(2k+1)$ and from the inequality on $F(\theta, e)$ (Eq. 8.18)

$$\frac{2\Gamma(2k+1)}{\pi F^{2k+1}(0, e)} \int_0^\pi d\theta > |S_{2k+1}|$$

The result for $\imath$ is then

$$\frac{1}{\imath} = \lim_{k \to \infty} \left| \frac{2\Gamma(2k+1)}{\Gamma(2k+2) F^{2k+1}(0, e)} \right|^{1/(2k+1)}$$

Or

$$\imath = F(0, e)$$

Let us now look back on what has been accomplished. We have shown that $E(M)$, expressed as a power series in M about $M = 0$, converges for $|M| < F(0, e)$. The missing point is the uniform convergence of the I_{2k} integral which we now provide: It is necessary to demonstrate that

$$\int_0^\infty \nu^{2k} \left\{ \frac{1}{\pi} \int_0^\pi \exp[-\nu F(\theta, e)]\, d\theta \right\} d\nu$$

$$= \frac{1}{\pi} \int_0^\pi \left\{ \int_0^\infty \nu^{2k} \exp[-\nu F(\theta, e)]\, d\nu \right\} d\theta$$

This will be true if the inner integral on the right-hand side converges uniformly in θ for $\theta \in [0, \pi]$. From Eq. 8.18 one can see that the integral can be majorized by the gamma function so that by de la Vallee Poussin's comparison test for convergence the inner integral on the right-hand side does converge uniformly. The last thing needed is to inquire into the permissibility of substituting one power series into another, performing the Cauchy multiplications, rearranging, and summing. From Pierpont (1912, Section 161) one can see that in the present circumstances this is permissible and that the radius of convergence of the f and g series is precisely given by Eq. 8.12.

One might wonder as to the relationship between this and the results of Moulton (1902; e.g., Section 100), wherein r and v are expressed in a mean anomaly Fourier series. These are known to be divergent for some values of M once $e \geqslant 0.6627434$. This number is the modulus of a complex root of $F^2(0, e) = 1$. The connection arises because of the nature of the Kapteyn series for $E(M)$ and the use of Lagrange's formula. The series for r and v are really power series in e whose coefficients happen to be trigonometric functions of the mean anomaly. If, for example, $\sin M$ is expanded in a power series in

M, its radius of convergence would clearly be at most $0.6 \cdots$. In fact it would be much less. It appears that the theorem given in the beginning of this section is the most general result one can obtain without special arguments, depending on the values of e and M.

Lagrange Series. Before proceeding to Moulton's proof, we elaborate further on the connection just mentioned. Lagrange showed that the solution of

$$x = y + pf(x)$$

for a parameter p and differentiable function f is of the form

$$x = y + pf(y) + \frac{p^2}{2!}\frac{\partial f^2(y)}{\partial y} + \frac{p^3}{3!}\frac{\partial^2 f^3(y)}{\partial y^2} + \frac{p^4}{4!}\frac{\partial^3 f^4(y)}{\partial y^3} + \cdots$$

It turns out that such series converge for $|p| < P$ where P is the smallest value of p such that there exists a double root for x (y fixed). In the Kepler problem

$$x = E, \qquad y = M, \qquad p = e, \qquad f(x) = \sin x$$

There will be a double root of $E - M - e \sin E = 0$ if the derivative with respect to E vanishes or $1 = e \cos E$. Hence

$$\cos E = \frac{1}{e}, \qquad \sin E = (1 - 1/e^2)^{1/2}$$

and Kepler's equation may be rewritten as

$$\cos^{-1}\left(\frac{1}{e}\right) = M + i(1 - e^2)^{1/2}$$

($i^2 = -1$). Taking the cosine of both sides and using the addition formula for the cosine and Euler's formula relating the trigonometric functions to the exponential function,

$$\frac{1}{e} = \cos[M + i(1 - e^2)^{1/2}]$$

$$= \cos M \cos[i(1 - e^2)^{1/2}] - \sin M \sin[i(1 - e^2)^{1/2}]$$

$$= \cos M \left\{ \frac{\exp[-(1 - e^2)^{1/2}] + \exp[(1 - e^2)^{1/2}]}{2} \right\}$$

$$- \sin M \left\{ \frac{\exp[-(1 - e^2)^{1/2}] - \exp[(1 - e^2)^{1/2}]}{2i} \right\}$$

If $M = 0$, then the smallest root of this equation is $e = 1$, but if $M = \pi/2$, the equation can be recast as

$$1 + (1 + \varepsilon^2)^{1/2} = \varepsilon \exp(1 + \varepsilon^2)^{1/2}$$

where $\varepsilon = ie$. But $F(0, e) = 0$ is (see Eq. 8.17)

$$\ln\left[\frac{1 + (1 - e^2)^{1/2}}{e}\right] = (1 - e^2)^{1/2}$$

An Alternative Proof. The general form of a Kapteyn series is

$$\sum_{n=0}^{\infty} a_n J_{n+\nu}[(n + \nu)z]$$

Such series are convergent and represent an analytic function of z throughout the domain in which

$$\left|\frac{z \exp(1 - z^2)^{1/2}}{1 + (1 - z^2)^{1/2}}\right| < \liminf_{n \to \infty} |a_n^{-(n+\nu)/2}|$$

This follows from Cauchy's root test, Weierstrass's M test, and a result of Kapteyn's Eq. 2.21 mentioned earlier,

$$|J_n(nz)| \le \left|\frac{z^n \exp[n(1 - z^2)^{1/2}]}{[1 + (1 - z^2)^{1/2}]^n}\right| \tag{2.21}$$

From this it follows that $|J_n(nz)| \le 1$ if $|z| \le 1$ and that

$$\left|\frac{z \exp(1 - z^2)^{1/2}}{1 + (1 - z^2)^{1/2}}\right| \le 1 \tag{8.19}$$

To discern the region delineated by the equality, set $z = \sigma e^{i\psi}$ and define χ via

$$\sigma \sinh \chi \cosh \chi = (\sinh^2 \chi + \sin^2 \psi)^{1/2}$$

Then, at equality for Eq. 8.19,

$$\sigma(\sinh^2 \chi + \sin^2 \psi)^{1/2} = \chi$$

$$\sigma^2 = 2\chi \operatorname{csch} 2\chi, \qquad \sin^2 \psi = (\chi \cosh \chi - \sinh \chi) \sinh \chi$$

As χ increases from 0 to $1.19967864\ldots$, $\sin^2 \psi$ increases from zero to unity and σ decreases from 1 to $0.66274342\ldots$. Hence, there is an oval region in the complex plane defining convergence.

Moulton's Proof

Moulton's (1903) original proof will be recounted for completeness. He first supposes that one can expand E as a power series in $M - M_0$ for $|M - M_0| < \imath$. To establish $\imath$, he relies on a theorem of Cauchy's, namely that the series in $M - M_0$ converges for all values of $M - M_0$ whose moduli are less than the distance (in the complex plane) from M_0 to the nearest singular point. The singular points of $E(M)$ are determined by looking for places where $|dE/dM| = \infty$ or $1 - e \cos E = 0$. Taking E to be complex and equal to $u + iw$, it is easy to show that this implies that

$$\sin u \sinh w = 0, \qquad e \cos u \cosh w = 1$$

The only solutions for this pair are $(0 \leqslant e < 1)$

$$u = 2n\pi, \qquad n = \ldots, -2, -1, 0, 1, 2, \ldots$$

$$w^2 = \cosh^{-1}\left(\frac{1}{e}\right) = \ln\left(\frac{1 + (1 - e^2)^{1/2}}{e}\right)$$

The corresponding values of $M = \xi + i\eta$ are

$$\xi = 2n\pi$$

$$\eta = y - e \sinh y$$

and $\imath = (M_0^2 + |\eta|^2)^{1/2}$ in M. Obviously the radius of convergence in t is $P_\imath/2\pi$. These equations are equivalent to those embodied in Eq. 8.12.

The Implications for Initial Orbital Determination

$F(0, e)$ is given in Table 5 for $e = 0(0.1)1.0$. Note that $F(0, 0.1)/F(0, 0.7) = 11$, which shows the dramatic drop in the permissible time span as $e \to 1$. If $n = 2$ rev/day, $e = 0.7$, then the maximum time for which the f and g series converge at perigee is $< 21\,min$. Table 5 also gives the maximum eccentricity such that the f and g series converge for $|T|/P \leqslant 0(0.05)0.5$. Although it is clear that for $M_0 = 0$, $\imath \to 0$ very rapidly as $e \to 1$, one will rarely be so unfortunate as to observe a satellite at perigee or an asteroid at perihelion. The most promising search patterns will find artificial satellites near $\delta = 0$ or asteroids near $b = 0$. The most common high-eccentricity satellites can be characterized by (roughly) $n = 2$ rev/day, $e = 1/\sqrt{2}$, $\omega = 3\pi/2$. Hence, if $\delta = 0$, $E = \pi/4$, $M_0 = (\pi - 2)/4$, and $P_\imath/(2\pi) = 38.3$ min. Of course, the radial velocity is now a maximum and the speed nearly a maximum so that their neglect is especially serious. The difficulties for main belt asteroids are minimal.

Table 5. Radius of Convergence of the f and g Series

e	$\imath = F(0, e)$	max time/P	e
0	∞	0	1^-
0.1	1.9982	0.05	0.5885
0.2	1.3126	0.10	0.4096
0.3	0.9199	0.15	0.2930
0.4	0.6503	0.20	0.2118
0.5	0.4509	0.25	0.1539
0.6	0.2986	0.30	0.1121
0.7	0.1814	0.35	0.0817
0.8	0.0931	0.40	0.0597
0.9	0.0313	0.45	0.0436
1.0	0	0.50	0.0318

It would seem that because one cannot know the orbital phase of the initial observations the best one can do is retain a reasonable number of terms in the f and g series, numerically investigate their divergence, and restrict the time span of the observations to a safe, small duration. Table 6 contains the relative percentage error in the partial sums of the f and g series for truncations after 5, 9, and 13 terms, at eccentricities $e = 0.1(0.2)0.9$ and at mean anomalies $M_0 = 0(60)180°$. When one turns to the rapidity of the convergence of the f and g series as a function of e, T, and M_0, one discovers that the series converge or they do not. What is meant by this tautology is that the partial sums of the f and g series for fixed e, T, and M_0 are either constant as a function of their order (and essentially equal to f and g) or they are meaningless. Numerical values for the third- through twelfth-order partial sums were computed before reaching this conclusion [and for eccentricities $= 0(0.1)0.9$].

Tests

I have computed more initial orbits on high-eccentricity objects using angles-only data than has anyone else. Gauss's method (with or without Gibb's refinement) is not an acceptable initial orbit determination procedure. This is because neither the orbital element set nor the proper time intervals (e.g., $t_2 - t_1$ and $t_3 - t_2$) are known that will guarantee not exceeding the radius of convergence of the f and g series. Hence, in the real world, one unwittingly runs over this boundary or is excessively cautious. The former circumstance is catastrophic. The latter, when coupled with the inevitable errors of observation and the real-time constraints at a telescope, results in a numerical mess—art and intuition, not science and logic. The only corrective for this is perfect data.

The direction of technological change—electro-optical cameras, laser radars, antisatellite satellites, space-borne surveillance systems, high-value and maneuverable military satellites, coherent radars, and the decreasing time of flight of submarine-launched ballistic missiles—all portend an increasing emphasis on rapid and accurate initial orbit determination. There is no method that will suffice for all orbits, over all data sets, in all observing scenarios. Thus, it becomes ever more important to search for new techniques, to delve into the physics and mathematics of old ones, and to understand all of their limitations. Note, however, one can never *prove* the superiority of one technique over another by a finite set of numerical experiments. The best that can be done is to partition orbital element space, or the space of observables, into discrete portions wherein the competing (and hopefully supplementary if not complementary) methods of initial orbit determination can be ranked on the basis of performance.

How can even this limited goal be accomplished? Considering the multiplicity of potential information available, a complete examination would involve a huge amount of computer time. Hence we shall separate the (laser) radar problem(s) from the angles-only case. The reason is simple—radars give distance, and distance estimation is what initial orbit determination is all about. The discussion will be confined to the high angular speed range of the spectrum,

Table 6. Convergence Properties of the f and g Series[a]

$e = 0.1$	$M_0 = 0°$ Relative % Error		$M_0 = 60°$ Relative % Error		$M_0 = 120°$ Relative % Error		$M_0 = 180°$ Relative % Error	
	f	g	f	g	f	g	f	g
$\tau = 0.1$								
$N = 5$	−897−5	477−3	−102−3	245−3	−102−3	619−4	157−4	969−4
9	−411−9	193−7	−441−9	−184−7	465−8	817−11	−943−9	229−7
13	−228−13	110−11	333−12	−116−11	−214−12	−853−12	328−13	−208−11
$\tau = 0.2$								
$N = 5$	−614−3	770−2	−352−2	350−2	−342−2	726−3	118−2	189−2
9	−449−6	497−5	−702−7	−467−5	248−5	226−6	−116−5	611−5
13	−430−9	454−8	215−8	103−8	−227−8	−517−9	626−9	−880−8
$\tau = 0.3$								
$N = 5$	−791−2	397−1	−305−1	160−1	−295−1	266−2	190−1	130−1
9	−291−4	129−3	344−5	−120−3	109−3	109−4	−975−4	167−3
13	−141−6	593−6	453−6	167−6	−501−6	−902−7	280−6	−128−5
$\tau = 0.4$								
$N = 5$	−544−1	129+0	−159+0	462−1	−162+0	601−2	228+0	609−1
9	−629−3	131−2	149−3	−121−2	188−2	157−3	−389−2	184−2
13	−958−5	190−4	224−4	625−5	−273−4	−357−5	380−4	−475−4
$\tau = 0.5$								
$N = 5$	−293+0	325+0	−682+0	104+0	−843+0	104−1	−252+1	240+0
9	−818−2	797−2	242−2	−743−2	240−1	124−2	113+0	126−1
13	−303−3	281−3	549−3	106−3	−848−3	−634−4	−293−2	−874−3

$e = 0.3$								
$\tau = 0.1$								
$N = 5$	−257−5	187−3	−172−4	−122−3	−297−5	−190−3	358−5	−341−3
9	−117−9	756−8	407−9	661−8	−116−8	362−8	231−9	412−9
13	−899−14	426−12	−890−11	389−9	−154−11	537−10	0	787−12
$\tau = 0.2$								
$N = 5$	−168−3	298−2	−503−3	−196−2	−637−4	−294−2	255−3	−565−2
9	−121−6	192−5	180−6	156−5	−578−6	818−6	259−6	161−6
13	−116−9	175−8	−207−10	−161−8	171−9	145−8	209−10	345−8
$\tau = 0.3$								
$N = 5$	−197−2	150−1	−358−2	−996−2	−281−3	−145−1	352−2	−305−1
9	−720−5	484−4	593−5	394−4	−225−4	190−4	177−4	691−5
13	−347−7	223−6	791−9	−203−6	310−7	182−6	507−8	489−6
$\tau = 0.4$								
$N = 5$	−117−1	468−1	−145−1	−316−1	−368−3	−451−1	270−1	−106+0
9	−134−3	474−3	689−4	387−3	−313−3	176−3	414−3	120−3
13	−203−5	686−5	208−6	−622−5	126−5	563−5	123−6	176−4
$\tau = 0.5$								
$N = 5$	−480−1	113+0	−436−1	−774−1	156−2	−110+0	172+0	−294+0
9	−133−2	276−2	455−3	228−2	−254−2	981−3	609−2	127−2
13	−491−4	971−4	684−5	−882−4	232−4	807−4	−932−5	310−3

$e = 0.5$								
$\tau = 0.1$								
$N = 5$	−145−5	121−3	954−5	−145−3	941−5	−143−3	−892−7	−263−3
9	−656−10	487−8	−565−9	−642−10	−117−9	−501−8	640−10	−103−7
13	−672−14	270−12	−123−13	−423−12	−257−13	−266−12	733−14	−388−12

Table 6—Continued

e = 0.5	$M_0 = 0°$ Relative % Error		$M_0 = 60°$ Relative % Error		$M_0 = 120°$ Relative % Error		$M_0 = 180°$ Relative % Error	
	f	g	f	g	f	g	f	g
$\tau = 0.2$								
N = 5	−935 − 4	192 − 2	302 − 3	−219 − 2	295 − 3	−216 − 2	−501 − 5	−435 − 2
9	−674 − 7	123 − 5	−270 − 6	−522 − 7	−527 − 7	−121 − 5	710 − 7	−273 − 5
13	−642 − 10	112 − 8	−144 − 9	127 − 8	−226 − 9	−241 − 9	114 − 9	−154 − 8
$\tau = 0.3$								
N = 5	−108 − 2	955 − 2	229 − 2	−105 − 1	223 − 2	−104 − 1	−397 − 4	−234 − 1
9	−390 − 5	308 − 4	−983 − 5	−204 − 5	−183 − 5	−295 − 4	471 − 5	−743 − 4
13	−188 − 7	141 − 6	−277 − 7	152 − 6	−424 − 7	−314 − 7	380 − 7	−210 − 6
$\tau = 0.4$								
N = 5	−615 − 2	296 − 1	969 − 2	−317 − 1	956 − 2	−317 − 1	−145 − 4	−808 − 1
9	−700 − 4	299 − 3	−126 − 3	−258 − 4	−227 − 5	−283 − 3	104 − 3	−811 − 3
13	−106 − 5	432 − 5	−118 − 6	453 − 5	−175 − 5	−993 − 6	261 − 5	−711 − 5
$\tau = 0.5$								
N = 5	−241 − 1	705 − 1	300 − 1	−741 − 1	303 − 1	−752 − 1	179 − 2	−224 + 0
9	−664 − 3	172 − 2	−916 − 3	−179 − 3	−161 − 3	−163 − 2	133 − 2	−548 − 2
13	−245 − 4	604 − 4	−217 − 4	627 − 4	−321 − 4	−145 − 4	803 − 4	−114 − 3

e = 0.7								
$\tau = 0.1$								
N = 5	−100 − 5	901 − 4	106 − 4	−976 − 4	812 − 5	−931 − 4	−721 − 6	−181 − 3
9	−514 − 10	362 − 8	319 − 9	−545 − 8	261 − 9	−479 − 8	−145 − 10	−886 − 8
13	−335 − 14	193 − 12	680 − 12	−904 − 10	410 − 12	−366 − 10	−561 − 15	−560 − 12
$\tau = 0.2$								
N = 5	−642 − 4	143 − 2	321 − 3	−145 − 2	248 − 3	−140 − 2	−483 − 4	−299 − 2
9	−460 − 7	913 − 6	154 − 6	−128 − 5	127 − 6	−113 − 5	−155 − 7	−235 − 5
13	−438 − 10	830 − 9	280 − 10	−948 − 9	755 − 10	−114 − 8	215 − 11	−233 − 8
$\tau = 0.3$								
N = 5	−731 − 3	708 − 2	233 − 2	−687 − 2	182 − 2	−673 − 2	−595 − 3	−160 − 1
9	−264 − 5	228 − 4	568 − 5	−307 − 4	468 − 5	−272 − 4	−959 − 6	−638 − 4
13	−127 − 7	105 − 6	541 − 8	−113 − 6	140 − 7	−136 − 6	811 − 9	−322 − 6
$\tau = 0.4$								
N = 5	−412 − 2	218 − 1	948 − 2	−204 − 1	752 − 2	−203 − 1	−376 − 2	−547 − 1
9	−466 − 4	220 − 3	731 − 4	−289 − 3	608 − 4	−257 − 3	−189 − 4	−697 − 3
13	−705 − 6	318 − 5	230 − 6	−339 − 5	577 − 6	−407 − 5	650 − 7	−111 − 4
$\tau = 0.5$								
N = 5	−158 − 1	517 − 1	282 − 1	−473 − 1	228 − 1	−478 − 1	−170 − 1	−150 + 0
9	−433 − 3	126 − 2	531 − 3	−163 − 2	449 − 3	−146 − 2	−203 − 3	−471 − 2
13	−159 − 4	442 − 4	423 − 5	−469 − 4	104 − 4	−563 − 4	235 − 5	−185 − 3

e = 0.9								
$\tau = 0.1$								
N = 5	−764 − 6	721 − 4	620 − 5	−520 − 4	543 − 5	−593 − 4	−665 − 6	−124 − 3
9	−342 − 10	290 − 8	283 − 9	−323 − 8	220 − 9	−360 − 8	−252 − 10	−586 − 8
13	−558 − 14	112 − 12	111 − 12	−200 − 10	553 − 11	−713 − 9	−559 − 15	−387 − 12
$\tau = 0.2$								
N = 5	−487 − 4	114 − 2	186 − 3	−803 − 3	164 − 3	−892 − 3	−443 − 4	−203 − 2
9	−348 − 7	729 − 6	135 − 6	−760 − 6	103 − 6	−686 − 6	−296 − 7	−155 − 5
13	−331 − 10	663 − 9	134 − 9	−868 − 9	105 − 9	−108 − 8	−228 − 10	−158 − 8

Table 6—Continued

$e = 0.9$	$M_0 = 0°$ Relative % Error		$M_0 = 60°$ Relative % Error		$M_0 = 120°$ Relative % Error		$M_0 = 180°$ Relative % Error	
	f	g	f	g	f	g	f	g
$\tau = 0.3$								
$N = 5$	$-551-3$	$565-2$	$134-2$	$-380-2$	$119-2$	$-428-2$	$-540-3$	$-108-1$
9	$-198-5$	$182-4$	$489-5$	$-182-4$	$377-5$	$-164-4$	$-168-5$	$-419-4$
13	$-951-8$	$834-7$	$244-7$	$-103-6$	$173-7$	$-877-7$	$-722-8$	$-216-6$
$\tau = 0.4$								
$N = 5$	$-308-2$	$174-1$	$539-2$	$-113-1$	$486-2$	$-129-1$	$-337-2$	$-366-1$
9	$-347-4$	$175-3$	$618-4$	$-170-3$	$483-4$	$-155-3$	$-334-4$	$-454-3$
13	$-525-6$	$253-5$	$973-6$	$-305-5$	$699-6$	$-260-5$	$-456-6$	$-745-5$
$\tau = 0.5$								
$N = 5$	$-117-1$	$410-1$	$158-1$	$-261-1$	$115-1$	$-304-1$	$-148-1$	$-987-1$
9	$-319-3$	$999-3$	$441-3$	$-955-3$	$350-3$	$-876-3$	$-365-3$	$-303-2$
13	$-117-4$	$350-4$	$169-4$	$-418-4$	$123-4$	$-358-4$	$-123-4$	$-122-3$

[a] An entry of $-123+4$ means -1.23×10^4.

thereby eliminating natural bodies from consideration. Thus, and not totally accidentally, the objects of interest are deep- (sic) space or high-altitude (sic) artificial satellites. Furthermore, this discussion will be restricted to non-parallax-type data acquisition.

In such circumstances on such objects three initial orbit determination procedures merit consideration. One is the Gauss–Gibbs method. This algorithm is included because of its reputation, not because I am promoting it. The principal method I use on the faster-moving subset of the deep-space artificial satellite population is my modification of Laplace's method (Chapter Seven). The third technique considered here is the Taff–Hall angular velocity method (i.e., the exact method presented in Chapter Seven). Now the latter two schemes presume the observational capability to rapidly acquire large amounts of high-quality data. This information will then be smoothed in some fashion and ultimately differentiated. Because of this, and the fact that the Gauss–Gibbs method is perforce restricted to three sets of angles-only data, a way has to be found to try and balance the scales. This is done as follows: A subset of orbital element sets are selected from the deep-space population (as it existed on a certain date). This subset of the actual population's element sets is used to generate topocentric position vectors for a particular geographical location (the observatory of Example 3.7, but that is not relevant). The time spacing between position vectors is 2 min—the baseline specification for the GEODSS network (see Chapter Five). From this set of passes from the subset of element sets chosen, 96 passes of 25 different satellites are selected. For the midpoint of each pass the radius of convergence of the f and g series is computed. In each case this figure is then reduced by 20% and eight-tenths of the radius of convergence, rounded down to the nearest whole 2 min, defines the total time span of the data. For the tests of the Gauss–Gibbs method the input data are topocentric right ascension, declination, and time at the beginning, middle, and end of the time interval just defined. For the tests of the other two methods the input data consisted of all the (A, Δ, t) triplets within the time interval unless there were more than ~12–15. In the case of the larger radii of convergence only every other [or every third, etc. (if need be)] position vector is regarded as known. In this fashion, while denying the Laplace-Taff and Taff–Hall techniques data that would be available in practice, I have tried not to tip the scales too much in their favor.

The above-mentioned passes and orbital element sets are weighted toward the stressing cases, that is, the high-eccentricity, fast-moving portion of the hundreds of passes originally generated. This, however, merely reflects the actual state of affairs (cf. Chapter Five). Also, the number of data points reduction scheme outlined above keep the total number of assumed data points to at most 16. The data have the same number of significant decimal digits— namely right ascensions to the nearest tenth of a second of time and declinations to the nearest whole second of arc. In order to complete the analysis, the entire test should be redone, preferably at two (but at least one) other levels of precision. This would provide sorely needed information about the robustness of the three procedures. A part of this has been completed and is discussed below.

Now that the ground rules have been outlined, what criteria should be used to measure the efficacy of the three competing techniques? I could define some six-dimensional norm in orbital element set space and compare each initial orbital element set to the real one (e.g., the one used to generate the pseudo-observational data). If the metric needed to define a meaningful norm were known and how large "bad" is or how small "good" is, then I would consider this alternative. Not being in possession of such knowledge I prefer to test the generated orbital element sets against what they will be used for (in the angles-only context—a nice touch of self-consistency), pointing a telescope. Therefore, the measure used is the angular error, on the topocentric celestial sphere, between the position from the computed orbital element set and the position from the real orbital element set. More than this, such pointing predictions ("look angles" if you must) have been generated at intervals of 0.5, 1, and 2 hr into the future (sometimes in the past; Newton's equations of motion are time reversible). Therefore, for each pass of each satellite there are nine angular errors as measured on the topocentric sky—an error from each of three initial orbit determination schemes at three different prediction intervals. This information has been reproduced in Table 7. The units are minutes of arc. Errors $>100'$ are indicated by ∞.

Gauss–Gibbs.   This test is not designed to highlight Gaussian orbit determination. Rather it was designed to see if the Gauss–Gibbs method could ever work well on fast-moving artificial satellites in eccentric orbits. (It should work

Table 7. Positional Errors in Initial Orbit Determination

Satellite	Information	Time Spans	High Precision		Low Precision		Angular Speed ("/sec)
			Gauss–Gibbs	Laplace–Taff	Gauss–Gibbs	Laplace–Taff	
SDC #	11926						
n	1.02	0^h5	$0{.}''14$	$0{.}''03$	∞	$0{.}''2$	
e	0.331	1.0	0.46	0.04	∞	0.7	32.10
i	$10°44$	2.0	0.58	0.13	∞	2.0	
SDC #	12137						
n	2.37	0.5	5.25	0.76	18.7	15.5	
e	0.724	1.0	4.50	1.20	4.0	1.6	66.66
i	$46°.73$	2.0	4.20	2.05	3.0	2.8	
SDC #	12679						
n	3.49	0.5	4.94	0.07	5.7	4.0	
e	0.626	1.0	4.15	0.03	3.3	3.9	32.00
i	$89°94$	2.0	2.34	0.21	8.2	8.4	
SDC #	12679						
n	3.49	0.5	6.02	0.78	6.4	2.3	
e	0.626	1.0	5.70	3.85	7.4	11.4	68.20
i	$89°94$	1.5	5.06	8.72	8.6	26.1	
SDC #	83781						
n	2.01	0.5	9.41	∞			
e	0.735	1.0	9.94	∞			112.80
i	$64°55$	2.0	13.47	∞			
SDC #	83871						
n	2.01	0.0	10.69	0.07	∞	1.1	
e	0.735	1.0	8.74	8.33	∞	4.9	136.73
i	$64°55$	2.0	9.79	21.57	∞	14.2	
SDC #	83878						
n	1.99	0.5	3.68	3.34	3.5	6.8	
e	0.740	1.0	4.72	6.30	4.4	11.0	113.89
i	$62°87$	2.0	7.44	9.21	6.7	17.2	
SDC #	83878						
n	1.99	0.5	6.75	6.75	7.1	15.5	
e	0.740	1.0	2.02	1.34	2.3	2.2	54.42
i	$62°87$	2.0	2.85	1.77	3.5	4.5	
SDC #	83878						
n	1.99	0.5	5.87	15.58	5.7	16.8	
e	0.740	1.0	2.74	4.63	2.7	5.0	35.50
i	$62°87$	1.5	4.59	8.98	4.5	9.7	
SDC #	83878						
n	1.99	0.5	2.59	0.42	2.5	0.6	
e	0.740	1.0	1.36	0.51	1.0	0.8	164.60
i	$62°87$	2.0	0.66	1.39	1.6	0.8	
SDC #	83878						
n	1.99	0.25	3.22	0.52	8.0	9.3	
e	0.740	1.0	4.18	0.63	1.7	3.0	68.97
i	$62°87$	2.0	18.71	1.96	1.3	10.4	

Table 7—Continued

Satellite	Information	Time Spans	High Precision		Low Precision		Angular Speed ("/sec)
			Gauss–Gibbs	Laplace–Taff	Gauss–Gibbs	Laplace–Taff	
SDC#	10167						
n	1.99	0^h5	1.98	0.58	2.0	2.8	
e	0.652	1.0	2.96	1.86	3.7	5.7	73.74
i	$64°28$	1.5	3.88	2.83	4.1	8.9	
SDC#	10167						
n	1.99	0.5	2.17	0.62	2.4	0.5	
e	0.652	1.0	1.28	0.57	0.2	0.9	42.10
i	$64°28$	2.0	3.44	3.22	2.8	4.2	
SDC#	12996						
n	2.04	0.5	2.68	3.74	2.7	3.1	
e	0.685	1.0	1.86	4.49	1.9	8.8	59.40
i	$61°35$	2.0	0.44	4.88	0.8	31.0	
SDC#	83746						
n	2.01	0.5	5.77	No			
e	0.660	1.0	7.80	No			132.92
i	$71°11$	2.0	18.27	No			
SDC#	83601						
n	2.13	0.5	9.20	17.29	No	23.9	
e	0.710	1.0	10.34	47.50	No	74.0	130.20
i	$64°16$	2.0	13.92	∞	No	∞	
SDC#	83744						
n	2.46	0.5	3.20	0.66	1.4	3.4	
e	0.715	1.0	7.83	1.88	6.2	17.2	8.10
i	$10°32$	2.0	∞	6.83	∞	∞	
SDC#	83744						
n	2.46	0.5	0.71	7.97	0.9	8.4	
e	0.715	1.0	1.05	11.69	1.5	17.3	158.51
i	$10°32$	2.0	1.83	16.65	13.3	38.6	
SDC#	83885						
n	2.27	0.5	3.86	0.80	2.5	2.2	
e	0.732	1.0	11.19	3.40	6.8	10.1	22.91
i	$27°40$	2.0	39.30	13.83	26.4	43.6	
SDC#	83885						
n	2.27	0.5	2.03	0.70	2.2	0.4	
e	0.732	1.0	25.37	5.84	1.5	8.2	10.08
i	$27°40$	2.0	∞	63.72	27.6	∞	
SDC#	898						
n	2.01	0.5	5.94	No			
e	0.643	1.0	6.22	∞			174.00
i	$71°28$	2.0	7.73	∞			
SDC#	898						
n	2.01	0.5	1.63	0.35	2.0	0.3	
e	0.643	1.0	3.12	2.11	3.5	0.9	8.50
i	$71°28$	2.0	13.66	15.61	12.3	4.7	

Table 7—Continued

Satellite	Information	Time Spans	High Precision		Low Precision		Angular Speed ("/sec)
			Gauss-Gibbs	Laplace-Taff	Gauss-Gibbs	Laplace-Taff	
SDC#	898						
n	2.01	$0^h.5$	4".54	0".11	4".4	1".2	
e	0.643	1.0	5.44	0.42	3.6	∞	19.83
i	71°.28	2.0	14.31	1.56	6.3	∞	
SDC#	83750						
n	4.22	0.5	2.84	0.65	2.8	1.8	
e	0.593	1.0	3.91	1.81	4.0	4.8	84.42
i	27°.14	2.0	10.63	4.19	11.3	11.2	
SDC#	83750						
n	4.22	0.5	3.74	2.08	5.0	5.1	
e	0.593	1.0	4.49	6.61	6.8	30.3	22.43
i	27°.14	2.0	78.31	42.26	87.5	∞	
SDC#	83750						
n	4.22	0.5	1.38	2.98	5.5	5.8	
e	0.593	1.0	10.99	11.24	14.6	12.4	59.63
i	27°.14	2.0	29.94	52.46	53.6	44.4	
SDC#	83887						
n	2.29	0.5	1.36	1.23	3.5	4.3	
e	0.723	1.0	1.66	3.65	6.3	13.5	66.80
i	47°.39	2.0	3.89	9.88	14.6	37.3	
SDC#	83887						
n	2.29	0.5	0.50	0.69	No	No	
e	0.723	1.0	4.02	2.67	No	No	146.29
i	47°.39	2.0	13.05	7.89	No	No	
SDC#	83887						
n	2.29	0.5	2.38	0.25	No	1.6	
e	0.723	1.0	3.80	1.25	No	6.5	8.91
i	47°.39	2.0	16.90	9.28	No	51.8	

on objects in nearly circular orbits.) However, for the first time proper cognizance of the essential restrictions that are part and parcel of Gaussian initial orbit determination have been included. The numbers will speak for themselves. The use of the Gauss–Gibbs angles-only method is not recommended. For the 29 passes detailed in Table 7 the average positional errors at 0.5, 1, and 2 hr (and the standard deviation about their means) were 3".95 ± 2".70, 5".58 ± 4".91, and 12".56 ± 16".01. For the other passes in the original sample the same quantities are 1".83 ± 2".34, 2".23 ± 4".03, and 6".25 ± 16".05. For the low-precision data the 29-pass Gauss–Gibbs set results are 4".52 ± 3".80, 4".16 ± 3".17, and 14".4 ± 21".33. Invisible in these statistics is the failure rate of the Gauss–Gibbs technique even though all of the analytically necessary restrictions have been observed. This rate was negligible for the high-precision data runs (both groups) but approached $\frac{1}{3}$ for the low-precision runs.

Laplace-Taff. For the 0.5, 1, and 2-hr prediction intervals, the ratio of the positional error of the Gauss–Gibbs method to that obtained from the Laplace–Taff method has been computed. The larger this number is, the worse Gaussian initial orbit determination fares relative to Laplacian initial orbit determination. The three averages, and their standard deviations about their means, are 8.0 ± 27, 13.2 ± 32, and 2.4 ± 3.2. The accuracy of the Laplacian data alone, in the same format as given for the Gauss–Gibbs technique, is $2\rlap{.}''65 \pm 4\rlap{.}''53$, $5\rlap{.}''15 \pm 9\rlap{.}''22$, and $12\rlap{.}''44 \pm 16\rlap{.}''48$. For the low-precision data Laplace–Taff runs the appropriate numbers are $5\rlap{.}''48 \pm 6\rlap{.}''23$, $7\rlap{.}''85 \pm 6\rlap{.}''92$, and $18\rlap{.}''98 \pm 16\rlap{.}''93$. The Laplace–Taff method failure rate on the poorer data is about half that of the Gaussian.

From my point of view the ratio numbers tell the real story. Laplace–Taff is an order of magnitude better than is Gauss–Gibbs without having to know half of the orbital element set before starting the computation. Higher data rates will improve its performance.

Taff-Hall. The Taff–Hall technique rests on the fact that the topocentric expression of angular momentum conservation and energy conservation is a quartet of equations involving the position $\mathbf{l}(A, \Delta)$, the angular velocity $\dot{\mathbf{l}}(A, \Delta, \dot{A}, \dot{\Delta})$, the topocentric distance R, and the topocentric radial velocity $\dot{R}$. The key is to use the fact that these are constants of the motion and write

$$\mathbf{L}_1 = \mathbf{L}_2, \qquad E_1 = E_2 \tag{8.20}$$

at two times t_1, t_2. There are now 12 unknowns in these four equations. In the radar case eight quantities (A, Δ, R, and $\dot{R}$ at both times) are measured so that the problem is well posed. In the high angular speed case eight quantities are known too (A, Δ, $\dot{A}$, and $\dot{\Delta}$ at both times).

The early (Taff and Hall 1977) solution method was a four-dimensional Newton–Raphson technique. This procedure was not robust, did not exploit the analytical simplicity of Eqs. 8.20, and did not work well. When I started this work I switched to a steepest descent method that was not robust, did exploit some of the analytical simplicity of Eqs. 8.20, and did not work well. Finally I fully exploited the analytical simplicity of Eqs. 8.20 to reduce the problem to a one-dimensional one. In particular, the system was reformulated so as (to appear) to be a single equation (for $E_1 - E_2$) in one unknown (R_2). From an assumed R_2, R_1 was computed (by explicit algebra) and then $\dot{R}_1$ and $\dot{R}_2$ (similarly). This required the use of three equations and the angular momentum conservation equations were utilized. Next this was used to compute $E_1 - E_2$ (which should vanish). If $E_1 - E_2 \neq 0$, then R_2 is incremented and the R_2 iteration repeated until $E_1 - E_2$ changed sign. I then homed in on the root by decreasing the R_2 step size by a factor of 10 and simultaneously changing direction. I used $\Delta R_2 = 0.01$ Earth radii, started at $R_2 = 1$, and went out to $R_2 = 10$ before declaring a failure. (I know I'm doing high-altitude satellite initial orbit determination.)

Of the 30 passes of 15 different artificial satellites tried, no root was found in 14 cases, in 10 cases an incorrect root was found, and in 6 cases both a root was found and it produced a physically sensible orbital element set. The reason for the first group appears to be the extraordinarily large slope in the $E_1 - E_2$ versus R_2 relationship. An explanation for the middle group is the existence of multiple roots and an a priori inability to guess the correct one (for if R_2 were then known, then . . .). The last group represents the "successes" but the orbital element sets are so poor that the pointing errors have not been formally computed. Without a clearer and deeper understanding of the sensitivities of this method it must be regarded as a failure. I suspect that it will do much better with a higher data rate than one observation per 2 min or that it is appropriate for use of moderate (as opposed to high) angular speed artificial satellites. As supporting evidence, the average angular speeds for the three Taff–Hall subgroups are 94"/sec, 85"/sec, and 40"/sec.

Summary

Since Gauss (1809) did not solve the angles-only initial orbit determination problem, and no one else has for all applications, initial orbit determination retains an element of art. For asteroids the Gauss–Gibbs technique is perfectly acceptable except for the few ($\leqslant 50$ so far) high-eccentricity or the few (~ 60) Earth-approaching objects. As mentioned above, these can almost always be distinguished as such at the telescope (or in the darkroom for those who observe photographically). Comets have been a special case since Olbers presented his parabolic initial orbit determination scheme in 1797. New major planets are a zero probability event. New moons are not but they and binary stars (see Chapter Thirteen) have very limited spatial excursions and, therefore, present no time critical cases. So, as has been the case in science and technology since the onset of the Holocene, war will fuel the engine of advancement.

Low-altitude artificial satellites, ballistic missiles, direct-ascent antisatellites, and so on, are all observable by radars (and destroyable by laser radars). So until "stealth" technology spreads from the bomber fleet this author takes the position that initial orbit determination for these objects is conceptually and practically relatively simple. High-altitude satellites, and the next generation of killer satellites, are all beyond the range of ground-based or space-borne radars. However, as discussed in Chapter Five, this population of objects can be grouped into four large classes. One can (almost invariably) place the pinpoint of light on one's video monitor into the correct class at the telescope. Having done so, the correct choice between Laplace-Taff, Taff–Hall, NSDC (see Chapter Eleven), or Gauss–Gibbs then becomes obvious. Even this element of "art" can be programmed if desirable.

ORBITAL ELEMENT SET CONSTRUCTION

How we go from the end results of the Gaussian, Laplacian, and so on, initial orbit determination procedures to an orbital element set has not been explained. Building on the foundations laid down in Chapter Two, this will now be rectified. Except for some mixed-data-type determinations, which as special cases will not be treated, the end result of initial orbit determination is location and velocity at some time (the epoch) or two locations at two different times. It is not analytically difficult to go from $\mathbf{r}$ and $\dot{\mathbf{r}}$ or $\mathbf{r}_1$ and $\mathbf{r}_2$ to $\mathbf{a} = (a, e, \omega, i, \Omega, M_0)$. It is more important in the computer age, when one loses direct control over the process, to appreciate the multiplicity of the solution set. Hence robust formulas are desirable and these require thought and not merely algebraic dexterity.

$\mathbf{r}$ and $\dot{\mathbf{r}}$

Assuming that both location and velocity are known at the same time, the semi-major axis a can be computed from the conservation of energy equation,

$$\mathscr{E} = -\frac{\mu}{2a} = \frac{\dot{\mathbf{r}} \cdot \dot{\mathbf{r}}}{2} - \frac{\mu}{|\mathbf{r}|}$$

One could then get the eccentricity from the norm of the angular momentum vector (per unit mass),

$$L^2 = \mu a (1 - e^2) = |\mathbf{r} \times \dot{\mathbf{r}}|^2$$

but this is decidedly not a robust procedure for nearly circular orbits. It is better to obtain e and v or e and E from the equation of the orbit and its derivative, namely,

$$e \cos v = \frac{L^2}{\mu r} - 1 = \frac{|\mathbf{r} \times \dot{\mathbf{r}}|^2}{\mu r} - 1$$

$$e \sin v = \frac{L(\mathbf{r} \cdot \dot{\mathbf{r}})}{\mu r}$$

or

$$e \cos E = 1 - \frac{r}{a}$$

$$e \sin E = \frac{\mathbf{r} \cdot \dot{\mathbf{r}}}{(\mu a)^{1/2}}$$

The mean anomaly at epoch then follows from Kepler's equation.

The three Euler angles that specify the orientation of the orbit in space remain to be fixed. The direction cosines of $\mathbf{L}$ fix the inclination i and the longitude of the ascending node Ω directly from $\mathbf{L} = \mathbf{r} \times \dot{\mathbf{r}}$. This is the most robust aspect of element set (indeed initial orbit) construction; only quadrant ambiguities resulting from the inverse trigonometric functions need to be attended to. The argument of periapse ω is best determined via the argument of latitude $u = \omega + v$,

$$r \cos u = \mathbf{r} \cdot (\cos \Omega, \sin \Omega, 0)$$

$$r \sin u = \mathbf{r} \cdot (0, 0, \csc i)$$

r_1 and r_2

Suppose that $t_2 > t_1$. Then, because the orbit lies in a plane, $r_1 \times r_2$ is in the same direction as is **L** (i.e., normal to the orbital plane). Therefore, both i and Ω can be simply obtained. The semi-major axis computation is more involved. Remember Lambert's theorem for an ellipse,

$$n(t_2 - t_1) = (\alpha - \sin \alpha) - (\beta - \sin \beta)$$

where

$$\sin\left(\frac{\alpha}{2}\right) = \left(\frac{r_1 + r_2 + c}{4a}\right)^{1/2}$$

$$\sin\left(\frac{\beta}{2}\right) = \left(\frac{r_1 + r_2 - c}{4a}\right)^{1/2}$$

defines α and β and n is the mean motion, $n^2 = \mu/a^3$. Since t_1 and t_2 are known and c is the chord length from the end of r_1 to the end of r_2 (i.e., $c^2 = r_1^2 + r_2^2 - 2r_1 \cdot r_2$), the above triplet represents an implicit set of equations for the determination of a.

In order to get e and ω, first compute u_1 and u_2 via

$$\cos u_j = r_j \cdot (\cos \Omega, \sin \Omega, 0)/r_j$$

then, by manipulation of $r = a(1 - e^2)/(1 + e \cos v) = p/[1 + e \cos(u - \omega)]$ $[p = a(1 - e^2)]$, one can derive the pair of equations

$$e \sin \omega = \frac{p(r_2 \cos u_2 - r_1 \cos u_1) + r_1 r_2 (\cos u_1 - \cos u_2)}{r_1 r_2 \sin(u_1 - u_2)}$$

$$e \cos \omega = \frac{p(r_2 \sin u_2 - r_2 \sin u_1) + r_1 r_2 (\sin u_1 - \sin u_2)}{r_1 r_2 \sin(u_2 - u_1)}$$

Squaring each equation and adding gives e^2 in terms of p. This equation has two positive roots, only one of which is physically meaningful. Finally division provides an expression for $\tan \omega$.

The last needed quantity is the mean anomaly at epoch. From either u_1 or u_2 and ω we get v_1 or v_2 and thence E_1 or E_2. The mean anomaly at epoch immediately follows from Kepler's equation.

Related Topics

p Iteration

There is an alternative method of obtaining the eccentricity when the initial data are r_1 and r_2. Observe that if v_1 and v_2 are the corresponding true anomalies, then

$$r_1 r_2 \cos(v_2 - v_1) = r_1 \cdot r_2$$

But $r = p/(1 + e \cos v)$, so by rearrangement,

$$e \cos v_j = \frac{p}{r_j} - 1, \qquad j = 1, 2$$

Moreover $\cos v_2 = \cos[(v_2 - v_1) + v_1]$, whence

$$e \sin v_1 = [(e \cos v_1) \cos(v_2 - v_1) - (e \cos v_2)] \csc(v_2 - v_1)$$

$$e \sin v_2 = [-(e \cos v_2) \cos(v_2 - v_1) + (e \cos v_1)] \csc(v_2 - v_1)$$

Hence $e \sin v_1$ and $e \sin v_2$ can be computed and then e as $[(e \sin v_j)^2 + (e \cos v_j)^2]^{1/2}$. Finally, $a = p/(1 - e^2)$.

Clearly the above development assumes a knowledge of p, which is not available. If it were, however, the computation could then be continued to obtain the corresponding eccentric anomalies E_j via

$$\sin E_j = \frac{r_j (1 - e^2)^{1/2}}{p} \sin v_j, \qquad \cos E_j = \left(\frac{r_j}{p}\right)(e + \cos v_j)$$

Finally, $t_2 - t_1$ can be calculated via Kepler's equation and this acts as a constraint on the above process (viewed as an iteration) when the value of p is guessed. After trying this with values of p and p', the function

$$f = n(t_2 - t_1) - (M_2 - M_1)$$

is known at two points (the M's are the mean anomalies corresponding to E_1 and E_2). Numerical differentiation then allows a refinement of the value of p by a Newton-Raphson algorithm.

v Iteration

The method discussed in the previous section fails for $t_2 - t_1 = P/2$, for then $E_2 - E_1 = M_2 - M_1 = v_2 - v_1 = \pi$. A way around this problem is to eliminate p via

$$r_1(1 + e \cos v_1) = p = r_2(1 + e \cos v_2)$$

so that the eccentricity is obtained from

$$e = \frac{r_2 - r_1}{r_1 \cos v_1 - r_2 \cos v_2}$$

But $v_2 - v_1 = \Delta v_{12}$ is known from $r_1 r_2 \cos \Delta v_{12} = \mathbf{r}_1 \cdot \mathbf{r}_2$, so if v_1 is guessed, e can be computed and then p and a. Again, the constraint on this process is the prediction of t_1 and t_2 and the subsequent enforcement of Kepler's equation.

The Jacobian

All of this analysis is useless unless the Jacobian's $\partial(\mathbf{r}, \dot{\mathbf{r}})/\partial(\mathbf{a})$ and $\partial(\mathbf{r}_1, \mathbf{r}_2)/\partial(\mathbf{a})$ are nonzero. It is well known that both vanish whenever the observer is located in the plane of the orbit. It follows from this analytical result that severe numerical difficulties exist if one is nearly in the orbital plane. Such is the case for most solar system initial orbit determination problems. If one were exactly in the orbital plane, then four independent measurements of the longitude coordinate would be necessary. My analytical intuition also makes me wary of the polar $(i = \pi/2)$, infinitely slow $(a = \infty)$, and parabolic $(e = 1)$ cases. Indeed, for polar orbits

$$\left| \frac{\partial(\mathbf{r}, \dot{\mathbf{r}})}{\partial(\mathbf{a})} \right| = \frac{\mu^2 e \sec^2 u}{2ar^4}$$

showing that there are potential numerical problems in initial orbit determination for the many nearly polar, low-altitude, artificial satellites launched each year.

PROBLEM

1. Reflect on the differences in approach between Laplace's method (which attacks the initial orbit determination problem through the differential equations of motion), Gauss's method (which works through the solution), and my method (which works through the first integrals of the equations of motion).

   a) Is there an avenue not yet exploited?

   b) Are there particular numerical algorithms associated with each inventor's approach that are unique?

   c) Can you invent a new numerical approach for each line of attack? If you come up with a good one write to me.
   See *Astron J.* **89**, 1426 (1984)..

2. Find a paper by Taff and Randall in a 1985 issue of *Celestial Mechanics.* Read it and compare with pp. 284–6 of the text.

chapter nine

Introduction to Perturbation Theory

In *my* view the problem is that when you really need perturbation theory (in celestial mechanics) it does not work very well (as applied in the past). The existing applications of it do not explain the Kirkwood gaps in the asteroid belt. They do not explain the fascinating discoveries in the Saturnian rings/moons system. They do not predict physically correct rates of change for the orbital elements of a satellite revolving about an oblate primary. Some of these failures may be attributable to faulty mathematics, some to the difference between a trajectory theory and a field theory. The reader is advised that the above is a minority opinion (and taken slightly out of context). It may be so much in the minority that it is unique. Future editorials will be as clearly delineated and the exact context of the above-mentioned specific failures are discussed below or in the next chapter.

The reader might now ask two questions: "Why should one bother to read this and the next chapter?" and "What's he got that's better?". The answer to the first question is multiplefold. First of all the presentation in this chapter is distinctly nontraditional, and the solutions presented here and later for the oblateness, atmospheric drag, and third-body perturbations have a rigorous mathematical basis. Hence, there is no doubt that they are correct (as far as they go). Also, in order to criticize intelligently one's forerunners one must thoroughly understand what they did and why they did it. No, as of this writing (January, 1985) I do not have anything better. The main reason for this is that it will not be until second order that a competing version of perturbation theory can be shown to be better. Solving, completely through second order, any interesting problem in a different formalism for perturbation theory is most definitely nontrivial (and extraordinarily tedious). I now turn to a brief survey of the physical problems of interest.

Within the solar system the principal perturbing forces are due to the presence of a third body (Earth–artificial satellite–Moon/Sun, Sun–minor planet–Jupiter, planet–moon–moon) or the nonspherical symmetry of the primary (Earth–artificial satellite, Jupiter/Saturn–moon). In neither case can the general problem be solved. There are, in both instances, special restrictions that allow an exact analytical solution. These were illustrated in Chapter Six.

The dominant force in either of these problems is the central $1/r^2$ attraction of the primary. In the real satellite problem (artificial or natural) both third-body and oblateness perturbations are significant forces. Moreover, for many artificial satellite orbits atmospheric drag is present too. Oblateness and third-body disturbances are conservative, but air resistance is not. Thus there are two forms of the perturbation equations; one form was developed by Lagrange and assumes that the perturbing force may be written as the gradient of a velocity-independent scalar. The other form was developed by Gauss and is completely general.

The outline of the remainder of this chapter is as follows; a short discussion of oblateness effects, third-body perturbations, atmospheric drag, and other (minor) perturbative forces. Then a discussion of the mathematical foundations of perturbation theory (within the context of celestial mechanics), including a compendium of forms for Lagrange's (planetary) equations and for Gauss's equations. Scattered throughout the next chapter are further discussions, in much more depth, of these subjects.

THE PRINCIPAL PERTURBATIONS

Oblateness Perturbations

The general form of the exterior potential for a body with axial symmetry can be written as

$$U(r, \theta, \phi) = -\frac{GM}{r}\left(1 - \sum_{n=2}^{\infty} \frac{J_n P_n(\sin \theta)}{r^n}\right)$$

Here the z axis is the axis of symmetry, $z = r \cos \theta$, M is the total mass, and r is the distance from the center of mass to an arbitrary point with spherical coordinates r, θ, ϕ. There is no longitudinal (i.e., ϕ) dependence because of the assumed symmetry (coupled with choosing the correct coordinate system). There is no $n = 1$ term in U (more accurately it vanishes) because the center of mass has been chosen as the origin. The *main problem* of artificial satellite theory is the solution of

$$m\ddot{\mathbf{r}} = -m\nabla U$$

where

$$U \equiv -\frac{GM}{r}\left(1 - \frac{J_2 P_2(\sin \theta)}{r^2}\right)$$

For the Earth $J_2 \simeq 10^{-3}$; $|J_3|, |J_4| \sim J_2^2$.

To discuss the perturbations of an arbitrary orbit one can use the work done by and the torque developed by the J_2 term in U to simply derive the secular first-order results. The total power developed by the potential is

$$\frac{d\mathscr{E}}{dt} = -\dot{\mathbf{r}} \cdot \nabla U$$

and over the course of one revolution the total work done is

$$\Delta\mathscr{E} = -\oint \dot{\mathbf{r}} \cdot \nabla U \, dt = -\oint \nabla U \cdot d\mathbf{r}$$

However, one cannot simply note that $\nabla \times \nabla U = \mathbf{0}$ and use Green's theorem to evaluate this. The sign, $\oint$, means integrate around one complete circuit of the *perturbed* orbit. One cannot assume that the terminal point is the same as the starting point because it is not known if the perturbed orbit is closed. One can, however, split U into $U_0 + J_2 U_2$, where $U_0 = -GM/r$, so that now

$$\Delta\mathscr{E} = \Delta\mathscr{E}_0 + \Delta\mathscr{E}_2 = -\oint [\nabla U_0 + J_2 \nabla U_2] \cdot \dot{\mathbf{r}} \, dt$$

To first order in J_2, the contour for the U_0 integration can be deformed to be the unperturbed orbit. Then the fact that the force generated by U_0 is conservative can be exploited, whence $\Delta\mathscr{E}_0 = 0$. Because $J_2 U_2$ is already first order in magnitude, we can again deform the contour and deduce that $\Delta\mathscr{E}_2 = 0$. But

$$\mathscr{E} = -\frac{GM}{2a}$$

whence

$$\Delta a = 0 \qquad \text{or} \qquad \frac{\Delta a}{P} = \left.\frac{da}{dt}\right)_{\text{sec}} = 0 \qquad (9.1)$$

where $(da/dt)_{\text{sec}}$ is the first-order secular rate of change of the semi-major axis owing to J_2.

The rate of change of the angular momentum owing to J_2 is given by

$$\frac{d\mathbf{L}}{dt} = \mathbf{r} \times \mathbf{F} = -\mathbf{r} \times \nabla U = -J_2 \mathbf{r} \times \nabla U_2$$

since U_0 represents a central force. By direct computation $(\mathbf{L} = \mathbf{r} \times \dot{\mathbf{r}})$ one can verify the formula

$$\dot{\mathbf{L}} = -3\left(\frac{GMJ_2 z}{r^5}\right)(y, -x, 0)$$

and therefore that, using the representation in Eq. 2.11 for $\mathbf{L}$,

$$\Delta\mathbf{L} = \oint \dot{\mathbf{L}} \, dt = \int_0^P \dot{\mathbf{L}} \, dt = \frac{-3GMJ_2 \pi \sin 2i}{2na^2(1-e^2)^{3/2}}(\cos \Omega, \sin \Omega, 0)$$

$$= (\Delta L_x, \Delta L_y, \Delta L_z)$$

Again $\oint dt$ means integrate over the perturbed spatial path, but because $\dot{\mathbf{L}}$ is of order J_2, the contour to $\int_0^P dt$ can be deformed with no additional errors of the first order. Using Eq. 2.11 again, $\Delta\mathbf{L}$ has the form

$$\Delta\mathbf{L} = (L+\Delta L)(\sin[\Omega+\Delta\Omega]\sin[i+\Delta i],$$
$$-\cos[\Omega+\Delta\Omega]\sin[i+\Delta i], \cos[i+\Delta i]) - \mathbf{L}$$
$$\approx \Delta L(\sin\Omega\sin i, -\cos\Omega\sin i, \cos i)$$
$$+ L\,\Delta\Omega(\cos\Omega\sin i, \sin\Omega\sin i, 0)$$
$$+ L\Delta i(\sin\Omega\cos i, -\cos\Omega\cos i, -\sin i)$$

Combining the above two expressions for $\Delta\mathbf{L}$ yields

$$\Delta L = \Delta L_x \sin\Omega\sin i - \Delta L_y \cos\Omega\sin i + \Delta L_z \cos i$$
$$L\,\Delta\Omega\sin i = \Delta L_x \cos\Omega + \Delta L_y \sin\Omega$$
$$L\,\Delta i = \Delta L_x \sin\Omega\cos i - \Delta L_y \cos\Omega\cos i - \Delta L_z \sin i$$

So, after dividing by P and setting $\Delta i/P = di/dt)_{\text{sec}}$

$$\left.\frac{di}{dt}\right)_{\text{sec}} = 0 \tag{9.2}$$

$$\left.\frac{d\Omega}{dt}\right)_{\text{sec}} = -\frac{3GMJ_2 n\cos i}{2(1-e^2)^2} \tag{9.3}$$

$$\left.\frac{da}{dt}\right)_{\text{sec}} = \frac{2ea}{1-e^2}\left.\frac{de}{dt}\right)_{\text{sec}}$$

where Eq. 2.4 has been used to derive the last formula from $dL/dt)_{\text{sec}} = 0$. Equation 9.1 and the last equation above together imply that

$$\left.\frac{de}{dt}\right)_{\text{sec}} = 0 \tag{9.4}$$

Therefore, the application of the equations for the time rates of change and of the energy and of the angular momentum simply and directly enables one to solve two-thirds of the perturbation problem. This approach also gives a direct physical insight into the cause of the perturbations as well as to their effects. The last two needed results are those for $d\omega/dt$ and dT/dt. They turn out to be (the analysis is in the next chapter)

$$\left.\frac{d\omega}{dt}\right)_{\text{sec}} = \frac{3GMJ_2 n(5\cos^2 i - 1)}{4(1-e^2)^2} \tag{9.5}$$

$$\left.\frac{dT}{dt}\right)_{\text{sec}} = -\frac{3GMJ_2(1-3\sin^2 i\sin^2\omega)}{2(1-e)^3} \tag{9.6}$$

Note that for $\cos^2 i = \frac{1}{5}$ ($i = 63°26'05''\!.82$ or $i = 116°33'54''\!.18$, or $\tan^{-1} i = 2$) there is no perigee advance due to J_2 (in first order theory!).

Third-Body Perturbations

This problem admits an analytical first-order solution directly for the time dependence of r and v (the true anomaly) instead of for the orbit $r(v)$. To set the stage the satellite's unperturbed orbit is circular, namely,

$$r = a, \qquad v = \frac{2\pi t}{P}, \qquad \text{with} \ \frac{4\pi^2 a^3}{P^2} = \mu = GM_\oplus$$

In the plane of the orbit lies another body, of mass M_p, also revolving about the Earth in a circular orbit (ψ is an arbitrary phase),

$$r_p = a_p, \qquad v_p = \frac{2\pi t}{P_p} + \psi, \qquad \frac{4\pi^2 a_p^3}{P_p^2} = \mu$$

However, $a_p \gg a$ or $P_p \gg P$ and we regard the perturbing body to be fixed for a few revolutions of the satellite. The exact potential for the problem is

$$U = -\frac{\mu}{r} - \frac{GM_p}{|\mathbf{r}_p - \mathbf{r}|}$$

which can be expressed as

$$U = -\frac{\mu}{r} - \frac{GM_p}{[a_p^2 + r^2 - 2ra_p \cos(v - v_p)]^{1/2}}$$

The planar equations of motion of the satellite are

$$\ddot{r} - r\dot{v}^2 = -\frac{\partial U}{\partial r}, \qquad r\ddot{v} + 2\dot{r}\dot{v} = -\frac{1}{r}\frac{\partial U}{\partial v}$$

If one uses the approximation of a stationary perturber, then they can be simplified to

$$\ddot{r} - r\dot{v}^2 = -\frac{\mu}{r^2} + \left(\frac{\mu\varepsilon}{a^2}\right)\cos(v - \psi)$$

$$r\ddot{v} + 2\dot{r}\dot{v} = -\left(\frac{\mu\varepsilon}{a^2}\right)\sin(v - \psi)$$

where the small parameter of the problem is ε,

$$\varepsilon = \frac{GM_p/a_p^2}{GM_\oplus/a^2}$$

We set

$$r = a + \varepsilon r_1, \qquad v = \frac{2\pi t}{P} + \varepsilon v_1$$

and substitute these expressions into the equations of motion, linearize with respect to ε, and deduce

$$\ddot{r}_1 - \frac{12\pi^2 r_1}{P^2} - \frac{4\pi a \dot{v}_1}{P} = \left(\frac{\mu}{a^2}\right)\cos\left(\frac{2\pi t}{P} - \psi\right)$$

$$a\ddot{v}_1 + \frac{4\pi \dot{r}_1}{P} = -\left(\frac{\mu}{a^2}\right)\sin\left(\frac{2\pi t}{P} - \psi\right)$$

As initial conditions we use

$$r(0) = a, \; v(0) = 0 \qquad\qquad r_1(0) = v_1(0) = 0$$
$$\text{or}$$
$$\dot{r}(0) = 0, \; \dot{v}(0) = \frac{2\pi}{P} \qquad\qquad \dot{r}_1(0) = \dot{v}_1(0) = 0$$

One can integrate the v_1 equation once, substitute that result into the r_1 equation, and then integrate it twice. We can then return to the v_1 equation to finish the solution. The results are

$$\frac{r}{a} = 1 + \varepsilon\left[B\cos\left(\frac{2\pi t}{P} + \phi\right) + \left(\frac{3\pi t}{P}\right)\sin\left(\frac{2\pi t}{P} - \psi\right) - 2\cos\psi \right]$$

$$v = \frac{2\pi t}{P} - \varepsilon\left\{ 2B\sin\left(\frac{2\pi t}{P} + \phi\right) - \left(\frac{6\pi t}{P}\right)\left[\cos\left(\frac{2\pi t}{P} - \psi\right) + \cos\psi\right] \right.$$

$$\left. + 2\sin\left(\frac{2\pi t}{P} - \psi\right) + 5\sin\psi \right\}$$

where the constants of integration B and ϕ are fixed by

$$B\cos\phi = 2\cos\psi \qquad \text{from } r_1(0) = 0, \; \dot{v}_1(0) = 0$$
$$B\sin\phi = -\tfrac{3}{2}\sin\psi \qquad \text{from } \dot{r}_1(0) = 0, \; v_1(0) = 0$$

This solution is valid for times $|t| \lesssim P/(2\pi\varepsilon)$.

Given the artificial nature of the problem, it seems best just to illustrate the motion for a few values of ε and ψ. Figures 36–39 show the unperturbed circular orbit as a full curve and the perturbed orbit as a series of equally spaced dots [$\Delta t = \frac{1}{36}$ of the unperturbed period] for $\psi = 0(90)360°$, $\varepsilon = 0.05$ from $t = 0$ until $t = P$. From the location of M_p and the direction of the initial velocity vector, the results can be interpreted. It is a numerical accident that for $\psi = 90°$ or $270°$ $v(P) = 0$.

Atmospheric Drag

The most commonly used (simple) expression for atmospheric drag is

$$\mathbf{F}_D = \frac{AC_D\rho}{2}|\mathbf{v} - \mathbf{v}_a|(\mathbf{v}_a - \mathbf{v})$$

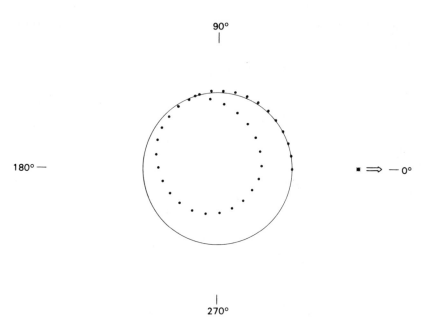

Figure 36. Unperturbed (solid) and perturbed (dotted) orbits for one unperturbed period. The perturbing third body is indicated by the arrow and the equal time spacing of the dots is $\frac{1}{36}$ of the unperturbed period.

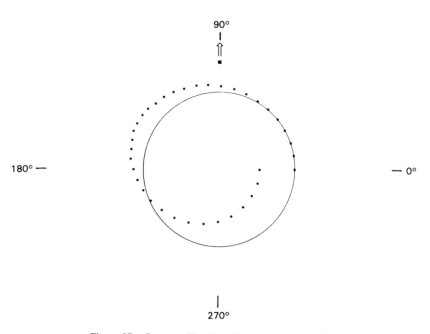

Figure 37. Same as Fig. 36 with the perturber at 90°.

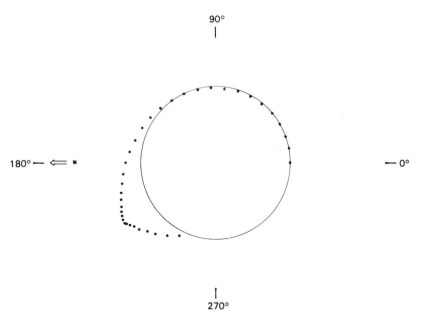

Figure 38. Same as Fig. 36 with the perturber at 180°.

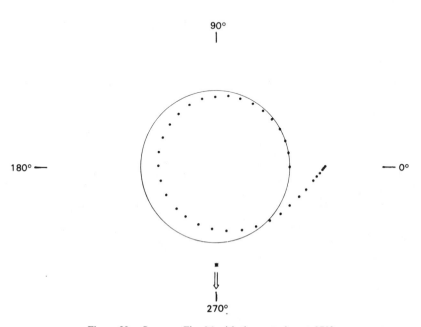

Figure 39. Same as Fig. 36 with the perturber at 270°.

where $\mathbf{v}$ is the satellite's inertial velocity, $\mathbf{v}_a$ is the inertial velocity of the atmosphere, ρ is the atmospheric density at the satellite's location, A is the effective cross-sectional area of the satellite, and C_D is the nondimensional drag coefficient, $C_d \simeq 2$. This expression is very difficult to deal with analytically. Moreover, ρ is dependent on spatial location and time, including diurnal and seasonal periodicities. Hence drag theories are semi-analytical at best.

What drag does is steal energy from the satellite's orbit. It takes work to push the atmosphere out of the way, and this work is performed at the expense of the orbital kinetic energy of the artificial satellite. As the satellite loses energy it drops $[\mathscr{E} = -\mu/2a$ so $\Delta\mathscr{E} = (\mu/2a^2)\,\Delta a$ and $\Delta\mathscr{E} < 0]$ thereby speeding up. The increase in speed plus the rising density of the lower layers of the atmosphere act to increase the drag force, which takes even more energy from the satellite's orbit, and so on.

For a high-altitude satellite atmospheric drag is a problem only if its orbit is highly elliptical. Then (a Molniya satellite would be a good example) the drag forces tend to act mostly at perigee. It should be remembered that the satellite is moving fastest at perigee, which both increases the magnitude of the drag force and decreases the amount of time over which the force can act. In the limit of an instantaneous perigee passage the drag force may be regarded as impulsive. The effect of this impulse is to decrease the apogee height. Over many revolutions the subsequent decrements in apogee distance will lead to a nearly circular orbit. This is one reason why almost all low-altitude satellites are in a nearly circular orbit—whether by design or not.

A simple energy analysis will reveal the main features. The kinetic energy per unit mass of the satellite at apogee is

$$K = GM_\oplus \left(\frac{1}{r} - \frac{1}{2a}\right)\bigg|_{r=a(1+e)}$$

Instead of the semi-major axis a and the eccentricity e we use the apogee and perigee distances R_a, R_P;

$$K = \frac{GM_\oplus R_a}{2(R_a + R_P)R_P}$$

If R_a changes by ΔR_a, then K changes by ΔK,

$$\Delta K = \frac{\partial K}{\partial R_a}\Delta R_a = \frac{GM_\oplus \Delta R_a}{2(R_a + R_P)^2}$$

This change is caused by the impulsive atmospheric drag force. This frictional force has done work equal to

$$\int \mathbf{F}_D \cdot d\mathbf{s}$$

during perigee passage. Substituting the expression for the drag force $\mathbf{F}_D$ given above, we get

$$m \, \Delta K = \tfrac{1}{2} A C_D \int \rho^2 v \, ds$$

The speed of the atmosphere has been ignored (which is a result of the diurnal rotation of the Earth). Since the impulsive approximation is being used,

$$m \, \Delta K \simeq \tfrac{1}{2} A C_D \rho_p v_p^2 \, \Delta s$$

or, using the above expression for ΔK,

$$\Delta R_a = (R_a + R_p) \frac{R_a}{R_p} \frac{A C_D}{m} \rho_p \, \Delta s$$

Now suppose that this process has gone on long enough that the orbit is nearly circular. Then the satellite's entire orbit is within the atmosphere, and the drag force acts continuously. By equating the work done in overcoming the drag force with the loss in energy (as above), we obtain

$$-F_D \, \Delta s = -\frac{G M_\oplus \, \Delta r}{2 r^2}$$

The difference between the element of the path length Δs and the radial distance Δr depends on the depression, the angular distance below the horizon. We call this d and let d be positive for a decrease in height. Then

$$\Delta r = -\sin d \, \Delta s$$

and

$$\sin d = \frac{A C_D \rho}{m r}$$

The final part of the flight path occurs when the component of gravitational acceleration forward is just balanced by the component of drag backward. This happens once d exceeds

$$\sin^{-1}\left(\frac{A C_D \rho}{2 m r}\right)$$

Thereafter the deceleration is very rapid, resulting in a spiralling in of the spacecraft.

Other Perturbations

Radiation Pressure

Electromagnetic radiation carries energy, momentum, and angular momentum with it. Hence it should not be surprising that radiation exerts a pressure. Its effects on artificial satellites are especially noticeable on the balloon-type satellites (e.g., Echo I). Since satellites may go in and out of eclipse, this force is time dependent and usually treated in the Gaussian formulation of the perturbation equations (see below). The magnitude of the solar radiation pressure force is given by

$$\frac{AP_\odot}{4\pi mcD^2}$$

where A is the effective cross-sectional area of the satellite, $P_\odot$ is the total radiated solar power, m is the satellite's mass, c is the speed of light in vacuo, and D is the satellite–Sun distance. For close Earth satellites $D \simeq 1$ A.U. The direction of this force is along the satellite–Sun line or nearly along the Earth–Sun line. A sufficient approximation is to say that the solar radiation pressure acts along the unit vector $\mathbf{n}$,

$$\mathbf{n} = (\cos\Lambda_\odot,\ \cos\varepsilon\,\sin\Lambda_\odot,\ \sin\varepsilon\,\sin\Lambda_\odot)$$

where $\Lambda_\odot$ is the geocentric longitude of the Sun, ε is the obliquity of the ecliptic, and $\mathbf{n}$ is in the geocentric equatorial coordinate system. There is no analytical formulation for the effects of solar radiation pressure on the orbital element set (but if there are no eclipses, then $\Delta a = 0$).

Relativistic Effects

The classical mechanics described thus far is generally good to one part in 10^8. General relativistic effects are dependent on the ratio of the gravitational potential to the square of the speed of light. Special relativistic effects depend on the square of the ratio (object speed/c). The Earth's orbital speed (~ 30 km/sec) is such that $v^2/c^2 \simeq 10^{-8}$ and its rotational speed (~ 0.5 km/sec) results in post-Newtonian approximations of 10^{-12}. The presence of the Moon and Sun affects the gravitational potential in the vicinity of the Earth by $\sim 10^{-8}$. Relativistic corrections crop up in time, radial velocity, length, and direction measurements at the part in 10^8 level. Indeed, second-order effects in Doppler corrections have now been measured. When necessary, these relativistic corrections can be regarded as small perturbations to the motion or in the data reduction.

Planetary Theory

The presentation of "planetary theory" has been systematically neglected in this text and will continue to be. The reasons are twofold—it is treated in depth in all of the classical references (Smart 1953; Plummer 1918; Brouwer and Clemence 1961; etc.), and I think that (beyond first-order results) it's mostly wrong. However, as this has been the classical realm of application for celestial mechanics, many of the subject's concepts and terminology derive from it. In particular, when a solution to the full equations of motion gave rise to certain types of terms, they were labeled as a function of their time dependence. When an explicit linear function of the time appeared in the expression for an orbital element, this was called a *secular term*. The implication is that of a steady, uniform increase (or decrease) in the appropriate quantity. If, instead, the time dependence were periodic, and with a period on the order of the fundamental period of the perturbed object, then these were called *short-period terms* or *short-period inequalities*. Clearly there is a huge difference in the evolution of an orbit if a perturbed quantity is undergoing short periodic variations rather than secular ones. Sometimes the coefficients of such short periodic terms would be functions of the various mean motions involved. For instance, in the planetary problem, a typical term in a perturbed orbital element would be of the form

$$\frac{mf}{j_1 n_1 + j_2 n_2} \sin[(j_1 n_1 + j_2 n_2)t + \phi]$$

where m is the mass of the perturber (in solar units and therefore $\ll 1$), f is some messy function of the eccentricities and semi-major axis ratios ($|f| \ll 1$ too usually), n_1 and n_2 are the mean motions of the object of interest and the perturber, j_1 and j_2 are integers telling you which term of a Fourier series all of this comes from, and ϕ is a phase dependent on the mean anomalies at epoch, the longitudes of the ascending nodes, and the arguments of perihelion. Now j_1 and j_2 can be positive or negative. Hence, if the mean motions are nearly commensurate, there exists some pair of j_1, j_2 such that $j_1 n_1 + j_2 n_2 \approx 0$. The numerical effect of such a term will be huge and its period will be long.

As an example, consider Jupiter (subscript 1) perturbed by Saturn (subscript 2). The period of this kind of term is clearly $2\pi/(j_1 n_1 + j_2 n_2)$. For $j_1 = 1 = j_2$ this is 8.5 years, for $j_1 = 1 = -j_2$ it is 19.9 years, and for $j_1 = 2$, $j_1 = 1$ it is 4.9 years. However, for $j_1 = -2$ and $j_2 = 5$ it is about 900 years. Such terms are called *small divisors* and are said to lead to *long-periodic terms*.

In the artificial satellite problem the result is the same, but the analytical cause is different. It turns out that some of the perturbations are periodic functions of the Euler angles (i, ω, Ω) of the orbit. If these suffer from secular perturbations, then trigonometric functions of them will be long-term periodic. Finally sometimes there are mixed or Poisson terms; these are of the form $t \times$ periodic function of t.

Resonances. The solar system is full of orbital resonances (Peale 1976). The satellite systems of Jupiter and Saturn show them; the ring/moon system of Saturn has them; the asteroid belt exhibits them with respect to Jupiter; Jupiter/Saturn, Neptune/Pluto, and so on exhibit them; the Earth–Moon and Mercury–Sun systems show revolution/rotation locking; and so on. Except for the rotational period = revolution period spin–orbit coupling due to tidal friction, none of these are really explained. Interpreted in terms of mean motion commensurability yes, explained no.

Solar System Stability. The question of the ultimate stability of the solar system has been a problem of great interest. Laplace took up the problem and demonstrated the absence of first-order secular terms in the semi-major axes of the planets. Poisson took up the attack to second order and others have gone further. The question is still unsettled and although powerful analytic techniques have been brought to bear (Jeffreys and Szebehely 1978) no clarification seems near. Numerical studies (Nacozy 1976) also seem limited in predictive ability but are interesting to read.

THE DEVELOPMENT OF PERTURBATION THEORY

One could attempt to continue to develop more tricks akin to the above in order to more fully discuss the disturbances caused by various perturbing forces. This would not (necessarily) be coherent, and it would be very difficult to deduce the detailed time behavior of the orbit's size, shape, and spatial orientation. We have, however, all learned analytical geometry and (presumably) feel comfortable with conic sections. As long as the perturbing forces are small, our intuition leads us to believe that the changes produced in $a, e, \omega, \ldots$ will be small too. Hence, long ago (Lagrange 1783) the utility of replacing the second-order system $m\ddot{\mathbf{r}} = \mathbf{F}$ with an equivalent system $\dot{a} = $, $\dot{e} = $, $\dot{\omega} = $, $\ldots$ was recognized. Furthermore, when the transformation is performed correctly, the new system of equations will be rigorously equivalent to the original one, will have vanishing right-hand sides when there are no perturbing forces, and will still allow us to deduce location and velocity in the usual (Keplerian) way. This transformation is discussed below. The only alternatives to this procedure are the straightforward numerical integrations of the equations of motion or the (messy and nontransparent) approximate analytical solutions of them. The former is only recently practical with accuracy over long time scales, it is inelegant, and it is no fun. The latter *is* messy and is not perspicuous. It does have some advantages all the same, and that is why I have tended to pursue it herein.

When the perturbing forces can be represented as the gradient of a velocity-independent scalar, the resulting form of the perturbation equations is known as Lagrange's (planetary) equations. For an arbitrary perturbing force Gauss (in 1814) constructed the appropriate generalization. The derivation of either set is usually labeled tedious. It seems to me more important for the reader to see the logic of the procedure than be intimately familiar with the details. Thus, I first discuss the why and then, separately for the Lagrangian and the Gaussian sets, the how.

The Transformation

Consider the case when the perturbed equations of motion take the form

$$\ddot{\mathbf{r}} + \nabla U = \nabla R(\mathbf{r}, t) \qquad (9.7)$$

where $U = -GM/|\mathbf{r}|$ and $|\nabla U| \gg |\nabla R|$. M, depending on the coordinate system, will be the sum of the masses, the mass of the primary, the reduced mass, or possibly some more complicated function of the masses. R is the negative of the perturbing potential (this symbol and sign convention are standard). If $\nabla R = 0$, then the solution of Eq. 9.7 would be in the form

$$\mathbf{r} = \mathbf{r}(\mathbf{a}, t) \tag{9.8}$$

where $\mathbf{a}$ is a six-dimensional constant vector. Equations 2.12 and 2.13 provide one explicit example of a possible form for $\mathbf{r}(\mathbf{a}, t)$. We shall regard $\mathbf{a}$ as the classical set of orbital elements a, e, ω, i, Ω, and T or some simple combination thereof. Since $\mathbf{a}$ is a constant, it follows that, in unperturbed motion, the velocity vector $\mathbf{v}$ is given by

$$\mathbf{v} = d\mathbf{r}/dt = \partial \mathbf{r}(\mathbf{a}, t)/\partial t \tag{9.9}$$

We wish to reformulate the equations of motion 9.7 into an equation for $\dot{\mathbf{a}}$ while keeping the form and (instantaneous) meaning of Eqs. 9.8 and 9.9. If this can be successfully performed, then $\mathbf{a}(t = t_0)$ represents the orbital elements of the ellipse that the object would move on were the perturbation suddenly removed at $t = t_0$. This ellipse is tangent to the actual path at $t = t_0$ and is known as the *osculating ellipse*. Similarly one speaks of the osculating orbital element set $\mathbf{a}(t_0)$. Furthermore, if we can enforce Eqs. 9.8 and 9.9, then we can compute the instantaneous location and velocity of the object by the usual Keplerian formulas. We shall now see how all of this can be rigorously accomplished.

In perturbed motion we have

$$d\mathbf{r}/dt = \partial \mathbf{r}/\partial t + \partial \mathbf{r}/\partial \mathbf{a} \cdot \dot{\mathbf{a}} \tag{9.10}$$

Hence, looking at Eq. 9.9, we need to force the second term in Eq. 9.10 to be zero. This is called the *condition of osculation* (i.e. Eq. 9.11b) and is what allows one to compute the velocity in the Keplerian fashion. (The representation 9.8 permits it for the location.) That we have this much freedom should be clear—a second differentiation of Eq. 9.10 and subsequent substitution into Eq. 9.7 provides only three equations for the six-dimensional vector $\mathbf{a}$.

Imposing the condition of osculation and then differentiating Eq. 9.10 once more yields

$$\frac{d^2\mathbf{r}}{dt^2} = \frac{\partial^2\mathbf{r}}{\partial t^2} + \left(\frac{\partial}{\partial \mathbf{a}} \frac{\partial \mathbf{r}}{\partial t} \right) \cdot \dot{\mathbf{a}}$$

Replacing the left-hand side of Eq. 9.7 with this expression, and remembering that $\partial^2\mathbf{r}/\partial t^2 + \nabla U = \mathbf{0}$ by construction, we deduce that

$$\dot{\mathbf{a}} \cdot \frac{\partial}{\partial \mathbf{a}} \frac{\partial \mathbf{r}}{\partial t} = \nabla R \tag{9.11a}$$

supplemented with the constraint

$$\dot{\mathbf{a}} \cdot \frac{\partial \mathbf{r}}{\partial \mathbf{a}} = 0 \tag{9.11b}$$

The system 9.11 represents six nonlinear, inhomogeneous, first-order, ordinary differential equations for the six unknowns $\mathbf{a}(t)$, which are rigorously equivalent to the system 9.7 as long as the Jacobian $\partial(\mathbf{r}, \mathbf{v})/\partial(\mathbf{a})$ does not vanish. One can show that the Jacobian does vanish only on a set of zero measure (but I do not recommend trying it). The last remaining step is to write Eqs. 9.11a and b in a more usable form.

The Details

Now that the basic ideas have been presented we turn to carrying out the necessary analytical manipulations in detail. The perturbed equations of motion are given in Eq. 9.7

$$\ddot{\mathbf{r}} + \nabla U = \nabla R(\mathbf{r}, t) \tag{9.7}$$

where $U = -GM/|\mathbf{r}|$ is the two-body potential. R is some scalar (the negative of the perturbing potential) whose gradient gives all of the other forces in the system. When no such scalar exists (i.e., the forces are nonconservative), a similar but analytically different procedure is necessary (see below). The solution of the unperturbed dynamical problem ($R = $ constant) is known and parameterized by some six-dimensional vector $\mathbf{a} = (a_1, a_2, \ldots, a_6)$. This vector, when $R = $ constant, is a constant. However, the existence of the perturbing forces represented by ∇R causes the motion to depart from the unperturbed motion, and we will allow the parameters $\mathbf{a}$ to vary (hence the phrase "variation of parameters"). It is important to remember that not only is the path through space given in the form

$$\mathbf{r} = \mathbf{r}(\mathbf{a}, t) \tag{9.8}$$

but that the velocity

$$\mathbf{v} = \frac{d\mathbf{r}}{dt} = \frac{\partial \mathbf{r}}{\partial t} \tag{9.9}$$

is also given in a similar form. Even though we allow the parameters to vary, since $|\nabla R| \neq 0$, we want to preserve these formal results, especially Eq. 9.9.

Consider just the x component. We write

$$x = x(\mathbf{a}, t)$$

so that when $\mathbf{a}$ is no longer a constant

$$\dot{x} = \frac{dx}{dt} = \frac{\partial x}{\partial t} + \frac{\partial x}{\partial \mathbf{a}} \cdot \dot{\mathbf{a}}$$

$$= \frac{\partial x}{\partial t} + \sum_{n=1}^{6} \frac{\partial x}{\partial a_n} \frac{da_n}{dt}$$

In a similar fashion we could compute $\ddot{x}$. If we did so and then substituted the results into the perturbed equations of motion (Eq. 9.7), it should be clear that we would find three equations for each a_n, that is, three relations involving $\dot{a}_n$ and $\ddot{a}_n$ for $n = 1, 2, \ldots, 6$. Remembering that we want to enforce Eq. 9.9, we remove these extra degrees of freedom by insisting that $\dot{\mathbf{a}}$ satisfies (for x, there are exactly analogous equations for y and z)

$$\sum_{n=1}^{6} \frac{\partial x}{\partial a_n} \cdot \frac{da_n}{dt} = \nabla_{\mathbf{a}} x \cdot \dot{\mathbf{a}} = 0 \qquad \text{(cf. Eq. 9.11b)}$$

Note that there are many different constraints that could be imposed to remove the extra degrees of freedom. Only the above choice (the condition of osculation) preserves the Keplerian formula for $\dot{\mathbf{r}}$.

Now we compute $d^2 x / dt^2$,

$$\ddot{x} = \frac{\partial^2 x}{\partial t^2} + \frac{\partial}{\partial \mathbf{a}}\left(\frac{\partial x}{\partial t} \cdot \dot{\mathbf{a}}\right)$$

$$= \frac{\partial^2 x}{\partial t^2} + \sum_{n=1}^{6} \frac{\partial^2 x}{\partial a_n \, \partial t} \frac{da_n}{dt}$$

The x component of Eq. 9.7 can now be written as

$$\frac{\partial^2 x}{\partial t^2} + \frac{\mu x}{r^3} + \sum_{n=1}^{6} \frac{\partial^2 x}{\partial a_n \, \partial t} \frac{da_n}{dt} \frac{\partial}{\partial t} = \frac{\partial R}{\partial x}$$

(with exactly analogous expressions for y and z). We remember that the first two terms sum to zero, for $x(\mathbf{a}, t)$ is precisely the solution of the two-body problem. Moreover we can replace $\partial x(\mathbf{a}, t)/\partial t$ with $\dot{x}$ by virtue of Eq. 9.11b. Hence the above can be written more compactly as

$$\nabla_{\mathbf{a}} \dot{x} \cdot \dot{\mathbf{a}} = \frac{\partial R}{\partial x} \qquad \text{(cf. Eq. 9.11a)}$$

If we add the y and z components to this result, then the six equations for the determination of the parameters $\mathbf{a}$ can be simply written as in Eqs. 9.11, namely,

$$\mathbf{a} \cdot \frac{\partial \dot{\mathbf{r}}}{\partial \mathbf{a}} = \nabla R \qquad\qquad (9.11a)$$

$$\dot{\mathbf{a}} \cdot \frac{\partial \mathbf{r}}{\partial \mathbf{a}} = 0 \qquad\qquad (9.11b)$$

In explicit component form these are

$$\frac{\partial x}{\partial a_1}\frac{da_1}{dt}+\frac{\partial x}{\partial a_2}\frac{da_2}{dt}+\frac{\partial x}{\partial a_3}\frac{da_3}{dt}+\frac{\partial x}{\partial a_4}\frac{da_4}{dt}+\frac{\partial x}{\partial a_5}\frac{da_5}{dt}+\frac{\partial x}{\partial a_6}\frac{da_6}{dt}=0$$

$$\frac{\partial y}{\partial a_1}\frac{da_1}{dt}+\frac{\partial y}{\partial a_2}\frac{da_2}{dt}+\frac{\partial y}{\partial a_3}\frac{da_3}{dt}+\frac{\partial y}{\partial a_4}\frac{da_4}{dt}+\frac{\partial y}{\partial a_5}\frac{da_5}{dt}+\frac{\partial y}{\partial a_6}\frac{da_6}{dt}=0$$

$$\frac{\partial z}{\partial a_1}\frac{da_1}{dt}+\frac{\partial z}{\partial a_2}\frac{da_2}{dt}+\frac{\partial z}{\partial a_3}\frac{da_3}{dt}+\frac{\partial z}{\partial a_4}\frac{da_4}{dt}+\frac{\partial z}{\partial a_5}\frac{da_5}{dt}+\frac{\partial z}{\partial a_6}\frac{da_6}{dt}=0$$

$$\frac{\partial \dot{x}}{\partial a_1}\frac{da_1}{dt}+\frac{\partial \dot{x}}{\partial a_2}\frac{da_2}{dt}+\frac{\partial \dot{x}}{\partial a_3}\frac{da_3}{dt}+\frac{\partial \dot{x}}{\partial a_4}\frac{da_4}{dt}+\frac{\partial \dot{x}}{\partial a_5}\frac{da_5}{dt}+\frac{\partial \dot{x}}{\partial a_6}\frac{da_6}{dt}=\frac{\partial R}{\partial x}$$

$$\frac{\partial \dot{y}}{\partial a_1}\frac{da_1}{dt}+\frac{\partial \dot{y}}{\partial a_2}\frac{da_2}{dt}+\frac{\partial \dot{y}}{\partial a_3}\frac{da_3}{dt}+\frac{\partial \dot{y}}{\partial a_4}\frac{da_4}{dt}+\frac{\partial \dot{y}}{\partial a_5}\frac{da_5}{dt}+\frac{\partial \dot{y}}{\partial a_6}\frac{da_6}{dt}=\frac{\partial R}{\partial y}$$

$$\frac{\partial \dot{z}}{\partial a_1}\frac{da_1}{dt}+\frac{\partial \dot{z}}{\partial a_2}\frac{da_2}{dt}+\frac{\partial \dot{z}}{\partial a_3}\frac{da_3}{dt}+\frac{\partial \dot{z}}{\partial a_4}\frac{da_4}{dt}+\frac{\partial \dot{z}}{\partial a_5}\frac{da_5}{dt}+\frac{\partial \dot{z}}{\partial a_6}\frac{da_6}{dt}=\frac{\partial R}{\partial z}$$

$$(9.11)$$

Formally the task is complete. The six equations of motion 9.11 are equivalent to the original equations of motion 9.7. In practice Eqs. 9.11 are not in the most convenient form for applications. It would be easier if we had something in the form $\dot{\mathbf{a}}=$ to deal with. We now turn to transforming Eqs. 9.11 into such a system.

Each of the explicitly displayed elements of Eqs. 9.11 are multiplied, in turn, by $-\partial \dot{x}/\partial a_n$, $-\partial \dot{y}/\partial a_n$, $-\partial \dot{z}/\partial a_n$, and $\partial x/\partial a_n$, $\partial y/\partial a_n$, $\partial z/\partial a_n$. All of these are then added. On the right-hand side the result is

$$\frac{\partial R}{\partial x}\frac{\partial x}{\partial a_n}+\frac{\partial R}{\partial y}\frac{\partial y}{\partial a_n}+\frac{\partial R}{\partial z}\frac{\partial z}{\partial a_n}=\nabla R\cdot\frac{\partial \mathbf{r}}{\partial a_n}=\frac{\partial R}{\partial a_n}$$

because we assumed that $R=R(\mathbf{r}, t)$ and disallowed (for now) velocity-dependent forces. The left-hand sides are a mess, and can be compactly written as

$$[a_1, a_1]\frac{da_1}{dt}+[a_1, a_2]\frac{da_2}{dt}+[a_1, a_3]\frac{da_3}{dt}+[a_1, a_4]\frac{da_4}{dt}$$

$$+[a_1, a_5]\frac{da_5}{dt}+[a_1, a_6]\frac{da_6}{dt}=\frac{\partial R}{\partial a_1}$$

$$[a_2, a_1]\frac{da_1}{dt}+[a_2, a_2]\frac{da_2}{dt}+[a_2, a_3]\frac{da_3}{dt}+[a_2, a_4]\frac{da_4}{dt}$$

$$+[a_2, a_5]\frac{da_5}{dt}+[a_2, a_6]\frac{da_6}{dt}=\frac{\partial R}{\partial a_2}$$

$$[a_3, a_1]\frac{da_1}{dt}+[a_3, a_2]\frac{da_2}{dt}+[a_3, a_3]\frac{da_3}{dt}+[a_3, a_4]\frac{da_4}{dt}$$

$$+[a_3, a_5]\frac{da_5}{dt}+[a_3, a_6]\frac{da_6}{dt}=\frac{\partial R}{\partial a_3}$$

$$[a_4, a_1]\frac{da_1}{dt}+[a_4, a_2]\frac{da_2}{dt}+[a_4, a_3]\frac{da_3}{dt}+[a_4, a_4]\frac{da_4}{dt}$$

$$+[a_4, a_5]\frac{da_5}{dt}+[a_4, a_6]\frac{da_6}{dt}=\frac{\partial R}{\partial a_4} \quad (9.12)$$

$$[a_5, a_1]\frac{da_1}{dt}+[a_5, a_2]\frac{da_2}{dt}+[a_5, a_3]\frac{da_3}{dt}+[a_5, a_4]\frac{da_4}{dt}$$

$$+[a_5, a_5]\frac{da_5}{dt}+[a_5, a_6]\frac{da_6}{dt}=\frac{\partial R}{\partial a_5}$$

$$[a_6, a_1]\frac{da_1}{dt}+[a_6, a_2]\frac{da_2}{dt}+[a_6, a_3]\frac{da_3}{dt}+[a_6, a_4]\frac{da_4}{dt}$$

$$+[a_6, a_5]\frac{da_5}{dt}+[a_6, a_6]\frac{da_6}{dt}=\frac{\partial R}{\partial a_6}$$

where the symbol $[a_m, a_n]$ is called a *Lagrange bracket*. It is defined by

$$[a_m, a_n]\equiv\frac{\partial x}{\partial a_m}\frac{\partial \dot{x}}{\partial a_n}-\frac{\partial x}{\partial a_n}\frac{\partial \dot{x}}{\partial a_m}+\frac{\partial y}{\partial a_m}\frac{\partial \dot{y}}{\partial a_n}-\frac{\partial y}{\partial a_n}\frac{\partial \dot{y}}{\partial a_m}+\frac{\partial z}{\partial a_m}\frac{\partial \dot{z}}{\partial a_n}-\frac{\partial z}{\partial a_n}\frac{\partial \dot{z}}{\partial a_m}$$

and has the useful properties of being both antisymmetric and time invariant. The antisymmetry is obvious and means that instead of 36 $(=6^2)$ different Lagrange brackets there are only $15=\binom{6}{2}$ potentially nonzero ones. Before showing that $\partial[a_n, a_m]/\partial t = 0$, we note that the last remaining step is to solve the six linear, inhomogeneous, equations in six unknowns (Eqs. 9.12). As there are really only six nonzero Lagrange brackets (for the two-body problem), it makes no sense to exhibit the formal solution via Cramer's rule. Finally, in a much more compact notation, Eqs. 9.12 may be written as

$$\sum_{n=1}^{6}[a_m, a_n]\frac{da_n}{dt}=\frac{\partial R}{\partial a_m}$$

The reason (for this discussion) that it is important that the Lagrange brackets be time independent is the implication that one can evaluate them at any place in the orbit. In particular one can choose an instant of time that simplifies the algebra. To see that the Lagrange bracket is time independent, consider one of them, say $[p, q]$. Writing only the x part (and remembering that the complete bracket contains comparable y and z terms),

$$[p, q] = \frac{\partial x}{\partial p}\frac{\partial \dot{x}}{\partial q} - \frac{\partial x}{\partial q}\frac{\partial \dot{x}}{\partial p}$$

Taking the derivative with respect to the time t

$$\frac{\partial}{\partial t}[p, q] = \frac{\partial^2 x}{\partial p\,\partial t}\frac{\partial \dot{x}}{\partial q} + \frac{\partial x}{\partial p}\frac{\partial^2 \dot{x}}{\partial q\,\partial t} - \frac{\partial^2 x}{\partial q\,\partial t}\frac{\partial \dot{x}}{\partial p} - \frac{\partial x}{\partial q}\frac{\partial^2 \dot{x}}{\partial p\,\partial t}$$

The right-hand member may be rewritten as

$$\frac{\partial}{\partial p}\left(\frac{\partial x}{\partial t}\frac{\partial \dot{x}}{\partial q} - \frac{\partial x}{\partial q}\frac{\partial \dot{x}}{\partial t}\right) - \frac{\partial}{\partial q}\left(\frac{\partial x}{\partial t}\frac{\partial \dot{x}}{\partial p} - \frac{\partial x}{\partial p}\frac{\partial \dot{x}}{\partial t}\right)$$

As discussed, $x, \dot{x}$ in this expression stand for the functions in elliptic motion with $a_1, \dots, a_6$ treated as constants. Thus, $\partial x/\partial t = \dot{x}$, $\partial \dot{x}/\partial t = \ddot{x}$, and

$$\frac{\partial}{\partial t}[p, q] = \frac{\partial}{\partial p}\left(\dot{x}\frac{\partial \dot{x}}{\partial q} - \ddot{x}\frac{\partial x}{\partial q}\right) - \frac{\partial}{\partial q}\left(\dot{x}\frac{\partial \dot{x}}{\partial p} - \ddot{x}\frac{\partial x}{\partial p}\right)$$

$$= \frac{\partial}{\partial p}\left(\frac{1}{2}\frac{\partial(\dot{x}^2)}{\partial q} + \frac{\partial U}{\partial x}\frac{\partial x}{\partial q}\right) - \frac{\partial}{\partial q}\left(\frac{1}{2}\frac{\partial(\dot{x}^2)}{\partial p} + \frac{\partial U}{\partial x}\frac{\partial x}{\partial p}\right)$$

since $\ddot{x} = -\partial U/\partial x$. Combining all of the (unwritten) y and z terms with these yields

$$\frac{\partial}{\partial t}[p, q] = \frac{1}{2}\frac{\partial^2 \dot{r}^2}{\partial p\,\partial q} + \frac{\partial^2 U}{\partial p\,\partial q} - \frac{1}{2}\frac{\partial^2 \dot{r}^2}{\partial q\,\partial p} - \frac{\partial^2 U}{\partial q\,\partial p} = 0$$

Q.E.D.

At this point in celestial mechanics books one usually reads how much work the evaluation of the 15 Lagrange brackets is. This is not true because there are two tricks (over and above evaluating them at $t = T$ so that $M = E = v = 0$) that can be used to shorten the labor. One (recognized, for instance, by Moulton in 1902) is to recall the explicit form for $\mathbf{r}(\mathbf{a}, t)$ given in Eq. 2.13a,

$$\mathbf{r} = S\mathbf{q}$$

Hence, there are $\binom{3}{2} = 3$ i, Ω, and ω brackets that are separate from the three a, e, and T brackets. In addition, there are $3 \times 3 = 9$ brackets wherein one element is a member of the angular triplet and one element is a member of the size, shape, and place triplet. Thus, if $S = (S_{ij})$ and $\mathbf{q} = (q_1, q_2, q_3)$

$$[a_m, a_n] = na^2(1 - e^2)^{1/2} \sum_{j=1}^{3}\left(\frac{\partial S_{j1}}{\partial a_m}\frac{\partial S_{j2}}{\partial a_n} - \frac{\partial S_{j2}}{\partial a_m}\frac{\partial S_{j1}}{\partial a_n}\right)$$

if a_m, a_n are both one of i, Ω, ω;

$$[a_m, a_n] = a(1-e) \sum_{j=1}^{3} \left(S_{j1} \frac{\partial S_{j1}}{\partial a_m} \frac{\partial \dot{q}_1}{\partial a_n} + S_{j2} \frac{\partial S_{j1}}{\partial a_m} \frac{\partial \dot{q}_2}{\partial a_n} \right)$$

$$- na \left(\frac{1+e}{1-e} \right)^{1/2} \sum_{j=1}^{3} \left(S_{j1} \frac{\partial S_{j2}}{\partial a_m} \frac{\partial \dot{q}_1}{\partial a_n} + S_{j2} \frac{\partial S_{j2}}{\partial a_m} \frac{\partial \dot{q}_2}{\partial a_n} \right)$$

if a_m is one of i, Ω, or ω and a_n is one of a, e, or T;

$$[a_m, a_n] = \left(\frac{\partial q_1}{\partial a_m} \frac{\partial \dot{q}_1}{\partial a_n} - \frac{\partial q_1}{\partial a_n} \frac{\partial \dot{q}_1}{\partial a_m} \right) \sum_{j=1}^{3} S_{j1}^2 + \left(\frac{\partial q_2}{\partial a_m} \frac{\partial \dot{q}_2}{\partial a_n} - \frac{\partial q_2}{\partial a_n} \frac{\partial \dot{q}_2}{\partial a_m} \right) \sum_{j=1}^{3} S_{j2}^2$$

$$+ \left[\left(\frac{\partial q_1}{\partial a_m} \frac{\partial \dot{q}_2}{\partial a_n} - \frac{\partial q_1}{\partial a_n} \frac{\partial \dot{q}_2}{\partial a_m} \right) + \left(\frac{\partial q_2}{\partial a_m} \frac{\partial \dot{q}_1}{\partial a_n} - \frac{\partial q_2}{\partial a_n} \frac{\partial \dot{q}_1}{\partial a_m} \right) \right] \sum_{j=1}^{3} S_{j1} S_{j2}$$

if a_m, a_n are both one of a, e, T.

The second trick is to exploit the fact that S is unitary $(S^{-1} = S^T)$ and, therefore, is composed of orthonormal row and column vectors. Whence,

$$\sum_{j=1}^{3} S_{j1} \frac{\partial S_{j1}}{\partial a_l} = \frac{1}{2} \sum_{j=1}^{3} \frac{\partial S_{j1}^2}{\partial a_l} = \frac{1}{2} \frac{\partial}{\partial a_l} \sum_{j=1}^{3} S_{j1}^2 = \frac{1}{2} \frac{\partial}{\partial a_l}(1) = 0$$

$$\sum_{j=1}^{3} \left(S_{j2} \frac{\partial S_{j1}}{\partial a_l} - S_{j1} \frac{\partial S_{j2}}{\partial a_l} \right) = \frac{\partial}{\partial a_l} \sum_{j=1}^{3} S_{j1} S_{j2} - 2 \sum_{j=1}^{3} S_{j1} \frac{\partial S_{j2}}{\partial a_l}$$

$$= \frac{\partial}{\partial a_l}(0) - 2 \sum_{j=1}^{3} S_{j1} \frac{\partial S_{j2}}{\partial a_l}$$

$$= -2 \sum_{j=1}^{3} S_{j1} \frac{\partial S_{j2}}{\partial a_l}$$

Exploiting the unitariness of S to the fullest extent eliminates most of the remaining work. One finds then, for the six nonzero Lagrange brackets,

$$[i, \Omega] = na^2(1-e^2)^{1/2} \sin i, \qquad [a, \omega] = -\frac{na}{2}(1-e^2)^{1/2}$$

$$[a, \Omega] = -\frac{na(1-e^2)^{1/2}}{2} \cos i, \qquad [e, \omega] = \frac{na^2 e}{(1-e^2)^{1/2}}$$

$$[e, \Omega] = \frac{na^2 e \cos i}{(1-e^2)^{1/2}}, \qquad [a, T] = \frac{a}{2}$$

We can now complete the final step, the inversion of Eqs. 9.12. The result is given immediately below.

Lagrangian Forms. Depending on taste and the problem at hand, one may prefer a form for **a** other than a, e, ω, i, Ω, and T. Especially for small eccentricity (ω indeterminate) or small inclination (Ω indeterminate) orbits, other forms for **a** are preferable. Listed below is a catalog of the more common forms. In all of the cases the abbreviations $L^2 = \mu a(1 - e^2)$, $e = \sin \phi$ have been used. (These are for reference, not casual reading.)

1. Using a, e, ω, i, Ω, and T $[M = n(t - T)]$, one has

$$\dot{a} = \frac{-2a^2}{\mu} \frac{\partial R}{\partial T}$$

$$\dot{e} = \frac{-L^2}{\mu^2 e} \frac{\partial R}{\partial T} - \frac{(1 - e^2)}{eL} \frac{\partial R}{\partial \omega}$$

$$\dot{\omega} = \frac{(1 - e^2)}{eL} \left(\frac{\partial R}{\partial e} - \frac{e \cot i}{1 - e^2} \frac{\partial R}{\partial i} \right)$$

$$\frac{di}{dt} = \frac{\csc i}{L} \left(\cos i \frac{\partial R}{\partial \omega} - \frac{\partial R}{\partial \Omega} \right)$$

$$\dot{\Omega} = \frac{\csc i}{L} \frac{\partial R}{\partial i}$$

$$\dot{T} = \frac{2a^2}{\mu} \frac{\partial R}{\partial a} + \frac{L^2}{\mu^2 e} \frac{\partial R}{\partial e}$$

2. We use a, e, ω, i, Ω, and $\chi = -nT$ $(M = nt + \chi)$. Only the $\dot{a}$, $\dot{e}$, and $\dot{T}$ equations change from set 1 into

$$\dot{a} = \frac{2}{na} \frac{\partial R}{\partial \chi}$$

$$\dot{e} = \frac{1 - e^2}{(\mu a e^2)^{1/2}} \frac{\partial R}{\partial \chi} - \left(\frac{1 - e^2}{\mu a e^2} \right)^{1/2} \frac{\partial R}{\partial \omega}$$

$$\dot{\chi} = -\frac{(1 - e^2)}{(\mu a e^2)^{1/2}} \frac{\partial R}{\partial e} - \frac{2}{na} \frac{\partial R}{\partial a}$$

3. Using a, e, ω, i, Ω, and M_0 $[M = n(t - t_0) + M_0]$, one finds

$$\dot{a} = \frac{2}{na} \frac{\partial R}{\partial M_0}$$

$$\dot{e} = \frac{1 - e^2}{na^2 e} \frac{\partial R}{\partial M_0} - \frac{(1 - e^2)^{1/2}}{na^2 e} \frac{\partial R}{\partial \omega}$$

$$\dot{\omega} = \frac{-\cot i}{L} \frac{\partial R}{\partial i} + \frac{(1 - e^2)^{1/2}}{na^2 e} \frac{\partial R}{\partial e}$$

$$\frac{di}{dt} = \frac{\cot i}{L} \frac{\partial R}{\partial \omega} - \frac{\csc i}{L} \frac{\partial R}{\partial \Omega}$$

$$\dot{\Omega} = \frac{\csc i}{L}\frac{\partial R}{\partial i}$$

$$\dot{M}_0 = -\frac{(1-e^2)}{na^2 e}\frac{\partial R}{\partial e} - \frac{2}{na}\frac{\partial R}{\partial a}$$

4. Using a, e, $\tilde{\omega}$, i, Ω, and $\varepsilon = \tilde{\omega} + \chi$ [$\tilde{\omega} = \omega + \Omega$, $M = nt + \chi$], it can be shown that

$$\dot{a} = \frac{2}{na}\frac{\partial R}{\partial \varepsilon}$$

$$\dot{e} = -\frac{\cot\phi}{na^2}\frac{\partial R}{\partial \tilde{\omega}} - \frac{\cos\phi\,\tan(\phi/2)}{na^2}\frac{\partial R}{\partial \varepsilon}$$

$$\dot{\tilde{\omega}} = \frac{\cot\phi}{na^2}\frac{\partial R}{\partial e} + \frac{\tan(i/2)\sec\phi}{na^2}\frac{\partial R}{\partial i}$$

$$\frac{di}{dt} = -\frac{\csc i\,\sec\phi}{na^2}\frac{\partial R}{\partial \Omega} - \frac{\tan(i/2)\sec\phi}{na^2}\left(\frac{\partial R}{\partial \tilde{\omega}} + \frac{\partial R}{\partial \varepsilon}\right)$$

$$\dot{\Omega} = \frac{\csc i\,\sec\phi}{na^2}\frac{\partial R}{\partial i}$$

$$\dot{\varepsilon} = -\frac{2}{na}\frac{\partial R}{\partial a} + \frac{\cos\phi\,\tan(\phi/2)}{na^2}\frac{\partial R}{\partial e} + \frac{\tan(i/2)\sec\phi}{na^2}\frac{\partial R}{\partial i}$$

5. Using a, e, ω, i, Ω, and M_0 with v (the true anomaly) as the independent variable [$M = n(t - t_0) + M_0$], the results are

$$\frac{da}{dv} = \frac{2}{n^2 a}\left(\frac{r}{a}\right)^2\frac{1}{(1-e^2)^{1/2}}\frac{\partial R}{\partial M_0}$$

$$\frac{de}{dv} = \frac{(1-e^2)^{1/2}}{n^2 a^2 e}\left(\frac{r}{a}\right)^2\frac{\partial R}{\partial M_0} - \frac{1}{n^2 a^2 e}\left(\frac{r}{a}\right)^2\frac{\partial R}{\partial \omega}$$

$$\frac{d\omega}{dv} = -\frac{\cot i}{n^2 a^2(1-e^2)}\left(\frac{r}{a}\right)^2\frac{\partial R}{\partial i} + \frac{1}{n^2 a^2 e}\left(\frac{r}{a}\right)^2\frac{\partial R}{\partial e}$$

$$\frac{di}{dv} = -\frac{\cot i}{n^2 a^2(1-e^2)}\left(\frac{r}{a}\right)^2\frac{\partial R}{\partial \omega} - \frac{\csc i}{n^2 a^2(1-e^2)}\left(\frac{r}{a}\right)^2\frac{\partial R}{\partial \Omega}$$

$$\frac{d\Omega}{dv} = \frac{\csc i}{n^2 a^2(1-e^2)}\left(\frac{r}{a}\right)^2\frac{\partial R}{\partial i}$$

$$\frac{dM_0}{dv} = -\frac{(1-e^2)^{1/2}}{n^2 a^2 e}\left(\frac{r}{a}\right)^2\frac{\partial R}{\partial e} - \frac{2}{n^2 a(1-e^2)^{1/2}}\left(\frac{r}{a}\right)^2\frac{\partial R}{\partial a}$$

6. Small eccentricity; $h = e \sin \tilde{\omega}, k = e \cos \tilde{\omega}$. Only the $\dot{e}, \dot{\tilde{\omega}}$ equations change from set 4 to

$$\dot{h} = \frac{\cos\phi}{(\mu a)^{1/2}}\frac{\partial R}{\partial k} + \frac{k\tan(i/2)\sec\phi}{(\mu a)^{1/2}}\frac{\partial R}{\partial i} - \frac{h\cos\phi\sec^2(\phi/2)}{2(\mu a)^{1/2}}\frac{\partial R}{\partial \varepsilon}$$

$$\dot{k} = -\frac{\cos\phi}{(\mu a)^{1/2}}\frac{\partial R}{\partial h} - \frac{h\tan(i/2)\sec\phi}{(\mu a)^{1/2}}\frac{\partial R}{\partial i} - \frac{k\cos\phi\sec^2(\phi/2)}{2(\mu a)^{1/2}}\frac{\partial R}{\partial \varepsilon}$$

7. Small inclination; $p = \sin i \sin \Omega$, $q = \sin i \cos \Omega$. Only the di/dt, $\dot{\Omega}$ equations change from set 4 to

$$\dot{p} = \frac{\cos i \sec\phi}{(\mu a)^{1/2}}\frac{\partial R}{\partial q} - \frac{p\cos i\sec^2(i/2)\sec\phi}{2(\mu a)^{1/2}}\left(\frac{\partial R}{\partial\tilde{\omega}} + \frac{\partial R}{\partial\varepsilon}\right)$$

$$\dot{q} = -\frac{\cos i \sec\phi}{(\mu a)^{1/2}}\frac{\partial R}{\partial p} - \frac{q\cos i\sec^2(i/2)\sec\phi}{2(\mu a)^{1/2}}\left(\frac{\partial R}{\partial\tilde{\omega}} + \frac{\partial R}{\partial\varepsilon}\right)$$

8. Small inclination; $p = \tan i \sin \Omega$, $q = \tan i \cos \Omega$. Only the di/dt, $\dot{\Omega}$ equations change from set 4 to

$$\dot{p} = \frac{\sec^3 i}{L}\frac{\partial R}{\partial q} - \frac{p\sec i\sec^2(i/2)}{2L}\left(\frac{\partial R}{\partial\tilde{\omega}} + \frac{\partial R}{\partial\varepsilon}\right)$$

$$\dot{q} = -\frac{\sec^3 i}{L}\frac{\partial R}{\partial p} - \frac{q\sec i\sec^2(i/2)}{2L}\left(\frac{\partial R}{\partial\tilde{\omega}} + \frac{\partial R}{\partial\varepsilon}\right)$$

9. Small eccentricity and inclination; $h = e \sin \tilde{\omega}, k = e \cos \tilde{\omega}, p = \sin i \sin \Omega$, $q = \sin i \cos \Omega$. The $\dot{a}$ and $\dot{\varepsilon}$ equations are unchanged from set 4. The rest are transformed into

$$\dot{h} = \frac{\cos\phi}{na^2}\frac{\partial R}{\partial k} + \frac{k\cot i\tan(i/2)\sec\phi}{na^2}\left(p\frac{\partial R}{\partial p} + q\frac{\partial R}{\partial q}\right)$$
$$- \frac{h\cos\phi\sec^2(\phi/2)}{2na^2}\frac{\partial R}{\partial\varepsilon}$$

$$\dot{k} = -\frac{\cos\phi}{na^2}\frac{\partial R}{\partial h} - \frac{h\cot i\tan(i/2)\sec\phi}{na^2}\left(p\frac{\partial R}{\partial p} + q\frac{\partial R}{\partial q}\right)$$
$$- \frac{k\cos\phi\sec^2(\phi/2)}{2na^2}\frac{\partial R}{\partial\varepsilon}$$

$$\dot{p} = \frac{\cos i\sec\phi}{na^2}\frac{\partial R}{\partial q} - \frac{p\cos i\sec^2(i/2)\sec\phi}{2na^2}\left(k\frac{\partial R}{\partial h} - h\frac{\partial R}{\partial k} + \frac{\partial R}{\partial\varepsilon}\right)$$

$$\dot{q} = -\frac{\cos i\sec\phi}{na^2}\frac{\partial R}{\partial p} - \frac{q\cos i\sec^2(i/2)\sec\phi}{2na^2}\left(k\frac{\partial R}{\partial h} - h\frac{\partial R}{\partial k} + \frac{\partial R}{\partial\varepsilon}\right)$$

10.   Small eccentricity; $h = e \sin \omega$, $k = e \cos \omega$ and $\sigma = \omega - nT$.

$$\dot{a} = 2\left(\frac{a}{\mu}\right)^{1/2} \frac{\partial R}{\partial \sigma}$$

$$\dot{h} = \frac{(1-e^2)}{L} \frac{\partial R}{\partial k} - \frac{k \cot i}{L} \frac{\partial R}{\partial i}$$

$$\dot{k} = -\frac{(1-e^2)}{L} \frac{\partial R}{\partial h} + \frac{k \cot i}{L} \frac{\partial R}{\partial i} - \frac{k(1-e^2)}{L[1+(1-e^2)^{1/2}]} \frac{\partial R}{\partial \sigma}$$

$$\frac{di}{dt} = L^{-1}\left[\cot i\left(k \frac{\partial R}{\partial h} - h \frac{\partial R}{\partial k} + \frac{\partial R}{\partial \sigma}\right) - \cos i \frac{\partial R}{\partial \Omega}\right]$$

$$\dot{\Omega} = \frac{\csc i}{L} \frac{\partial R}{\partial i}$$

$$\dot{\sigma} = \frac{1-e^2}{L[1+(1-e^2)^{1/2}]}\left(h \frac{\partial R}{\partial h} + k \frac{\partial R}{\partial k}\right) - 2\left(\frac{a}{\mu}\right)^{1/2} \frac{\partial R}{\partial a} - \frac{\cot i}{L} \frac{\partial R}{\partial i}$$

A Modification.   Without a modification the equations of set 2, in planetary applications, have a defect. The disturbing function will be expanded in a series of periodic terms. The elements a, e, and i appear in the coefficients, and the elements χ, ω, and Ω in the arguments. Element χ always appears with nt in the form $nt + \chi$, and n is a function of a, $n^2 a^3 = \mu$. The orbital element a is therefore present both explicitly in the coefficients and, through n, implicitly in the arguments. The derivative $\partial R/\partial a$ appears in $d\chi/dt$ only. This part of $d\chi/dt$ may be written as

$$-\frac{2}{na} \frac{\partial R}{\partial a} = -\frac{2}{na}\left(\frac{\partial R}{\partial a}\right) - \frac{2}{na} \frac{\partial R}{\partial \chi} t \frac{dn}{da}$$

in which the parentheses designate that part of $\partial R/\partial a$ that arises from the explicit presence of a in the coefficients of the periodic terms. Hence, the time appears as a factor in the coefficients of periodic terms. The expression may be written in the simpler form

$$-\frac{2}{na} \frac{\partial R}{\partial a} = -\frac{2}{na}\left(\frac{\partial R}{\partial a}\right) - t \frac{dn}{da} \frac{da}{dt}$$

$$= -\frac{2}{na}\left(\frac{\partial R}{\partial a}\right) - t \frac{dn}{dt}$$

If this is substituted into the equation for $d\chi/dt$, then

$$\frac{d\chi}{dt} = -\frac{2}{na}\left(\frac{\partial R}{\partial a}\right) - t \frac{dn}{dt} - \frac{(1-e^2)}{na^2 e} \frac{\partial R}{\partial e}$$

The presence of t outside of the trigonometric arguments may be avoided by defining X such that

$$\frac{dX}{dt} = \frac{d\chi}{dt} + t\frac{dn}{dt}$$

The significance of this change can be seen by differentiating the mean anomaly

$$M = nt + \chi$$

$$\frac{dM}{dt} = n + t\frac{dn}{dt} + \frac{d\chi}{dt}$$

$$= n + \frac{dX}{dt}$$

Integrating gives

$$M = \int n\,dt + X$$

This is frequently written as

$$M = \rho + X$$

wherein

$$\frac{d\rho}{dt} = n, \qquad \frac{d^2\rho}{dt^2} = \frac{dn}{dt} = -\frac{3}{2}\frac{n}{a}\frac{da}{dt} = -\frac{3}{a^2}\frac{\partial R}{\partial \chi}$$

The simplest way to achieve the desired form is to consider R a function of a, e, i, M, ω, and Ω. Then a is present in the coefficients only and the parentheses with $\partial R/\partial a$ are unnecessary. The variation of parameters equations are unchanged provided χ is understood to be X, that is, the relation between M and χ is

$$M = \int n\,dt + \chi$$

The partial differential quotient $\partial R/\partial \chi$ may be replaced by (or should be understood to mean) $\partial R/\partial M$, and one has to complete the above set with

$$M = \rho + \chi$$

$$\frac{d^2\rho}{dt^2} = -\frac{3}{a^2}\frac{\partial R}{\partial \chi} = -\frac{3}{a^2}\frac{\partial R}{\partial M}$$

The elimination of the time in the coefficients of trigonometric terms has been achieved at the cost of introducing a double integration for the mean anomaly.

It may appear that two superfluous constants of integration have been introduced, but this is not so. Since ρ and χ appear together in the form $\rho + \chi$, the additive constants in ρ and χ merge. Next,

$$\frac{d\rho}{dt} = n$$

which is the osculating mean motion, related to a by $n^2 a^3 = \mu$. Hence, the second constant of integration in the integration of the $d^2\rho/dt^2$ formula is not free but must be compatible with the constant introduced in the integration of da/dt.

The Remaining Details

When we wrote Eq. 9.7, we assumed that the perturbing forces could be written as the gradient of a velocity-independent scalar $R(r, t)$. This is not always true—consider the case of atmospheric resistance. Gauss developed an extension of Lagrange's planetary equations to deal with the general problem. The standard method is to resolve the perturbing force into three mutually perpendicular components. $\mathcal{R}$ is the component in the direction of the radius vector (positive in the direction of increasing radial distance), $\mathcal{T}$ is the component perpendicular to $\mathcal{R}$ in the orbital plane (positive in the direction of increasing longitude), and $\mathcal{W}$ is the component perpendicular to the orbital plane (positive in the direction in which the orbital motion appears counterclockwise).

Let r be the radius vector, λ the longitude, and z the coordinate perpendicular to the orbital plane (i.e., cylindrical coordinates). Then

$$\mathcal{R} = \frac{\partial R}{\partial r}, \qquad \mathcal{T} = \frac{1}{r}\frac{\partial R}{\partial \lambda}, \qquad \mathcal{W} = \frac{\partial R}{\partial z}$$

Let a be any one of the six orbital elements; then

$$\frac{\partial R}{\partial a} = \frac{\partial R}{\partial r}\frac{\partial r}{\partial a} + \frac{\partial R}{\partial \lambda}\frac{\partial \lambda}{\partial a} + \frac{\partial R}{\partial z}\frac{\partial z}{\partial a}$$

In order to obtain the expressions for $\partial R/\partial a$ in terms of $\mathcal{R}$, $\mathcal{T}$, and $\mathcal{W}$, it is first necessary to find the expressions for the partial derivatives of r, λ, and z in elliptic motion with respect to the elements. Rather than use the a, e, ω, i, Ω, and T set this time it turns out to be more convenient to use the a, e, $\tilde{\omega} = \omega + \Omega$ (longitude of periapse), i, Ω, and $\varepsilon = M + \tilde{\omega} - nT = \tilde{\omega} - nT$ set,

$$\frac{\partial r}{\partial a} = \frac{r}{a}, \qquad \frac{\partial \lambda}{\partial e} = \frac{2 + e \cos v}{1 - e^2}\sin v$$

$$\frac{\partial r}{\partial e} = -a \cos v, \qquad \frac{\partial \lambda}{\partial e} = \frac{a^2}{r^2}(1 - e^2)^{1/2} = \frac{(1 + e \cos v)^2}{(1 - e^2)^{3/2}}$$

$$\frac{\partial r}{\partial \varepsilon} = \frac{ae}{(1 - e^2)^{1/2}}\sin v, \qquad \frac{\partial \lambda}{\partial \tilde{\omega}} = 1 - \frac{\partial \lambda}{\partial \varepsilon}$$

$$\frac{\partial r}{\partial \tilde{\omega}} = -\frac{\partial r}{\partial \varepsilon}, \qquad \frac{\partial \lambda}{\partial \Omega} = -1 + \cos i$$

Also, with $u = v + \omega$,

$$dz = r \sin u \, di - r \cos u \sin i \, d\Omega$$

These relations allow one to calculate

$$\frac{\partial R}{\partial a} = \frac{r}{a}\frac{\partial R}{\partial r}$$

$$\frac{\partial R}{\partial e} = -a\cos v\frac{\partial R}{\partial r} + \left(\frac{1}{1-e^2} + \frac{a}{r}\right)\sin v\frac{\partial R}{\partial \lambda}$$

$$\frac{\partial R}{\partial i} = r\sin u\frac{\partial R}{\partial z}$$

$$\frac{\partial R}{\partial \varepsilon} = \frac{ae\sin v}{(1-e^2)^{1/2}}\frac{\partial R}{\partial r} + \frac{a^2}{r^2}(1-e^2)^{1/2}\frac{\partial R}{\partial \lambda}$$

$$\frac{\partial R}{\partial \tilde{\omega}} = -\frac{ae\sin v}{(1-e^2)^{1/2}}\frac{\partial R}{\partial r} + \left[1 - \frac{a^2}{r^2}(1-e^2)^{1/2}\right]\frac{\partial R}{\partial \lambda}$$

$$\frac{\partial R}{\partial \Omega} = -r\sin i\cos u\frac{\partial R}{\partial z} - 2\sin^2(i/2)\frac{\partial R}{\partial \lambda}$$

From these we deduce

$$\frac{da}{dt} = \frac{2}{n(1-e^2)^{1/2}}\left(\mathcal{R}e\sin v + \mathcal{T}\frac{p}{r}\right)$$

$$\frac{de}{dt} = \frac{(1-e^2)^{1/2}}{na}[\mathcal{R}\sin v + \mathcal{T}(\cos E + \cos v)]$$

$$\frac{di}{dt} = \frac{1}{na^2(1-e^2)^{1/2}}\mathcal{W}r\cos u$$

$$\frac{d\Omega}{dt} = \frac{\csc i}{na^2(1-e^2)^{1/2}}\mathcal{W}r\sin u$$

$$\frac{d\tilde{\omega}}{dt} = \frac{(1-e^2)^{1/2}}{nae}\left[-\mathcal{R}\cos v + \mathcal{T}\left(\frac{r}{p}+1\right)\sin v\right] + 2\frac{d\Omega}{dt}\sin^2(i/2)$$

$$\frac{d\varepsilon}{dt} = -\frac{2r}{na^2}\mathcal{R} + \frac{e^2}{1+(1-e^2)^{1/2}}\frac{d\tilde{\omega}}{dt} + 2(1-e^2)^{1/2}\frac{d\Omega}{dt}\sin^2(i/2)$$

where $p = a(1-e^2)$. Other sets are collected below for reference.

Gaussian Forms.

1. We use a, e, ω, i, Ω, and $\chi = -nT$ $(M = nt + \chi)$ and discover

$$\dot{a} = \frac{2e\sin v}{n(1-e^2)^{1/2}}\mathcal{R} + \frac{2a(1-e^2)^{1/2}}{nr}\mathcal{T}$$

$$\dot{e} = \frac{(1-e^2)^{1/2}\sin v}{na}\mathcal{R} + \frac{(1-e^2)^{1/2}}{na^2 e}\left[\frac{ap}{r} - r\right]\mathcal{T}$$

$$\dot{\omega} = -\frac{(1-e^2)^{1/2} \cos v}{nae} \mathscr{R} + \frac{(1-e^2)^{1/2}}{nae} \sin v \left(1 + \frac{r}{p}\right) \mathscr{T} - \frac{r \sin u \cot i}{L} \mathscr{W}$$

$$\frac{di}{dt} = \frac{r \cos u}{L} \mathscr{W}$$

$$\dot{\Omega} = \frac{r \csc i \sin u}{L} \mathscr{W}$$

$$\dot{\chi} = -\frac{1}{na} \left[\frac{2r}{a} - \frac{(1-e^2) \cos v}{e}\right] \mathscr{R} - \frac{(1-e^2) \sin v}{nae} \left(1 + \frac{r}{p}\right) \mathscr{T}$$

2.   We use a, e, $\tilde{\omega}$, i, Ω, and $\varepsilon = \tilde{\omega} + \chi$ [$\tilde{\omega} = \omega + \Omega$, $M = nt + \chi$] to deduce

$$\dot{a} = \frac{2}{n(1-e)^{1/2}} \left(e\mathscr{R} \sin v + \frac{p\mathscr{T}}{r}\right)$$

$$\dot{e} = \frac{(1-e^2)^{1/2}}{na} \left[\mathscr{R} \sin v + \left(\frac{e + \cos v}{1 + e \cos v} + \cos v\right) \mathscr{T}\right]$$

$$\dot{\tilde{\omega}} = 2\dot{\Omega} \sin^2(i/2) + \frac{(1-e^2)^{1/2}}{nae} \left[-\mathscr{R} \cos v + \left(1 + \frac{r}{p}\right) \mathscr{T} \sin v\right]$$

$$\frac{di}{dt} = \frac{r \cos u}{L} \mathscr{W}$$

$$\dot{\Omega} = \frac{r \csc i \sin u}{L} \mathscr{W}$$

$$\dot{\varepsilon} = -\frac{2r}{na^2} \mathscr{R} + \frac{e^2}{1 + (1-e^2)^{1/2}} \dot{\tilde{\omega}} + 2\dot{\Omega}(1-e^2)^{1/2} \sin^2\left(\frac{i}{2}\right)$$

3.   An alternative form ($e = \sin \phi$) is

$$\dot{a} = 2\left(\frac{a^3}{\mu}\right)^{1/2} [\mathscr{R} \tan \phi \sin v + \mathscr{T} \sec \phi(1 + e \cos v)]$$

$$\dot{e} = \left(\frac{a}{\mu}\right)^{1/2} \cos \phi [\mathscr{R} \sin v + \mathscr{T}(\cos v + \cos E)]$$

$$\dot{\tilde{\omega}} = \frac{\csc \phi \sec \phi}{(\mu a)^{1/2}} [-a\mathscr{R} \cos^2 \phi \cos v + r\mathscr{T}(2 + e \cos v) \sin v$$
$$+ r\mathscr{W} \tan(i/2) \sin \phi \sin u]$$

$$\frac{di}{dt} = \frac{r\mathscr{W} \sec \phi \cos u}{(\mu a)^{1/2}}$$

$$\dot{\Omega} = r\mathscr{W} \csc i \sec \phi \sin u / (\mu a)^{1/2}$$

$$\dot{\varepsilon} = -\frac{2r\mathscr{R}}{(\mu a)^{1/2}} + 2\dot{\tilde{\omega}} \sin^2(\phi/2) + 2\dot{\Omega} \cos \phi \sin^2\left(\frac{i}{2}\right)$$

Also

$$\dot{p} = 2r\left(\frac{a}{\mu}\right)^{1/2} \mathscr{T} \cos \phi$$

$$\dot{n} = -\frac{3 \sec \phi}{a}[\mathscr{R} \sin \phi \sin v + \mathscr{T}(1 + e \cos v)]$$

$$\dot{\chi} = \frac{\csc \phi}{(\mu a)^{1/2}}[(a \cos^2 \phi \cos v - 2r \sin \phi)\mathscr{R} - r\mathscr{T}(2 + e \cos v)]$$

THE MISAPPLICATIONS OF PERTURBATION THEORY

Power Series and the Order of the Theory

Within classical mechanics one deals with equations of motion of the form

$$\mathbf{F} = m\mathbf{a}$$

If $\mathbf{v} = d\mathbf{r}/dt$, then an equivalent first-order system is

$$\mathbf{v} = \frac{d\mathbf{r}}{dt}, \qquad \mathbf{F} = m\frac{d\mathbf{v}}{dt}$$

If $\mathbf{F}$ is conservative, $\mathbf{F} = -\nabla V$, then we can rewrite these as

$$\mathbf{v} = \frac{d\mathbf{r}}{dt}, \qquad m\frac{d\mathbf{v}}{dt} = -\nabla V$$

Whenever V is such that this equation cannot be explicitly solved, it is normal practice to separate it into two parts. One part, represented by U, is such that these equations of motion are exactly soluble, whereas the other part (represented by $-R$ in the standard sign convention) makes the problem intractable. The solution of

$$\mathbf{v} = d\mathbf{r}/dt, \qquad m\, d\mathbf{v}/dt = -\nabla U$$

has six arbitrary constants associated with it, say $\mathbf{a}$. The parameterization of the solution, $\mathbf{r} = \mathbf{r}(\mathbf{a}, t)$, is then used, as discussed above, to formulate an equation for $\dot{\mathbf{a}}$. Since $|\nabla R|$ is small (in some sense), we can write ∇R as $\varepsilon \mathbf{f}$ where ε is a small parameter. One further expects that the changes in the norm of $\mathbf{a}$ will be small too. Hence, we can keep the concept of an ellipse whose size, shape, and orientation may be slowly changing with time. Should it happen that these changes are periodic, using the coordinates of the osculating ellipse will prove to be a smart choice indeed. Before becoming too optimistic, let us examine the structure of the variation of parameters equations. Thus, we are led to study the solution of equations of the form (or systems of such equations)

$$\frac{da}{dt} = \varepsilon f(t, a), \qquad a(t_0) = a_0 \tag{9.13}$$

The development below is independent of the dimensionality of a and it is presented for the one-dimensional case.

We know of one rigorous method of approximately solving a set of ordinary differential equations—Picard's method of successive substitutions (the mathematical details were presented in Chapter One). One constructs the sequence $\{a_n(t)\}$ for Eq. 9.13, where

$$a_n(t) = a_0 + \varepsilon \int_{t_0}^{t} f[t', a_{n-1}(t')] \, dt', \qquad n = 1, 2, \ldots$$

The sequence $\{a_n - a_{n-1}\}$ converges uniformly and absolutely to a function $a(t)$ that is continuous and satisfies Eq. 9.13. In addition, the solution of Eq. 9.13 is unique and stable with respect to volume perturbations (e.g., small changes in f) and surface perturbations (e.g., small changes in a_0). That's it in a nutshell.

Actually carrying out the Picard process is enormously difficult in general. To my knowledge it has never been carried out beyond $n = 1$ in celestial mechanics. The reasons for this are twofold: The orbital element set is an extraordinarily poor reference frame that exploits none of the symmetries of R (or U) and the interesting problems of celestial mechanics (atmospheric drag, oblateness perturbations, and third-body effects) are not simple. In an effort to deal with these difficulties other forms of approximation have been invented. One is to try and obtain a solution to Eq. 9.13 as a power series in ε. For instance, one would calculate a_1 correctly via

$$A_1(t) = a_1(t) = a_0 + \varepsilon \int_{t_0}^{t} f(t_1, a_0) \, dt_1$$

but instead of computing a_2 from the rigorous expression

$$a_2(t) = a_0 + \varepsilon \int_{t_0}^{t} f[t_1, a_1(t_1)] \, dt_1$$

we replace a_1 by its expression,

$$a_2(t) = a_0 + \varepsilon \int_{t_0}^{t} f\left[t_1, a_0 + \varepsilon \int_{t_0}^{t_1} f(t_2, a_0) \, dt_2 \right] dt_1$$

and then expand f in a power series in ε $[\partial f(t, a) = \partial f(u, v)/\partial v|_{u=t, v=a}]$

$$a_2(t) \simeq a_0 + \varepsilon \int_{t_0}^{t} \left\{ f(t_1, a_0) + \varepsilon \, \partial f(t_1, a_0) \left[\int_{t_0}^{t_1} f(t_2, a_0) \, dt_2 \right] \right\} dt_1$$

Therefore,

$$A_2(t) = a_0 + \varepsilon \int_{t_0}^{t} f(t_1, a_0) \, dt_1 + \varepsilon^2 \int_{t_0}^{t} \partial f(t_1, a_0) \left[\int_{t_0}^{t_1} f(t_2, a_0) \, dt_2 \right] dt_1$$

If one continues this policy to higher orders—expanding f at each step so that for A_3 we start with the rigorous expression for a_3

$$a_3(t) = a_0 + \varepsilon \int_{t_0}^{t} f[t_1, a_2(t_1)] \, dt_1$$

and then approximate a_2, and so on; namely,

$$a_3 = a_0 + \varepsilon \int_{t_0}^{t} f\left\{ t_1, a_0 + \varepsilon \int_{t_0}^{t_1} f[t_2, a_1(t_2)] \, dt_2 \right\} dt_1$$

$$\approx a_0 + \varepsilon \int_{t_0}^{t} \left(f(t_1, a_0) + \varepsilon \, \partial f(t_1, a_0) \left\{ \int_{t_0}^{t_1} f[t_2, a_1(t_2)] \, dt_2 \right\} \right) dt_1$$

$$= a_0 + \varepsilon \int_{t_0}^{t} f(t_1, a_0) \, dt_1 + \varepsilon^2 \int_{t_0}^{t} \partial f(t_1, a_0)$$

$$\times \left\{ \int_{t_0}^{t_1} f\left[t_2, a_0 + \varepsilon \int_{t_0}^{t_2} f(t_3, a_0) \, dt_3 \right] dt_2 \right\} dt_1$$

$$\approx a_0 + \varepsilon \int_{t_0}^{t} f(t_1, a_0) \, dt_1 + \varepsilon^2 \int_{t_0}^{t} \partial f(t_1, a_0)$$

$$\times \left(\int_{t_0}^{t_1} \left\{ f(t_2, a_0) + \varepsilon \, \partial f(t_2, a_0) \left[\int_{t_0}^{t_2} f(t_3, a_0) \, dt_3 \right] \right\} dt_2 \right) dt_1$$

Thus,

$$A_3 = a_0 + \varepsilon \int_{t_0}^{t} f(t_1, a_0) \, dt_1 + \varepsilon^2 \int_{t_0}^{t} \partial f(t_1, a_0) \left[\int_{t_0}^{t_1} f(t_2, a_0) \, dt_2 \right] dt_1$$

$$+ \varepsilon^3 \int_{t_0}^{t} \partial f(t_1, a_0) \left\{ \int_{t_0}^{t_1} \partial f(t_2, a_0) \left[\int_{t_0}^{t_2} f(t_3, a_0) \, dt_3 \right] dt_2 \right\} dt_1$$

$$= A_2 + \varepsilon^3 \text{ term}$$

or, in general,

$$A_n(t) = A_{n-1} + \varepsilon^n \int_{t_0}^{t} \partial f(t_1, a_0) \cdots \left(\int_{t_0}^{t_{n-3}} \partial f(t_{n-2}, a_0) \left\{ \int_{t_0}^{t_{n-2}} \partial f(t_{n-1}, a_0) \right. \right.$$

$$\left. \left. \times \left[\int_{t_0}^{t_{n-1}} f(t_n, a_0) \, dt_n \right] dt_{n-1} \right\} dt_{n-2} \right) \cdots dt_1$$

Then one can prove that the sequence $\{A_n\}$ converges absolutely and uniformly to a continuous function of t that satisfies the initial condition. The series *does not* satisfy the differential equation. Hence it is useless.

If instead of substituting, expanding, substituting, and so on, one substitutes and then expands, one gets a different result (beyond a_2 or A_2). For instance, we return to the exact expression for a_3,

$$a_3(t) = a_0 + \varepsilon \int_{t_0}^{t} f[t_1, a_2(t_1)]\, dt_1$$

$$= a_0 + \varepsilon \int_{t_0}^{t} f\left\{ t_1, a_0 + \varepsilon \int_{t_0}^{t_1} f[t_2, a_1(t_2)]\, dt_2 \right\} dt_1$$

$$= a_0 + \varepsilon \int_{t_0}^{t} f\left\{ t_1, a_0 + \varepsilon \int_{t_0}^{t_1} f\left[t_2, a_0 + \varepsilon \int_{t_0}^{t_2} f(t_3, a_0)\, dt_3 \right] dt_2 \right\} dt_1$$

This is still rigorous. Now we expand from the inside outward

$$\tilde{A}_3(t) \simeq a_0 + \varepsilon \int_{t_0}^{t} f\Bigg(t_1, a_0$$

$$+ \varepsilon \int_{t_0}^{t_1} \left\{ f(t_2, a_0) + \varepsilon\, \partial f(t_2, a_0)\left[\int_{t_0}^{t_2} f(t_3, a_0)\, dt_3 \right] \right\} dt_2 \Bigg) dt_1$$

$$\simeq a_0 + \varepsilon \int_{t_0}^{t} f\Bigg\{ t_1, a_0 + \varepsilon \int_{t_0}^{t_1} f(t_2, a_0)\, dt_2$$

$$+ \varepsilon^2 \int_{t_0}^{t_1} \partial f(t_2, a_0)\left[\int_{t_0}^{t_2} f(t_3, a_0)\, dt_3 \right] dt_2 \Bigg\} dt_1$$

$$\simeq a_0 + \varepsilon \int_{t_0}^{t} \Bigg(f(t_1, a_0)$$

$$+ \left\{ \varepsilon \int_{t_0}^{t_1} f(t_2, a_0)\, dt_2 + \varepsilon^2 \int_{t_0}^{t_1} \partial f(t_2, a_0) \right.$$

$$\times \left. \left[\int_{t_0}^{t_2} f(t_3, a_0)\, dt_3 \right] dt_2 \right\}$$

$$\times \partial f(t_1, a_0) + \frac{\varepsilon^2}{2}\left[\int_{t_0}^{t_1} f(t_2, a_0)\, dt_2 \right]^2 \partial^2 f(t_1, a_0) \Bigg) dt_1$$

through all terms of order ε^2 $[\partial^2 f(t, a) = \partial^2 f(u, v)/\partial v^2|_{u=t,v=a}]$. This is not equal to a_3 or A_3, but the technique represents a perfectly acceptable approximation scheme (on the surface). Note too that the A_n expansion only required that $\partial f/\partial a$ exists but the $\tilde{A}_n$ one will require that all partial derivatives of f with respect to a exist and be continuous. Most of the "second-order" approximations in celestial mechanics are of one of the incorrect forms exhibited above. They are *not* rigorous nor are they successful. As with the f and g series and Gauss's method of orbit determination, this is one case where the mathematics must come before the physics and not the other way around.

Let us step back from the mathematical details of the fray and adopt the usual physicist's point of view. A new physical theory is formulated that purports to be a better, simpler, and more complete description of nature. We physicists rush to try it out by interpreting old experiments, devising new ones, or tackling novel problems. Frequently, we come to a complex problem beyond our analytical (or numerical) capabilities to solve. Perhaps this disputed point represents a critical test of the new formalism so that a solution, even an approximate one, is desirable. Historically, at such junctures in the development of or refinement of new physical theories, the scientific community rushes ahead using an intuitively obvious, attractive, or reasonable perturbation technique. As long as only a first-order result is sought, the correct answer has usually been obtained, an experiment is explained, a result predicted, or a burgeoning edifice falls. (Only if you had a good acquaintance with the history of physics could you name more than a few of these—phlogiston, the ether, Weber's electrodynamics, etc.) Later, sometimes much later, the mathematical physicists or mathematicians move in, clean up the heuristic perturbation theory, and progress to the second or higher order can be made. The perturbation theory developed for quantum electrodynamics would serve as an excellent illustration of this type of solidifying, t-crossing and i-dotting stage of refinement. (But contrast this with the current status of quantum chromodynamics.) In over three hundred years this has not yet happened, in a computationally useful fashion, for celestial mechanics.

The Method of Averaging

Return to Eq. 9.11a. When we replace $\mathbf{r}$ and $\mathbf{v}$ by their Keplerian formulas, then on the right-hand side there are two types of functional dependence upon the time. One type is explicit due to any time dependence of the perturbing force (think of radiation pressure as the disturbing force acting on an artificial satellite that passes in and out of the Earth's shadow) and the (former) appearance of $\mathbf{r}$ and $\mathbf{v}$. The other type of time dependence is implicit since $\mathbf{a}$ itself is time dependent. Now in many cases of interest the explicit time dependence of $\mathbf{f}$ is periodic, with the same period as the object's revolution period about a primary (P). This is true, for instance, when discussing perturbations due to the Earth's oblateness for artificial satellites. Because the rate of change of $\mathbf{a}$ is small by hypothesis, many revolutions may occur before $|\int \mathbf{f}(\mathbf{a}, t)\, dt|$ is appreciable. Hence, this arguments runs, replace the solution of Eq. 9.11a with the following pair

$$\frac{d\mathbf{a}_{\text{sec}}}{dt} = \left\langle \frac{d\mathbf{a}}{dt} \right\rangle = \frac{1}{P} \int_0^P \mathbf{f}(\mathbf{a}, \tau)\, d\tau \tag{9.14}$$

$$\frac{d\mathbf{a}_{\text{per}}}{dt} = \frac{d\mathbf{a}}{dt} - \frac{d\mathbf{a}_{\text{sec}}}{dt} = \mathbf{f}(\mathbf{a}, t) - \frac{1}{P} \int_0^P \mathbf{f}(\mathbf{a}, \tau)\, d\tau \tag{9.15}$$

where it is understood that in the integrands $\mathbf{a}$ is to be regarded as a constant.

The averaged equations of motion Eq. 9.14 provides the long-term or secular changes in **a** whereas Eq. 9.15 yields the short-term or periodic changes in **a**.

If some of the elements of **a** change on a time scale of $\sim P/|\varepsilon|$, for a perturbation of "magnitude" or "strength" $|\varepsilon| \ll 1$, then by averaging yet again over this period one separates the short-term periodic variations from the long-term ones $(\sim P/|\varepsilon|^2)$. Such theoretical structures are known as doubly averaged theories. Kozai (1959) exploits the method of averaging in order to solve the main problem of artificial satellite theory (see Chapter Ten for details). Various developments of the method of averaging (Lorell, Anderson, and Lass 1964; Lorell and Liu 1971; McClain 1977) all refer back to Kryloff and Bogoliubov (1937) and Bogoliubov and Mitropolsky (1961) for a theoretical basis. If this method has a rigorous foundation, I cannot find it. As an example consider the following (Bogoliubov and Mitropolsky 1961, p. 41):

> We note that in the case represented by equation (1.1) we might establish the convergence of expansions (1.2), (1.3) under very general conditions for the function $f(x, dx/dt)$.
>
> However, since in the future we will have to deal with cases in which similar expansions apparently diverge, we will not tie up the development of our method of the construction of asymptotic approximations with any proof of convergence.

I find it difficult to be so cavalier.

Arnold (1978) discusses this topic from a more sophisticated mathematical point of view than can be presented here. The essential result (p. 292) is

> We note that this principle [that of averaging] is neither a theorem, an axiom, nor a definition, but rather a physical proposition, i.e., a *vaguely* [emphasis added] formulated and, strictly speaking, untrue assertion. Such assertions are often fruitful sources of mathematical theorems.

But not yet in this case.

Editorial

Beyond first-order results I know of no useful result from perturbation theory in celestial mechanics because all of the higher-order results have no firm mathematical basis. Frequently the second approximation produces nonsensical results (the critical inclination, small divisors, etc.). There is no good analytical long-term perturbation theory. Why? I believe the problem is simply stated: Classical mechanics is a *trajectory* theory. Return to pages 1 and 2 and look at the central element under discussion; it is $\mathbf{r}(t)$. I have no doubt about Newton's laws of motions or his law of gravitation. Rather there is no efficient method of dealing with nontrivial gravitational problems.

What would be better? A *field* theory. More precisely a linear field theory (think of the Schrödinger equation, Poisson's equation, the wave equation, etc.). A second-order linear field theory. With such a theoretical basis perturbation theory reduces to eigenfunction expansions of Riemann's P equation (see Whittaker and Watson 1902). There is a rigorous, multifaceted approach to such problems (see, e.g., Morse and Feshbach 1953).

There is a field theoretic approach to classical mechanics known as Hamilton–Jacobi theory. I have purposefully avoided it in this book for two reasons. First, its introduction requires a nontrivial amount of space and the introduction of a whole new set of terms (Lagrangians, action angle variables, Hamiltonians, canonical transformations, Poisson brackets, etc.). Second, I did not see anything practical (in the applications sense) following its presentation. The Hamilton–Jacobi equation is nonlinear and it is first order in the space (and time) variables. Thus no luck here.

There is a more sophisticated approach to mechanics in general known as quantum mechanics. It is a field theory, it is linear, and it is second order (in the space variables). Furthermore, the correspondence principle assures us that in the limit of large quantum numbers (or as Planck's constant approaches zero) one will recover classical mechanics. Therefore, why not solve within the context of quantum mechanics the main problem of artificial satellite theory (i.e., an external field with quadrupole moments) or the three-body problem (i.e., the helium atom minus the Pauli exclusion principle)? Then take the large quantum number limit and see what you get. I believe that this course of action is doomed to failure because the zeroth-order approximation to the Schrödinger equation is the Hamilton–Jacobi equation (this deep connection is not an accident; see Goldstein (1980) for more). And the Hamilton–Jacobi equation is nonlinear and first order.

Not having a method that allows the solution of complex problems (or even simple problems) is no excuse for using unjustified techniques.

chapter ten

More on Perturbation Theory

This chapter has two very different parts. The first concentrates on a deeper investigation into the perturbations of an artificial satellite's motion due to the oblateness effects of the Earth, the third-body perturbations owing to the Sun and the Moon, and the frictional effects of atmospheric drag. These discussions are as complete as can be found in a variety of disparate reference sources. The concentration will be on the artificial satellite problem because the planetary problem is already discussed at length in the standard texts. Also, the artificial satellite problem is of immediate computational interest to a larger audience. The last section of this chapter discusses some misapplications of perturbation theory via the method of averaging or the improper use of power series expansions. This is done by using as a model the perturbed simple harmonic oscillator. Both anisotropic and anharmonic perturbations are discussed. Since these perturbed systems are exactly soluble in closed form, the pitfalls and shortcomings of several of the usual mathematical techniques utilized in perturbation theory are evident. The deep connection between the isotropic, three-dimensional simple harmonic oscillator and the Keplerian two-body problem was discussed in Chapter Six.

OBLATENESS PERTURBATIONS

Equation 1.25 gives the value of the potential through J_2,

$$U = -\frac{GM}{r}\left(1 - \frac{J_2 P_2(\sin \delta)}{r^2}\right) \qquad (1.25)$$

where δ is the declination. Now the effects of the perturbing potential

$$R = \frac{GMJ_2}{2r^3}(1 - 3\sin^2 \delta)$$

will be investigated on the orbital element set. The effects of higher-order terms in the potential will also be discussed. This will include both axially symmetric terms (e.g., J_3, J_4) and nonaxially symmetric terms (e.g., J_{21}, J_{22}) especially with regard to equatorial orbits. Kozai's (1959) analysis will be treated in depth because it is a widely used result and serves as an excellent straw man for the second part of this chapter. First, the first-order variations in the orbital element set due to R will be computed.

Perturbations Owing to J_2

The orbital element set used is a, e, ω, i, Ω, and T. Then, from set 1 of the Lagrangian forms in Chapter Nine,

$$\frac{da}{dt} = -\left(\frac{2a^2}{\mu}\right)\frac{\partial R}{\partial T}$$

$$\frac{de}{dt} = -\frac{a(1-e^2)}{\mu e}\frac{\partial R}{\partial T} - \left(\frac{1-e^2}{\mu ae^2}\right)^{1/2}\frac{\partial R}{\partial \omega}$$

$$\frac{di}{dt} = \frac{\csc i}{L}\left(\cos i\frac{\partial R}{\partial \omega} - \frac{\partial R}{\partial \Omega}\right)$$

$$\frac{d\Omega}{dt} = \frac{\csc i}{L}\frac{\partial R}{\partial i} \qquad\qquad (10.1)$$

$$\frac{d\omega}{dt} = \left(\frac{1-e^2}{\mu ae^2}\right)^{1/2}\left(\frac{\partial R}{\partial e} - \frac{e\cot i}{1-e^2}\frac{\partial R}{\partial i}\right)$$

$$\frac{dT}{dt} = \frac{2a^2}{\mu}\frac{\partial R}{\partial a} + \frac{a(1-e^2)}{\mu e}\frac{\partial R}{\partial e}$$

where $n^2a^3 = \mu = GM$. r and δ in the perturbing potential are replaced by their expressions in terms of the usual variables,

$$\sin\delta = \sin i \sin u, \qquad u = v + \omega$$

$$r = \frac{a(1-e^2)}{1+e\cos v} = a(1-e\cos E)$$

The next step is tedious analysis. We actually need to compute the indicated derivatives of R and, holding $a, \ldots, T$ constant on the right-hand sides of Eqs. 10.1, integrate with respect to the time to obtain the first-order approximations to the instantaneous values of $a, \ldots, T$. Note that what makes these first-order approximations is *not* the appearance of J_2 to the first order but rather the holding to $a, \ldots, T$ as constants.

(Because my secretary would never stand for it if I tried to reproduce all of these computations I will only outline the computation of the first-order changes.)

First-Order Changes

Starting with the equation for da/dt we need $\partial R/\partial T$,

$$\frac{\partial R}{\partial T} = -\frac{3\mu J_2}{2r^4}(1 - 3\sin^2\delta)\frac{\partial r}{\partial T} - \frac{3\mu J_2}{r^3}\sin\delta\cos\delta\frac{\partial\delta}{\partial T}$$

$$= \frac{3\mu nae J_2 \sin v}{2(1-e^2)^{1/2}r^4}(1 - 3\sin^2\delta) + \frac{3\mu na^2(1-e^2)^{1/2}}{r^5}J_2\sin^2 i \sin u \cos u$$

Next da/dt is integrated with respect to t from $t = T$ (perigee passage) to $t = T + P$ ($P = $ period $= 2\pi/n$). It is simpler to do this after a change of variable from the time to the true anomaly,

$$\int_{t_0}^{t} f(t')\,dt' = \int_{v_0}^{v} f[t(v')]\frac{dt(v')}{dv'}\,dv' = \int_{v_0}^{v} \frac{r^2(v')f[t(v')]}{L}\,dv'$$

where $L^2 = \mu a(1 - e^2)$. The result is

$$\Delta a = \int_{t_0}^{t} \frac{da(t')}{dt'}\,dt' = \frac{\mu J_2}{a(1-e^2)^3}(1 + e\cos v)^3(1 - 3\sin^2 i \sin^2 u)\big|_{v_0}^{v}$$

Hence, if $v_0 = 0$, $v = 2\pi$, then (cf. Eq. 9.1)

$$(\Delta a)_{\text{sec}} = 0$$

Turning to the eccentricity,

$$\frac{de}{dt} = -\frac{a(1-e^2)}{\mu e}\frac{\partial R}{\partial T} - \left(\frac{1-e^2}{\mu a e^2}\right)^{1/2}\frac{\partial R}{\partial\omega}$$

$$= \frac{1-e^2}{2ea}\frac{da}{dt} - \left(\frac{1-e^2}{\mu a e^2}\right)^{1/2}\frac{\partial R}{\partial\omega}$$

Since

$$\frac{\partial R}{\partial\omega} = -\frac{3\mu J_2}{r^3}\sin^2 i \sin u \cos u = -\frac{3\mu J_2}{2r^3}\sin^2 i \frac{\partial}{\partial v}(\sin^2 u)$$

the evaluation of Δe is relatively straightforward;

$$\Delta e = \left(\frac{1-e^2}{2ea}\right)\Delta a + \frac{3\mu J_2 \sin^2 i}{2ea^2(1-e^2)}\left[\sin^2 u + \frac{2e}{3}(\sin\omega\sin^3 u - \cos\omega\cos^3 u)\right]\Bigg|_{u_0}^{u}$$

where $u_0 = v_0 + \omega$, $u = v + \omega$. For $v_0 = 0$, $v = 2\pi$ (as in Eq. 9.4)

$$(\Delta e)_{\text{sec}} = 0$$

The simplest of the orbital elements to deal with next is inclination,

$$\frac{di}{dt} = \frac{\csc i}{L}\left(\cos i \frac{\partial R}{\partial\omega} - \frac{\partial R}{\partial\Omega}\right)$$

But $\partial R/\partial\Omega = 0$ because of the axial symmetry of U and

$$\frac{\partial R}{\partial\omega} = -\left(\frac{\mu ae^2}{1-e^2}\right)^{1/2}\left[\frac{de}{dt} - \frac{(1-e^2)}{2ea}\frac{da}{dt}\right]$$

so

$$\Delta i = -\frac{e\cot i}{1-e^2}\left[\Delta e - \frac{(1-e^2)}{2ea}\Delta a\right]$$

$$= -\frac{3\mu J_2 \sin i \cos i}{2a^2(1-e^2)^2}\left[\sin^2 u + \frac{2e}{3}(\sin\omega\sin^3 u - \cos\omega\cos^3 u)\right]\Bigg|_{u_0}^{u}$$

In particular (cf. Eq. 9.2)

$$\Delta i)_{\text{sec}} = 0$$

Another relatively straightforward element to deal with is the longitude of the ascending node,

$$\frac{d\Omega}{dt} = \frac{\csc i}{L}\frac{\partial R}{\partial i}, \qquad \frac{\partial R}{\partial i} = -\frac{3\mu J_2}{r^3}\sin i \cos i \sin^2 u$$

After integration,

$$\Delta\Omega = -\frac{3\mu J_2 \cos i}{a^2(1-e^2)^2}\left[\frac{u-\sin u\cos u}{2} + \frac{e}{3}(\cos\omega\sin^3 u + \sin\omega\cos^3 u)\right.$$

$$\left. - e\sin\omega\cos u\right]\Bigg|_{u_0}^{u}$$

or, as in Eq. 9.3,

$$\Delta\Omega)_{\text{sec}} = -\frac{3\pi\mu J_2 \cos i}{a^2(1-e^2)^2}$$

Turning next to the change in the argument of perigee

$$\frac{d\omega}{dt} = \left(\frac{1-e^2}{\mu ae^2}\right)^{1/2}\left(\frac{\partial R}{\partial e} - \frac{e\cot i}{1-e^2}\frac{\partial R}{\partial i}\right)$$

$$= \left(\frac{1-e^2}{\mu ae^2}\right)^{1/2}\left(\frac{\partial R}{\partial e} - \frac{e\cos i}{1-e^2}\frac{d\Omega}{dt}\right)$$

so

$$\Delta\omega + \Delta\Omega\cos i = \left(\frac{1-e^2}{\mu ae^2}\right)^{1/2}\int_{t_0}^{t}\frac{\partial R(t')}{\partial e}\,dt'$$

Now

$$\frac{\partial R}{\partial e} = \frac{3\mu aJ_2}{2r^4}(1-3\sin^2\delta)\cos v - \frac{3\mu aJ_2\sin^2 i}{r^4(1-e^2)}\left(\frac{r}{a}+1-e^2\right)\sin v\sin u\cos u$$

and the resulting expression is

$$\Delta\omega + \Delta\Omega \cos i = \frac{3\mu J_2}{2a^2 e(1-e^2)^2}$$

$$\times\left(-\sin v + e(v+\sin v \cos v) + e^2\left(\sin v - \frac{\sin^3 v}{3}\right) - 3\sin^2 i\right.$$

$$\times\left\{\sin^2 \omega\left[\sin v - \frac{\sin^3 v}{3} + \left(\frac{e}{16}\right)(12v + 8\sin 2v + \sin 4v)\right.\right.$$

$$\left. + \left(\frac{e^2}{240}\right)(150\sin v + 25\sin^3 v + 3\sin 5v)\right]$$

$$+ 2\sin\omega \cos\omega\left(\frac{-\cos^3 v}{3} - \frac{e}{2}\cos^4 v - \frac{e^2}{3}\cos^5 v\right)$$

$$+ \cos^2 \omega\left[\frac{\sin^3 v}{3} + \left(\frac{e}{4}\right)\left(v - \frac{\sin 4v}{4}\right)\right.$$

$$\left.\left. + \left(\frac{e^2}{15}\right)(3\sin^3 v \cos^2 v + 2\sin^3 v)\right]\right\}$$

$$-2\sin^2 i\left\{(\cos^2\omega - \sin^2\omega)\left[\frac{\sin^3 v}{3} + \left(\frac{e}{8}\right)\left(v - \frac{\sin 4v}{4}\right)\right]\right.$$

$$\left. + \sin\omega \cos\omega\left(\cos v - \frac{4\cos^3 v}{3}\right) - \left(\frac{e}{4}\right)(\sin 4v + \cos^4 v)\right\}$$

$$-2\sin^2 i\left\{(\cos^2\omega - \sin^2\omega)\left[\frac{\sin^3 v}{3} + \left(\frac{e}{4}\right)\left(v - \frac{\sin 4v}{4}\right)\right.\right.$$

$$\left. + \left(\frac{e^2}{15}\right)(3\sin^3 v \cos^2 v + 2\sin^3 v)\right]$$

$$+ \sin\omega \cos\omega\left[\cos v - \frac{2\cos^3 v}{3} - \left(\frac{e}{2}\right)(\sin^4 v + \cos^4 v)\right.$$

$$\left.\left.\left. - \left(\frac{e^2}{15}\right)(5\cos^3 v + 6\cos^5 v)\right]\right\}\right)\Big|_{v_0}^{v}$$

When $v_0 = 0$, $v = 2\pi$, this reduces to (as in Eq. 9.5)

$$\Delta\omega)_{sec} = \frac{3\pi\mu J_2(5\cos^2 i - 1)}{2a^2(1-e^2)^2}$$

Finally we come to T and restrict the discussion to the secular part (see below for an explanation):

$$\frac{dT}{dt} = \frac{2a^2}{\mu}\frac{\partial R}{\partial a} + \frac{a(1-e^2)}{\mu e}\frac{\partial R}{\partial e}$$

But from the above result

$$(\Delta\omega + \Delta\Omega \cos i)_{\text{sec}} = \left(\frac{1-e^2}{\mu a e^2}\right)^{1/2} \int_T^{T+P} \frac{\partial R(t')}{\partial e}\, dt'$$

$$= \frac{3\pi\mu J_2}{2a^2(1-e^2)^2}(2-3\sin^2 i)$$

so

$$\Delta T)_{\text{sec}} = \frac{2a^2}{\mu}\int_T^{T+P} \frac{\partial R(t')}{\partial a}\, dt' + \frac{3\pi\mu J_2(2-3\sin^2 i)}{2na^2(1-e^2)^{3/2}}$$

Now

$$\frac{\partial R}{\partial a} = -\frac{3\mu J_2}{2r^4}(1-3\sin^2\delta)\frac{\partial r}{\partial a} - \frac{3\mu J_2}{r^3}\sin^2 i\frac{\partial u}{\partial a}\sin u \cos u$$

$$= -\frac{3\mu J_2}{2r^4}(1-3\sin^2\delta)\left[\frac{r}{a} - 3ae\frac{(E-e\sin E)}{2r}\sin E\right]$$

$$+ \frac{9\mu a(1-e^2)^{1/2}}{2r^5}J_2\sin^2 i \sin u \cos u(E-e\sin E)$$

After much computation, one finds that (cf. Eq. 9.6)

$$\Delta T)_{\text{sec}} = -\frac{3\pi\mu J_2(1-3\sin^2 i \sin^2\omega)}{na^2(1-e)^3}$$

The analytical difficulty of calculating $\Delta T(t)$ should be apparent to the reader from the form of $\partial R/\partial a$—it is in the mixture of E terms and trigonometric functions of E, especially the $(1-e\cos E)^{-7}$ term—definite integrals of such expressions come much easier than do indefinite integrals. Part of the problem can be alleviated by a change of variable from t directly to E and more can be mitigated by the use of nT or M as the independent variable. Geyling and Westerman (1971) discuss this problem (pp. 164–175) and give explicit expressions in terms of v. Ehricke's second volume (1962) gives alternative forms (pp. 172–182) and yet more are in Sterne's (1960) text (pp. 116–127).

Equatorial Oblateness

A more general expression than Eq. 1.25 for U is the form given in Eq. 1.22. One should remember that in the generalized spherical harmonic expansion for the potential, J_2 is really J_{20}, and that there are longitude (e.g., right ascension) dependent terms arising from J_{21} and J_{22}. In this section we want to consider these terms or a potential of the form

$$U = -\frac{GM}{r}\left[1 + \frac{J_2}{2r^2}(1-3\sin^2\delta) + \frac{J_{22}}{2r^2}\cos^2\delta\cos 2\lambda\right]$$

where λ is the geographic longitude measured from the major axis of the elliptical equatorial cross section. $|J_{22}|$ is $\sim J_2^2$ and the zero of λ is $\sim 35°$ W. In order to investigate the motion of an artificial satellite, it must be remembered that the Earth rotates underneath an inertially fixed point. Hence, in an inertial reference frame (as throughout most of this book the motion of the center of mass of the Earth both about the Earth–Moon barycenter and the revolution of the Earth–Moon barycenter about the solar system barycenter are neglected)

$$U = -\frac{GM}{r}\left[1 + \frac{J_2}{2r^2}(1 - 3\sin^2\delta) + \frac{J_{22}}{2r^2}\cos^2\delta\cos 2(\alpha - \omega t) \right]$$

where ω is the sidereal rotation rate of the Earth.

The equations of motion are

$$\ddot{r} - r\dot{\delta}^2 - r\dot{\alpha}^2\cos^2\delta = -\frac{GM}{r^2} + \frac{3GMJ_2}{2r^4}(3\sin^2\delta - 1)$$

$$-\frac{3GMJ_{22}}{2r^4}\cos^2\delta\cos 2(\alpha - \omega t)$$

$$\frac{d}{dt}(r^2\dot{\alpha}\cos^2\delta) = -\frac{GMJ_{22}}{r^3}\cos^2\delta\sin 2(\alpha - \omega t)$$

$$\frac{d}{dt}(r^2\dot{\delta}) - r^2\dot{\alpha}^2\sin\delta\cos\delta = \frac{GM}{r^3}\sin\delta\cos\delta[3J_2 + J_{22}\cos 2(\alpha - \omega t)]$$

Clearly it will facilitate things if t is the mean sidereal time rather than Ephemeris Time so we assume this to be the case. The net effect is to absorb the factor 1.0027379093 into the constants. Now we make the following change of variables

$$\tau = \omega t, \qquad \psi = \alpha - N\omega t$$

where N is a constant yet to be chosen. Also we set $a^3 = GM/\omega^2$. Then, with primes denoting derivatives with respect to τ,

$$r'' - r(\delta')^2 - (N + \psi')^2 r\cos^2\delta = -\frac{a}{r^2} + \frac{3aJ_2}{2r^4}(3\sin^2\delta - 1)$$

$$-\frac{3aJ_{22}}{2r^4}\cos^2\delta\cos 2[\psi + (N - 1)\tau]$$

$$[(N + \psi')r^2\cos^2\delta]' = -\frac{aJ_{22}}{r^3}\cos^2\delta\sin 2[\psi + (N - 1)\tau]$$

$$(r^2\delta')' - r^2(N + \psi')^2\sin\delta\cos\delta = \frac{a\sin\delta\cos\delta}{r^3}$$

$$\times \left\{ 3J_2 + J_{22}\cos 2[\psi + (N - 1)\tau] \right\}$$

We further specialize to a nearly circular, nearly equatorial orbit. If the radius of the orbit is R, then $\rho = r - R$ will be a small quantity as will ψ and δ (the former because we can choose N such that $\langle \dot{\psi} \rangle = 0$ and the latter because the inclination is small). We linearize the equations of motion in terms of these variables to obtain

$$\rho'' - \left(N^2 + \frac{2a}{R^3} + \frac{6aJ_2}{R^5} \right) \rho - 2NR\psi' = \frac{3aJ_{22}}{R^4} \psi \sin[2(N-1)\tau]$$

$$- \frac{3aJ_{22}}{2R^4} \cos[2(N-1)\tau]$$

$$+ \frac{6aJ_{22}}{R^5} \rho \cos[2(N-1)\tau]$$

$$R^2\psi'' + 2NR\rho' = -\frac{2aJ_{22}\psi}{R^3} \cos[2(N-1)\tau] - \frac{aJ_{22}}{R^3} \sin[2(N-1)\tau]$$

$$+ \frac{3aJ_{22}}{R^4} \rho \sin[2(N-1)\tau]$$

$$\delta'' + \left(N^2 + \frac{3aJ_2}{R^5} \right) \delta = -\frac{aJ_{22}\delta}{R^5} \cos[2(N-1)\tau]$$

Now since $J_2 \ll 1$ and $|J_{22}| \ll J_2$, it follows that

$$N^2 R = \frac{a}{R^2} + \frac{3aJ_2}{2R^4} \approx \frac{a}{R^2}$$

so the linearized equations of motion can be approximated by the set

$$\rho'' - 3N^2\rho - 2NR\psi' = -\frac{3aJ_{22}}{2R^4} \cos[2(N-1)\tau]$$

$$R\psi'' + 2N\rho' = -\frac{aJ_{22}}{R^4} \sin[2(N-1)\tau]$$

$$\delta'' + \left(N^2 + \frac{3aJ_2}{R^5} \right) \delta = 0$$

At this stage one can see that the motion in declination has separated from the radial and azimuthal motions. Moreover, it is simple harmonic with frequency $\approx N$ and amplitude equal to the unperturbed inclination. The remaining coupled set is equivalent to a driven harmonic oscillator. To see this, observe that the equation for psi is directly integrable (once) to

$$R\psi' + 2N\rho = \frac{aJ_{22}}{2R^4(N-1)} \cos[2(N-1)\tau] + \text{const}$$

If this is substituted into the radial equation of motion, the resulting differential equation is linear, second order, and ordinary for ρ with a periodic forcing term; namely,

$$\rho'' + N^2\rho = \frac{(3-N)}{2R^4(N-1)} aJ_{22} \cos[2(N-1)\tau] + \text{const}$$

Thus, one should expect the usual resonance effects whenever N is such that $|2(N-1)| = |N|$ (e.g., $N = 2, \frac{2}{3}$). Now we specify that N is chosen such that in unperturbed motion

$$N\omega = \langle \dot{\alpha} \rangle$$

Then for a geosynchronous orbit $N = 1$, $N = 2$ for a 12-hr orbit and $N = 24/P$ for a P-hr orbit. I can now interpret the resonances to be at 12 and at 36 hr.

The particular solutions of the simplified equations of motion that exhibit these effects are

$$\left. \begin{aligned} \rho &= \frac{aJ_{22}(N-3)\cos[2(N-1)\tau]}{2(N-1)(N-2)(3N-2)R^4} \\[2ex] \psi &= \frac{aJ_{22}(N^2-2N+4)\sin[2(N-1)\tau]}{4(N-1)^2(N-2)(3N-2)R^5} \end{aligned} \right\} \quad N \neq 1, 2, \tfrac{2}{3}$$

The complete solution to the equations of motion is obtained by adding to these particular solutions the general solutions of the homogeneous versions thereof,

$$\rho = A \sin N\tau + B \cos N\tau + C$$

$$\psi = \left(\frac{2A}{R}\right) \cos N\tau - \left(\frac{2B}{R}\right) \sin N\tau - \frac{3NC}{2R}\tau + D$$

where A, B, C, and D are the four arbitrary constants of integration. If $C = D = 0$, then the mean values of ρ and δ both vanish.

The case $N = 1$. Despite the $N - 1$ divisor in the above formula, this does not represent a true resonance as the $N = 2$ and $\frac{2}{3}$ cases do. It is left to the reader to show that if $N = 1$, then there are four possible stationary points for circular orbits (in a coordinate system rotating with the Earth). These are, not surprisingly, at the extensions of the ends of the principal axes of the elliptical section of the equator. Furthermore, the reader should be able to demonstrate that the ones on the minor axis are stable and that the ones on the major axis are unstable.

The case $N = 2$ or $\frac{2}{3}$. The form of the particular solution is ($N = 2$)

$$\rho = \frac{aJ_{22}}{8R^4}(\cos 2\tau + \tau \sin 2\tau)$$

$$\psi = \frac{aJ_2}{8R^5}(2\tau \cos 2\tau - \sin 2\tau)$$

For $N = \frac{2}{3}$

$$\rho = -\frac{21\,aJ_{22}}{8R^4}\left[\tau \sin\left(\frac{2\tau}{3}\right) + 3\cos\left(\frac{2\tau}{3}\right)\right]$$

$$\psi = \frac{3\,aJ_{22}}{8R^5}\left[57\sin\left(\frac{2\tau}{3}\right) - 14\tau\cos\left(\frac{2\tau}{3}\right)\right]$$

Both of these orbits are unstable.

Kozai's Theory

The launch of the first artificial satellite in 1957 was the second element in the resurgence of celestial mechanics (the other was the development of the digital electronic computer). Soon thereafter in *The Astronomical Journal* articles on celestial mechanics appeared with a special emphasis on the oblate primary/atmospheric drag set of perturbations. A paper by Kozai (1959) has attracted particular attention over the years and has been used by many because not only are the terms owing to J_2 included but the J_3 and J_4 terms are too. I will use Kozai's analysis as a prototype of a misapplication of perturbation theory. Obviously, in order to do this, his theory must be understood in some detail. In order to facilitate comparison with Kozai's original paper and other discussions of it (which have also tended to follow his notation) I'll use his notation. The principal difference is that Kozai writes the potential as

$$U = \frac{GM}{r}\left[1 + \frac{A_2}{r^2}\left(\frac{1}{3} - \sin^2 \delta\right) + \frac{A_3}{r^3}\left(\frac{5}{2}\sin^2 \delta - \frac{3}{2}\right)\sin \delta \right.$$

$$\left. + \frac{A_4}{r^4}\left(\frac{3}{35} + \frac{1}{7}\sin^2 \delta - \frac{1}{4}\sin^2 2\delta\right) + \cdots\right]$$

instead of using J's and Legendre polynomials ($A_2 = 3J_2/2$, $A_3 = -J_3$, $A_4 = -55J_4/8$). Kozai also frequently uses the parameter p,

$$p = a(1 - e^2)$$

To begin, Kozai notes that A_2 is a small quantity compared to unity for the Earth. He goes on to note that *numerically* $|A_3| \simeq |A_4| \simeq A_2^2$ and so, *numerically*, are of the second order. This is true (for the Earth) but irrelevant to the meaning of *order* in Picard's method of successive substitutions. The use of this misleading word in the context of approximately trying to solve the variation of parameter equations might lead to the belief that those equations have been solved to second order by including terms proportional to A_3 and A_4. This is false. Kozai will use *second order* in yet another context—when terms proportional to A_2^2 appear. This too lacks rigor and adds a third meaning to "second order." Since the reader knows what second order should mean, in this section only I'll write using the incorrect terminology but do so with quote marks as a reminder.

Kozai defines the disturbing function R by

$$R = U - \frac{GM}{r}$$

$$= GM\left[\frac{A_2}{a^3}\left(\frac{a}{r}\right)^3\left[\frac{1}{3} - \frac{1}{2}\sin^2 i + \frac{1}{2}\sin^2 i \cos 2(v+\omega)\right]\right.$$

$$+ \frac{A_3}{a^4}\left(\frac{a}{r}\right)^4\left[\left(\frac{15}{8}\sin^2 i - \frac{3}{2}\right)\sin(v+\omega) - \frac{5}{8}\sin^2 i \sin 3(v+\omega)\right]\sin i$$

$$+ \frac{A_4}{a^5}\left(\frac{a}{r}\right)^5\left[\frac{3}{35} - \frac{3}{7}\sin^2 i + \frac{3}{8}\sin^4 i + \sin^2 i\left(\frac{3}{7} - \frac{1}{2}\sin^2 i\right)\cos 2(v+\omega)\right.$$

$$\left.\left. + \frac{1}{8}\sin^4 i \cos 4(v+\omega)\right]\right]$$

and replaces r and δ by their usual expressions;

$$r = \frac{a(1-e^2)}{1+e \cos v}, \qquad \sin \delta = \sin i \sin(v+\omega)$$

The true anomaly v is related to the mean anomaly M via

$$\frac{dv}{dM} = \frac{a^2}{r^2}(1-e^2)^{1/2}$$

At this point another false set of concepts is introduced as a computational device. Kozai looks at the functional form of R and, *knowing* the first-order secular results for this problem (Eqs. 9.1–9.6), isolates three types of terms. That part of R which is explicitly independent of M, Ω, and ω he calls the secular part. The remainder of R either depends on M alone and varies rapidly (e.g., with the unperturbed period $P = 2\pi/n$) or depends on ω and Ω and varies slowly (e.g., with period $\sim P/|A_2|$ because of the first-order secular perturbations in ω and Ω). He calls the M-dependent terms the short-periodic ones and the ω- and Ω-dependent terms the long-periodic ones. Of course there are no Ω terms in this particular illustration. Note that if he had included the J_{21} and J_{22} terms this would have introduced right ascension terms into R, hence longitude of the ascending node dependence (explicitly). By this same "reasoning" additional long-periodic terms would have been present. Note too that by the time ω (or Ω) has varied appreciably so much time has elapsed (it takes $\sim P/|A_2|$ for this to happen) that no first-order result is meaningful any longer. He further claims that the long-periodic terms originate from the second-order (meaning here the A_3, A_4 terms) terms in R, and to be consistent, one must keep secular and short-periodic terms to second order too (here meaning A_2^2 terms). Kozai then shows that if one averages with respect to the mean anomaly, R can be written as

$$R = R_1 + R_2 + R_3 + R_4$$

with

$$R_1 = GM \frac{A_2}{a^3} \left(\frac{1}{3} - \frac{1}{2} \sin^2 i \right) (1 - e^2)^{-3/2}$$

$$R_2 = GM \frac{A_4}{a^5} \left(\frac{3}{35} - \frac{3}{7} \sin^2 i + \frac{3}{8} \sin^4 i \right) (1 - e^2)^{-7/2} \left(1 + \frac{3}{2} e^2 \right)$$

$$R_3 = GM \left[\frac{3}{2} \frac{A_3}{a^4} \sin i \left(\frac{5}{4} \sin^2 i - 1 \right) e (1 - e^2)^{-5/2} \sin \omega \right.$$
$$\left. + \frac{A_4}{a^5} \sin^2 i \left(\frac{9}{28} - \frac{3}{8} \sin^2 i \right) e^2 (1 - e^2)^{-7/2} \cos 2\omega \right]$$

$$R_4 = GM \frac{A_2}{a^3} \left(\frac{a}{r} \right)^3 \left\{ \left(\frac{1}{3} - \frac{1}{2} \sin^2 i \right) \left[1 - \left(\frac{r}{a} \right)^3 (1 - e^2)^{-3/2} \right] + \frac{1}{2} \sin^2 i \cos 2(v + \omega) \right\}$$

where R_1, R_2, R_3, and R_4 are the first-order secular, the second-order secular, the long-periodic, and the short-periodic parts of the disturbing function. This result follows since (Tisserand 1889)

$$\overline{\left(\frac{a}{r} \right)^3} = \frac{1}{2\pi} \int_0^{2\pi} \left(\frac{a}{r} \right)^3 dM = (1 - e^2)^{-3/2}$$

$$\overline{\left(\frac{a}{r} \right)^3 \sin 2v} = \overline{\left(\frac{a}{r} \right)^3 \cos 2v} = 0$$

$$\overline{\left(\frac{a}{r} \right)^4 \cos v} = e(1 - e^2)^{-5/2}$$

$$\overline{\left(\frac{a}{r} \right)^4 \sin v} = \overline{\left(\frac{a}{r} \right)^4 \cos 3v} = \overline{\left(\frac{a}{r} \right)^4 \sin 3v} = 0$$

$$\overline{\left(\frac{a}{r} \right)^5} = (1 - e^2)^{-7/2}(1 + \frac{3}{2} e^2)$$

$$\overline{\left(\frac{a}{r} \right)^5 \cos 2v} = \frac{3}{4} e^2 (1 - e^2)^{-7/2}$$

$$\overline{\left(\frac{a}{r} \right)^5 \sin 2v} = \overline{\left(\frac{a}{r} \right)^5 \cos 4v} = \overline{\left(\frac{a}{r} \right)^5 \sin 4v} = 0$$

The next step is to write the variation of parameters equations;

$$\frac{da}{dt} = \frac{2}{na} \frac{\partial R}{\partial M}$$

$$\frac{de}{dt} = \frac{1 - e^2}{na^2 e} \frac{\partial R}{\partial M} - \frac{(1 - e^2)^{1/2}}{na^2 e} \frac{\partial R}{\partial \omega}$$

$$\frac{d\omega}{dt} = -\frac{\cot i}{na^2(1-e^2)^{1/2}}\frac{\partial R}{\partial i} + \frac{(1-e^2)^{1/2}}{na^2 e}\frac{\partial R}{\partial e}$$

$$\frac{di}{dt} = \frac{\cot i}{na^2(1-e^2)^{1/2}}\frac{\partial R}{\partial \omega}$$

$$\frac{d\Omega}{dt} = \frac{\csc i}{na^2(1-e^2)^{1/2}}\frac{\partial R}{\partial i}$$

$$\frac{dM}{dt} = n - \frac{1-e^2}{na^2 e}\frac{\partial R}{\partial e} - \frac{2}{na}\frac{\partial R}{\partial a}$$

$(n^2 a^3 = GM$ as usual) and solve with R set equal to R_1, R_2, R_3, or R_4. Except for some complications arising from the misleading use of the expression "second order," this is relatively straightforward. The time t is related to M or v via

$$dt = \frac{dt}{dM}\,dM = \frac{1}{n}\left(\frac{r}{a}\right)^2 \frac{dv}{(1-e^2)^{1/2}}$$

The results are

$$da_s = \frac{A_2}{a}\left\{\frac{2}{3}\left(1-\frac{3}{2}\sin^2 i\right)\left[\left(\frac{a}{r}\right)^3 - (1-e^2)^{-3/2}\right] + \left(\frac{a}{r}\right)^3 \sin^2 i \cos 2(v+\omega)\right\}$$

$$de_s = \frac{1-e^2}{e}\frac{A_2}{a^2}\left\{\frac{1}{3}\left(1-\frac{3}{2}\sin^2 i\right)\left[\left(\frac{a}{r}\right)^3 - (1-e^2)^{-3/2}\right]\right.$$
$$\left. + \frac{1}{2}\left(\frac{a}{r}\right)^3 \sin^2 i \cos 2(v+\omega)\right\}$$
$$- \frac{\sin^2 i}{2e}\frac{A_2}{ap}\left[\cos 2(v+\omega) + e\cos(v+2\omega) + \left(\frac{e}{3}\right)\cos(3v+2\omega)\right]$$

$$d\omega_s = \frac{A_2}{p^2}\left\{\left(2-\frac{5}{2}\sin^2 i\right)(v-M+e\sin v)\right.$$
$$+ \left(1-\frac{3}{2}\sin^2 i\right)\left[\frac{1}{e}(1-e^2/4)\sin v + \frac{1}{2}\sin 2v + \frac{e}{12}\sin 3v\right]$$
$$- \frac{1}{e}\left[\frac{1}{4}\sin^2 i + \left(\frac{1}{2}-\frac{15}{16}\sin^2 i\right)e^2\right]\sin(v+2\omega)$$
$$+ \frac{e}{16}\sin^2 i \sin(v-2\omega) - \frac{1}{2}\left(1-\frac{5}{2}\sin^2 i\right)\sin 2(v+\omega)$$
$$+ \frac{1}{e}\left[\frac{7}{12}\sin^2 i - \frac{1}{6}\left(1-\frac{19}{8}\sin^2 i\right)e^2\right]\sin(3v+2\omega)$$
$$+ \frac{3}{8}\sin^2 i \sin(4v+2\omega) + \frac{e}{16}\sin^2 i \sin(5v+2\omega)\right\}$$

$$di_s = \frac{1}{4}\frac{A_2}{p^2}\sin 2i\left[\cos 2(v+\omega)+e\cos(v+2\omega)+\frac{e}{3}\cos(3v+2\omega)\right]$$

$$d\Omega_s = -\frac{A_2}{p^2}\cos i\left[v-M+e\sin v-\frac{1}{2}\sin 2(v+\omega)\right.$$

$$\left.-\frac{e}{2}\sin(v+2\omega)-\frac{e}{6}\sin(3v+2\omega)\right]$$

$$e\,dM_s = \frac{A_2}{p^2}(1-e^2)^{1/2}\left\{-\left(1-\frac{3}{2}\sin^2 i\right)\left[\left(1-\frac{e^2}{4}\right)\sin v+\frac{e}{2}\sin 2v+\frac{e^2}{12}\sin 3v\right]\right.$$

$$+\sin^2 i\left[\frac{1}{4}\left(1+\frac{5}{4}e^2\right)\sin(v+2\omega)-\frac{e^2}{16}\sin(v-2\omega)\right.$$

$$-\frac{7}{12}\left(1-\frac{e^2}{28}\right)\sin(3v+2\omega)-\frac{3}{8}e\sin(4v+2\omega)$$

$$\left.\left.-\frac{e^2}{16}\sin(5v+2\omega)\right]\right\}$$

Since

$$\overline{\cos jv} = \left[\frac{-e}{1+(1-e^2)^{1/2}}\right]^j[1+j(1-e^2)^{1/2}]$$

the mean values of these short-periodic perturbations are not zero (except for a). Their mean values with respect to M are

$$\overline{de_s} = \frac{A_2}{p^2}\sin^2 i\frac{1-e^2}{6e}\overline{\cos 2v}\cos 2\omega$$

$$\overline{d\omega_s} = \frac{A_2}{p^2}\left\{\left(\frac{1}{8}+\frac{1-e^2}{6e^2}\overline{\cos 2v}\right)\sin^2 i+\frac{1}{6}\overline{\cos 2v}\cos^2 i\right\}\sin 2\omega$$

$$\overline{di_s} = -\frac{1}{12}\frac{A_2}{p^2}\overline{\cos 2v}\sin 2i\cos 2\omega$$

$$\overline{d\Omega_s} = -\frac{1}{6}\frac{A_2}{p^2}\overline{\cos 2v}\cos i\sin 2\omega$$

$$\overline{dM_s} = -\frac{A_2}{p^2}(1-e^2)^{1/2}\left(\frac{1}{8}+\frac{1+e^2/2}{6e^2}\overline{\cos 2v}\right)\sin^2 i\sin 2\omega$$

Expressions for the mean anomaly and the argument of perigee are rather complicated so it is better to combine the four elements a, e, ω, and M into the radius vector r and the argument of latitude $L=v+\omega$; namely,

$$\frac{dr}{a} = \frac{e}{(1-e^2)^{1/2}}\sin v\,dM+\frac{r}{a}\frac{da}{a}-\cos v\,de$$

$$dv = \frac{a^2}{r^2}(1-e^2)^{1/2}\,dM + \left(1+\frac{r}{p}\right)\frac{a}{r}\sin v\,de$$

We set (a_0 = the unperturbed semi-major axis)

$$da = -a_0\frac{A_2}{p^2}\left(1-\frac{3}{2}\sin^2 i\right)(1-e^2)^{1/2} + da_s$$

$$d\omega = d\omega_s - \frac{3}{8}\frac{A_2}{p^2}\sin^2 i\sin 2\omega, \qquad de = de_s$$

$$dM = dM_s + \frac{3}{8}\frac{A_2}{p^2}(1-e^2)^{1/2}\sin^2 i\sin 2\omega$$

Then the deviations of the magnitude of the radius vector and the argument of latitude from those computed by the mean orbital element set can be calculated:

$$\frac{dr}{a} = \frac{1}{3}\frac{A_2}{ap}\left(1-\frac{3}{2}\sin^2 i\right)\left\{-1-\frac{1}{e}[1-(1-e^2)^{1/2}]\cos v + \frac{r}{a}\frac{1}{(1-e^2)^{1/2}}\right\}$$

$$+\frac{1}{6}\frac{A_2}{ap}\sin^2 i\cos 2(v+\omega)$$

$$dL = \frac{A_2}{p^2}\left(\left(2-\frac{5}{2}\sin^2 i\right)(v-M+e\sin v) + \left(1-\frac{3}{2}\sin^2 i\right)\right.$$

$$\times\left\{\frac{2}{3e}\left[1-\frac{e^2}{2}-(1-e^2)^{1/2}\right]\sin v\right.$$

$$+\left.\frac{1}{6}[1-(1-e^2)^{1/2}]\sin 2v\right\}$$

$$-\left(\frac{1}{2}-\frac{5}{6}\sin^2 i\right)e\sin(v+2\omega)-\left(\frac{1}{2}-\frac{7}{12}\sin^2 i\right)\sin 2(v+\omega)$$

$$\left.-\frac{e}{6}\cos^2 i\sin(3v+2\omega)\right)$$

Kozai gets the secular perturbations of the first order by setting $R = R_1$; the familiar results are

$$\bar{\omega} = \omega_0 + \frac{A_2}{p^2}\bar{n}\left(2-\frac{5}{2}\sin^2 i\right)t$$

$$\bar{\Omega} = \Omega_0 - \frac{A_2}{p^2}\bar{n}t\cos i$$

$$\bar{M} = M_0 + \bar{n}t$$

$$\bar{n} = n_0 + \frac{A_2}{p^2}n_0\left(1-\frac{3}{2}\sin^2 i\right)(1-e^2)^{1/2}$$

He then introduces the quantity $\bar{a}$,

$$\bar{a} = a_0\left[1 - \frac{A_2}{p^2}\left(1 - \frac{3}{2}\sin^2 i\right)(1-e^2)^{1/2}\right]$$

so that

$$\bar{n}^2\bar{a}^3 = GM\left[1 - \frac{A_2}{p^2}\left(1 - \frac{3}{2}\sin^2 i\right)(1-e^2)^{1/2}\right]$$

As might be expected the long-periodic perturbations are a bit more difficult to obtain, but by use of the constancy of the z component of the angular momentum Kozai succeeds:

$$da_1 = 0$$

$$di_1 = \overline{di_s} - \frac{A_2}{p^2}\frac{e^2 \sin 2i}{8(4-5\sin^2 i)}\left(\frac{14-15\sin^2 i}{6} - \frac{A_4}{A_2^2}\frac{18-21\sin^2 i}{7}\right)\cos 2\omega$$

$$- \frac{3}{4}\frac{A_3}{A_2 p}e\cos i\sin\omega$$

$$de_1 = -\frac{1-e^2}{e}\tan i\, di_1$$

$$= \overline{de_s} + \frac{A_2}{ap}\frac{e\sin^2 i}{4(4-5\sin^2 i)}\left(\frac{14-15\sin^2 i}{6} - \frac{A_4}{A_2^2}\frac{18-21\sin^2 i}{7}\right)\cos 2\omega$$

$$+ \frac{3}{4}\frac{A_3}{A_2 a}\sin i\sin\omega$$

$$d\Omega_1 = \overline{d\Omega_s} - \frac{A_2}{p^2}\frac{e^2\cos i}{2(4-5\sin^2 i)}$$

$$\times\left[\left(\frac{7-15\sin^2 i}{6} - \frac{A_4}{A_2^2}\frac{9-21\sin^2 i}{7}\right)\right.$$

$$\left.+ \frac{5\sin^2 i}{2(4-5\sin^2 i)}\left(\frac{14-15\sin^2 i}{6} - \frac{A_4}{A_2^2}\frac{18-21\sin^2 i}{7}\right)\right]\sin 2\omega$$

$$+ \frac{3}{4}\frac{A_3}{A_2 p}e\cot i\cos\omega$$

$$d\omega_1 = \overline{d\omega_s} - \frac{3}{8}\frac{A_2}{p^2}\sin^2 i\sin 2\omega$$

$$- \frac{A_2}{p^2}\left\{\frac{1}{4-5\sin^2 i}\left[\frac{14-15\sin^2 i}{24}\sin^2 i - e^2\frac{28-158\sin^2 i+135\sin^4 i}{48}\right.\right.$$

$$\left.\left. - \frac{A_4}{A_2^2}\left(\frac{18-21\sin^2 i}{28}\sin^2 i - e^2\frac{36-210\sin^2 i+189\sin^4 i}{56}\right)\right]\right.$$

$$-\frac{e^2 \sin^2 i(13-15\sin^2 i)}{(4-5\sin^2 i)^2}\left(\frac{14-15\sin^2 i}{24}\right.$$

$$\left.\left.-\frac{A_4}{A_2^2}\frac{18-21\sin^2 i}{28}\right)\right\}\sin 2\omega$$

$$+\frac{3}{4}\frac{A_3}{A_2 p}\frac{\sin^2 i-e^2\cos^2 i}{e\sin i}\cos\omega$$

where he used (Tisserand 1889)

$$\overline{\left(\frac{a}{r}\right)^6}=(1-e^2)^{-9/2}\left(1+3e^2+\frac{3}{8}e^4\right)$$

$$\overline{\left(\frac{a}{r}\right)^6\cos v}=2e(1-e^2)^{-9/2}\left(1+\frac{3}{4}e^2\right)$$

$$\overline{\left(\frac{a}{r}\right)^6\cos 2v}=\frac{3}{2}e^2(1-e^2)^{-9/2}\left(1+\frac{1}{6}e^2\right)$$

$$\overline{\left(\frac{a}{r}\right)^6\cos 3v}=\frac{e^3}{2}(1-e^2)^{-9/2}$$

$$\overline{\left(\frac{a}{r}\right)^6\cos 4v}=\frac{e^4}{16}(1-e^2)^{-9/2}$$

$$\overline{\left(\frac{a}{r}\right)^6\cos jv}=0,\qquad\text{for }j>4$$

$$\overline{\left(\frac{a}{r}\right)^3}=(1-e^2)^{-3/2}$$

$$\overline{\left(\frac{a}{r}\right)^3\cos v}=\frac{e}{2}(1-e^2)^{-3/2}$$

$$\overline{\left(\frac{a}{r}\right)^3\cos jv}=0,\qquad\text{for }j>1$$

Kozai gives up on the mean anomaly perturbations dM, arguing that (remember it is 1959) atmospheric perturbations will cause decay before these become significant. In sum

$$a=\bar a+da_s,\qquad \bar a=a_0\left[1-\frac{A_2}{p^2}\left(1-\frac{3}{2}\sin^2 i\right)(1-e^2)^{1/2}\right]$$

$$e=\bar e+de_s-\overline{de_s}+de_1$$

$$i=\bar i+di_s-\overline{di_s}+di_1$$

$$\omega=\omega_0+\dot\omega t+d\omega_s-\overline{d\omega_s}+d\omega_1$$

$$\Omega=\Omega_0+\dot\Omega t+d\Omega_s-\overline{d\Omega_s}+d\Omega_1$$

$$M = M_0 + \bar{n}t + dM_s$$

$$\bar{n} = n_0\left[1 + \frac{A_2}{p^2}\left(1 - \frac{3}{2}\sin^2 i\right)(1 - e^2)^{1/2}\right]$$

$$n_0^2 a_0^3 = GM$$

The Critical Inclination

In the formulas above the factor $4 - 5\sin^2 i$ or $2 - \frac{5}{2}\sin^2 i$ occur in the numerator of some first-order secular or short-period changes but in the *denominator* of some second-order or long-period changes. The zeros of $4 - 5\sin^2 i$ are at $\sin i = \pm(\frac{4}{5})^{1/2}$, $\cos i = \pm 1/\sqrt{5}$, or $\tan i = \pm 2$. The numerical values are $63°26'05''816$ and $116°33'54''184$ and these are known (unfortunately) as the critical inclinations. It is a prediction of first-order perturbation theory that the argument of perigee, for an orbit about an oblate primary, will not advance or retard if the inclination of the orbit is equal to one of these values. It just works out that way. Similarly, if $i = 90°$, then $\Delta\Omega)_{sec} = 0$, but at least that looks physically meaningful.

It is *not* a prediction of second-order perturbation theory that there are infinitely large or infinitely rapid changes in some of the orbital elements at the critical inclinations. No one has ever solved this problem, in this representation, to second order. It is, however, a prediction of a misapplication of perturbation theory that such infinities do occur. There has been much heat concerning this point in the literature. Arguments are made one way or the other, librations are introduced, expansions in $\sqrt{J_2}$ instead of J_2 are utilized, and so on. The correct resolution of these unphysical predictions is not to rely on bad mathematics.

Numerical Considerations

Hori and Kozai (1975) gave a paper at a 1974 symposium on artificial satellite dynamics held in São Paulo. They dealt mainly with perturbations due to oblateness but also discussed, for example, drag. They summarized in particular the number of terms necessary to yield a given accuracy for the satellite's motion. Quoting from their paper (with permission) the results were as follows:

The size of disturbing forces of various sources and the perturbations produced are summarized in Table 8 where n is the mean motion of the satellite in degrees per day, A/M the area-to-mass ratio in cgs unit, and k_2 (=0.3) is the Love's number. The factors in the second column represent also the orders of magnitude for short-period perturbations, while the third column shows those for long-period and secular perturbations after 1000 revolutions of satellites.

In order to compute the positions and velocities of satellites with an accuracy of 10^{-6} (10 m and 1 cm/sec) it is necessary to evaluate the first order short-period perturbations due to J_2 and J_{22} in the geopotential and those due to the luni-solar terms for the stationary satellite as shown in the fourth column. As for the secular and the long-period perturbations, the first order theory is sufficient for the

Table 8

Disturbing Force	Factor	lp & s	10⁻⁶ Accuracy sp	10⁻⁶ Accuracy lp & s	10⁻⁸ Accuracy sp	10⁻⁸ Accuracy lp & s
			Orders of Perturbation Needed			
J_2	10^{-3}	1	1	2	2	3
$J_n\ (n>2)$	10^{-6}	10^{-3}	no	1	1	$1, J_2, J_n$
Nonzonal	$10^{-5}\,(J_{22}),\ 10^{-6}$ (others)	10^{-4}	1	1	1	$1 \times J_2,\ 2$
Lunisolar	$3.1/n^2 \sim 10^{-7}$ (close sat.) $\sim 10^{-5}$ (station. sat.)	10^{-4} / 10^{-2}	no / 1	1 / 2	1	$3, \ldots$
Tides	$3.1\left(\dfrac{R}{a}\right)^5 \dfrac{k_2}{n^2} \sim 10^{-8}$	10^{-5}	no	1	no	1
Solar radiation	$4.6 \times 10^{-8}\left(\dfrac{a}{R}\right)^2 \dfrac{A}{M} \sim 10^{-8}$	10^{-2} (mean anomaly) / 10^{-5} (other elements)	no	2	no	3
Air drag	$\left(\dfrac{A}{M}\right)\zeta \sim 10^{-8}$	10^{-2} (mean anomaly) / 10^{-5} (other elements)	no	2 / 1	no	3 / 1

sp = short-period perturbations.

lp & s = long-period and secular perturbations.

The third column shows orders of magnitude for the long-period and secular perturbations after 1000 revolutions.

perturbations caused by nonzonal and higher zonal harmonics, and also by luni-solar, tidal, solar radiation, and air-drag forces. The second order theory is required, however, for the perturbations due to J_2, and luni-solar attractions (for the stationary satellite), and those due to the solar radiation and the air-drag (in the computation of the mean anomaly). These are shown in the fifth column of the table. In addition to the disturbing forces given in the table, there are radiation effects from the surface of the earth, and also the effects of the motion of the earth's equator (precession and nutation).

When an accuracy of 10^{-8} (10 cm and 0.1 mm/sec) is needed, analytical theories become much more complicated as shown in the last two columns of the table. Even the computations of short-period perturbations require all the disturbing terms larger than 10^{-8}, and the formulations of secular and long-period perturbations require all terms larger than 10^{-11} in the transformed disturbing function which is derived by eliminating short-period terms. Numerical integration of these transformed equations of motion can be adopted. Here the step of integration is as long as one day even for close satellites since short-period terms have been eliminated. This semi-analytical procedure can adequately handle satellite data with the accuracy of 10^{-8}.

LUNISOLAR PERTURBATIONS

As can be seen, the detailed discussion of a problem within the context of perturbation theory is very complicated. Part of this is because of the complexity of even a simple-sounding problem—try to solve, analytically, for the motion of a charged particle in a dipole magnetic field—the other part is on account of the fact that the orbital element set is a poor coordinate system. Perturbations due to a third body are even more troublesome than are perturbations due to oblateness effects because the perturbing force is now time dependent. Thus, while there exists a simple set of first-order results for the dominant oblateness perturbations (Eqs. 9.1–9.6), there is nothing comparable in this case.

Lunisolar perturbations of an artificial satellite are representative of the more general problem of third-body perturbations. Third-body perturbations open up the whole field of the expansion of the disturbing function—an analytically straightforward, tedious, and unrewarding task ideally suited to algebraic manipulators. So before discussing the satellite case in particular, we want to adopt a more general perspective. Next is a quick summary of the simpler analytical results available, including a discussion of the solar perturbations of the lunar orbit.

The N-Body Problem

We return to the general N-body equations of motion (Eq. 1.15):

$$m_n \ddot{\mathbf{r}}'_n = -G m_n \sum_{\substack{l=1 \\ l \neq n}}^{N} m_l \frac{\mathbf{r}'_n - \mathbf{r}'_l}{|\mathbf{r}'_n - \mathbf{r}'_l|^3}$$

or

$$m_n \ddot{\mathbf{r}}'_n = -\nabla' U$$

with

$$U = -\frac{G}{2} \sum_{\substack{l,n=1 \\ l \neq n}}^{N} \frac{m_l m_n}{r'_{ln}}, \qquad \mathbf{r}'_{ln} = \mathbf{r}'_l - \mathbf{r}'_n$$

Suppose that the Nth body is chosen to be the new origin of a system of noninertial coordinates. These coordinates are denoted by the same symbol without primes,

$$\mathbf{r}'_n = \mathbf{r}_n + \mathbf{r}'_N, \qquad n = 1, 2, \ldots, N-1$$

Clearly $\nabla U = \nabla' U$ so the new equations of motion are

$$\ddot{\mathbf{r}}_n + \ddot{\mathbf{r}}'_N = -\frac{1}{m_n} \nabla_{\mathbf{r}_n} U$$

and, since $\mathbf{r}_N = \mathbf{0}$,

$$\ddot{\mathbf{r}}'_N = -G \sum_{l=1}^{N-1} \frac{m_l \mathbf{r}_l}{(r'_{lN})^3}$$

The potential energy U can be written as the sum of two terms,

$$U = -G m_N \sum_{n=1}^{N-1} \frac{m_n}{r'_{nN}} - \frac{G}{2} \sum_{\substack{l,n=1 \\ l \neq n}}^{N-1} \frac{m_l m_n}{r'_{ln}}$$

$$= -G m_N \sum_{n=1}^{N-1} \frac{m_n}{r'_{nN}} + U'$$

and the equations of motion can now be written as

$$\ddot{\mathbf{r}}_n + G(m_n + m_N) \frac{\mathbf{r}_n}{r_{nN}^3} = -\frac{1}{m_n} \nabla_{\mathbf{r}_n} U' - G \sum_{\substack{l=1 \\ l \neq n}}^{N-1} \frac{m_l \mathbf{r}_l}{r_{lN}^3}$$

If we define the disturbing function R_{nl} as

$$R_{nl} = G \left(\frac{1}{r_{nl}} - \frac{\mathbf{r}_n \cdot \mathbf{r}_l}{r_{lN}^3} \right)$$

then the equations of motion can be more compactly written as

$$\ddot{\mathbf{r}}_n + G(m_n + m_N) \frac{\mathbf{r}_n}{r_{nN}^3} = \sum_{\substack{l=1 \\ l \neq n}}^{N-1} m_l \nabla_{\mathbf{r}_n} R_{nl}$$

The usefulness of this form becomes clearer if one remembers that in the solar system $M_\odot \gg m_{\text{planet}}$. Hence, the center of mass of the solar system is almost at the Sun's center and a heliocentric coordinate system is almost an inertial one. So when N is the Sun and l, n refer to two planets, we may write

$$\ddot{\mathbf{r}}_1 + G(m_1 + M_\odot)\frac{\mathbf{r}_1}{r_1^3} = m_2 \nabla_{\mathbf{r}_1} R_{12}$$

$$\ddot{\mathbf{r}}_2 + G(m_2 + M_\odot)\frac{\mathbf{r}_2}{r_2^3} = m_1 \nabla_{\mathbf{r}_2} R_{12}$$

$$R_{12} = G\left(\frac{1}{r_{12}} - \frac{\mathbf{r}_1 \cdot \mathbf{r}_2}{r_2^3}\right)$$

$$r_{12} = |\mathbf{r}_1 - \mathbf{r}_2|, \qquad r_1 = |\mathbf{r}_1|, \qquad r_2 = |\mathbf{r}_2|$$

and the Sun's location is $\mathbf{0}$.

Development of the Disturbing Function

Classically there are three steps involved in the expansion of the perturbation function (see any old text such as Brown and Shook 1933): (1) R_{12} is expanded as a power series in the square of the sine of the mutual half-inclination of the orbital planes (which is a very small quantity for solar system problems); (2) the coefficients of this series are developed into power series in the two eccentricities (usually both very small in solar system problems—only some asteroids have large eccentricities); and (3) the coefficients of these new series are developed into Fourier series in the angular variables (e.g., the two mean longitudes, the two arguments of perihelion, and the two longitudes of the ascending nodes). We will illustrate this procedure, at least partially, in outline. Let

$$\mathbf{r}_1 \cdot \mathbf{r}_2 = r_1 r_2 \cos \chi$$

and I be the mutual inclination of the two orbital planes. Further, let γ_1 and γ_2 be the angular distances from the planet's ascending nodes to the intersection of their orbital planes (measured eastward as usual). Then in terms of their true anomalies v_1, v_2, arguments of perihelion ω_1, ω_2, longitudes of the ascending node Ω_1, Ω_2, and arguments of latitude u_1, u_2 ($=v_1 + \omega_1$, $v_2 + \omega_2$),

$$\begin{aligned}
\cos \chi &= \cos(u_1 - \gamma_1) \cos(u_2 - \gamma_2) \\
&\quad + \sin(u_1 - \gamma_1) \sin(u_2 - \gamma_2) \cos I \\
&= \cos(u_1 - u_2 + \gamma_2 - \gamma_1) \\
&\quad - 2 \sin(u_1 - \gamma_1) \sin(u_2 - \gamma_2) \sin^2(I/2)
\end{aligned}$$

$$u_1 - \gamma_1 = v_1 + \omega_1 - \gamma_1, \qquad u_2 - \gamma_2 = v_2 + \omega_2 - \gamma_2$$

Spherical trigonometry applied to the triangle with vertices at the two ascending nodes and orbital intersection point yields (i_1, $i_2 =$ the two planetary inclinations relative to the ecliptic; see Fig. 40)

$$\sin \gamma_1 \sin I = \sin i_2 \sin(\Omega_1 - \Omega_2)$$

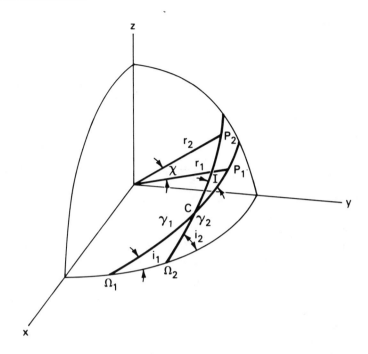

Figure 40. Geometry for the development of the disturbing function.

$$\sin \gamma_2 \sin I = \sin i_1 \sin(\Omega_1 - \Omega_2)$$

$$\cos \gamma_1 \sin I = \sin i_1 \cos i_2 - \cos i_1 \sin i_2 \cos(\Omega_1 - \Omega_2)$$

$$\cos \gamma_2 \sin I = -\cos i_1 \sin i_2 + \sin i_1 \cos i_2 \cos(\Omega_1 - \Omega_2)$$

$$\cos I = \cos i_i \cos i_2 + \sin i_1 \sin i_2 \cos(\Omega_1 - \Omega_2)$$

Therefore

$$R_{12} = [r_1^2 + r_2^2 - 2r_1 r_2 \cos(u_1 - u_2 + \gamma_2 - \gamma_1)]^{-1/2}$$

$$\times \left[1 + \frac{4r_1 r_2 \sin(u_1 - \gamma_1) \sin(u_2 - \gamma_2) \sin^2(I/2)}{r_1^2 + r_2^2 - 2r_1 r_2 \cos(u_1 - u_2 + \gamma_2 - \gamma_1)} \right]^{-1/2}$$

$$- \frac{r_1}{r_2^2} \left[\cos(u_1 - u_2 + \gamma_2 - \gamma_1) - 2 \sin(u_1 - \gamma_1) \sin(u_2 - \gamma_2) \sin^2\left(\frac{I}{2}\right) \right]$$

To expand this result by the binomial theorem, we need to be assured that the second term in the second square root above is less than unity (in absolute value). It is clearly less than $4r_1 r_2 \sin^2(I/2)/(r_1 - r_2)^2$, which will only be appreciable for a close approach. Let us not dwell on this point but proceed to the expansion. The leading terms are, with $c = \cos(u_1 - u_2 + \gamma_2 - \gamma_1)$,

$$R_{12} = (r_1^2 + r_2^2 - 2r_1 r_2 c)^{-1/2}$$

$$- 2r_1 r_2 (r_1^2 + r_2^2 - 2r_1 r_2 c)^{-3/2} \sin(u_1 - \gamma_1) \sin(u_2 - \gamma_2) \sin^2\left(\frac{I}{2}\right)$$

$$+ 6r_1^2 r_2^2 (r_1^2 + r_2^2 - 2r_1 r_2 c)^{-5/2} \sin^2(u_1 - \gamma_1) \sin^2(u_2 - \gamma_2) \sin^4\left(\frac{I}{2}\right) + \cdots$$

$$+ \left(\frac{r_1}{r_2^2}\right)\left[-c + 2\sin(u_1 - \gamma_1) \sin(u_2 - \gamma_2) \sin^2\left(\frac{I}{2}\right)\right]$$

Now it starts to get messy. The two heliocentric distances r_1 and r_2 are replaced by $a_1(1 - e_1 \cos E_1)$ and $a_2(1 - e_2 \cos E_2)$, respectively. Since R_{12} is a homogeneous function of r_1 and r_2 (or of a_1 and a_2) of degree -1, the expansion of the disturbing function in powers of the two eccentricities turns out to be equivalent to the expansion (say, there is an obvious $1 \leftrightarrow 2$ symmetry) of

$$\frac{e_1 \cos E_1 - e_2 \cos E_2}{1 + e_2 \cos E_2}$$

The next step is the representation of terms of the form $(a_1^2 + a_2^2 - 2a_1 a_2 c)^{-k/2}$ for odd integral k. Since $c = \cos(u_1 - u_2 + \gamma_2 - \gamma_1)$, a Fourier series expansion in multiples of the argument $u_1 - u_2 + \gamma_2 - \gamma_1$ seems appropriate. Clearly the general term is of the form

$$f(a_1, a_2, e_1, e_2) \cos(j_1 u_1 + j_2 u_2 + k_1 \gamma_1 + k_2 \gamma_2)$$

and the integers j_1, j_2, k_1, and k_2 range over all values. Enough.

Lunar Theory. As might be imagined, the perturbations of the Moon by the Sun is an area ripe for the application of these techniques. Of course, the notion that the Moon revolves about the Earth is erroneous. The Earth and the Moon revolve about their common center of mass. The mean distance between the Earth and the Moon is 384,400 km. The mass of the Moon is 1/81.30 of the mass of the Earth. The center of mass of the system is 4728 km from the center of the Earth (about three-fourths of the way to the surface).

Describing the motion of the Earth–Moon system is complex. First of all, the center of mass revolves around the Sun once per year. The Earth and the Moon both revolve about their common center of mass once in 27.3 days. As a result the celestial longitude of any object exhibits fluctuations with a period of 27.3 days. These periodic fluctuations in longitude were, in fact, the most reliable source for determining the Moon's mass until spacecraft flew close to the Moon.

The orbital period of the Moon is slowly increasing as a consequence of the fact that the distance between the Earth and Moon is increasing. The slow recession of the Moon can be explained by the fact that the tidal bulge in the Earth's oceans (raised by the Moon) is carried eastward by the Earth's rotation.

This shifts the center of mass of the Earth to the east of the line joining the centers of mass of the Earth and Moon and gives the Moon a small acceleration in the direction of its orbital motion. This in turn causes the Moon to speed up and, therefore, to spiral slowly outward. Energy is lost in this process (known as tidal friction; see Chapter Thirteen).

As noted above, the mean value of the semi-major axis is 384,400 km. The average time for the Moon to make one complete revolution around the Earth relative to the stars is 27.31661 days. Due to solar perturbations the sidereal period may vary by as much as 7 hr. The mean eccentricity of the Moon's orbit is 0.054900489. Small periodic changes in the eccentricity occur at intervals of 31.8 days. This effect, called *evection*, was discovered more than 2000 years ago by Hipparchus and can result in perturbations of the lunar geocentric longitude up to 75'. The Moon's orbit is inclined to the ecliptic (the plane of the Earth's orbit) by about 5°8'. The line of nodes, which is the intersection of the Moon's orbital plane with the ecliptic, rotates westward, making one complete revolution in 18.6 years (the Saros cycle). The average time for the Moon to go around in its orbit from node to (the same) node is 27.21222 days.

The inclination of the Moon's orbit to the ecliptic actually varies between 4°59' and 5°18'; its mean value is 5°8'. The Earth's equator is inclined to the ecliptic by 23°27', and except for the slow precession of the Earth's axis of rotation with a period of 26,000 years, the equatorial plane is relatively stationary. The angle between the celestial equator and the Moon's orbital plane varies because of the rotation of the Moon's line of nodes. When the Moon's ascending node is in the direction to the vernal equinox, the inclination of the Moon's orbit to the equator is a maximum (e.g., the sum of 5°8' and 23°27', or 28°35'). When the ascending node is at the vernal equinox, the inclination of the Moon's orbit to the equator is given by the difference, 18°19'. Thus, the inclination relative to the equator varies between 18°19' and 28°35' with a period of 18.6 years. Both the slight variation in inclination relative to the ecliptic and the regression of the line of nodes were first observed by Flamsteed (about 1670).

The line of apsides rotates in the direction of the Moon's orbital motion causing ω to change by 360° in about 8.9 years. There are several more perturbations of the lunar orbit which go by the names of the *annual equation* (amplitude ~11'), the *variation* (39'.5), and the *parallactic inequality* (2'). Moulton's (1902) text contains a good introductory discussion of this material, including an analytical derivation of several of them (pp. 337–362). The classic text is by Brown (1896). A recent summary of lunar theory is Henrard (1982).

Artificial Satellites

Geyling and Westerman (1971) have directly solved the Lagrangian variation of parameter equations when the third body is *stationary* in an inertial reference frame. This is an obvious approximation to make (cf. Chapter Nine) when the period of the perturbing body is much longer than that of the body being perturbed. More than that, they have done it explicitly both in terms of the true anomaly and the eccentric anomaly. The expressions are worth repeating here to illustrate the true degree of messiness of the situation, even for a fixed perturbing third body. So with the perturbing object (Sun or Moon) at $\mathbf{r}_p$ with mass M_p, and correcting a few misprints (copyright 1971 Bell Telephone Laboratories, reprinted by permission),

$$R = GM_P\left(\frac{1}{r} - \frac{xx_p + yy_p + zz_p}{r_p^3}\right)$$

for the disturbing function, where $r^2 = (x - x_p)^2 + (y - y_p)^2 + (z - z_p)^2$. Forming the gradient of R, one should find that

$$\frac{\partial R}{\partial x} = \frac{GM_P}{r_p^5}[x(3x_p^2 - r_p^2) + 3yx_p y_p + 3zx_p z_p], \ldots$$

The following abbreviations are used:

$$s_1 = \mathbf{S}_1 \cdot \mathbf{r}_p, \qquad s_2 = \mathbf{S}_2 \cdot \mathbf{r}_p$$
$$S = \sin v, \qquad C = \cos v, \qquad \phi^2 = 1 - e^2$$

where $\mathbf{S}_1$ and $\mathbf{S}_2$ are the first two column vectors of the matrix S in Eq. 2.12b. Then, with $\Delta a = a(t) - a(t_0) = a(v) - a(v_0)$, and so on,

$$\Delta a = \frac{a^4 M_p}{r_p^5 M_\oplus}\phi^4\left[\frac{(3/e^2)(s_2^2 - s_1^2)(1 + 2eC)}{(1 + eC)^2} + \frac{(3s_2^2 - r_p^2) + 6s_1 s_2 SC}{(1 + eC)^2}\right]\Big|_{v_0}^{v}$$

$$\Delta e = \frac{(1 - e^2)}{2ae}\Delta a - \frac{3a^3 M_p \phi}{r_p^5 M_\oplus}$$

$$\times\left(\frac{\phi}{6e(1 + eC)^3}\left\{(s_2^2 - s_1^2)\frac{\phi^6(1 + 3eC)}{e^2}\right.\right.$$

$$- s_1 s_2 S[e(8 + 9e^2 - 2e^4) - 3(2 - 9e^2 - 3e^4)C$$

$$\left.\left.- 3(2 - 9e^2 - 8e^4)C^2]\right\} + \frac{5}{2}es_1 s_2 E\right)\Big|_{v_0}^{v}$$

$$\Delta i = \frac{a^3 M_p}{2r_p^5 M_\oplus}\left\{\frac{1}{(1 + eC)^3}\left[\frac{\phi^3(1 + 3eC)}{2e^2}\frac{\partial}{\partial i}(s_2^2 - s_1^2)\right.\right.$$

$$\left.- As_1 S\frac{\partial s_2}{\partial i} + Bs_2\phi^2 S\frac{\partial s_1}{\partial i}\right]$$

$$\left.+ \frac{3}{\phi}\left[s_1(1 + 4e^2)\frac{\partial s_2}{\partial i} - s_2\phi^2\frac{\partial s_1}{\partial i}\right]E\right\}\Big|_{v_0}^{v}$$

$$\Delta\Omega = \frac{a^3 M_p}{r_p^3 M_\oplus}\frac{\csc i}{2}\left\{\frac{\phi^6(1 + 3eC)}{e^3(1 + eC)^3}\frac{\partial}{\partial i}(s_1 s_2) - \frac{S}{2(1 + eC)^3}\left(A\frac{\partial s_1^2}{\partial i} + B\phi^2\frac{\partial s_2^2}{\partial i}\right)\right.$$

$$\left.+ \frac{3}{2\phi}\left[(1 + 4e^2)\frac{\partial s_1^2}{\partial i} + \phi^2\frac{\partial s_2^2}{\partial i}\right]E\right\}\Big|_{v_0}^{v}$$

$$\Delta\omega + \Delta\Omega \cos i = \frac{a^3 M_p}{r_p^5 M_\oplus}$$

$$\times\left(\frac{\phi^2}{2e(1+eC)^2}\left\{(r_p^2-3s_1^2)S[2+e^2+e(1+2e^2)C]-\frac{3\phi^4}{e}s_1s_2\right\}\right.$$

$$+\frac{\phi^2}{2e(1+eC)^3}\left\{(s_2^2-s_1^2)S[2+e^2+3e(1+e^2)C+(5e^2-2)C^2]\right.$$

$$\left.+s_1s_2\frac{2\phi^4}{e^3}(2-e^2+6eC+6e^2C^2)\right\}$$

$$\left.+\frac{3\phi}{2}(4s_1^2-s_2^2-r_p^2)E\right)\Bigg|_{v_0}^{v}$$

$$\Delta\chi + \phi[\Delta\Omega \cos i + \Delta\omega] + t\,\Delta n$$

$$= -\frac{3}{2}\frac{a^3 M_p}{r_p^5 M_\oplus}\phi^4 n\,\Delta t\frac{r_p^2-3(s_1 S+s_2 C)^2}{(1+eC)^2}-\frac{7}{2}\frac{a^3 M_p}{r_p^3 M_\oplus}\phi$$

$$\times\left\{\frac{S}{2(1+eC)^3}\left[\frac{e(18-5e^2+2e^4)+3e^2(9+e^2)C+e^2(11+4e^2)C^2}{3}\right.\right.$$

$$\left.-\frac{s_1^2}{r_p^2}A-\frac{\phi^2 s_2^2}{r_p^2}B\right]$$

$$\left.+\frac{\phi^6(1+3eC)}{e^2(1+eC)^3}\frac{s_1 s_2}{r_p^2}-\frac{1}{2\phi}\left[2+3e^2-\frac{3s_1^2}{r_p^2}(1+4e^2)-\frac{3s_2^2}{r_p^2}\phi^2\right]E\right\}\Bigg|_{v_0}^{v}$$

where Kepler's equation is $M = \chi + nt$ and

$$A = e(13+2e^2)-3(1-9e^2-2e^4)C-e(1-10e^2-6e^4)C^2$$

$$B = e(5-2e^2)+3(1+e^2)C+e(1+2e^2)C^2$$

As an alternative, let

$$s = \sin E, \qquad c = \cos E$$

Then

$$\Delta a = 2\frac{a^4 M_p}{r_p^3 M_\oplus}\left\{ec+\frac{e^2}{2}s^2-\frac{3}{r_p^2}\left[s_1^2\left(\frac{s^2}{2}+ec\right)+s_1 s_2\phi s(e-c)-s_2^2\phi^2\frac{s^2}{2}\right]\right\}\Bigg|_{E_0}^{E}$$

$$\Delta e = \frac{(1-e^2)\,\Delta a}{2ae}-\frac{3a^3 M_p\phi}{r_p^5 M_\oplus e}\left\{(s_2^2-s_1^2)\phi\left[\frac{1+e^2}{2}s^2+ec+\frac{e}{3}c^3\right]\right.$$

$$+s_1 s_2\left[\frac{5}{2}e^2 E+\frac{2+e^2}{2}sc-\left(\frac{7}{3}e+\frac{4}{3}e^3\right)s\right.$$

$$\left.\left.-\left(\frac{2}{3}e-\frac{e^2}{3}\right)sc^2\right]\right\}\Bigg|_{E_0}^{E}$$

$$\Delta i = \left[\frac{\Delta a}{2a} - \frac{e\,\Delta e}{\phi^2} \right] \cot i - \frac{3a^3 M_p}{r_p^5 M_\oplus} \frac{\csc i}{\phi}$$

$$\times \left\{ s_1 \frac{\partial s_1}{\partial \Omega} \left[\frac{1+4e^2}{2} E - (3+e^3)s + \frac{1+2e^2}{2} sc + \frac{e}{3} s^3 \right] \right.$$

$$+ \frac{\partial(s_1 s_2)}{\partial \Omega} \phi \left[\frac{1+e^2}{2} s^2 + ec + \frac{e}{3} c^3 \right]$$

$$\left. + s_2 \frac{\partial s_2}{\partial \Omega} \phi^2 \left[\frac{E}{2} - \frac{sc}{2} - \frac{e}{3} s^3 \right] \right\} \Bigg|_{E_0}^{E}$$

$$\Delta \Omega = \frac{3a^3 M_p}{r_p^5 M_\oplus} \frac{\csc i}{\phi} \left\{ s_1 \frac{\partial s_1}{\partial i} \left[\frac{1+4e^2}{2} E - (3e+e^3)s + \frac{1+2e^2}{2} sc + \frac{e}{3} s^3 \right] \right.$$

$$+ \frac{\partial(s_1 s_2)}{\partial i} \phi \left[\frac{1+e^2}{2} s^2 + ec + \frac{e}{3} c^3 \right]$$

$$\left. + s_2 \frac{\partial s_2}{\partial i} \phi^2 \left[\frac{E}{2} - \frac{sc}{2} - \frac{e}{3} s^3 \right] \right\} \Bigg|_{E_0}^{E}$$

$$\Delta \omega + \Delta \Omega \cos i = \frac{a^3 M_p}{r_p^3 M_\oplus} \frac{\phi}{e} \left((1+e^2)s - \frac{3}{2} eE - \frac{e}{2} sc - \frac{3}{r_p^2} \left\{ s_1^2 \left[\frac{s^3}{3} + (1+e^2)s - 2eE \right] \right. \right.$$

$$+ \frac{s_1 s_2}{\phi} \left[(3e^2-2)c - \frac{e+e^3}{2} c^2 + \frac{2-e^3}{3} c^3 \right]$$

$$\left. \left. + s_2^2 \left[\frac{eE}{2} - \frac{esc}{2} - \frac{s^3}{3} \right] \right\} \right) \Bigg|_{E_0}^{E}$$

$$\Delta \chi + (1-e^2)(\Delta \omega + \Delta \Omega \cos i) + t\,\Delta n$$

$$= \frac{a^3 M_p}{r_p^3 M_\oplus} \left\{ \left(2 + \frac{15}{4} e^2 + 3ec + \frac{3}{2} e^2 s \right) E + \frac{7}{6} e^3 s^3 \right.$$

$$- \left(9e + 2e^3 + \frac{3}{2} e^3 s^2 + 3e^2 c \right) e + \frac{21}{4} e^2 sc - \frac{3}{r_p^2}$$

$$\times \left(s_1^2 \left[\left(\frac{1}{4} + \frac{11}{2} e^2 + \frac{3}{2} s^2 + 3ec \right) E + \frac{7+14e^2}{4} sc \right. \right.$$

$$\left. + \frac{7}{6} es^3 - \left(9e + 2e^3 + \frac{3}{2} es^2 + 3e^3 c \right) s \right]$$

$$+ s_1 s_2 \phi \left[3s(e-c)E + \frac{7}{2}(1+e^2)s^2 + \frac{7}{3} ec^3 + 3es(c-e)s + 7ec \right]$$

$$\left. \left. + s_2^2 \phi^2 \left[\frac{7-6s^2}{4} E - \frac{7}{4} sc - \frac{7e}{6} s^3 + es^3 \right] \right] \right) \Bigg|_{E_0}^{E}$$

The reader may want to verify the identity of these two forms. On the other hand . . .

Lunisolar Secular Changes

Cook (1962) too has obtained approximate secular results for first-order changes assuming that the perturbing body is relatively distant. with $\eta = GM_p/r_p^3$ and λ, μ, ν the direction cosines to the perturbing body (i.e., a fixed third body again),

$$\Delta e = -\frac{15\pi\eta e(1-e^2)^{1/2}}{n^2}\left[\lambda\mu\cos 2\omega - \frac{1}{2}(\lambda^2-\mu^2)\sin 2\omega\right]$$

$$\Delta\Omega = \frac{3\pi\eta\nu\csc i}{2n^2(1-e^2)^{1/2}}[5\lambda e^2\sin 2\omega + \mu(2+3e^2-5e^2\cos 2\omega)]$$

$$\Delta i = \frac{3\pi\eta\nu}{2n^2(1-e^2)^{1/2}}[\lambda(2+3e^2+5e^2\cos 2\omega)+5\mu e^2\sin 2\omega]$$

$$\Delta\omega + \Delta\Omega\cos i = \frac{3\pi\eta(1-e^2)^{1/2}}{n^2}\left\{5\left[\lambda\mu\sin 2\omega + \frac{1}{2}(\lambda^2-\mu^2)\cos 2\omega\right]\right.$$
$$-1+\frac{3}{2}(\lambda^2+\mu^2)+\frac{1}{2r_p}\frac{5a}{e(1-e^2)^{1/2}}$$
$$\left.\times(\lambda\cos\omega+\mu\sin\omega)\left[1-\frac{5}{4}(\lambda^2+\mu^2)\right]\right\}$$

$$\Delta a = 0$$

The direction cosines are computed from the orbital data as follows:

$$\lambda = \cos(\Omega-\Omega_p)\cos u_p + \cos i_p\sin u_p\sin(\Omega-\Omega_p)$$

$$\mu = \cos i[-\sin(\Omega-\Omega_p)\cos u_p + \cos i_p\sin u_p\cos(\Omega-\Omega_p)]$$
$$+ \sin i\sin i_p\sin u_p$$

$$\nu = \sin i[\cos u_p\sin(\Omega-\Omega_p)-\cos i_p\sin u_p\cos(\Omega-\Omega_p)]$$
$$+ \cos i\sin i_p\sin u_p$$

where $u_p = \omega_p + v_p$.

For the Sun, $\Omega_p = 0$, $i_p = \varepsilon$, $u_p = L$, where L is the geometric mean longitude measured in the ecliptic from the equinox of date, and ε is the mean obliquity of the ecliptic. For the Moon the necessary elements are

$$\cos i_p = \cos\varepsilon\cos i_{\mathbb{C}} - \sin\varepsilon\sin i_{\mathbb{C}}\cos L_\Omega$$

$$\sin\Omega_p = \frac{\sin i_{\mathbb{C}}\sin L_\Omega}{\sin i_p}$$

$$u_p = L_{\mathbb{C}} - L_\Omega + \csc\left(\frac{\sin\varepsilon\sin L_\Omega}{\sin i_p}\right)$$

where $i_{\mathbb{C}}$ is the inclination of the lunar orbit to the ecliptic and L_{Ω} is the longitude of the mean ascending node of the lunar orbit on the ecliptic measured from the mean equinox of date. $L_{\mathbb{C}}$ is the mean longitude, measured in the ecliptic from the mean equinox of the date to the mean ascending node of the orbit, and then along the orbit. The effects of the Moon on a satellite are about twice those of the Sun.

ATMOSPHERIC DRAG

Atmospheric drag is the most complex and the most difficult of the important artificial satellite perturbations. It is complex because the functional form of the force law is not known (the formula used in Chapter Nine is a semi-empirical approximation) and the atmosphere is variable. The nature of this variability includes departures from spherical symmetry, lunisolar tides, diurnal heating, the strength of the solar wind, and so on. Moreover, the atmosphere is moving as the rotating Earth carries it along. Hence, this perturbation is hard to handle analytically or numerically with any accuracy. We shall limit the discussion to the static, exponential atmosphere, the usual drag formula, and nearly circular orbits. As discussed in Chapter Nine, one of the effects of drag is to circularize orbits.

The magnitude of the drag force is

$$F_D = \frac{A\rho C_D}{2m} v^2$$

where A is the effective cross-sectional area of the body, ρ is the atmospheric density at the geocentric distance r, C_D is the drag coefficient (~ 2), v is the satellite–atmosphere relative speed, and m is the mass of the satellite. The force acts, when the atmosphere is regarded as static, opposite to the satellite's velocity vector or in the negative tangential direction. The Gaussian equations for the variation of parameters are (my use of v for the true anomaly has finally caught up with me; in this section *only* v = speed, ν = true anomaly)

$$\frac{da}{dt} = -\left(\frac{AC_D\rho}{m}\right)\frac{v^2(1 + 2e\cos\nu + e^2)^{1/2}}{n(1 - e^2)^{1/2}}$$

$$\frac{de}{dt} = -\left(\frac{AC_D\rho}{m}\right)\frac{v^2(1 - e^2)^{1/2}(e + \cos\nu)}{na(1 + 2e\cos\nu + e^2)^{1/2}}$$

$$\frac{d\omega}{dt} = -\left(\frac{AC_D\rho}{m}\right)\frac{v^2(1 - e^2)^{1/2}\sin\nu}{nae(1 + 2e\cos\nu + e^2)^{1/2}}$$

$$\frac{di}{dt} = 0$$

$$\frac{d\Omega}{dt} = 0$$

$$\frac{dM}{dt} = n + \left(\frac{AC_D\rho}{m}\right)\frac{v^2(1-e^2)\sin\nu}{nae(1+e\cos\nu)}\left[\frac{1+e\cos\nu+e^2}{(1+2e\cos\nu+e^2)^{1/2}}\right]$$

In addition to the absence of first-order changes in the inclination and the longitude of the ascending node, the changes in the argument of perigee and mean anomaly (or time of perigee passage) appear to be periodic. This is assumed because each is proportional to $\sin\nu$ (this is patently false but is the usual procedure).

A form for $v^2 = \dot{r}^2 + r^2\dot{\nu}^2$ is

$$v^2 = \left(\frac{\mu}{p}\right)(1+2e\cos\nu+e^2)$$

where $p = a(1-e^2)$. After making the change of variable from t to E, one has

$$\frac{da}{dE} = -\left(\frac{AC_D\rho}{m}\right)a^2\left[\frac{(1+e\cos E)^3}{1-e\cos E}\right]^{1/2}$$

$$\frac{de}{dE} = -\left(\frac{AC_D\rho}{m}\right)a\left(\frac{1+e\cos E}{1-e\cos E}\right)^{1/2}(1-e^2)\cos E$$

The next step is to compute the secular rates of change of these two quantities assuming both an exponential model for the atmosphere,

$$\rho = \rho(0)\,e^{-r/R}$$

(where R is the atmospheric scale height) and that the eccentricity is small. These two approximations allow one to express the first-order secular changes of the semi-major axis and of the eccentricity in terms of modified Bessel functions of the first kind. Because we have already extracted most of the physics (the orbit is circularized) we go no further [see Escobal (1965), pp. 207–213].

THE SIMPLE HARMONIC OSCILLATOR

There is no problem in celestial mechanics that perturbation theory provides a satisfactory solution to. Saturn's rings, the Kirkwood gaps, Jupiter's satellites, the Sun–Jupiter–Saturn system, and so on, all exhibit resonances, unusual structures, or some other peculiarity not yet deduced from celestial mechanics. Celestial mechanics may provide *explanations* (as in the fact that the Kirkwood gaps occur at simple multiples of Jupiter's mean motion) but not *predictions.* Why is this state of affairs so poor? I believe that the reason is twofold (at least). First, celestial mechanics is a trajectory theory and not a field theory. The field theories with good perturbation techniques rely on linear second differential equations not nonlinear first-order ones (e.g., the Hamilton–Jacobi equation). Second, the mathematical nature of the problem in celestial mechanics is compounded by a poor choice of coordinate system. It does no good to express perturbation theory in terms of the orbital element set because we will then immediately and intuitively understand the result, if we can never solve any problems in this coordinate system. This becomes compounded when our frustration with our failures leads us astray in the use of unjustifiable analytical techniques. Examples of the latter include unjustified power series expansions or the method of averaging.

The reader who has skipped Chapter Nine might not be aware of the shortcomings and pitfalls of the misapplications of perturbation theory in celestial mechanics. This section will serve to inform the reader of the situation by a carefully constructed sequence of illustrations. Because the nature of the demonstration is by counterexample, it follows that the counterexample must itself be exactly soluble. Moreover, it is clear that no such meaningful example is likely to come from the normal realm of problems dealt with in celestial mechanics. However, it would be preferable to deal with a system as close to the two-body problem as possible. The existence of Bertrand's (1873) theorem is used to find the suitable connection and the alternative physical model—the three-dimensional, isotropic, simple harmonic oscillator (see Chapter Six). Bertrand's theorem simply states that of all possible central-force potential functions only two give rise to bounded, closed orbits. One is the two-body potential $(-k/r)$ and the other is the isotropic, simple harmonic oscillator $(kr^2/2)$. A perturbation theory for simple harmonic orbits is developed and a set of first-order differential equations is derived. These are exactly analogous to Lagrange's variation of parameter equations in the two-body problem.

Next we attempt to use the variation of parameters equations when an anisotropic perturbation is applied. This very closely follows Kozai's (1959) artificial satellite theory in format. The purpose is not to criticize Kozai but rather to illustrate the invalidity of a technique widely used in celestial mechanics. This section concludes with the development of a new set of variation of parameter equations utilizing a different set of constants (that is, other than the orbital element set). In this representation the perturbation equations are exactly soluble to all orders—indeed, the whole system is exactly soluble in closed form. The penultimate section introduces anharmonic perturbations. These too, in the right basis, allow a set of perturbation equations for which one can obtain the solution to all orders (explicitly) as well as in closed form. In the final section nonconservative perturbations due to drag and "third-body" forces for the harmonic oscillator are discussed. This is very brief as are some remarks concerning additional representations. I have proved my points and our energies should be devoted to productive rearrangements of perturbation theory for the two-body problem.

A Perturbation Theory

Because I want to consider more complicated physics than $U = kr^2/2$, and my analytical capabilities are not up to solving the more realistic problems, a perturbation theory is needed. Since our knowledge of analytical geometry is almost intuitive I'll feel more comfortable if the perturbation theory is expressed in terms of the orbital element set $\mathbf{a} = (a, e, \omega, i, \Omega, T)$ or some function thereof. Hence, confining myself to a conservative disturbing force $+m\nabla R$ (for the moment), I shall derive the perturbation equations equivalent to

$$m\ddot{\mathbf{r}} = -m\nabla U + m\nabla R$$

where U is the zeroth-order potential. I set $\mathbf{r} = \mathbf{r}(\mathbf{a}, t)$ and deduce that

$$\nabla_{\mathbf{a}} \frac{\partial \mathbf{r}}{\partial t} \cdot \dot{\mathbf{a}} = \nabla R$$

Along the way the condition of osculation on $\mathbf{r}$ was imposed,

$$\dot{\mathbf{r}} = \frac{\partial \mathbf{r}}{\partial t} \qquad \text{or} \qquad \nabla_{\mathbf{a}} \mathbf{r} \cdot \dot{\mathbf{a}} = 0$$

so that both locations and velocities can be computed from the usual (i.e., unperturbed) formulas. Finally I rearrange via Lagrange brackets and obtain

$$\sum_{k=1}^{6} [a_l, a_k]\dot{a}_k = \frac{\partial R}{\partial a_l} \tag{10.2}$$

where the Lagrange bracket $[a_l, a_k]$ is defined by

$$[a_l, a_k] = \left(\frac{\partial x}{\partial a_l} \frac{\partial \dot{x}}{\partial a_k} - \frac{\partial x}{\partial a_k} \frac{\partial \dot{x}}{\partial a_l} \right) + (x \rightarrow y) + (x \rightarrow x)$$

Note that $d[a_l, a_k]/dt = 0$ so that the Lagrange brackets can be evaluated at any convenient place in the orbit. I choose $M = 0$ ($t = T$). I further note that there are $\binom{6}{2} = 15$ independent brackets because a Lagrange bracket is antisymmetric and the most there could be is $6^2 = 36$. The last point of interest is to remember Eq. 6.33 for $\mathbf{r}$ and realize that there are three types of Lagrange brackets:

Type I: both a_l, a_k are one of i, ω, Ω of which there are $\binom{3}{2} = 3$ such.
Type II: a_l is one of i, ω, Ω but a_k is one of a, e, T (or M) of which there are $3^2 = 9$ such.
Type III: both a_l, a_k are one of a, e, T (or M) of which there are $\binom{3}{2} = 3$.

So far this is (formally) just like the two-body problem. In fact all of the brackets not involving a are the same as therein and the whole set of nonzero results is

$$[a, M] = -an(2 - e^2) = -[M, a]$$

$$[a, \Omega] = -2bn \cos i = -[\Omega, a]$$

$$[a, \omega] = -2bn = -[\omega, a]$$

$$[e, M] = a^2 ne = -[M, e]$$

$$[e, \Omega] = \frac{a^2 ne \cos i}{(1 - e^2)^{1/2}} = -[\Omega, e]$$

$$[e, \omega] = \frac{a^2 ne}{(1 - e^2)^{1/2}} = -[\omega, e]$$

$$[i, \Omega] = abn \sin i = -[\Omega, i]$$

Alternatively $[T, e] = a^2 n^2 e$, $[T, a] = an^2 e^2$.

The last step is to put the results into Eq. 10.2 and explicitly solve for $\dot{a}$ to obtain

$$\frac{da}{dt} = \frac{1}{ane^2} \frac{\partial R}{\partial M} - \frac{(1 - e^2)^{1/2}}{ane^2} \frac{\partial R}{\partial \omega}$$

$$\frac{de}{dt} = \frac{2(1 - e^2)}{a^2 ne^3} \frac{\partial R}{\partial M} - \frac{(2 - e^2)(1 - e^2)^{1/2}}{a^2 ne^3} \frac{\partial R}{\partial \omega}$$

$$\frac{d\omega}{dt} = \frac{(1 - e^2)^{1/2}}{a^2 ne^3} \left[(2 - e^2) \frac{\partial R}{\partial e} + ae \frac{\partial R}{\partial a} - \frac{e^3 \cot i}{1 - e^2} \frac{\partial R}{\partial i} \right]$$

$$\frac{di}{dt} = \frac{\csc i}{na^2 (1 - e^2)^{1/2}} \left(\cos i \frac{\partial R}{\partial \omega} - \frac{\partial R}{\partial \Omega} \right)$$

$$\frac{d\Omega}{dt} = \frac{\csc i}{na^2 (1 - e^2)^{1/2}} \frac{\partial R}{\partial i}$$

$$\frac{dM}{dt} = \frac{-1}{nae^2} \frac{\partial R}{\partial a} - \frac{2(1 - e^2)}{a^2 ne^3} \frac{\partial R}{\partial e}$$

A New Representation

The above "solution" of the perturbation problem leaves much to be desired—like an answer. The problem is the basis of the representation $a = (a, e, \omega, i, \Omega, T)$. We go back to the original, unperturbed equations of motion $(U = kr^2/2)$,

$$m\ddot{\mathbf{r}} = \mathbf{F} = -\nabla U = -k\mathbf{r}$$

The general solution is

$$\mathbf{r} = \mathbf{A} \cos nt + \mathbf{B} \sin nt$$

where $n^2 = k/m$. If we add a perturbing force $+m\nabla R$, then the values of $\mathbf{A}$ and $\mathbf{B}$ will no longer be constant. I now develop a perturbation theory for $\mathbf{a} = (\mathbf{A}, \mathbf{B})$. The work is done and is given by Eq. 10.2. Once again there are only 15 independent Lagrange brackets to evaluate, since 21 of them obviously vanish. They are also of three types

Type I: both a_l, a_k are one of the elements of $\mathbf{A}$ of which there are $\binom{3}{2} = 3$ such.

Type II: a_l is one of the elements of $\mathbf{A}$ but a_k is one of the elements of $\mathbf{B}$ of which there are $3^2 = 9$ such.

Type III: both a_l, a_k are one of the elements of **B** of which there are $\binom{3}{2} = 3$. All type I and III brackets vanish as do the "nondiagonal" type II. The only nonzero ones are

$$[A_x, B_x] = [A_y, B_y] = [A_z, B_z] = n$$

and their negatives. The simplicity of this result already makes one suspect that this is a very nice basis indeed.

The perturbed equations of motion are

$$m\ddot{\mathbf{r}} = -k\mathbf{r} - (\kappa - k)(0, 0, z)$$

Let $\kappa = k(1 + \varepsilon)$ with $|\varepsilon| \ll 1$. Then $R = -\varepsilon n^2 z^2 / 2$ or

$$R = \frac{-\varepsilon n^2}{2}(A_z \cos nt + B_z \sin nt)^2$$

The variation of parameters equations are simply

$$\dot{A}_x = \dot{A}_y = \dot{B}_x = \dot{B}_y = 0$$

$$\dot{A}_z = \varepsilon n(A_z \cos nt + B_z \sin nt) \sin nt$$

$$\dot{B}_z = -\varepsilon n(A_z \cos nt + B_z \sin nt) \cos nt$$

Much simpler than this it would be difficult to imagine. Moreover, because these equations are first-order, linear, ordinary differential equations with polynomial or exponential coefficients, Picard's method of successive approximations can be carried out, *explicitly*, to all orders.

Let the zeroth-order approximation be denoted by $A_z(0)$, $B_z(0)$. These are simply related to the initial values for z and $\dot{z}$ in the unperturbed case,

$$z(0) = A_z(0), \qquad \dot{z}(0) = nB_z(0)$$

Using primes to indicate the order in Picard's method, we have, successively,

$$\dot{A}'_z = \varepsilon n[A_z(0) \cos nt + B_z(0) \sin nt] \sin nt$$

$$\dot{B}'_z = -\varepsilon n[A_z(0) \cos nt + B_z(0) \sin nt] \cos nt$$

$$A'_z = A_z(0) + \left(\frac{\varepsilon}{2}\right) A_z(0) \sin^2 nt + \left(\frac{\varepsilon}{2}\right) B_z(0)(nt - \sin nt \cos nt)$$

$$B'_z = B_z(0) - \left(\frac{\varepsilon}{2}\right) B_z(0) \sin^2 nt - \left(\frac{\varepsilon}{2}\right) A_z(0)(nt + \sin nt \cos nt)$$

$$\dot{A}''_z = \varepsilon n(A'_z \cos nt + B'_z \sin nt) \sin nt$$

$$\dot{B}''_z = -\varepsilon n(A'_z \cos nt + B'_z \sin nt) \cos nt$$

$$A''_z = A'_z - \left(\frac{\varepsilon^2}{8}\right) A_z(0)[n^2 t^2 + (\sin nt - 2nt \cos nt) \sin nt]$$

$$+ \left(\frac{\varepsilon^2}{16}\right) B_z(0)(3 \sin 2nt - 4nt - 2nt \cos 2nt)$$

$$B_z'' = B_z' + \left(\frac{\varepsilon^2}{8}\right) B_z(0)[-n^2 t^2 + (3 \sin nt - 2nt \cos nt) \sin nt]$$

$$+ \left(\frac{\varepsilon^2}{16}\right) A_z(0)(\sin 2nt - 2nt \cos 2nt)$$

$$\dot{A}_z''' = n(A_z'' \cos nt + B_z'' \sin nt) \sin nt$$

$$\dot{B}_z''' = -n(A_z'' \cos nt + B_z'' \sin nt) \cos nt$$

and so on. Short-period terms, secular terms, and mixed terms can be seen in the above equations. Long-period terms will never appear, however. Another advantage of the **A, B** representation over the orbital element set representation should now be clear. Since Picard's method can be rigorously applied to all orders, all we need do is sum the series to get closed-form results for A_z and B_z. Of course, for this problem a much simpler course is available, for there is an even better basis than the **A, B** one.

Define α and β as

$$\alpha = A_z \cos nt + B_z \sin nt, \qquad \beta = -A_z \sin nt + B_z \cos nt \qquad (10.3)$$

They satisfy

$$\dot{\alpha} = n\beta, \qquad \dot{\beta} = -n\alpha(1+\varepsilon)$$

or

$$\ddot{\alpha} = -n^2(1+\varepsilon)\alpha$$

The solution to this can be written by inspection,

$$\alpha = C \cos \nu t + D \sin \nu t$$

where C and D are the arbitrary constants of integration and

$$\nu^2 = n^2(1+\varepsilon)$$

Also, $\beta = \dot{\alpha}/n = (\nu/n)(-C \sin \nu t + D \cos \nu t)$. We can recover A_z and B_z from Eqs. 10.3,

$$A_z = \alpha \cos nt - \beta \sin nt, \qquad B_z = \alpha \sin nt + \beta \cos nt$$

or

$$A_z = C\left[\cos nt \cos \nu t + \left(\frac{\nu}{n}\right) \sin nt \sin \nu t\right]$$

$$+ D\left[\cos nt \sin \nu t - \left(\frac{\nu}{n}\right) \sin nt \cos \nu t\right]$$

$$B_z = C\left[\sin nt \cos \nu t - \left(\frac{\nu}{n}\right) \cos nt \sin \nu t\right] \qquad (10.4)$$

$$+ D\left[\sin nt \sin \nu t + \left(\frac{\nu}{n}\right) \cos nt \cos \nu t\right]$$

We can also identify C and D as

$$C = A_z(0), \qquad D = \left(\frac{n}{\nu}\right) B_z(0) = \frac{B_z(0)}{(1+\varepsilon)^{1/2}}$$

Complete success—the exact, analytical solution of the perturbation equations.

It is appropriate now to reflect on what has been accomplished. First, a basis (a set of arbitrary constants) for the development of a perturbation theory has been found that yields a much simpler set of variation of parameter equations than one would have expected. Second, this system can be solved by actually implementing Picard's method of successive approximations. Hence, based on the discussion in Chapter One, we now know that a solution exists, it is continuous, it is unique, and so forth. Moreover, *for this problem*, the nth-order successive approximation in Picard's method is of order ε^n. This coincidence is due to the linearity of the differential equations for **A** and **B**. (See the next section for a counterexample to this result as a general proposition.) Third, the full set of variation of parameters equations is exactly soluble in closed form. This reflects the fact that the original problem was analytically tractable (since $\alpha \equiv z$, $\beta \equiv \dot{z}/n$) *and* the choice of the correct representation. Finally, should one choose to expand the direct solution for A_z, B_z in Eqs. 10.4 in a Taylor series in ε, one would recover, order by order, the successive steps of the Picard scheme.

One last point. An argument used in favor of the orbital element basis for perturbation theory was that doing so made the interpretation of the results easier. I claim that the form of Eqs. 10.4 makes it crystal clear that the three-dimensional anisotropic, simple harmonic oscillator is space filling.

The Anharmonic, One-Dimensional, Harmonic Oscillator

Anharmonic perturbations will now be discussed, and since the harmonic oscillator problem separates in rectangular coordinates, one dimension is sufficient. The unperturbed equations of motion are

$$m\ddot{x} = -kx, \qquad n^2 = \frac{k}{m}$$

whose general solution is

$$x = A \cos nt + B \sin nt$$

The perturbed equations of motion are

$$m\ddot{x} = -kx - \eta k x^3$$

so $R = -\eta n^2 x^4 / 4$ (we could have chosen a quadratic form for the anharmonic term, but this choice simplifies the analysis somewhat). The basis for the perturbation theory is $\mathbf{a} = (A, B)$. The only nonzero Lagrange bracket is $[A, B] = -[B, A] = n$. The variation of parameter equations are

$$\dot{A} = -\left(\frac{1}{n}\right)\frac{\partial R}{\partial B}, \qquad \dot{B} = \left(\frac{1}{n}\right)\frac{\partial R}{\partial A}$$

Since $R = -\eta n^2 x^4/4$,

$$R = -\left(\frac{\eta n^2}{4}\right)(A \cos nt + B \sin nt)^4$$

or

$$\dot{A} = \eta n(A \cos nt + B \sin nt)^3 \sin nt$$

$$\dot{B} = -\eta n(A \cos nt + B \sin nt)^3 \cos nt$$

Let A', B' be the first-order set of results from Picard's method; namely,

$$A'(t) = A(0) + \left(\frac{\eta}{32}\right)[8A^3(0)(1 - \cos^4 nt) + 3A^2(0)B(0)(4nt - \sin 4nt)$$

$$+ 24A(0)B^2(0) \sin^4 nt + B^3(0)(12nt - 8 \sin 2nt + \sin 4nt)]$$

$$B'(t) = B(0) - \left(\frac{\eta}{32}\right)[A^3(0)(12nt + 8 \sin 2nt + \sin 4nt) + 24A^2(0)B(0)$$

$$\times (1 - \cos^4 nt) + 3A(0)B^2(0)(4nt - \sin 4nt) + 8B^3(0) \sin^4 nt]$$

Clearly the next approximation, obtained from

$$\dot{A}'' = \eta n(A' \cos nt + B' \sin nt)^3 \sin nt$$

$$\dot{B}'' = -\eta n(A' \cos nt + B' \sin nt)^3 \cos nt$$

results in powers of η of order 4, not just order 2. Just as clearly an anharmonic perturbation of the equations of motion of the form $-\eta k x^{p-1}$ results in A'', B'' containing terms of the form η^p. I hope that this result clearly illustrates the differences between the order (in the sequence of successive approximations) of the solution and the functional form or highest power present of a small parameter.

The perturbation equations can be solved analytically. Define α and β by

$$\alpha = A \cos nt + B \sin nt$$

$$\beta = -A \sin nt + B \cos nt$$

Then, because of the condition of osculation,

$$\dot{\alpha} = n\beta, \qquad \dot{\beta} = -n\alpha - \eta n \alpha^3$$

or

$$\ddot{\alpha} = -n^2 \alpha - \eta n^2 \alpha^3$$

A first integral is

$$\dot{\alpha}^2 = n^2(L^2 - \alpha^2) + \left(\frac{\eta n^2}{2}\right)(L^4 - \alpha^4)$$

if the initial conditions are $x(0) = L$, $x(0) = 0$. Other initial conditions do not yield more transparent solutions or allow for additional analytical points of interest.

The full solution for α is

$$\alpha = L \operatorname{cn}[nt(1+\eta L^2)^{1/2}, \gamma]$$

$$\gamma^2 = \frac{L^2}{2L^2+2/\eta}, \qquad \gamma \in [0, 1]$$

where cn is the cosine amplitude Jacobian elliptic function. Since $\beta = \dot{\alpha}/n$,

$$\beta = -L(1+\eta L^2)^{1/2} \operatorname{sn}[nt(1+\eta L^2)^{1/2}, \gamma] \operatorname{dn}[nt(1+\eta L^2)^{1/2}, \gamma]$$

where sn is the sine amplitude Jacobian elliptic function and dn is the delta amplitude Jacobian elliptic function. The solutions for A and B are [$\mu = nt(1+\eta L^2)^{1/2}$]

$$\frac{A}{L} = \operatorname{cn}(\mu, \gamma) \cos nt + (1+\eta L^2)^{1/2} \operatorname{sn}(\mu, \gamma) \operatorname{dn}(\mu, \gamma) \sin nt$$

$$\frac{B}{L} = \operatorname{cn}(\mu, \gamma) \sin nt - (1+\eta L^2)^{1/2} \operatorname{sn}(\mu, \gamma) \operatorname{dn}(\mu, \gamma) \cos nt$$

As $\eta \to 0$

$$\gamma^2 \to \frac{\eta L^2}{2}$$

$$\operatorname{sn}(\mu, \gamma) \to \sin \mu - \left(\frac{\gamma}{2}\right)^2 \cos \mu \, (\mu - \sin \mu \cos \mu)$$

$$\operatorname{cn}(\mu, \gamma) \to \cos \mu + \left(\frac{\gamma}{2}\right)^2 \sin \mu \, (\mu - \sin \mu \cos \mu)$$

$$\operatorname{dn}(\mu, \gamma) \to 1 - \left(\frac{\gamma^2}{2}\right)^2 \sin^2 \mu$$

By some laborious algebra one can show that the first-order Taylor series expansions of A and B exactly match the expressions for A' and B' [$A(0) = L$, $B(0) = 0$]. Clearly, the statement is not true for the second-order Taylor series because A'' and B'' contain terms proportional to η^4. It is unclear if the fourth-order Taylor series expansions of A and B would match A'' and B'' (I have only a finite amount of patience but I would bet on it, however).

Lastly, the original equations of motion are also exactly soluble. For $x(0) = L$, $\dot{x}(0) = 0$, the solution is

$$x = L \operatorname{cn}(\mu, \gamma) \to L \cos nt \qquad \text{as } \eta \to 0$$

and for $x(0) = 0$, $\dot{x}(0) > 0$, the solution is

$$x = \frac{(L^2+2/\eta)^{1/2} \operatorname{sn}(\mu, \gamma)}{\operatorname{dn}(\mu, \gamma)}$$

where $\max(x) = L$. As $\eta \to 0$, x approaches $L \sin nt$. The general solution is neither a linear combination of these two (it is *an*harmonic) nor is it worth displaying.

Concluding Remarks

Another way of writing the general solution to the harmonic oscillator equations of motion (one dimensional for the moment) is

$$x = C \cos(nt + \phi)$$

The Lagrange bracket $[C, \phi] = -nC$ and the perturbation equations are

$$-nC\dot{\phi} = \frac{\partial R}{\partial C}, \qquad nC\dot{C} = \frac{\partial R}{\partial \phi}$$

But R is time dependent for either anharmonic or anisotropic perturbations. This obviously complicates the solution and this alternative has not been explored in depth. Also, the analog of Gauss's equation for nonconservative perturbations has not been derived. As an example, if there is a drag term,

$$m\ddot{x} = -kx - 2m\gamma\dot{x}$$

then the general solution can be written as (assuming γ is small compared to n, $n^2 = k/m$ still)

$$x = (A \cos \nu t + B \sin \nu t)\, e^{-\gamma t}$$

where $\nu^2 = n^2 - \gamma^2$. It would be interesting to pursue this problem.

An even more interesting generalization would be the simple harmonic oscillator subject to drag and an external, periodic force. The equation of motion would then be (say)

$$m\ddot{x} = -kx - 2m\gamma\dot{x} + F \cos(\omega t + \theta)$$

As discussed, the steady-state solution involves a resonance. The phase of x is offset from that of the external force by ψ,

$$\tan \psi = \frac{2\gamma\omega}{n^2 - \omega^2}$$

and the amplitude of x is $(F/m)/[(n^2 - \omega^2)^2 + 4\gamma^2\omega^2]^{1/2}$. When the atmospheric drag is small, there is a resonance at $\omega = n - \gamma^2/n$. (I wonder what classical perturbation theory would do with this.)

PROBLEMS

1. Consider the motion of an artificial satellite about an oblate Earth. Show that, in the mean, its ground trace will repeat for orbital frequency n if

$$\dot{\omega} + \dot{M} + n(\dot{\Omega} - \dot{\tau}) = 0$$

where the dotted quantities denote first-order secular rates of change and τ is the mean sidereal time. If the perturbation is solely due to J_2, derive a formula for the semi-major axis in terms of the other elements.

2. Prove the remarks made in the text concerning the instability of the 12 and 36-hr orbit for an elliptical equatorial cross section of the Earth.

3. Prove the remarks made in the text concerning the existence and stability of the 24-hr orbits for an elliptical equatorial cross section of the Earth.

4. For the one-dimensional harmonic oscillator perturbed by drag, deduce the "Gaussian" variation of parameter equations using $\mathbf{a} = (A, B)$ as the "orbital element set." Solve for the motion.

chapter eleven

Differential Correction

No representation of the forces acting on an artificial satellite, moon, asteroid, comet, or planet is exact. This is true either because some of the intrinsic properties of the body are not perfectly known, the locations of perturbing bodies are not precisely calculated, or some of the assumed properties of the perturbers are inaccurate. On the other side of the coin are our observations of such objects. Here again there is a lack of exactitude. Thus, no representation of the orbit of any solar system object is "correct." As new observations (with their new errors) become available and (somewhat less frequently) new theories of motion* are constructed, one should continuously refine the osculating orbital element set for each object. This is a correction process. As one's knowledge of the forces is usually pretty good and one's data not too badly corrupted with errors, it follows that the current set of osculating orbital elements is fairly close to the true set. Hence only differential (i.e., first order) forms of the corrections are entertained. Thus the title of this chapter.

There is one more element of a differential correction beyond the theory of motion and the observations—the method of correction. The most common mode is that of linearized least squares. This is outlined in detail in the second section of this chapter. Sequential methods based on Kalman filters and different probabilistic bases (such as the principle of maximum likelihood) are also in current vogue. An example of the formulation of a maximum likelihood based differential correction process for angular and angular rate data is outlined in the last section.

This chapter commences with a presentation of a completely analytical theory of motion and a second-order differential correction process applied to near-stationary artificial satellites. The analysis includes both distance and radial velocity data. This problem illustrates several things; (1) that there still are problems left for which an analytical theory is appropriate and within

---

* A theory of motion refers to an analytical, semi-analytical, or numerical representation of an object's ephemeris. All such theories are approximations to a full treatment of the motion and the term does not imply a competing structure of mechanics.

ordinary human patience, (2) the use of a nonlinear convergence scheme to a set of coupled, transcendental equations (the method of steepest descent), (3) a result successfully demonstrated both in numerical experiments and in the real world, and (let's hope) (4) a novel and transparent introduction to an obtuse subject where art counts as much as science.

After presenting the near-stationary differential correction process, differential correction in a more general form is expounded upon. As an illustration, a very powerful and pretty set of computer programs known as ANODE (ANalytical Orbit DEtermination) is discussed. This software and its analytical structure were created by Ramaswamy Sridharan and William P. Seniw (Sridharan and Seniw 1979, 1980) of Lincoln Laboratory's Surveillance Techniques (Millstone Hill Radar) Group. Finally, we close with the maximum likelihood differential corrector developed by Taff and Hall (1980).

NEAR-STATIONARY DIFFERENTIAL CORRECTOR

The Near-Stationary Differential Corrector (NSDC) was suggested to me by John M. Sorvari in late 1978. We developed it shortly thereafter (Taff and Sorvari 1979, 1982) and tested it in the late winter of 1979 (Taff 1979b). What did we do? We exploited the fact that for any orbital motion there exists a unique coordinate system in which the object is stationary. Consider, for example, an asteroid being observed from the Earth. Its apparent motion is complicated by the Earth's noncentral position in the solar system, the observer's diurnal parallax, the Earth's motions, and the minor planet's heliocentric motion. To eliminate these effects we first transform into a heliocentric coordinate system, then rotate into the asteroid's orbital plane, next align with the line of apsides, then rotate the coordinate system with the asteroid's instantaneous angular velocity, and lastly use the asteroid's instantaneous heliocentric distance as the unit of distance. This coordinate system rotates and pulsates with a period equal to the asteroid's orbital period, but the asteroid is fixed. Clearly to perform this type of coordinate system transformation one needs to know the orbit and the observer's location in space. However, if one had an approximate set of orbital elements for the asteroid, then one could construct a coordinate system in which the asteroid was *nearly* stationary. If, in addition, one were at the origin of this coordinate system, then the description of the departures from stationary motion would not only be small but would also be easily modeled.

We are at the origin of any topocentric coordinate system. There also happens to be a populous class of artificial satellites that are growing in number and importance, that are frequently maneuvered, and that are naturally nearly stationary in any topocentric coordinate system. The geocentric orbits of these satellites are also simple, having low inclination, small eccentricity, and mean motion $\simeq \dot{\tau} (\equiv 1.0027379093$ rev/day). There are no meaningful perturbations acting on them over a time span of a period ($\simeq 1$ mean sidereal day) so that the analysis will be the simplest, analytically, for this orbital type. Moreover, there are *two* natural topocentric coordinate systems in which the analysis can be performed and all "small" quantities of the theory have comparable magnitudes. There are also many practical problems having to do with searching for such satellites that require essentially instantaneous orbital element set construction. For all of these reasons the exploration of this new concept of differential correction is most efficiently performed for this type of orbit. The theoretical aspects of the near-stationary artificial satellite problem and its third-order solution for angles-only data in the topocentric equatorial coordinate system are developed immediately below. Also included are the distance and radial velocity results (to second order). By way of contrast, the analysis is also carried out, through terms of the second order, in the horizon system.

Formulation

A satellite whose motion will be modeled by this theory, for a maximum time duration $T_{max} > 0$, is defined to be a near-stationary satellite if

$$i \leq 0.15,\ e \leq 0.15,\ \left| \frac{n}{\dot{\tau}} - 1 \right| \leq 0.15,\ \text{and}\ |(n - \dot{\tau})T_{max}| \leq 0.15 \qquad (11.1)$$

The satellite's inclination is i, its mean motion is n, and its eccentricity is e. The mean sidereal time is τ and $\dot{\tau} = d\tau/dt$ where t is Ephemeris Time. T_{max} is the maximum valid duration of the theory. The numerical value of 0.15 is a stationary satellite's mean equatorial horizontal parallax and has been taken to be the upper limit for a small quantity in the theory. The theory developed below is complete through all third-order terms of this magnitude. It turns out, as with many other orbital analysis techniques, that the mean motion is a difficult quantity to accurately compute. However, it will also develop that as $n \to \dot{\tau}$ the practical limits on i and e rise to $\sim 25°$ and ~ 0.25, respectively.

Using standard notation to describe the Keplerian orbit (ω = argument of perigee, Ω = longitude of the ascending node, v = true anomaly, and r = geocentric distance), the starting point is

$$h = \tau - \Omega - \tan^{-1}(\cos i \tan u), \qquad \delta = \sin^{-1}(\sin i \sin u)$$

$$r = \frac{a(1 - e^2)}{1 + e \cos v}, \qquad u = v + \omega \qquad (11.2)$$

Here $h(H)$ is the geocentric (topocentric) hour angle, $\delta(\Delta)$ is the geocentric (topocentric) declination, and $r(R)$ is the geocentric (topocentric) distance. Once the appropriate approximate expressions have been developed for the geocentric variables, we will transform to the topocentric coordinate system via

$$\tan(H-h) = \frac{p \sin h}{1 - p \cos h}$$

$$\tan(\delta - \Delta) = \frac{q \sin(\gamma - \delta)}{1 - q \cos(\gamma - \delta)} \qquad (11.3)$$

$$R = r \sin(\delta - \gamma) \csc(\Delta - \gamma)$$

where (cf. Eqs. 3.23, 3.24, and 3.25)

$$p = \left(\frac{\rho}{r}\right) \cos \phi' \sec \delta, \qquad q = \left(\frac{\rho}{r}\right) \sin \phi' \csc \gamma$$

$$\tan \gamma = \tan \phi' \cos\left(\frac{H-h}{2}\right) \sec\left[h - \frac{H-h}{2}\right]$$

The observer's geocentric distance is ρ and ϕ' is his geocentric latitude.

Parameterization

When developing the formulas it turns out that the Keplerian orbital element set is not a convenient framework to use. Instead we shall use

$$N = n - \dot{\tau}, \qquad \lambda = \tau - \dot{\tau}T - (M_0 + \omega + \Omega)$$

$$E_c = e \cos M_0, \qquad E_s = e \sin M_0 \qquad (11.4)$$

$$I_c = i \cos(\omega + M_0), \qquad I_s = i \sin(\omega + M_0)$$

where M_0 is the value of the mean anomaly at the epoch t_0 and $T = t - t_0$. [That is, the mean anomaly has been written in the form $M = n(t - t_0) + M_0$.] We also abbreviate some frequently occurring combinations of the parameters besides $m = \dot{\tau}T$,

$$E = e \sin(m + M_0), \qquad E' = \frac{\dot{E}}{\dot{\tau}} = e \cos(m + M_0)$$

$$ \qquad (11.5)$$

$$I = i \sin(\omega + m + M_0), \qquad I' = \frac{\dot{I}}{\dot{\tau}} = i \cos(\omega + m + M_0)$$

Note that $d\lambda/dt \equiv 0$ so that the new parameters are constants.

Method. The procedure is straightforward: Express the true anomaly as a power series in e and NT, substitute into Eqs. 11.2, replace trigonometric functions of i by their Maclaurin series, and then expand everything through third order in e, i, and NT (cf. Eq. 11.1) using Eqs. 11.4 and 11.5. The results are $(\dot{\tau}^2 a_0^3 = GM_\oplus)$

$$h = \lambda - (NT+2E) - 2NTE' - \frac{5EE'}{2} + \frac{II'}{.2} - \frac{5NT(E'^2 - E^2)}{2}$$

$$+ \frac{(NT+2E)(I'^2 - I^2)}{2} + EN^2 T^2 + \frac{13E^3}{3} - 3E(E_c^2 + E_s^2)$$

$$\delta = I + I'(NT+2E) - \frac{I(NT+2E)^2}{2} + 2(NT+2E)E'I' - \frac{3EE'I'}{2} \qquad (11.6)$$

$$+ \frac{I^3}{6} - \frac{I(I_c^2 + I_s^2)}{6}$$

$$\frac{r}{a_0} = \left(1 - \frac{2N}{3\dot{\tau}}\right)(1 - E') + (NT+E)E + \frac{5N^2}{9\dot{\tau}^2}$$

We now repeat this type of analysis by using Eqs. 11.6 in Eqs. 11.3 and regarding ρ/a_0 as a small parameter. The results are

$$\frac{(H-h)\sec\phi'}{\rho/a_0} = \left[1 + \frac{2N}{3\dot{\tau}} + E' - \frac{(NT+2E)^2}{2}\right.$$

$$+ \frac{2NE'}{3\dot{\tau}} - NTE + E'^2 - E^2 - \frac{N^2}{9\dot{\tau}^2} + \frac{I^2}{2}\right]\sin\lambda$$

$$- \left\{(NT+2E) + 2NTE' + \frac{5EE'}{2} - \frac{II'}{2} + (NT+2E)\left[\frac{2N}{3\dot{\tau}}\right.\right.$$

$$\left.\left. + E'\right]\right\}\cos\lambda + \left(\frac{\rho}{a_0}\right)\left[\frac{1}{2} + \frac{2N}{3\dot{\tau}} + E'\right]\cos\phi'\sin 2\lambda$$

$$- \left(\frac{\rho}{a_0}\right)(NT+2E)\cos\phi'\cos\phi'\cos 2\lambda$$

$$+ \left(\frac{\rho}{a_0}\right)^2 \cos^2\phi'\sin 3\lambda/3$$

$$\frac{\Delta - \delta}{\rho/a_0} = -\left[1 + \frac{2N}{3\dot{\tau}} + E' + \frac{2NE'}{3\dot{\tau}} - NTE + E'^2\right.$$

$$- E^2 - \frac{N^2}{9\dot{\tau}^2} - \frac{I^2}{2}\right]\sin\phi' + \left\{I\left[1 + \frac{2N}{3\dot{\tau}} + E'\right]\right.$$

$$\left. + I'(NT+2E)\right\}\cos\phi'\cos\lambda - I(NT+2E)\cos\phi'\sin\lambda$$

$$- \left(\frac{\rho}{a_0}\right)\left\{1 + 2\left[\frac{2N}{3\dot{\tau}} + E'\right]\right\}\cos\phi'\sin\phi'\cos\lambda$$

$$
+ \left(\frac{\rho}{a_0}\right)\left(\frac{I}{2}\right)(2\cos^2\lambda - \sin^2\lambda)\cos^2\phi' + \left(\frac{\rho}{a_0}\right)[(NT
$$

$$
+ 2E)\cos\phi'\sin\lambda - I\sin\phi']\sin\phi' + \left(\frac{\rho}{a_0}\right)^2\left[\frac{\sin^2\phi'}{3}\right.
$$

$$
\left. + \frac{\cos^2\phi'(\sin^2\lambda - 2\cos^2\lambda)}{2}\right]\sin\phi' \qquad (11.7)
$$

$$
\frac{R}{a_0} = 1 - \left[\frac{2N}{3\dot{\tau}} + E'\right] - \left(\frac{\rho}{a_0}\right)\cos\phi'\cos\lambda + \frac{2NE'}{3\dot{\tau}} + NTE + E^2
$$

$$
+ \frac{5N^2}{(3\dot{\tau})^2} + \left(\frac{\rho}{a_0}\right)[(NT + 2E)\cos\phi'\sin\lambda - I\sin\phi']
$$

$$
+ \left[\frac{(\rho/a_0)^2}{2}\right](\sin^2\phi' + \cos^2\phi'\sin^2\lambda) \qquad (11.8)
$$

$$
\frac{\dot{R}}{a_0} = (N + \dot{\tau})E + (NT + 2E)\dot{\tau}E' + \left(\frac{\rho}{a_0}\right)[(N + 2\dot{\tau}E')\cos\phi'\sin\lambda
$$

$$
- \dot{\tau}I'\sin\phi'] - \frac{2NE}{3}
$$

Optimization.   We can now formulate the optimization problem. Suppose that we have N observations of topocentric hour angle and declination $\{\mathscr{H}_n, \mathscr{D}_n, t_n\}$, $n = 1, 2, \ldots, N$. We define [the factor of $\cos^2 \Delta_n$ multiplying $(\mathscr{H}_n - H_n)^2$ is effectively always unity for near-stationary satellites]

$$
S = \sum_{n=1}^{N} [(\mathscr{H}_n - H_n)^2 + (\mathscr{D}_n - \Delta_n)^2]
$$

and demand that S be a minimum with respect to N, λ, E_c, E_s, I_c, and I_s. H_n and Δ_n are expressions 11.7 evaluated at time T_n, and

$$
T_n = t_n - \langle t \rangle, \qquad \langle t \rangle = \sum_{n=1}^{N} \frac{t_n}{N} \equiv t_0
$$

The problem formulated is a nonlinear least-squares problem that takes cognizance of the fact that there are only six independent parameters. The method of steepest descent (explained below) seems to be the most promising way to proceed. In order to be consistent with the idea of a differential correction, we should start with a historical point $\{N, \lambda, E_c, E_s, I_c, I_s\}$ for each case. It turns out, however, that a single very simple point worked satisfactorily for most cases, namely,

$$
\lambda = \sum_{n=1}^{N} \frac{\mathscr{H}_n}{N}, \qquad \text{all other parameters} = 0
$$

Note in particular that setting E_c, E_s, I_c, and I_s equal to zero does not prejudice the values for ω, Ω, and M_0. This is important because there is little forcing changes in these angles in the gradient or Hessian matrix of S.

Now what can we anticipate for the numerical results? Because the first-order terms will carry most of the weight, we concentrate on them. From the second of Eqs. 11.6 and 11.7 both the inclination and $\omega + M_0$ can be well determined. From the first of Eqs. 11.6 and 11.7 one sees that $\omega + \Omega + M_0$ will be fixed, but that an eccentricity–mean motion swap is possible. This follows because as $T \to 0$ the coefficient of T in h is $-(N + 2\dot{r}e \cos M_0)$, and there will be no way to distinguish which part of this quantity is being contributed by $N \neq 0$ and which part is being contributed by $e \neq 0$. (When $T \to 0$, the constant part of $h \to \lambda - 2e \sin M_0$ but λ is of the zeroth order whereas e is of the first order. Hence, the numerical separation problem should not be as severe.) This difficulty (especially for N) can be partially ameliorated if the units of N are rev/day instead of rad/day because the factor of 2π accentuates small changes in N. When computations are done using the same data but with N in the different units, there can be a noticeable difference in the results. Equations 11.8 are appropriate for a radar and have not been tested by me.

Tests of the Theory

It was originally thought that this development would remove the necessity for a traditional differential correction procedure. Such procedures frequently have artificial singularities for $e = 0$ or $i = 0$. Hence, the second set of tests described below were performed first. It was also tested in real time during an artificial satellite search. Happily it worked extremely well. Table 9 lists (1) the total time duration of the first three observation sequences, (2) the number of observations performed in each of these observing sessions, (3) the time interval between the successive observing intervals, and (4) the ratio of the time difference between the start of the last and first observing intervals to the sum of the first and second observing durations. The last quantity can be taken as a measure of the "gain" of the technique. Quite clearly a gain of 10 can be achieved. In the intervening time other tasks can be performed by the telescope. No satellites were lost on account of the use of the technique. In addition, the actual orbital element sets converged quickly toward the true values (within the limitations discussed above and in the next section). The precision of the data was 10–15″.

To ascertain the limits of the theory, a series of longer duration numerical tests were performed on 11 different satellites. For each satellite one positional measurement per half hour was supplied for an entire night. The data had a precision of $\approx 20''$. The appropriate fits were made and the theory was used to predict the position at the beginning of the next evening ($T_{\max} \approx 0.8$ day). This was compared against the actual position at the time. The results are in Table 10 for the (1) second-order theory with N in rad/day, (2) the third-order theory with N in rad/day, and (3) the third-order theory with N in rev/day.

Table 9. Short-Time Observational Tests

Satellite Number	Observing Time Duration	Number of Observations	Successive Observing Time Interval	Ratio
1	13^m3	8	40^m4	10.8
	2.7	3	126.9	
	2.2	3		
2	5.9	6	31.7	7.7
	3.4	3	38.0	
	2.0	3		
3	4.1	5	22.9	5.4
	1.9	3	8.7	
	2.6	3		
4	7.4	5	24.6	7.2
	0.8	2	39.6	
	17.6	4		
5	4.0	5	23.0	8.7
	1.2	2	20.9	
	0.9	2		
6	7.1	5	163.0	25.6
	2.0	2	72.4	
	1.4	3		
7	4.3	6	18.0	9.5
	1.3	3	33.6	
	1.7	3		
8	5.3	6	12.7	3.3
	1.8	3	9.1	
	1.9	3		
9	3.7	5	21.0	10.8
	2.0	3	39.5	
	2.2	3		
10	4.7	5	54.7	10.4
	8.3	3	78.8	
	8.3	3		

Also listed is the Space Defense Center satellite identification number, the satellites' inclinations, eccentricities, and N values.

In general, the second-order theory yields the same results as does the third-order theory. The largest difference is for the International Ultraviolet Explorer satellite (10637), which clearly shows both the importance of the higher-order terms (in this instance) and that the limits for i and e in Eq. 11.1 can be considerably extended as $N \to 0$. It is also clear that as N departs from zero the accuracy very rapidly degrades. In fact, almost all of the error is in the hour angle (remember that the declination coordinate has no first-order secular term, see Eq. 11.7).

Table 10. Long-Time Numerical Tests

Satellite Number	i	e	$N = n - \dot{\tau}$ rev/day	Total Positional Error		
				Second Order rad/day	Third Order rad/day	Third Order rev/day
6278	0°00	0.000	−0.0000	0°10	0°1	0°0
83594	1.85	0.001	−0.0000	0.20	0.2	1.1
83598	6.36	0.003	−0.011	4.1	3.9	0.1
1317	11.44	0.000	+0.000	0.1	0.3	0.2
83517	1.56	0.105	0.015	5.5	6.3	14.6
83589	6.60	0.134	0.020	7.6	7.0	10.5
4632	9.41	0.144	0.200	70.8	69.9	35.0
83546	1.77	0.005	0.091	32.8	32.1	17.3
3623	7.86	0.011	−0.105	38.7	37.9	21.2
73505	14.21	0.138	0.014	4.9	3.4	10.2
10637	28.41	0.238	+0.000	4.6	1.0	20.8

In order to understand the relationship between the two different third-order results, we must remember that if N is in rev/day, there is a multiplier of 2π exaggerating $N \neq 0$ values. When N is larger this is good (satellites 4632, 83546, and 3623). When N is small but i and e large, this is bad (73505 and 10637). We conclude that a more traditional differential correction procedure could be much more profitably used than this one over these time spans. Of course, the initial orbital element set for the traditional procedure should be one obtained from this method.

The Horizon System

Because the satellite is nearly stationary, and the topocentric horizon system is centered on the observer, this appears to be an excellent coordinate system in which to work. The parallax correction in azimuth is also extremely simple. The drawback, in my view, is the requirement of undoing the direction cosines.

In addition to geocentric hour angle and declination (h, δ) we need geocentric zenith distance (z) and azimuth (A). A is measured from the south positive westward. Finally, if ϕ is the observer's astronomical latitude (earlier Φ was used for this quantity, but in this application the deflection of the vertical can be neglected), then

$$\sin z \sin A = \cos \delta \sin h$$

$$\sin z \cos A = \sin \phi \cos \delta \cos h - \cos c\phi \sin \delta$$

$$\cos z = \cos \phi \cos \delta \cos h + \sin \phi \sin \delta$$

Using Eqs. 11.2, this triplet can be converted to

$$\sin z \sin A = \cos u \sin(\tau - \Omega) - \cos i \sin u \cos(\tau - \Omega)$$

$$\sin z \cos A = [\cos u \cos(\tau - \Omega) + \cos i \sin u \sin(\tau - \Omega)] \sin \phi$$
$$- \cos \phi \sin i \sin u$$
$$\cos z = [\cos u \cos(\tau - \Omega) + \cos i \sin u \sin(\tau - \Omega)] \cos \phi$$
$$+ \sin \phi \sin i \sin u$$

We proceed as outlined above and find (through second order)

$$\sin z \sin A = \sin \lambda - (NT + 2E) \cos \lambda - \left(\frac{N^2 T^2}{2}\right) \sin \lambda$$

$$- 2NT(E \sin \lambda + E' \cos \lambda) - 2E^2 \sin \lambda - \left(\frac{5EE'}{2}\right) \cos \lambda$$

$$+ \left[\frac{(I_c^2 + I_s^2)^{1/2} I}{2}\right] \cos(\tau - \Omega)$$

$$\sin z \cos A = [\cos \lambda + (NT + 2E) \sin \lambda] \sin \phi - I \cos \phi$$

$$- \left(\frac{N^2 T^2}{2}\right) \sin \phi \cos \lambda + 2NT(E' \sin \lambda - E \cos \lambda) \sin \phi$$

$$- INT \cos \phi - 2EI' \cos \phi$$

$$- \left\{2E^2 \cos \lambda - \left(\frac{5EE'}{2}\right) \sin \lambda + [I(I_c^2 + I_s^2)^{1/2}] \sin(\tau - \Omega)\right\} \sin \phi$$

$$\cos z = [\cos \lambda + (NT + 2E) \sin \lambda] \cos \phi + I \sin \phi - \left(\frac{N^2 T^2}{2}\right) \cos \phi \cos \lambda$$

$$+ 2NT(E' \sin \lambda - E \cos \lambda) \cos \phi + NTI' \sin \phi + 2IE \sin \phi$$

$$- \left\{2E^2 \cos \lambda - \left(\frac{2EE'}{2}\right) \sin \lambda + \left[\frac{I(I_c^2 + I_s^2)^{1/2}}{2}\right] \sin(\tau - \Omega)\right\} \cos \phi$$

If $\mathscr{Z}$, $\mathscr{A}$ are the topocentric zenith distance and azimuth, then

$$\tan(\mathscr{A} - A) = \frac{P \sin A}{1 - P \cos A}, \qquad \tan(\mathscr{Z} - z) = \frac{Q \sin(z - \Gamma)}{1 - Q \cos(z - \Gamma)}$$

where (cf. Eqs. 3.19, 3.20, and 3.21)

$$P = \left(\frac{\rho}{r}\right) \sin(\phi - \phi') \csc z, \qquad Q = \left(\frac{\rho}{r}\right) \cos(\phi - \phi') \sec \Gamma$$

$$\tan \Gamma = \tan(\phi - \phi') \cos\left[\frac{\mathscr{A} + A}{2}\right] \sec\left[\frac{\mathscr{A} - A}{2}\right]$$

But

$$\phi - \phi' = \eta \sin 2\phi' + (\eta^2/2) \sin 4\phi' + \cdots$$

where

$$\eta = f + \frac{f^2}{2} + \cdots$$

and f is the flattening of the Earth, $1/297.25$. Therefore, through second order,

$$\mathscr{A} = A$$

$$\mathscr{L} = z + \left(\frac{\rho}{a_0}\right)\left[1 + \frac{2N}{3\dot{\tau}} + E + \left(\frac{\rho}{a_0}\right)\cos z\right]\sin z$$

Hence,

$$\sin \mathscr{L} \sin \mathscr{A} = \left[1 - \left(\frac{\rho}{a_0}\right)^2 \sin^2 \mathscr{L}\right]\sin z \sin A + \left(\frac{\rho}{a_0}\right)\left[1 + \frac{2N}{3\dot{\tau}}\right.$$

$$\left. + E + \left(\frac{\rho}{a_0}\right)\cos \mathscr{L}\right]\sin z \sin A \cos z$$

$$\sin \mathscr{L} \cos \mathscr{A} = \left[1 - \left(\frac{\rho}{a_0}\right)^2 \sin^2 \mathscr{L}\right]\sin z \cos A + \left(\frac{\rho}{a_0}\right)\left[1 + \frac{2N}{3\dot{\tau}}\right.$$

$$\left. + E + \left(\frac{\rho}{a_0}\right)\cos \mathscr{L}\right]\sin z \cos A \cos z$$

$$\cos \mathscr{L} = \cos z - \left(\frac{\rho}{a_0}\right)\left[1 + \frac{2N}{3\dot{\tau}} + E + 2\left(\frac{\rho}{a_0}\right)\cos \mathscr{L}\right]\sin^2 z$$

Not only must the direction cosines be eliminated to obtain explicit expressions for $\mathscr{L}$ and $\mathscr{A}$, but the problem is implicit. Hence, we opted for the (relative) simplicity of the equatorial coordinate system.

Why Does It Work?   To try to answer this question let us compare a traditional differential correction procedure with the one presented here. Because no one would propose developing a sophisticated, complex algorithm (not to mention the successful coding of it for an electronic digital computer) for use over such short arcs, it would be designed for longer arcs. As the length of the arc increases, the significance of the perturbing effects of the nonsphericity of the Earth, the presence of the Sun and the Moon, of air drag, of solar radiation pressure, and so on, all increase. Hence, the physics of a traditional differential correction procedure is much more elaborate and complex than that used herein. To make this investment profitable one needs precise, rigorously reduced data. In contradistinction, this technique should be able to produce the same results with simplified physics and with poorer quality, unreduced data. (This is *not* a testimonial to inaccurate data acquisition or reduction.) One also formulates the physics in an inertial reference frame (almost) rather than a noninertial reference frame. Therefore, the coordinate transformation handled here, explicitly and analytically, is handled implicitly and numerically. This complexity forces, as a practical matter, the search for the orbital elements (or geocentric initial conditions) to be an iterative linear one with the attendant numerical computation of the various partial derivatives that one needs. On the other hand we use a second-order solution technique and proceed exactly (within the constraints of the order of the theory) because we do it analytically. Another consequence of using a more correct physical model of the situation is the desire to simplify the computations as much as possible. This leads to the use of various analytical devices (e.g., averaging in first or second order). This frequently leads to artificial analytical singularities, typically at zero inclination, zero eccentricity, or the critical inclinations ($5 \cos^2 i = 1$). Because the model used here is pure Keplerian motion, there are no artificial singularities owing to the use of analytical devices (and certainly not for $e = 0$, $i = 0$, or at the critical inclinations!). Moreover, one does not expend the effort necessary to design a sophisticated, complex differential correction procedure for one type of orbit. This generality of the traditional methods, coupled with the vicissitudes of orbital analysis, means that the art of orbital analysis is frequently as important as is the science of orbital analysis for them to work successfully. In the near-stationary case human intervention is almost superfluous.* Finally, the analysis herein is performed in a coordinate system that makes the motion nearly stationary. This is not a feature of traditional procedures. These points are summarized in Table 11.

It would appear, over the arcs with which we are concerned here, that only the first four lines of Table 11 agree. Of these, it is probably the near-stationary aspect of the motion that is of pre-eminent significance. By accident (for the

Table 11. Comparison of Traditional and Present Differential Correction Procedures

Traditional DC	NSDC
Not near-stationary	Near-stationary
Implicit geometry, numerical	Explicit geometry, analytical
Numerical computation of partial derivatives	Analytical computation of partial derivatives
First-order, linear, solution technique	Second-order, nonlinear, solution technique
For all types of orbits	For one type of orbit
Artificial singularities	No singularities
Complex, accurate physics	Simple, approximate physics
Can involve considerable art	No art
Accurate data	Unreduced data

* An iterative fitting for the mean motion may help when n significantly departs from $\bar{n}$. Solving the problem using both the rad/day and rev/day and then performing a new solution is the maximum art that we could envision.

purpose of this discussion, not for the practical uses of near-stationary artificial satellites), there is a natural, topocentric coordinate system in which some real artificial satellites' motions are nearly stationary. If it is the nearly stationary aspect that really counts, then a major change in the short-time differential corrections of orbits may be at hand.

A Non-Near-Stationary Orbit. If we look back at the formulas presented earlier, we will see that explicit use of $\dot{n} \simeq \tau$ has been made twice. Once was in computing the satellite's parallax; the other instance was in the constancy of λ. If we consider any low inclination, small eccentricity orbit, with mean motion n_0, then the near-stationary constraints 11.1 would be modified to

$$i \leqslant \pi_0, \ e \leqslant \pi_0, \ \left| \frac{n}{n_0} - 1 \right| \leqslant \pi_0, \text{ and } |n - n_0| T_{\max} \leqslant \pi_0$$

where π_0 is the (mean) equatorial horizontal parallax of a satellite with mean motion $n_0 \{ \pi_0 = \sin^{-1}[R_\oplus (n_0^2 / GM_\oplus)^{1/3}] \}$. Then, if we redefine the parameters of the theory by $(m = n_0 T)$,

$$N = n - n_0, \qquad \lambda = \tau - n_0 T - (M_0 + \omega + \Omega)$$

$$E' = \frac{\dot{E}}{n_0}, \qquad I' = \dot{I}/n_0$$

all of the above analysis will remain valid. The theory is still of order π_0^3, but earlier comments concerning poor data or perturbations may need modification. Roughly, J_2 will become important when $J_2 \pi_0^2 \sim \pi_0^3$, lunar perturbations will become important when $M_{\mathrm{C}} \pi_{\mathrm{C}}^2 / (M_\oplus \pi_0^2) \sim 1$, and solar perturbations will become important when $M_\odot \pi_\odot^2 / (M_\oplus \pi_0^2) \sim 1$. This extension is not only trivial, it is not very important.

Consider instead the general problem of making any artificial satellite nearly stationary. (The heliocentric parallax of the Earth complicates all other astronomical problems and does not add anything. As the problem appears intractable anyhow we ignore it for now.) One must, at least, have estimates for i, Ω, n, and e. It seems clear that the argument of latitude $(u = v + \omega)$ is the variable to use. From

$$\cos u = \cos \delta \cos(\alpha - \Omega)$$

$$\sin u = \cos \delta \cos i \sin(\alpha - \Omega) + \sin i \sin \delta$$

we can write that

$$u(i, \Omega, \alpha, \delta) = u[i, \Omega, \alpha(r, A, \Delta), \delta(r, A, \Delta)]$$

$$= u\{i, \Omega, \alpha[r(n, e), A, \Delta], \delta[r(n, e), A, \Delta]\}$$

Let i_0, Ω_0, n_0, and e_0 be the initial values for i, Ω, n, and e. Then, to first order in $\Delta i = i - i_0$, and so on,

$$u \simeq u_0 + \frac{\partial u}{\partial i}\bigg|_0 \Delta i + \frac{\partial u}{\partial \Omega}\bigg|_0 \Delta\Omega + \left(\frac{\partial u}{\partial \alpha}\frac{\partial \alpha}{\partial r} + \frac{\partial u}{\partial \delta}\frac{\partial \delta}{\partial r}\right)\frac{\partial r}{\partial n}\bigg|_0 \Delta n$$

$$+ \left(\frac{\partial u}{\partial \alpha}\frac{\partial \alpha}{\partial r} + \frac{\partial u}{\partial \delta}\frac{\partial \delta}{\partial r}\right)\frac{\partial r}{\partial e}\bigg|_0 \Delta e$$

Here u_0 is not really $u(i_0, \Omega_0, n_0, e_0)$ since the dependence of the true anomaly on the mean motion and the eccentricity has not been taken into account. Thus, with $v_0 = v(n_0, e_0)$

$$u_0 \simeq v_0 + \omega_0 + \frac{\partial u}{\partial i}\bigg|_0 \Delta i + \frac{\partial u}{\partial \Omega}\bigg|_0 \Delta\Omega + \left[\frac{\partial v}{\partial n} + \left(\frac{\partial u}{\partial \alpha}\frac{\partial \alpha}{\partial r} + \frac{\partial u}{\partial \delta}\frac{\partial \delta}{\partial r}\right)\frac{\partial r}{\partial n}\right]\bigg|_0 \Delta n$$

$$+ \left[\frac{\partial v}{\partial e} + \left(\frac{\partial u}{\partial \alpha}\frac{\partial \alpha}{\partial r} + \frac{\partial u}{\partial \delta}\frac{\partial \delta}{\partial r}\right)\frac{\partial r}{\partial e}\right]\bigg|_0 \Delta e$$

Before, the three angles of the problem, ω, Ω, and M_0 [= the mean anomaly at the epoch time, *not* the initial guess for $M(T)$] naturally appeared in analytically convenient forms. ω and Ω (by default) still do, but M_0 does not. My numerical experience with the near-stationary satellite problem augers very poorly for the determination of M_0. Hence, either the formulation presented here needs modification or a more clever choice of variables is required.

One might inquire as to the advisability of again dealing directly with topocentric coordinates. Since the topocentric coordinate system is not now the unique one referred to earlier, such an analysis would not likely be of much benefit.

The Method of Steepest Descent

In this section we review the standard technique for solving a nonlinear estimation problem and introduce the method of steepest descent. Suppose that we have some function g of the unknown parameters $\mathbf{x}$ and that we wish to minimize

$$G = \sum_{j=1}^{N} g_j^2(\mathbf{x}) \tag{11.9}$$

The standard technique (e.g., generalized Newton–Raphson) to find the value of $\mathbf{x}$ (i.e., $\mathbf{x}_m$) that minimizes G is to first expand $g_j(\mathbf{x})$ about some guess for $\mathbf{x}_m$, say $\mathbf{x}_0$. Thus, to first order,

$$G = \sum_{j=1}^{N} [g_j(\mathbf{x}_0) + \nabla_\mathbf{x} g_j(\mathbf{x})|_{\mathbf{x}=\mathbf{x}_0} \cdot (\mathbf{x} - \mathbf{x}_0)]^2 \tag{11.10}$$

As given by Eq. 11.10, G is now a quadratic function of $\mathbf{x} - \mathbf{x}_0$ and the normal equations are formed as is usual (i.e., demand that $\nabla_\mathbf{x} G = 0$). After solving the normal equations for $\mathbf{x} - \mathbf{x}_0$ the new guess for $\mathbf{x}_m$ is $\mathbf{x}_n = (\mathbf{x} - \mathbf{x}_0) + \mathbf{x}_0$. If necessary, this iteration is repeated until $|\mathbf{x} - \mathbf{x}_0|$ is sufficiently small and $|\nabla_\mathbf{x} G|$ is sufficiently small too.

Now we return to Eqs. 11.9 and 11.10 but continue the expansion of $G(\mathbf{x} + \lambda \mathbf{X})$ to second-order terms. If $\Gamma(\mathbf{x})$ is the Hessian matrix of G (the matrix of second partial derivatives), then

$$G(\mathbf{x} + \lambda \mathbf{X}) = G(\mathbf{x}) + \nabla_\mathbf{x} G(\mathbf{x}) \cdot \lambda \mathbf{X} + (\lambda^2/2) \mathbf{X} \cdot \Gamma(\mathbf{x}) \cdot \mathbf{X}$$

Since we are looking for a minimum of G, we take $\mathbf{X}$ to be in the direction of the maximum rate of decrease of G, namely,

$$\mathbf{X} = -\nabla_\mathbf{x} G(\mathbf{x})$$

With this value for $\mathbf{X}$ we can find the value of λ to use by insisting that $G[\mathbf{x} - \lambda \nabla_\mathbf{x} G(\mathbf{x})]$ be a minimum with respect to λ. The result is

$$\lambda = |\nabla_\mathbf{x} G(\mathbf{x})|^2 / [\nabla_\mathbf{x} G(\mathbf{x}) \cdot \Gamma(\mathbf{x}) \cdot \nabla_\mathbf{x} G(\mathbf{x})] \tag{11.11}$$

Thus, starting from a guess $\mathbf{x} = \mathbf{x}_0$ for $\mathbf{x}_m$, we compute the new guess from

$$\mathbf{x}_n = \mathbf{x}_0 - \lambda \nabla_\mathbf{x} G(\mathbf{x})|_{\mathbf{x} = \mathbf{x}_0} \tag{11.12}$$

with λ evaluated via Eq. 11.11 at $\mathbf{x} = \mathbf{x}_0$. If $G(\mathbf{x})$ has a unique minimum (e.g., $\mathbf{x}_m$) in any closed region and the metric defined by $\Gamma(\mathbf{x})$ has a positive upper bound in this region, then this sequence of iterations does converge to $\mathbf{x}_m$. For most practical problems this technique is difficult to apply because one needs accurate values for $\Gamma(\mathbf{x})$. In the case of NSDC we have analytic expressions for all of the second derivatives and we implemented Eqs. 11.11 and 11.12.

ANODE

Differential Correction in General

Let us start by taking a generalized look at the differential correction process. We have a set of observables consisting of location and velocity in some coordinate system. Say $\mathbf{R}$ and $\mathbf{V}$ to distinguish them from the theoretical values $\mathbf{r}$ and $\mathbf{v}$. The latter come from the theoretical structure (the theory of motion) that has been developed to predict location and velocity with some set of initial conditions, say $\mathbf{r}_0$ and $\mathbf{v}_0$. The prediction can be symbolized by two propagators,

$$\mathbf{r}(t) = \chi(\mathbf{r}_0, \mathbf{v}_0, t, t_0; \mathbf{p}), \qquad \mathbf{v}(t) = \psi(\mathbf{r}_0, \mathbf{v}_0, t, t_0; \mathbf{p})$$

where the vector functions χ and ψ contain the parameters of the theory $\mathbf{p}$. Typical elements of χ or ψ could be the values of the coefficients in the geopotential expansion, the atmospheric density, the observer's location, and so on. Ideally

$$\mathbf{R}(t) = \mathbf{r}(t), \qquad \mathbf{V}(t) = \mathbf{v}(t) \; \forall t$$

for whatever subset of location and velocity is observed. In practice this is not the case, and the discrepancies allow one to form the equations of condition. This is done by supposing that small changes in $\mathbf{r}_0$, $\mathbf{v}_0$, and $\mathbf{p}$ are necessary to reduce the residuals. Hence we write

$$\mathbf{R}(t) = \mathbf{r}(t) + \frac{\partial \boldsymbol{\chi}}{\partial \mathbf{r}_0} \cdot \Delta \mathbf{r}_0 + \frac{\partial \boldsymbol{\chi}}{\partial \mathbf{v}_0} \cdot \Delta \mathbf{v}_0 + \frac{\partial \boldsymbol{\chi}}{\partial \mathbf{p}} \cdot \Delta \mathbf{p} + \boldsymbol{\varepsilon}_R$$

$$\mathbf{V}(t) = \mathbf{v}(t) + \frac{\partial \boldsymbol{\psi}}{\partial \mathbf{r}_0} \cdot \Delta \mathbf{r}_0 + \frac{\partial \boldsymbol{\psi}}{\partial \mathbf{v}_0} \cdot \Delta \mathbf{v}_0 + \frac{\partial \boldsymbol{\psi}}{\partial \mathbf{p}} \cdot \Delta \mathbf{p} + \boldsymbol{\varepsilon}_V$$

(this is purely formal). The residuals that we cannot account for are symbolized by the epsilon terms. If we adapt the least-squares point of view, then we wish to adjust $\mathbf{r}_0$, $\mathbf{v}_0$, and $\mathbf{p}$ so as to minimize

$$\sum (\boldsymbol{\varepsilon}_R^2 + \boldsymbol{\varepsilon}_V^2)$$

where the sum is over all of the observations available (appropriately weighted). Notice too that by the appropriate use of a covariance matrix for the data any mixture of any sequence of observables can be utilized in the differential correction process. Furthermore we can alternatively keep various elements of $\mathbf{r}_0$, $\mathbf{v}_0$, or $\mathbf{p}$ fixed on the assumption that we know them especially accurately (or just to see what happens to the various fits to the data). This, in brief, is the concept and method of the differential correction process.

The Conceptual Structure of ANODE

The Millstone Hill Radar (run by M.I.T.'s Lincoln Laboratory) has been active since February 1975 as a contributing sensor to the deep-space satellite detection and tracking network of the North American Aerospace Defense Command (Space Track). The primary mission of the radar has been the detection and tracking of deep-space satellites. (Satellites whose orbital periods are greater than 220 min have been arbitrarily classified as in the deep-space regime.) The computer system at the Millstone Hill radar was upgraded in August 1977 and a new software system design was made at that time. One of the components of the system is an on-line, real-time, analytic orbit determination program. The purpose of the program is threefold: (1) It is intended to act as a real-time monitor on the tracking performance of the radar; (2) it is designed to function as a rapid orbit estimator available interactively to the analyst; and (3) it is a preprocessor for an accurate but computationally expensive numerical orbit estimation program.

ANODE is the *AN*alytic Orbit *DE*termination program that has been developed at Millstone to serve these functions. Special attention in the theory, structure, and functioning of ANODE as an orbit propagation and estimation program has been paid to its computer implementation and efficiency, as it operates in the real-time environment of a tracking radar. The high-altitude satellites of concern to the Millstone radar are perturbed by departures from spherical symmetry in the geopotential, lunisolar attractions, and atmospheric drag.

A substantial body of theory has been developed over the years for the evolution of orbits of artificial satellites using a variety of techniques. The most commonly used technique has been the method of averages. Formulations using the method of averages differ in their choice of orbital elements. Sridharan and Seniw (1979, 1980) decided to adopt the Keplerian elements of the orbit as the basis for the development of the software. The method of averages proceeds to develop (see Chapters Nine and Ten) a theory of motion by eliminating the fastest periodic variable in the system of differential equations. Let

$$\frac{d\mathbf{a}}{dt} = \mathbf{f}(\mathbf{a}, \mathbf{F})$$

where $\mathbf{a}$ is the set of orbital elements and $\mathbf{F}$ is the perturbing force. Typically the mean anomaly M is the fastest variable in the system of equations with a period P, the orbital period of the satellite. The averaged rate of change $\mathbf{a}$ is defined as

$$\left\langle \frac{d\mathbf{a}}{dt} \right\rangle = \frac{1}{P} \int_0^P \mathbf{f}(\mathbf{a}, \mathbf{F}) \, d\tau$$

This averages the fastest variable out of the system of equations. To render the theory complete the short-periodic variations must also be computed from

$$\Delta \mathbf{a}_{sp} = \int^t \left[\mathbf{f}(\mathbf{a}, \mathbf{F}) - \left\langle \frac{d\mathbf{a}}{d\tau} \right\rangle \right] d\tau$$

The resulting singly averaged dynamical system for the orbit exhibits a periodic dependence on ω, the argument of perigee. Hence, conventional orbital theory proceeds to average over the period of the argument of perigee, P_ω, thus resulting in a secular or doubly averaged theory

$$\left\langle \left\langle \frac{d\mathbf{a}}{dt} \right\rangle \right\rangle = \frac{1}{P_\omega} \int_0^{P_\omega} \left\langle \frac{d\mathbf{a}}{d\tau} \right\rangle d\tau$$

Periodic terms from this step of averaging are called the long-periodics and are given by $(P_\omega \sim P/|J_2|)$

$$\Delta \mathbf{a}_{LP} = \int^t \left[\left\langle \frac{d\mathbf{a}}{d\tau} \right\rangle - \left\langle \left\langle \frac{d\mathbf{a}}{d\tau} \right\rangle \right\rangle \right] d\tau$$

Thus, if one uses the doubly averaged results, a complete theory would consist of the doubly averaged or secular terms, the long-periodic terms, and the short-periodic terms.

Several different formulations are available in the literature for an orbital theory in the Keplerian elements. The formulation by Liu (1974) was adopted for ANODE because (1) it includes the zonal harmonics of the geopotential with terms owing to J_{20}, J_{20}^2, J_{30}, and J_{40}, (2) both the singly averaged and the doubly averaged results are given, and (3) expressions are given explicitly for the rates of change of the Keplerian elements. Two points about theories such as this are worthy of note. First, depending on the elements and the method used, a theory may have associated singularity problems. These will be discussed below. Second, any theory that is developed by "hand algebra" tends to be inefficient in terms of its computer implementation. In particular, advantage is rarely taken of commonality of expressions to reduce the computations involved. Every effort was made in the implementation of the theory in ANODE to identify common expressions and precompute them before computing any of the secular or periodic corrections to the elements. The effort was probably incomplete, but it still resulted in a significant saving of computer time. The most computationally compact form of the equations can be achieved only by a rederivation of the theory using computer algebra. (See below for a brief discussion of algebraic manipulators.)

Liu's theory is restricted to the effects of the first few terms in the zonal harmonics of the geopotential. However, in high-altitude orbits, the perturbations due to lunisolar gravitation are significant—particularly so in near-stationary orbits where the gravitational perturbations of the Moon can be a tenth to a half as large as that of the principal oblateness term (J_{20}) of the Earth. An interesting technique called "intermediate" averaging was developed during the course of this work in order to efficiently capture lunisolar perturbations in the theory. The theoretical basis of this technique is described below. Finally, satellites in high eccentricity orbits about the Earth are subject to atmospheric drag as their perigee heights decrease. An old suggestion of King–Hele was adopted to account for drag effects. Namely, since satellites in high eccentricity orbits are in a significant atmospheric drag regime (defined as <500 km elevation) for a very short interval of time (5–15 min) relative to the orbital period, it is feasible to represent the drag effect as an impulse applied at perigee.

Major users within the Department of Defense community have adopted the WGS-72 model of geopotential constants as the standard for all orbital computations. The same set is used for ANODE also. However, this model does not list any adopted values of the ratios of the masses of the Sun and the Moon to that of the Earth. Hence, these constants are taken from the list of recommended constants of the International Astronomical Union (see the Appendix).

The Analytical Structure of ANODE

The distinctive feature of the theory for ANODE lies in the double averaging of the differential equations—once over the period of the orbit and then over the period of the Moon. This technique is called intermediate averaging to distinguish it from the conventional doubly averaged theories and to denote the time scales inherent in the averaging technique. The alternative technique to analytical theories is the accurate, relatively compact, but computationally expensive method of the direct numerical integration of the differential equations of motion [see below for a brief review of this and Dallas (1970) for a bibliography of analytical theories].

By far, the most common technique that has been applied in the development of analytical theories is that of averaging over the time scales inherent in the differential equations. An exposition of these techniques can be found in Lorell et al. (1964). The characteristic features of the theories of motion so developed are: (1) They are expressed as analytical formulas for the rates of change of mean elements, (2) the theories are generally either to first or to second order in the expansion of the geopotential (higher-order theories do exist and are employed occasionally), (3) the first averaging is over the mean anomaly of the satellite (which is the fastest periodic variable in the system) with the necessary short-periodic terms also being formulated, and (4) the second averaging is over the period of the argument of perigee (which is typically the second fastest variable for low-altitude orbits) and the necessary long-periodic terms are also derived. As mentioned above, a complete theory would then consist of the doubly averaged secular rates of change of the mean elements along with the long-periodic and the short-periodic terms.

The Gaussian form of the variational equations was used to develop an alternative theory for high-altitude satellite orbits. The equations are;

$$\frac{da}{dt} = 2\frac{a}{\varepsilon}\left(\frac{p}{\mu}\right)^{1/2}\left[eF_1 \sin v + F_2 \frac{p}{r}\right]$$

$$\frac{de}{dt} = -\frac{r}{ae}\left(\frac{p}{\mu}\right)^{1/2} F_2 + \frac{\varepsilon}{2ae}\frac{da}{dt}$$

$$\frac{d\Omega}{dt} = \frac{r \csc i}{(\mu p)^{1/2}} F_3 \sin u$$

$$\frac{di}{dt} = \frac{r}{(\mu p)^{1/2}} F_3 \cos u$$

$$\frac{d\omega}{dt} = \frac{1}{e}\left(\frac{p}{\mu}\right)^{1/2}\left[-F_1 \cos v + F_2\left(1+\frac{r}{p}\right)\sin v\right] - \frac{d\Omega}{dt}\cos i$$

and

$$\frac{dM}{dt} - n = -\frac{2r}{(\mu a)^{1/2}} F_1 - \varepsilon^{1/2}\left(\frac{d\Omega}{dt}\cos i + \frac{d\omega}{dt}\right)$$

where $\varepsilon = 1 - e^2$, $p = a\varepsilon$, $\mu = GM_\oplus$, and F_1, F_2, F_3 are the components of the perturbing force along the instantaneous radial, transverse, and normal directions to the satellite orbit.

Certain features of the above equations are worthy of note. The Gaussian form has been altered slightly in the equations for e, ω, and M in order to take advantage of commonality of terms (see Chapter Nine). Also, the differential equations are linear in the perturbing forces. Finally, singularities at $e = 0$ or $i = 0$ exist; these will be commented on later.

The perturbations considered here result from the zonal harmonic terms in the geopotential and the point mass gravitational effects of the Sun and the Moon. The perturbing force due to the zonal harmonics can be written as

$$\mathbf{F}_n = \frac{\mu J_n}{r^2}\left(\frac{R_\oplus}{r}\right)^n [(n+1)P_n\mathbf{e}_r - P_n' \sin i \cos u\mathbf{e}_t - P_n' \cos i\mathbf{e}_n]$$

where n = degree of the zonal harmonic
         J_n = nth zonal harmonic coefficient
         $R_\oplus$ = equatorial radius of the Earth
         P_n = nth order Legendre polynomial in the argument $\sin \delta$
         δ = geocentric latitude of the satellite
         $P_n' = dP_n/d(\sin \delta)$
    $\mathbf{e}_r, \mathbf{e}_t, \mathbf{e}_n$ = instantaneous unit vectors along the radial, transverse, and nor-
         mal directions

The perturbing force due to a third body (the Moon or the Sun) can be written as

$$\mathbf{F}_q = \frac{\mu_k}{r_k^2}\left(\frac{r}{r_k}\right)^q [-P_q'\mathbf{e}_r + P_{q+1}'\mathbf{e}_k]$$

where q = order of the term in the Legendre polynomial expansion of the
         perturbing force
    μ_k = gravitational constant of the kth body (i.e., G times its mass)
    r_k = geocentric distance to the kth body
    P_q = qth-order Legendre polynomial in the argument $\mathbf{e}_r \cdot \mathbf{e}_k$
    $\mathbf{e}_k$ = instantaneous geocentric unit vector to the third body

Geopotential zonal harmonic terms up to order 4 will be included in the theory, but only the first term in the expansion of the third-body perturbation will be considered. Neglected terms are at least an order of magnitude smaller, for satellites in half-synchronous orbits, and at least a factor of 4 smaller for synchronous satellites.

The Averaging. The first step in developing the theory is to average the differential equations over the fastest variable on the right-hand side of the differential equations (the mean anomaly of the satellite). The results of Liu (1974) for the singly averaged equations resulting from the geopotential were adopted. Liu tabulates both the averaged rates and the short-periodic terms, the latter being computed only to order J_2.

The third-body perturbations must also be averaged over the period of the satellite. The relevant results for the averaged effects are taken from Sridharan and Renard (1975). The short-periodic terms caused by the third body are in Sridharan and Seniw (1979).

The singly averaged equations deduced from the geopotential perturbation show a periodic dependence on the argument of perigee. However, the rate of change of the argument of perigee in high-altitude orbits is less than $1°/$day. Thus, P_ω is of the order of a year or longer. The singly averaged equations, as a result of the presence of third-body perturbations, exhibit dependence on the position of the third body and hence on its period. For the Moon the period is only 29 days. Hence, the fastest variable left in the singly averaged equations is the mean anomaly of the Moon. Mathematically,

$$\left\langle \frac{d\mathbf{a}}{dt} \right\rangle = \mathbf{f}_G(\mathbf{a}) + \mathbf{f}_T(\mathbf{a}, M_T)$$

where

$\langle \; \rangle =$ singly averaged rates
$\mathbf{f}_G =$ terms from the geopotential
$\mathbf{f}_T =$ terms from the third body
$M_T =$ mean anomaly of the third body
$\mathbf{a} =$ orbital element set of the satellite's orbit

The next step in the method of intermediate averaging is

$$\left\langle\!\!\left\langle \frac{d\mathbf{a}}{dt} \right\rangle\!\!\right\rangle = \mathbf{f}_G(\mathbf{a}) + \frac{1}{P_{\!\mathbb{C}}} \int_0^{P_{\!\mathbb{C}}} \mathbf{f}_T(\mathbf{a}, M_T) \, d\tau$$

where $P_{\mathbb{C}}$ is the Moon's period of revolution about the Earth. Neither the geopotential terms nor the third-body terms caused by the Sun are affected by this step. The theory would not be complete without the intermediate periodic terms whose basic period is that of the Moon. These terms have been derived and are also given in the above-mentioned references.

The original differential equations are singular at $e = 0$ or $i = 0$, and these singularities carry over. As the eccentricity goes to zero, the argument of the perigee becomes ill defined. Hence, the short-periodic corrections in the mean anomaly and in the argument of perigee become large. However, these two corrections are of nearly equal magnitude and opposite in sign. Thus the position of the satellite in the orbit, as defined by $\omega + M$, does not exhibit any singularity. Unless the eccentricity becomes so small that the precision of computation affects results (an unlikely case), the theory remains valid.

As the inclination goes to zero, the line of nodes, and hence the value of Ω, becomes ill defined. The theory exhibits this singularity in the secular terms. Once more the change in the value of Ω is nearly equal and opposite in sign to that of ω. Thus, the position of perigee can be defined using the value of $\omega + \Omega$. Another solution is also possible: The reference plane can be changed from the equator to, say, the ecliptic, or preferably a plane containing the Earth's polar axis. Very little needs to be changed in the theory to accommodate such a plane change. Finally, the classic problem of singularity at critical inclination does not exist in this theory as no averaging is carried out over the period of the argument of perigee.

The Computational Structure of ANODE

ANODE is an orbit estimation program. Thus it has to have access to orbital elements and observational data, it has to compute pointing and residuals at sundry observation times, and it has to estimate corrections to the orbital elements in order to fit the data. This section is devoted to a computational description of the structure of the program that performs these functions.

Table 12 is an outline of the structure of ANODE. The input data to the orbit processor consists of a set of orbital elements and a sequence of observational data points (i.e., the satellite "metric" data base). These data may be from any type of instrument. The data consist of observation times associated with values for one or more of right ascension, declination, distance, or radial velocity. The orbit estimation is carried out in terms of the mean Keplerian elements. (Mean elements are calculated by subtracting the periodic corrections from the osculating elements. The definition of the mean elements is inextricably tied to the theory.)

Let the mean elements at time t_0 be $\mathbf{a}_0$. Define $\dot{\mathbf{a}}_0$ to be equal to the secular (or averaged) rates of change. Let the times of observation be $t_1, t_2, \ldots, t_N$. Then at time t_k

$$\mathbf{a}_k = \mathbf{a}_0 + (t_k - t_0)\dot{\mathbf{a}}_0, \qquad k = 1, 2, \ldots, N$$

The osculating elements at t_k can be calculated by

$$\mathbf{a}_{\mathrm{osc},k} = \mathbf{a}_k + (\delta \mathbf{a})_p$$

where $(\delta \mathbf{a})_p$ are the periodic corrections calculated for the location at the instant t_k. Furthermore, the differentials of the elements at t_k with respect to the elements at epoch t_0 are

$$\frac{\partial \mathbf{a}_k}{\partial \mathbf{a}_0} = \mathcal{I} + (t_k - t_0) \frac{\partial \dot{\mathbf{a}}_0}{\partial \mathbf{a}_0}$$

where $\mathcal{I}$ is the 6×6 unit matrix.

The next step is the calculation of the residuals. The computed pointing at time t_k is given by

$$\mathbf{C}_k = \mathbf{C}(\mathbf{a}_{\mathrm{osc},k}, \boldsymbol{\rho})$$

where $\mathbf{C}_k$ is the pointing in location and radial velocity, $\mathbf{a}_{\mathrm{osc},k}$ is the set of osculating elements at $t = t_k$, and $\boldsymbol{\rho}$ is the observer's location. Given $\mathbf{P}_k$, the actual observations, the residuals are

$$\mathbf{Y}_k = \mathbf{P}_k - \mathbf{C}_k \qquad \text{for } k = 1, 2, \ldots, N$$

A complete vector of residuals is thus built up. This vector has a maximum size of $4N$. The differentials of the residuals with respect to the mean elements at epoch t_0 are computed according to

$$\frac{\partial \mathbf{Y}_k}{\partial \mathbf{a}_0} = -\frac{\partial \mathbf{C}_k}{\partial \mathbf{a}_0}$$

Table 12. Outline of the Structure of ANODE

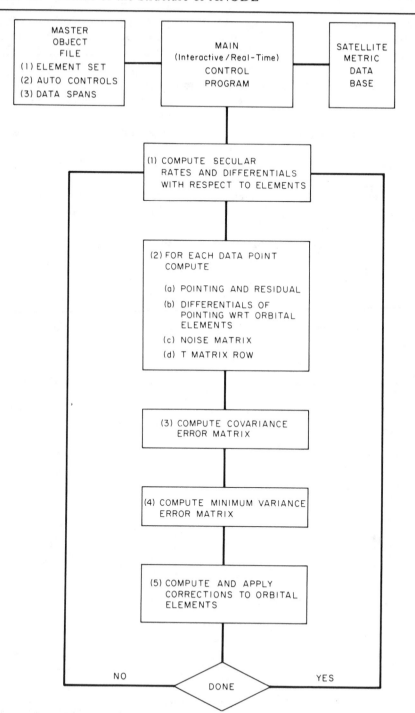

Note that the $\{\mathbf{P}_k\}$ are actual data and hence $\partial\mathbf{P}_k/\partial\mathbf{a}_0 = 0$. Furthermore,

$$\frac{\partial\mathbf{C}_k}{\partial\mathbf{a}_0} = \left(\frac{\partial\mathbf{C}_k}{\partial\mathbf{a}_{osc,k}}\right)\left(\frac{\partial\mathbf{a}_{osc,k}}{\partial\mathbf{a}_k}\right)\left(\frac{\partial\mathbf{a}_k}{\partial\mathbf{a}_0}\right)$$

where $\mathbf{a}_{osc}$ denotes the osculating orbital element set. But $\partial\mathbf{a}_{osc,k}/\partial\mathbf{a}_k$ are functions of the periodic corrections only; hence this matrix can, for simplicity, be set equal to the unit matrix. This approximation is allowed only in the partial differential calculation above, and its only negative effect would be to increase the number of iterations required to reach a solution. This approximation will not affect the final results. The positive effect of the approximation lies in the saving of the inordinate amount of computer time that would be needed for the calculation of the differentials of the periodic corrections. Consequently,

$$\frac{\partial\mathbf{Y}_k}{\partial\mathbf{a}_0} = -\frac{\partial\mathbf{C}_k}{\partial\mathbf{a}_0} = -\frac{\partial\mathbf{C}_k}{\partial\mathbf{a}_k}\cdot\frac{\partial\mathbf{a}_k}{\partial\mathbf{a}_0} = -\mathscr{T}$$

where the right-hand side can be computed as a function of the osculating (or the mean) elements under the same caveat as stated above. The resulting matrix of the derivatives of the residuals with respect to the epoch elements has a size of $4N \times 6$.

A set of corrections to the orbital elements at epoch t_0 can now be computed using a minimum variance technique. Let

$$\mathbf{Y} = \frac{\partial\mathbf{Y}}{\partial\mathbf{a}_0}\cdot\Delta\mathbf{a} + \mathbf{n} = -\mathscr{T}\,\Delta\mathbf{a}_0 + \mathbf{n}$$

wherein $\mathbf{n}$ represents the errors of the observations. The minimum variance estimate for the corrections $\Delta\mathbf{a}_0$ is given by

$$\Delta\mathbf{a}_0 = \mathscr{W}\mathbf{Y}$$

where $\mathscr{W} = (\mathscr{T}^T\mathscr{R}^{-1}\mathscr{T})^{-1}\mathscr{T}^T\mathscr{R}^{-1}$, $\mathscr{R} = \mathbf{n}\mathbf{n}^T$, and $\mathbf{Y} = \mathbf{P} - \mathbf{C}$.

Under the usual normality assumptions about the distribution of the errors of observation, the minimum variance estimate is also the weighted maximum likelihood estimate. The covariance error matrix of the estimate $\Delta\mathbf{a}_0$ is

$$\mathscr{S} = (\mathscr{T}^T\mathscr{R}^{-1}\mathscr{T})^{-1}$$

The new estimate of the elements at epoch is given by

$$\mathbf{a}_0 = \mathbf{a}_0 + \Delta\mathbf{a}_0$$

The estimate of the elements at epoch can be iteratively improved by going through the above steps repeatedly (as shown in Table 12). At each iteration the estimate of the mean elements at epoch from the previous iteration is used for all calculations. Convergence can be assessed in at least two ways. An analytically pleasing technique is to calculate the sum of the squares of the weighted residuals and look at the reduction in its value from iteration to iteration. For the purposes of ANODE a simpler alternative was chosen. The size of the correction $\Delta\mathbf{a}_0$ is monitored, and convergence is assumed to have occurred when the norm is sufficiently small.

Cautions.   Various cautionary steps have to be taken during the estimation process. First, multiplicative constants ("sigma" multipliers) for the weights of the observations have to be used so that all observations are initially accepted. These sigma multipliers have to be reduced in a logical fashion as the iterations proceed, and convergence has to be linked to those values too. Second, logic has to be provided for eliminating observations that have large errors. Third, the size of the corrections applied at any iteration has to be bounded. It must be remembered that a nonlinear problem is being solved by an iterative linear approximation process. Hence, large corrections may drive the solution away from the correct root. Fourth, lack of convergence has to be detected. The software structure of ANODE provides for all of these eventualities.

The Implementation of ANODE

ANODE has been designed to function both as an interactive program, driven by command inputs as supplied by the operator, and also as a real-time processor in the Millstone Hill Radar software system. In the real-time mode all necessary command input parameters are extracted from the Master Object File entry for the desired satellite. A command input design philosophy was chosen in the interactive mode. This provides the analyst with flexible control of the parameters determining the differential correction process. Moreover, any future modification can be easily implemented. Any desired action is carried out by entering a command name followed by the appropriate parameters, and all of the inputs can be entered in a free format. After decoding the command and the parameter input, a branch is made to a section in the main routine that is designed to handle that particular command. In the case of a relatively simple procedure, such as displaying the elements, the main routine contains all of the necessary code to carry out that command. However, if the command involves a more complex procedure, or if the real-time program needs the same capability (e.g., extracting and storing an element set from the Master Object File), a branch is made to a subroutine designed to handle it. Therefore, the only major difference between the real-time and interactive versions lies in the front-end command decoding routine.

All real variables and constants in the program are stored as double precision (48-bit, 11-decimal digits) words, and all floating-point operations and intrinsic functions (sin, cos, etc.) are carried out using the full precision available. Specific efficiency considerations will not be detailed here, but these considerations extend both to minimizing the total size of the program and reducing its execution time. Any special assembly language routines that are used to increase execution speed have a standard FORTRAN IV counterpart that ensures portability from one computer to another, leaving only obvious differences, such as word size and input/output incompatibilities, to transferring the program to other machines.

ANODE accesses two major data bases—the Master Object File and the Metric Data Base. The Master Object File (MOF) is a random-access file in which the current information on every satellite is stored. The orbital elements contained in the MOF entry for a given satellite are read in and stored in a common block and then are used as the initial orbital elements to commence the fit. When a fit has successfully converged, the new orbital elements can be inserted into the MOF.

The desired observational data for a satellite are accessed by a dynamic logical file number assignment. The records are read sequentially from the beginning of the file in the Metric (i.e., observational) Data Base. When an observation meets the specified input criteria, the measured pointing, time, site, and estimated error (weight) of that particular observation are decoded and stored. All of this information is kept in a common block too.

The subroutine that actually calculates the differential corrections applied to the orbital elements is entered next. The necessary control inputs that are passed to this routine are three in number; (1) the maximum number of iterations to perform, (2) a flag indicating whether the initial orbital elements or the elements calculated in the previous iterations are to be used at the start of this current cycle, and (3) a flag to indicate that the iterations should proceed in an "automatic" mode, that is, whether the sigma multipliers (multiplicative factors of the observational errors) should be reduced if the iteration has converged at that level. Once the sigma multipliers have reached their final minimum values and correct convergence has been achieved, control is returned to the main program in this mode. Finally, at the beginning of the first iteration an average time of the observations is calculated, and the initial orbital elements are propagated to that time. This is done to minimize the effects of uneven distribution of the observations over the time span.

The iteration cycle can be broken down into five major steps; (1) the calculation of the secular rates of change and their differentials with respect to the current elements, (2) the matrix of the differentials of the pointing with respect to the current elements and the differences between the measured and computed pointing (i.e., the residuals), (3) the covariance error matrix, (4) the minimum variance estimation matrix, and (5) the corrections to the elements. The details of the calculations in each step will be outlined next.

Step 1. The calculation of the secular rates of change ($\dot{\mathbf{a}}_0$) and the differentials of these rates with respect to the elements is performed first. A special subroutine does all of the necessary computations. Both the secular rates and the differentials are derived from either the singly or doubly averaged formulas depending on the value of the inclination. If the inclination lies between 52° and 75°, singly averaged formulas will be used; otherwise, the doubly averaged equations will be employed.

Step 2. In order to calculate the matrix $\mathcal{T}$ of the differentials of the pointing with respect to the current elements, the elements need to be updated to the time of observation, the pointing from the site in question needs to be computed, and the differentials of the computed pointing with respect to the updated elements also need to be calculated. A call is made to a special subroutine for these computations. The input arguments are the current elements and the update time (the time of the observation). After updating the elements for their secular rates of change, the periodic variations are calculated and then applied to the elements by a call to yet another subroutine. All of the necessary rotation and translation transformations to go from the location in the orbital coordinate system to pointing from the given site are done next. The location of the site is in another common block. If the current site is not the observation site, a call to yet another subroutine is made to recompute these values for the site of the observation before the call to the $\mathcal{T}$ matrix subroutine is made.

After the return, the residuals are calculated for each of the data points of the observation. The datum point will be rejected if its norm exceeds the variance allowed for that quantity (the sigma multipliers referred to above). If the observation is accepted, then the inverse of the error vector element is stored directly in $\mathcal{R}$ for that residual. Finally the differential matrix row of the $\mathcal{T}$ matrix of that residual, of that observation type is filled in.

Step 3. Once the $\mathcal{T}$ and $\mathcal{R}$ matrices have been computed it is a simple matter to compute the covariance error matrix, $\mathcal{S}$. Let

$$\mathcal{S}^{-1} = \mathcal{T}^T \mathcal{R}^{-1} \mathcal{T}$$

Since $\mathcal{R}^{-1}$ has been stored directly as $\mathcal{R}$, and the $\mathcal{T}$ matrix is really $\mathcal{T}^T$ of the analytic derivation, a call is made next to a matrix inversion subroutine that computes the inverse of $\mathcal{S}^{-1}$. If it exists, this is the covariance error matrix $\mathcal{S}$. It should be pointed out that it is not always possible to invert the matrix $\mathcal{S}^{-1}$ accurately (within the word length of the computer), and thus the iteration procedure may not converge. The singular matrix problem is usually the result of insufficient or poorly spaced data and is most apparent in the case of a single track on a near-stationary satellite.

Step 4. After the covariance error matrix $\mathcal{S}$ has been computed, the minimum variance estimation matrix $\mathcal{W}$ is calculated next; analytically

$$\mathcal{W} = \mathcal{S}\mathcal{T}^T \mathcal{R}^{-1}$$

Step 5. All that is left of the complete computation are the corrections to the elements. Once these have been calculated, a check is made to determine whether or not any of the corrections exceed the maximum allowable changes that have been specified by the analyst. If any of the corrections do exceed their bounds, a scale factor is computed to reduce the offender (and all of the other element corrections appropriately) to be within bounds. If the new value of inclination is negative, then set $i = -i$, $\Omega = \Omega + \pi$, and $\omega = \omega - \pi$.

Iteration Control. The first aspect of iteration control is the assurance that the present iteration has not rejected too many of the data points. This check is made after all the observations have been processed in step 2 of the cycle. During that step, a running count had been kept of the number of data point types (right ascension, declination, distance, and radial velocity) that have been accepted and rejected. Call these N_{acc} and N_{rej}. In the automatic mode of iteration control if any of the data types i have

$$\frac{N_{acc}(i)}{N_{acc}(i) + N_{rej}(i)} < \begin{cases} 0.5 & \text{if this is the first iteration with} \\ & \text{the present sigma multipliers} \\ 0.25 & \text{if not} \end{cases}$$

then all the sigma multipliers will be increased by dividing by its corresponding reduction factor. This step of the cycle will then be repeated with these new, looser values.

If the nonautomatic mode has been selected by the analyst, then the check is made only on the total number of points accepted and rejected and not on the individual types; so that if

$$\frac{S_{acc}}{S_{acc} + S_{rej}} < 0.5$$

where

$$S_{acc} = \sum_{i=1}^{4} N_{acc}(i), \qquad S_{rej} = \sum_{i=1}^{4} N_{rej}(i)$$

then the sigma multipliers will be increased and the step repeated.

After the corrections have been computed and been applied to the orbital elements, subject to the constraints detailed in step 5, the iteration cycle will be repeated with the corrected elements and the same sigma multipliers if the nonautomatic mode was specified. In this mode the iteration cycle will be repeated until the maximum number of cycles has been completed. However, in the automatic fit mode, after the corrections have been applied, a check is made to see whether or not the corrections have converged at the present level of sigma multiplier values. Convergence is assumed only if all the following requirements are met: (1) In all cases convergence in the change in i, e, and a is demanded; (2) for an eccentric ($e > 0.01$), inclined ($i > 1°$) orbit convergence is demanded for Ω, ω, and the mean anomaly at epoch too; (3) for a low eccentricity ($e \leqslant 0.01$), inclined ($i > 1°$) orbit convergence is demanded for $\omega + M_0$ and Ω; and (4) for a low eccentricity ($e \leqslant 0.01$), low inclination orbit ($i \leqslant 1°$) convergence is demanded for $\omega + \Omega + M_0$. If any of these criteria are not met, the iteration will be repeated with the corrected elements and current sigma multipliers (assuming that the maximum number of iterations has not been reached). On the other hand, if all these conditions are satisfied, then the sigma multipliers will be reduced (if they are not at their specified minimum values). The iteration cycle will now be repeated.

Once the final desired values of the sigma multipliers have been achieved, and the convergence criteria of the orbital element corrections have also converged at a level of 0.001 times the specified criteria, final convergence is assumed. Control is then returned to the main calling program.

Numerical Integration

Both of the differential correctors presented above rely on the analytical or semi-analytical integration of the equations of motion. Such techniques are referred to (in the celestial mechanics literature) as *general perturbations*. *Special perturbations* refers to the full numerical solution of the appropriate differential equations. The advent of large-scale, very fast, electronic digital computers has revolutionized numerical integration capabilities. There is another difference—one needs to be clever to generate a good general perturbation scheme, but this is not necessarily true of a good special perturbation scheme.

The principal question about numerical integration schemes is "Which is the best one?" Defining "best" is not trivial since it must deal with the rates of change of the location and velocity, the sensitivity of the algorithm to small errors, and the total time span of the integrations. Given a definition, the factors that influence one's decision include the effects of truncation errors, the effects of roundoff errors, the intrinsic stability of the finite-difference scheme being used, the ability to change the step size of the independent variable, and the ability to use large step sizes for the independent variable to reduce the overall computing time. Note that a change of variable from time to eccentricity anomaly (i.e., a regularizing transformation) will do more in the latter direction than will the most clever software.

It is facts such as these, which contrast analytic ability with numerical ability, plus the existence of a flood of numerical methods books that limit this section. No eighth-order Adams–Bashforth multistep predictor–corrector method with a time step in excess of the local radius of convergence of the f and g series will overcome a lack of analytical insight by the programmer. Finally, you can rarely prove the superiority of a numerical integration method. The best you can do is to compare a subset of the ones available (Runge–Kutta, Milne, Gill, Adams–Moulton, Euler–Cauchy, Gauss–Jackson, Bowie, or Obrechkoff) on a subset of problems of interest. The one that turns out to be the best in one test will not be the best in another. Some nice examples of this is the work of Black (1973) or Henrici (1962, 1964, especially the Appendix of 1963). Finally there are two classic methods in the field that should be mentioned for the sake of completeness.

Cowell's Method. Cowell's method, which he developed early in this century for application to Jupiter's eighth satellite, is to write down the differential equations of motion in rectangular coordinates

$$\mathbf{v} = \frac{d\mathbf{r}}{dt}, \qquad \frac{d\mathbf{v}}{dt} = \frac{\mathbf{F}}{m}$$

and integrate them step by step. **F** includes all of the forces acting on the satellite. Simple, brute force, numerical integration—the most general of all techniques—is slow and will eventually be overrun by the gradual accumulation of roundoff errors. A big boost in its popularity was the publication of an extensive computation of the rectangular coordinates of the seven major planets by the British Nautical Almanac Office (since outdated).

Encke's Method. These tables also did much to advance the more sophisticated method created by Encke in 1857. Herein one uses a reference orbit, the orbit the body would follow were all of the perturbing influences absent. After a while the differences between the coordinates in the real orbit and the osculating orbit become so large that a new osculating orbit is created by a process known as rectification. The use of the osculating orbit makes Encke's method about three times as fast (in execution speed) as Cowell's method for an artificial satellite and usually ten times as fast in the case of a planet.

Algebraic Manipulators

Presumably the reader is familiar with the use of digital electronic computers as a computational aid. In this role such machines perform large amounts of arithmetic operations (actually only binary additions) very rapidly. Also, I expect that the reader is less familiar with the use of analog electronic computers or human computers in such a mode. Both of these have been superseded by the digital machines. Finally, I predict that the reader is not used to the concept of utilizing these machines to perform formal analytical operations. Examples of this would be obtaining $\partial^3 u/\partial x^2\, \partial y$ for $u = \sin(3xy\, e^{x^2})$, the expansion of $(x+3y+9z^2)^4$ as a multinomial, or obtaining the indefinite integral of some polynomial. Such routine operations can now be performed on digital electronic computers without the numerical substitution of values for the variables appearing in the various expressions. The programs necessary to do this are known (generically) as algebraic manipulators.

An impediment to the development of a perturbation theory is the amount of analysis necessary to carry it through beyond the first order. In planetary theory the typical term in the disturbing function is of the form

$$\sum_{jklm} A_{jklm} \cos[(j+k+l)\lambda_1 + (m-k+l)\lambda_2 - j\omega_1 - m\omega_2]$$

where all four sum indices run from $-\infty$ to $+\infty$, λ_1 and λ_2 are the two mean heliocentric longitudes of the planets, and ω_1, ω_2 are their arguments of perihelion. The expansion coefficients $\{A\}$ are themselves threefold power series in the two eccentricities e_1 and e_2 and the square of the sine of the half inclination of their orbital planes, namely,

$$A_{jklm} = \sum_{pqr} A_{jklm}^{pqr} e_1^p e_2^q \sin^{2r}(I/2)$$

The expansion coefficients A_{jklm}^{pqr} are in turn transcendental functions of the ratio of semi-major axes. Clearly, working out such expansions is tedious and of limited potential for intellectual enrichment. Hence, astronomers reasoned that if such computations could be performed by a robot, then much dull labor would be saved (as well as the need to check it). Remember that we are discussing man-decades of labor in this context. Astronomers also pressed for the development of such facilities within digital computers.

I cannot discuss list or symbol processing in detail in this text. Nor can I expertly weigh FORMAC (*FOR*mula *MA*nipulation *C*omputer) against TRIGMAN (*TRIG*onometrical *MAN*ipulator) or ALPAK against formula ALGOL. Parts of these are assembly language or machine dependent. They are also likely to evolve before this text gets into print. Useful references are Tobey, Bobrow, and Zilles (1965), Tobey (1965), Sammet (1966), Jeffreys (1971), Davis (1973), and Pavelle et al. (1981). Recently software developed at M.I.T. for Digital Equipment machinery under the trademark name of MACSYMA has become available.

I mention the subject because I want the reader to be aware of a potentially useful tool. I also remind the reader that any child can keep the CPU of the largest machines continuously going—it takes a bit more thought to have it compute something interesting or useful.

A MAXIMUM LIKELIHOOD CORRECTOR

If the set $\{q_k\}$ represents a set of mutually exclusive exhaustive propositions, p an additional proposition, and H the available knowledge, then the conditional probability of both one $q = q_r$ and p being true given H is

$$P(q_r, p|H) = P(p|H)P(q_r|p, H) = P(q_r|H)P(p|q_r, H)$$

This follows from the calculus of probabilities. Using the fact that $\{q_k\}$ is exhaustive implies that

$$P(p|H) = \sum_k P(q_k|H)P(p|q_k, H)$$

so

$$P(q_r|p, H) = \frac{P(q_r, p|H)}{\sum_k P(q_k, p|H)} = \frac{P(q_r|H)P(p|q_r, H)}{\sum_k [P(q_k|H)P(p|q_k, H)]} \qquad (11.13)$$

The relationship in Eq. 11.13 is known as Bayes' theorem and the quantity $P(p|q_r, H)$ is called the likelihood. Bayes' postulate states that, given no knowledge, one assigns equal values to $\{P(q_r|H)\}$.

Since

$$P(q_r|p, H) \propto P(q_r|H)L(p|q_r, H)$$

an alternative to Bayes' postulate is to choose that hypothesis q_r that maximizes the likelihood L. Viewed another way, the Principle of Maximum Likelihood requires one to maximize the probability of the observed event. Moreover, the Principle of Maximum Likelihood is independent of the parameterization of the problem; for example, if L is a function of some parameter θ and Θ is that value of it that maximizes L, then $L[f(\theta)]$ is also maximized at $F = f(\Theta)$.

In general the likelihood function is the product of the frequency functions of the N observations. This fact will be used below in Eq. 11.14. Uniqueness of the solution for the maximum of L can be proved whenever sufficient statistics exist. Maximum likelihood estimators are, in general, consistent, efficient, asymptotically normal, and biased. More details can be found in Kendall and Stuart (1969). Now let us consider the construction of a differential corrector using this as the fundamental statistical assumption rather than that of least squares. Assume that the observational data consists of angles and angular rates.

Theory of Motion

The geocentric location vector is $\mathbf{r} = r(\cos \delta \cos \alpha, \cos \delta \sin \alpha, \sin \delta)$ where r is the distance, δ is the declination, and α is the right ascension. The topocentric location vector is $\mathbf{R} = R(\cos \Delta \cos A, \cos \Delta \sin A, \sin \Delta)$. The observer's location is given by $\boldsymbol{\rho} = \rho(\cos \phi' \cos \tau, \cos \phi' \sin \tau, \sin \phi')$ where ρ is his geocentric distance, ϕ' is his geocentric latitude, and τ is the local mean sidereal time. The fundamental geometrical relationship is

$$\mathbf{r} = \mathbf{R} + \boldsymbol{\rho}$$

The fundamental physics is given by

$$\ddot{\mathbf{r}} = \mathbf{f}$$

where $\mathbf{f}$ is the total force per unit mass and overdots denote differentiation with respect to Ephemeris Time. The force model for $\mathbf{f}$ may include non-spherical terms from the principal body, third-body disturbing functions, atmospheric drag, solar radiation pressure, and so on. Because our application will be to artificial satellites over short time spans, only the Keplerian force due to the Earth and the oblateness of the Earth (as measured by J_2) are considered here. Hence,

$$\mathbf{f} = -\nabla U$$

$$U = -\left(\frac{\mu}{r}\right)\left[1 - \frac{J_2}{r^2}P_2(\sin \delta)\right]$$

In dimensionless units $J_2 = 1.0827 \times 10^{-3}$. The formulation presented below can handle arbitrarily complex functions for $\mathbf{f}$. Finally, we shall sometimes express $\mathbf{r}$ and $\mathbf{R}$ in rectangular coordinates; $\mathbf{r} = (x, y, z)$, $\mathbf{R} = (X, Y, Z)$.

Statement of the Problem

A total of $N(>2)$ observations have been made at the ordered times $\{t_k\}$, $k = 1, 2, \ldots, N$. A complete observation consists of a determination of topocentric right ascension (A), declination (Δ), and their rates $(\dot{A}, \dot{\Delta})$. By topocentric we mean corrected for general precession, nutation, astronomical refraction, annual aberration, and so on, excluding planetary aberration, parallactic refraction, and diurnal parallax. A vector of observations at the time $t = t_k$, $\mathbf{m}(t_k)$, is given by

$$\mathbf{m}(t_k) = \mathbf{m}_k = \begin{vmatrix} A(t_k) \\ \dot{A}(t_k) \\ \Delta(t_k) \\ \dot{\Delta}(t_k) \end{vmatrix}$$

An incomplete observation is one wherein at least one element of $\mathbf{m}$ has not been determined.

Except for the errors inherent in $\mathbf{m}(t_k)$, had we known the six initial conditions $\mathbf{a}(t_0)$ at some epoch $t = t_0$, $\mathbf{m}(t_k)$ could have been predicted from the propagator $\mathbf{F}[\mathbf{a}(t_0), t_k, t_0] = \mathbf{F}_k$. The propagator includes all of the physics embodied in $\mathbf{f}$ as well as the conversion from geocentric rectangular coordinates to topocentric spherical coordinates. If the vector of errors at time $t = t_k$ is symbolized by $\mathbf{n}(t_k) = \mathbf{n}_k$, then

$$\mathbf{m}(t_k) = \mathbf{F}[\mathbf{a}(t_0), t_k, t_0] + \mathbf{n}(t_k)$$

The problem is to estimate the initial conditions $\mathbf{a}(t_0)$ from the observations $\{\mathbf{m}(t_k)\}$ given a reasonable set of assumptions concerning the measurement errors $\{\mathbf{n}(t_k)\}$.

We make the standard assumptions that (1) $\{\mathbf{n}(t_k)\}$ are randomly distributed, (2) $\mathbf{n}(t_k)$ is not correlated with $\mathbf{n}(t_l)$ if $k \neq l$, and (3) $\{\mathbf{n}(t_k)\}$ are unbiased. With these assumptions, the conditional probability that the noise at time $t = t_k$ has the value $\mathbf{n}(t_k)$ is given by

$$g[\mathbf{n}(t_k)|\mathbf{a}(t_0)] = \frac{\exp[-\frac{1}{2}\mathbf{n}_k^T \mathcal{M}_k^{-1}\mathbf{n}_k]}{(2\pi)^2|\det(\mathcal{M}_k)|^{1/2}}$$

where $\mathcal{M}_k = \mathcal{M}(t_k)$ is the covariance matrix of observations (see below). The likelihood of the sample is simply the product of g's;

$$L_N[\mathbf{a}(t_0)] = L[\mathbf{n}_1, \mathbf{n}_2, \mathbf{n}_3, \ldots, \mathbf{n}_N | \mathbf{a}(t_0)] = \prod_{k=1}^{N} g[\mathbf{n}_k | \mathbf{a}(t_0)] \qquad (11.14)$$

The Principle of Maximum Likelihood dictates that one choose $\mathbf{a}(t_0)$ such that $L_N[\mathbf{a}(t_0)]$ is a maximum.

Since $L_N > 0$, if L_N is maximized so is $\ln(L_N)$. We deal with the latter instead of the former because of its computational simplicity. The procedure used to find the correct value of $\mathbf{a}(t_0)$ is to linearize L_N about some guess for it, say $\mathbf{a}'(t_0)$, find the change in $\mathbf{a}'(t_0)$ which increases L_N the most, and then repeat the iteration procedure until the proposed change becomes sufficiently small. To be successful this procedure requires a good initial guess.

If we use Taylor's series to first order to expand the propagator at time $t = t_k$ about the estimate $\mathbf{a}'(t_0)$ for $\mathbf{a}(t_0) = \mathbf{a}_0$, then we obtain

$$\mathbf{F}[\mathbf{a}, t_k, t_0)] \simeq \mathbf{F}[\mathbf{a}'_0, t_k, t_0] + \nabla_{\mathbf{a}}\mathbf{F}[\mathbf{a}, t_k, t_0]\|_{\mathbf{a}=\mathbf{a}'_0} \cdot (\mathbf{a} - \mathbf{a}'_0)$$

In a more condensed notation we can write this as

$$\mathbf{F}_k \simeq \mathbf{F}'_k + \nabla\mathbf{F}'_k \cdot (\mathbf{a} - \mathbf{a}'_0) \tag{11.15}$$

The substitution of Eq. 11.15 into Eq. 11.14 makes L_N a function of $\mathbf{a}$. We require that the gradient of L_N with respect to $\mathbf{a}$ vanish. This yields an equation for the change to be applied to $\mathbf{a}'(t_0)$ whose solution is

$$\Delta\mathbf{a}'(t_0) = \left[\sum_{k=1}^{N} (\nabla\mathbf{F}'_k)^T \mathcal{M}_k^{-1}(\nabla\mathbf{F}'_k)\right]^{-1} \cdot \left[\sum_{k=1}^{N} (\nabla\mathbf{F}'_k)^T \mathcal{M}_k^{-1}(\mathbf{m}_k - \mathbf{F}'_k)\right]$$

In practice fewer than 30 iterations are required to reduce $|\Delta\mathbf{a}'(t_0)|$ to less than 10^{-3}. Once the maximum likelihood estimate of the initial conditions is obtained, the location and velocity at any time can be computed via the propagator $\mathbf{F}$.

The Computation of $\nabla_{\mathbf{a}}\mathbf{F}(\mathbf{a}, t_k, t_0)$

After the forces to be included in $\mathbf{f}$ have been specified, the main computational problems are integrating the equations of motion to determine the $\{\mathbf{F}_k\}$ and in calculating the gradient of the $\{\mathbf{F}_k\}$. As pointed out above, all of the relevant forces should be incorporated into $\mathbf{f}$. In addition, the numerical integration of the equations of motion should also be accurately performed. However, $\nabla\mathbf{F}_k$ need not be computed with especial precision because the linearization embodied in Eq. 11.15 partially vitiates the effect of higher-order terms. Thus, we set $J_2 = 0$ (but only use this assumption in Eq. 11.19 below; otherwise the analysis of this section is perfectly general).

We regard $\mathbf{a}(t_0)$ as the geocentric location and velocity vector at the time $t = t_0$. Furthermore, we assume $t_0 < t_1$ for simplicity. [If $\mathbf{a}(t_0)$ is to be regarded as the classical orbital elements of the osculating ellipse at time $t = t_0$ (or some combination of them), the analysis is still valid but an additional step is necessary.] Remembering that $\mathbf{F}_k$ produces error-free values of $\mathbf{m}(t_k)$,

$$\nabla_{\mathbf{a}_0}\mathbf{F}_k = \begin{pmatrix} \partial A_k/\partial x_0 & \partial A_k/\partial y_0 & \partial A_k/\partial z_0 & \partial A_k/\partial \dot{x}_0 & \partial A_k/\partial \dot{y}_0 & \partial A_k/\partial \dot{z}_0 \\ \partial \dot{A}_k/\partial x_0 & \cdot & \cdot & \cdot & \cdot & \vdots \\ \partial \Delta_k/\partial x_0 & \cdot & \cdot & \cdot & \cdot & \\ \partial \dot{\Delta}_k/\partial x_0 & \cdot & \cdot & \cdot & \partial \dot{\Delta}_k/\partial \dot{z}_0 \end{pmatrix} \tag{11.16}$$

Let $\mathbf{r}(t_k) = \mathbf{r}_k$ and $\dot{\mathbf{r}}(t_k) = \dot{\mathbf{r}}_k$ be the geocentric location and velocity vectors at the time $t = t_k$. Then, applying the chain rule of partial differentiation to each element of Eq. 11.16 in the form

$$\frac{\partial A_k}{\partial x_0} = \nabla_{\mathbf{a}_k}A_k \cdot \frac{\partial \mathbf{a}_k}{\partial x_0} + \nabla_{\dot{\mathbf{r}}_k}A_k \cdot \frac{\partial \dot{\mathbf{r}}_k}{\partial x_0}$$

results in $\nabla_{\mathbf{a}_0}\mathbf{F}_k$ being written as the product of two matrices. Calling these $\mathcal{B}_k$ and $\mathcal{C}_k$ we can symbolize this as

$$\nabla_{\mathbf{a}_0}\mathbf{F}_k = \mathcal{B}_k\mathcal{C}_k$$

with

$$\mathcal{B}_k = \begin{vmatrix} \partial Ak/\partial x_k & \partial A_k/\partial y_k & \partial A_k/\partial z_k & \partial A_k/\partial \dot{x}_k & \partial A_k/\partial \dot{y}_k & \partial A_k/\partial \dot{z}_k \\ \partial \dot{A}_k/\partial x_k & \cdot & \cdot & \cdot & \cdot & \vdots \\ \partial \Delta_k/\partial x_k & \cdot & \cdot & \cdot & \cdot & \\ \partial \dot{\Delta}_k/\partial x_k & \cdot & \cdot & \cdot & \cdot & \partial \dot{\Delta}_k/\partial \dot{z}_k \end{vmatrix} \quad (11.17)$$

and

$$\mathcal{C}_k = \begin{vmatrix} \partial x_k/\partial x_0 & \partial x_k/\partial y_0 & \partial x_k/\partial z_0 & \partial x_k/\partial \dot{x}_0 & \partial x_k/\partial \dot{y}_0 & \partial x_k/\partial \dot{z}_0 \\ \partial y_k/\partial x_0 & & & & & \\ \partial z_k/\partial x_0 & \cdot & \cdot & \cdot & \cdot & \cdot \\ \partial \dot{x}_K/\partial x_0 & \cdot & \cdot & \cdot & \cdot & \cdot \\ \partial \dot{y}_K/\partial x_0 & \cdot & \cdot & \cdot & \cdot & \cdot \\ \partial \dot{z}_K/\partial x_0 & & & & & \partial \dot{z}_K/\partial \dot{z}_0 \end{vmatrix}$$

The matrix $\mathcal{B}_k$ represents the transformation from geocentric rectangular coordinates to topocentric spherical coordinates at the time $t = t_k$. Hence, using the chain rule again, it too can be written as a product $\mathcal{B}_k = \mathcal{D}_k\mathcal{E}_k$. $\mathcal{D}_k$ involves the rectangular coordinate to spherical coordinate conversion, and $\mathcal{E}_k$ involves the geocentric coordinate to topocentric coordinate conversion. From the geometrical relationships it follows that $\mathcal{E}_k$ is just the identity matrix. Therefore, $\mathcal{B}_k$ is given by Eq. 11.17 with uppercase letters replacing the lowercase letters in the denominator of each element of the array. Explicitly, dropping the k subscript, if

$$\mathbf{s} = (X, Y, 0), \qquad s^2 = X^2 + Y^2, \qquad R^2 = s^2 + Z^2$$

then,

$$\mathcal{B} = \begin{vmatrix} -Y/s^2 & X/s^2 & 0 & 0 & 0 & 0 \\ [\dot{Y}(Y^2 - X^2) + & [\dot{X}(Y^2 - X^2) - & 0 & -Y/s^2 & X/s^2 & 0 \\ 2XY\dot{X}]/R^2 & 2XY\dot{Y}]/R^2 & & & & \\ -XZ/(sR^2) & -YZ/(sR^2) & s/R^2 & 0 & 0 & 0 \\ \partial\dot{\Delta}/\partial X & \partial\dot{\Delta}/\partial Y & \partial\dot{\Delta}/\partial Z & -XZ/(sR^2) & -YZ/(sR^2) & s/R^2 \end{vmatrix}$$

with

$$\nabla_{\mathbf{R}}\dot{\Delta} = -\mathbf{s}(\dot{Z} - \dot{R}\sin\Delta)/s^3 - (\nabla_{\mathbf{R}}\dot{R})\sin\Delta/s - (\nabla_{\mathbf{R}}\Delta)\dot{R}\cos\Delta/s$$

$$\nabla_{\mathbf{R}}\dot{R} = \frac{R\dot{\mathbf{R}} - \dot{R}\mathbf{R}}{R^2}$$

$$\nabla_{\mathbf{R}}\Delta = (-XZ, -YZ, s^2)/(sR^2)$$

The matrix $\mathscr{C}_k$ represents the dependence of the geocentric location and velocity at time $t = t_k$ on the initial conditions $\mathbf{a}(t_0)$. If we use the chain rule k times, then the $\{\mathscr{C}_k\}$ can be recursively computed from

$$\mathscr{C}_k = \prod_{n=1}^{k} c_n = c_k \mathscr{C}_{k-1} \tag{11.18}$$

The matrix c_n represents the dependence of the geocentric location and velocity at time $t = t_n$ on the same quantities at time $t = t_{n-1}$. Hence, each one is essentially diagonal. A convenient and fairly accurate evaluation of them can be made by using Taylor's series to second order and the equations of motion with $J_2 = 0$. If $\Delta t_n = t_n - t_{n-1}$, then this is written as

$$\mathbf{r}_n \simeq \mathbf{r}_{n-1} + \Delta t_n \dot{\mathbf{r}}_{n-1} - \frac{\mu (\Delta t_n)^2 \mathbf{r}_{n-1}}{2 r_{n-1}^3}$$

$$\dot{\mathbf{r}}_n \simeq \dot{\mathbf{r}}_{n-1} - \frac{\mu \Delta t_n \mathbf{r}_{n-1}}{r_{n-1}^3} - \mu (\Delta t_n)^2 \frac{(r_{n-1} \dot{\mathbf{r}}_{n-1} - 3 \dot{r}_{n-1} \mathbf{r}_{n-1})}{r_{n-1}^4} \tag{11.19}$$

It turns out that c_n can be partitioned into four 3×3 arrays each of which is symmetric, namely,

$$c_n = \begin{pmatrix} c_n^{(1)} & c_n^{(2)} \\ c_n^{(3)} & c_n^{(4)} \end{pmatrix}$$

Explicitly, if $\mathscr{I}$ is the 3×3 unit matrix and the $n-1$ subscript is temporarily dropped,

$$c_n^{(1)} = \mathscr{I} + \frac{\mu (\Delta t)^2}{(2 r^5)} \begin{vmatrix} 3x^2 - r^2 & 3xy & 3xz \\ 3xy & 3y^2 - r^2 & 3yz \\ 3xz & 3yz & 3z^2 - r^2 \end{vmatrix}$$

$$c_n^{(2)} = \Delta t_n \mathscr{I}$$

$$c_n^{(3)} = \frac{2(c_n^{(1)} - \mathscr{I})}{\Delta t_n} + \frac{\mu (\Delta t_n)^2}{2 r^6}$$

$$\times \begin{vmatrix} r^2 \dot{r} + 2x\dot{x}r - 5x^2 \dot{r} & r(\dot{x}y + x\dot{y}) - 5xy\dot{r} & r(\dot{x}z + x\dot{z}) - 5xz\dot{r} \\ r(\dot{x}y + x\dot{y}) - 5xy\dot{r} & r^2 \dot{r} + 2y\dot{y}r - 5y^2 \dot{r} & r(\dot{y}z + y\dot{z}) - 5yz\dot{r} \\ r(\dot{x}z + x\dot{z}) - 5xz\dot{r} & r(\dot{y}z + y\dot{z}) - 5yz\dot{r} & r^2 \dot{r} + 2z\dot{z}r - 5z^2 \dot{r} \end{vmatrix}$$

$$c_n^{(4)} = c_n^{(1)}$$

This completes the computation of $\nabla_{\mathbf{a}} F(\mathbf{a}, t_k, t_0)$.

Using the Orbital Elements for $\mathbf{a}(t_0)$. If one prefers to regard $\mathbf{a}(t_0)$ as some combination of the elements of the osculating ellipse at time $t = t_0$, then the above analysis is still valid but incomplete. Equation 11.18 has to be rewritten as $\mathscr{C}_k = \prod_{n=0}^{k} c_n = c_k \mathscr{C}_{k-1}$, where c_0 represents the dependence of $\mathbf{r}_0$ and $\dot{\mathbf{r}}_0$ on the orbital element set.

The Covariance Matrix.   If the angles and angular rates are obtained independently, then

$$\mathcal{M}(t_k) = \text{diag}[\text{var}(A_k), \text{var}(\dot{A}_k), \text{var}(\Delta_k), \text{var}(\dot{\Delta}_k)]$$

If. however, the rates are obtained by differentiating the angles, then $\mathcal{M}_k$ is no longer diagonal. Since $\Delta t_k = t_k - t_{k-1}$ is much too long a time span to use when approximating instantaneous values, we assume that there are observations of right ascension and declination close to and on either side of (A_k, Δ_k). Let these be at the times T_1^k, T_2^k, and T_3^k where

$$t_{k-1} < T_1^k < T_2^k = t_k < T_3^k < t_{k+1}$$

and

$$T_3^k - T_1^k \ll \min(t_k - t_{k-1}, t_{k+1} - t_k)$$

Then, if u_k is one of A_k, Δ_k and $\dot{u}_k$ is the corresponding rate, $\dot{u}_k = du(T)/dT|_{T=t_k}$,

$$\dot{u}_k = \frac{(T_2^k - T_3^k)u(T_1^k)}{(T_1^k - T_2^k)(T_1^k - T_3^k)} + \frac{[2T_2^k - (T_1^k + T_3^k)]u(T_2^k)}{(T_2^k - T_1^k)(T_2^k - T_3^k)} + \frac{(T_2^k - T_1^k)u(T_3^k)}{(T_3^k - T_1^k)(T_3^k - T_2^k)}$$

$$(11.20)$$

Equation 11.20 was obtained by differentiating a centered, three-point, interpolating polynomial for $u(T)$ and evaluating the result at T_2^k.

The other elements of $M(T_k)$ can now be computed from

$$\text{var}(\dot{u}_k) = \sum_{n=1}^{3} \left[\frac{\partial \dot{u}_k}{\partial u(T_n^k)}\right]^2 \text{var}[u(T_n^k)]$$

and

$$\text{cov}(u_k, \dot{u}_k) = \text{sgn}\left[\frac{2T_2^k - (T_1^k + T_3^k)}{(T_2^k - T_1^k)(T_2^k - T_3^k)}\right] \text{var}(u_k) \text{var}(\dot{u}_k)$$

When one can set $\text{var}[u(T_n^k)] = \text{var}(u_k)$ for $n = 1, 2, 3$, these quantities will be their smallest if $T_2^k = (T_1^k + T_3^k)/2$; namely,

$$\text{var}(\dot{u}_k) = \frac{2\,\text{var}(u_k)}{(T_3^k - T_1^k)^2}$$

$$\text{cov}(u_k, \dot{u}_k) = 0$$

Note that the assumption that $\text{var}[u(T_n^k)] = \text{var}(u_k)$ is very different from assuming that $\mathcal{M}_k = \mathcal{M}, \forall k \in [1, N]$ and that, even if this assumption were true, the absolute minimum of the variance of $\dot{u}(T)$ does not necessarily occur when $T = T_2^k$.

chapter twelve

Stellar Dynamics

Up until now this book has been concerned with bodies within the solar system—planets, asteroids, or satellites. In this chapter clusters of stars (large and small), aggregations of stars by the billions, and finally conglomerations of these entities will be discussed. No matter how large the distance scale, whether the few hundreds of A.U. necessary to traverse the solar system or binary star systems, the millions of A.U. required to cross a cluster of stars, the ten billions of A.U. across which galaxies stretch, or the trillions of A.U. spanned by clusters of galaxies, it is believed that these systems evolve under the influence of the self-gravitation of their parts. Thus, in a deceptively simple way, we still need no new physics.

This field of astrophysics has been jazzed up lately. Whereas quasars, black holes, pulsars, and neutron stars previously dominated the astronomical fashion scene, more recently massive galactic halos, tidal stripping, the gravothermal catastrophe, and galactic cannibalism have become the chic topics for computationally oriented theoreticians. With all of this sex and violence among celestial bodies the potential for puns is astronomical. I promise no more. Of more importance to the reader is that this general area of astrophysics is my specialty. Thus, since I believe that I know at least as much about large-scale gravitational problems as about the small-scale (read two-body) ones, this chapter is offered as an overall summary of the foundations (?) of stellar dynamics.

Before going a single step further, let me make my opinions crystal clear—I believe that there is no individual or set of theoretical, analytical, numerical, or physical reasons to believe that 70 years of stellar dynamics have provided the astronomical community with any real understanding of large-scale gravitational systems. All of the techniques described below that are used to study these problems suffer from fatal flaws. Thus, a large amount of space will be devoted to explaining the physical, mathematical, and astronomical nature of the various senarios, but very little space will be devoted to a discussion of the results of these computations. References to such discussion will be provided though (Michie 1964; Aarseth and Lecar 1975; and King 1980).

Now that my point of view has been established, I shall outline the remainder of this chapter. Immediately below I define the fundamental physical objects in this field of study and the astrophysical contexts in which they occur. Following that is a brief overview of the methods of attack used to study their structure and evolution. The bulk of the chapter is a more leisurely exploration of these methods, especially their mathematical and physical foundations. At the end of this chapter a special self-gravitating system is discussed in detail. This (somewhat artificial) system possesses several unique features. I believe that with the proper attention it will illuminate many of the deeply interconnected lines of research now left dangling in stellar dynamics (Taff 1977).

OVERVIEW

The Celestial Objects

A star is a self-gravitating, self-luminous, ellipsoidal (usually oblate spheroidal or spherical), localized accumulation of matter. For the purposes of this text a star may be replaced by a particle of the appropriate mass. As one scans across the nighttime sky, one finds stars grouped together in three levels of organization.* A *galactic cluster* or *open cluster* is a loose, irregular grouping of 50–1000 stars. Naked-eye examples are the Pleiades, the Hyades, and Orion. Typically these stars are young (on an astronomical time scale) and the clusters contain gas and dust. These constituents are remnants of the birth of the stars from the protocluster cloud. Some stars in such clusters are frequently very hot and bright.

One level up in complexity are groupings of stars known as *globular clusters.* These are organized, regular, indeed spherically symmetrical aggregates of 10^5–10^7 stars. A naked-eye example is the globular cluster in Hercules known as M13 (it is the thirteenth object in Charles *M*essier's catalog). Globular cluster stars are old, not very bright or hot, and globular clusters contain little (if any) gas or dust. Figures 41 and 42 show typical galactic and globular clusters.

The largest single grouping of stars is known as a *galaxy.* A big galaxy has $\sim 10^{11}$ stars in it. Our galaxy, known as the Milky Way Galaxy, is this large. Within the Milky Way there are ~ 150 globular clusters and thousands of galactic clusters. Galaxies basically come in two types (morphologically). Some appear to be quite flat except for a central spherical bulge. Because the central region is called the nucleus, the bulge is referred to as the nuclear bulge. Within the disc is a mixture of stars, gas, and dust; the arrangement of this material can be quite striking (cf. Figs. 43–45). In particular there are lenticular systems devoid of any obvious structure, the spiral galaxies with their (usually) two spiral arms winding through the disc, and the barred spirals that possess

---

* Be warned that this discussion is general—not wrong, but not detailed or complete.

Figure 41. The naked-eye visible galactic or open cluster known as the Pleiades (M45).

Figure 42. The naked-eye visible globular cluster in Hercules (M13).

a central bar from which the spiral arms emanate. The other principal type of galaxy is elliptical in shape (Fig. 46). Ellipticals contain no gas or dust, no young stars, and no spiral arms.

As a rule, galaxies are not isolated in space. They too occur in groupings. Sometimes only a few galaxies are together—the Milky Way Galaxy is a part of a ~25 member band known as the Local Group. Hundreds of galaxies found together mimic the open clusters in that they tend to be loose and irregular in structure. They also tend to contain a fairly high proportion of spiral and lenticular galaxies. Really large clusters of galaxies contain thousands or tens of thousands of members (mainly ellipticals); these objects are also spherically symmetrical (see Fig. 47).

Spend some time examining the accompanying photographs. The simple regular structure of globular clusters, spiral galaxies, elliptical galaxies, and rich clusters of galaxies (this is harder to see) cries out for a simple explanation. It is believed to be Newtonian gravity amongst thousands to billions of constituents—the stars. Aside from spiral structure, that's what this chapter is about. Spiral structure theories are neither compelling, satisfactory, nor simple. The most frequently used framework was developed by C. C. Lin in 1964. Lin and his students (F. H. Shu and F. W. K. Mark) have furthered this sketch

Figure 43. The naked-eye visible galaxy Andromeda (M31).

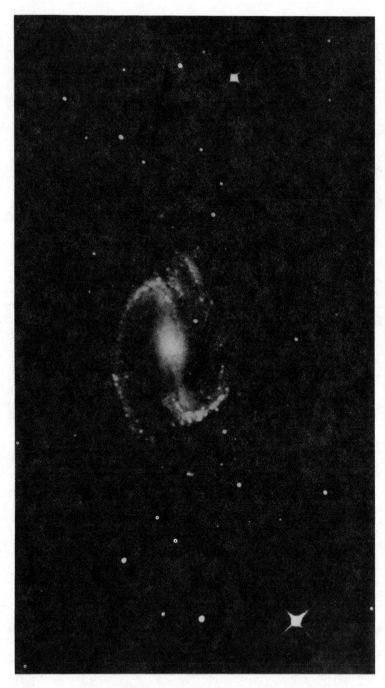

Figure 44. A barred spiral galaxy (NGC 1300).

Figure 45.   The great spiral galaxy in Ursa Major (M81).

quite far in the last 20 years. The essential concept is that the spiral arms are not static material structures, but rather waves in the underlying density distribution. Because spiral structure is less directly connected to celestial mechanics than are the other topics discussed, it is left out. There is a good review of the topic by A. Toomre (1977). Massive halos are a theoretical construct (a massive spheroidal component of spiral galaxies which so far is invisible) invented by Ostriker and co-workers to stabilize certain classes of models of disc systems. They too will be left behind at this point.

Stellar Dynamics

As treated herein, stellar dynamics is the study of the structure and evolution of symmetrical large-scale stellar systems such as globular clusters. By extension, elliptical galaxies and rich clusters of galaxies are also included. Galactic clusters have many fewer members than do the aforementioned systems. Moreover, because of their low mass and location in the plane of the disc of the Galaxy (globular clusters form a spherically symmetric subsystem about the galactic center), they are subject to external forces that can

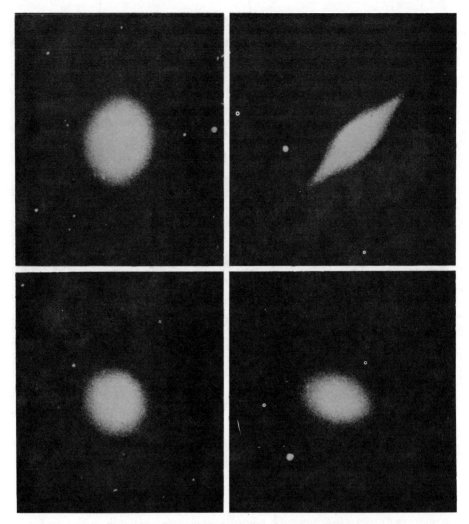

Figure 46. Four elliptical galaxies of varying ellipticity.

significantly perturb their internal structure. The distinction is not rigid, but since they do contain relatively few stars, the first approaches toward attempting to understanding the structure of open clusters were to simply integrate the N-body equations of motion directly,

$$m_n \ddot{\mathbf{r}}_n = -G m_n \sum_{\substack{l=1 \\ l \neq n}}^{n} \frac{m_l(\mathbf{r}_l - \mathbf{r}_n)}{|\mathbf{r}_l - \mathbf{r}_n|^3}, \qquad n = 1, 2, \ldots, N \qquad (12.1)$$

This defines the N-body problem and illustrates the fundamental approach to the N-body system. Clearly, were such a course of action successful, then one would have a completely dynamical specification of the evolution of such a system. Unfortunately, no compelling evidence exists that a successful numeri-

Figure 47. A cluster of galaxies in the constellation Ursa Major. The arrow points to the brightest one (NGC 4884).

cal solution of Eq. 12.1 has even been carried out. Moreover, much evidence to the contrary does exist (discussed in part below).

As in statistical mechanics (and even more so in thermodynamics) one becomes convinced that six values of location and velocity for N particles are not needed at every instant of time. What would one do with it? One would average it in some fashion to derive quantities you can deal with—the density, the pressure tensor, the entropy, and so on. So, since one cannot solve Eq. 12.1 in general, we settle for a macroscopic description. Statistics can be introduced into this problem in two ways. One is known (in this field) as the

Monte Carlo technique, the other as the fluid dynamical or kinetic theory technique. Now we show how one can be led to these developments.

Imagine a star in a globular cluster. As it moves along its orbit it feels two very different types of forces. Always in the background is the general, smoothed-out force field of the remainder of the cluster as a whole. Once in a while, however, it passes quite near another star belonging to the cluster. While these stars engage in their hyperbolic dance the cluster background acts as an audience. This "collision" takes place on a very short time scale compared with a crossing time—the average time it takes to traverse the cluster as a whole. It is also much more violent, as measured by the energy interchanged, than the gradual ebbing and flowing of kinetic and potential energies over the course of a whole orbit. Finally, these two-body interactions take place randomly—in time, in space, and in the parameters describing the interaction (relative velocity, distance of closest approach, etc.). With the Monte Carlo technique this fact is recognized and exploited. Every star is not followed for every step of its orbit. An individual "test star" is followed and randomly subjected to two-body encounters. If these random operations are properly performed and the averages of enough moments agree with the theoretically calculated ones, then one should have a computational method for describing the whole evolution of a globular cluster or of a cluster of galaxies. Calculations of this sort were carried out in America and in Europe during the seventies. At best they demonstrate internal self-consistency (not external since they do not agree) and circular reasoning.

The statistical mechanical or kinetic theory approach is very difficult to carry out in practice, so analytically averaged forms of it have been used. These fluid dynamical models have been pursued in both their hydrodynamical manifestations to perform evolutionary calculations and in their thermodynamical contexts to provide stability analyses. This type of computation is especially alluring but has no physical basis because of the special properties of the gravitational (*not* the $1/r^2$) force. In particular, the deduction of the existence of the gravothermal catastrophe, perhaps the most famous prediction of this type of work, is mathematically unjustifiable.

We continue to probe stellar dynamics because we look at the pictures (Figures 41–47), we look at Eqs. 12.1, and we say that it must be simple. Globular clusters and spiral galaxies are too beautiful not to succumb to our understanding or our analytical techniques or our digital computers. We insist on it.

N-BODY CALCULATIONS

Von Hoerner (1960) was an early practitioner of *N*-body calculations. To do this, one clearly needs a computer, a computing budget (preferably large), an integration scheme for Eqs. 12.1, and a set of initial conditions. Again, it must be stressed that the (deceptive) simplicity of Eqs. 12.1 led various astronomers to a direct numerical attack on the *N*-body problem. [We of course knew of the work of Bruns (1887) and Poincaré (1896), which precluded certain types of analytical solutions to the problem.] Multiple numerical integrations using different sets of masses, initial conditions, and values of *N* might allow one to gain an understanding of the behavior of self-gravitating systems. If one could delineate the important aspects of such systems, then perhaps one would really begin to interpret star clusters. Finally, such *N*-body integrations appear to be free from simplifying assumptions and this makes them even more attractive.

Limitations of memory space and computing budgets also limit the value of *N* that can be entertained. It is believed that interesting results require computations of a duration of order *N* crossing times and that the number of calculations per crossing time is of order N^2. Hence, early "numerical experiments" were run for small values of *N* (<50). More recently, $N = 500$ or 1000 has become feasible in three dimensions, $N = 10^{4-5}$ in two. The principal limiting factor is the calculation of the right-hand side of Eqs. 12.1. If one just uses his favorite Runge–Kutta technique on Eqs. 12.1, one encounters two very different types of problems. First of all, it is computationally expensive to compute the indicated sum. A common way around this inconvenience is to develop a polynomial in the time for the acceleration, say

$$\mathbf{a}_n = \sum_{j=1}^{J} \mathbf{A}_j (t - t_0)^j$$

The expansion coefficients are given by a set of divided backward differences with constant coefficients for weights [see Wielen (1967) for one such scheme]. The radius of convergence of this series is usually quite limited, but the integration of it is analytically trivial. Typically $J = 3$ or 4 and an additional iteration is performed using a one higher-order interpolating polynomial (based on the new location and velocity) to correct the errors of the extrapolating polynomial.

The second type of difficulty is directly connected with the size of the time step employed. Suppose that the time scale for the evolution or relaxation of the cluster as a whole (t_{relax}) is of order *N* times as long as the crossing time (t_{cross}). Suppose further that the typical crossing time is of order *N* times as long as the duration of a typical collision (t_{coll}). Then, for large *N*, one has the very strong inequalities

$$t_{\mathrm{relax}} \gg t_{\mathrm{cross}} \gg t_{\mathrm{coll}} \qquad (12.2)$$

The evolution of the cluster occurs on a time scale of order t_{relax}. The individual, violent perturbations of the stellar orbits take place on the t_{coll} time scale. One's computing budget demands that the time step be a fraction of t_{cross}. Thus, such procedures automatically smooth out the detailed interplay among stars, for in these systems one also has 2, 3, 4, ..., body collisions.

Three-body collisions can lead to the formation of binary subsystems within the *N*-body system. If such a pair is widely separated (a "soft binary"), then one can argue that it will be torn apart by the tidal forces exerted by the rest of the cluster on a time scale of order t_{cross}. Hence, its formation and description may be of no importance to the dynamics of the cluster as a whole. If, on the other hand, a tightly bound (or "hard binary") is formed, then it can act as an energy sink soaking up kinetic energy (as a unit) in collisions with other stars. Moreover, in a three-body-collision leading to the formation of a tightly bound binary, the third star must gain kinetic energy (since the energy of the subsystem is conserved—the cluster as a whole is static over the t_{coll} time scale). Therefore, it could entirely escape from the cluster during this process. Thus hard binaries permanently affect the evolution of a star cluster. The apparent solution to this problem is to use a smaller time step, but that is computationally expensive. Your favorite Runge–Kutta (or whatever) technique applied to Eqs. 12.1 is really a system of equations based on Taylor series expansions of $\mathbf{r}$ and $\dot{\mathbf{r}}$. By now we are intimately familiar with these under the nomenclature of the f and g series (see Chapter Eight). The same convergence problems encountered in Gaussian initial orbit determination will also plague the numerical integration scheme chosen.

Regularization

There is a way around the difficulty posed by the inequalities 12.2. One method employed in the past (by some workers in the field) has been to solve a different problem altogether. Instead of having the stars in their clusters interact via a $1/r^2$ force they changed it to a $1/(r^2 + a^2)$ force where $a > 0$. This softens the effects as $r \to 0$. Of course this is no longer an N-body calculation either. There is a correct method that deals with this difficulty while not simultaneously eating up all of the computing time in the treatment of close approaches. To illustrate it I follow the treatment given by Bettis and Szebehely (1971).

Returning to a one dimensional two-body problem, the equation of motion is

$$\frac{d^2x}{dt^2} = -\frac{G(m_1 + m_2)}{x^2} \tag{12.3}$$

where x is the separation between the two stars of mass m_1 and m_2. The energy integral is

$$\frac{(dx/dt)^2}{2} = G(m_1 + m_2)\left(\frac{1}{x} - \frac{1}{x_0}\right) \tag{12.4}$$

where it is assumed that their relative speed at separation x_0 was zero. Now both the equation of motion and its first integral exhibit a singularity as $x \to 0$ (i.e., a close approach or collision). No normal numerical integration of either equation can be successfully carried through the collision. The trick, first developed by Sundmann (1912), is to make a change of variable from t to τ that removed the singularity or *regularized* the equations of motion. This kind of idea has been extended with an eye on the numerical solution of the N-body system. We shall not discuss the details here, but in order to illustrate the technique we continue with a discussion of the one dimensional system.

A new time variable τ is defined by

$$d\tau = \frac{dt}{x} \qquad (12.5)$$

and the units are normalized such that $G(m_1 + m_2) = 1$. Notice that as x decreases, the time step of a numerical integration of Eqs. 12.3 or 12.4 should be decreased. From the definition of τ in Eq. 12.5 a constant might be used for $\Delta\tau$ as $x \to 0$ because of a compensation of these two effects. Even the best of regularization techniques will still not allow penetration (numerically) of the singularity. However, it will allow one to very closely approach $x = 0$ in a more economical fashion while simultaneously producing more accurate numerical results.

Let us further investigate τ. Define

$$x' = \frac{dx}{d\tau} = x\dot{x}$$

The energy integral Eq. 12.4 can now be written as [remember that $G(m_1 + m_2) = 1$ now too]

$$(x')^2 = 2x - \frac{2x^2}{x_0} \qquad (12.6)$$

Note that as $x \to 0$, $\dot{x} \to \infty$ but $x' \to 0$. Finally, Eq. 12.3 is transformed via

$$x'' = x^2\ddot{x} + x\dot{x}^2 = \frac{dx'}{d\tau} = x\frac{d(x\dot{x})}{dt}$$

so that

$$\ddot{x} = \frac{x'' - (x')^2/x}{x^2}$$

whence

$$x'' - (x')^2/x + 1 = 0 \qquad (12.7)$$

Now we know that $(x')^2/x = 2 - 2x/x_0$ from the conservation of energy equation. Hence, the transformed equation of motion, Eq. 12.7, is regular at $x = 0$. If we had lacked such knowledge (say in the two- or three-dimensional version wherein things are not so simple) and directly integrated Eq. 12.7 numerically, then we would still face difficulties. Thus, one considers more complex regularization procedures in general than Eq. 12.5—including the introduction of transformations of both the independent and dependent variables.

Equation 12.6 is used to simplify Eq. 12.7. The result is

$$x'' + \frac{2x}{x_0} - 1 = 0 \tag{12.8}$$

or the equation of a harmonic oscillator. If $\tau(0) = 0$, then the solution of Eq. 12.8 is

$$x = \frac{x_0}{2}\left\{1 + \cos\left[\left(\frac{2}{x_0}\right)^{1/2}\tau\right]\right\}, \quad x' = -\left(\frac{x_0}{2}\right)^{1/2}\sin\left[\left(\frac{2}{x_0}\right)^{1/2}\tau\right]$$

I can now integrate Eq. 12.5 to obtain the $t(\tau)$ relationship. The result is

$$t = \int_0^\tau x(\tau')\,d\tau' = \frac{x_0}{2}\left\{\tau + \left(\frac{x_0}{2}\right)^{1/2}\sin\left[\left(\frac{2}{x_0}\right)^{1/2}\tau\right]\right\}$$

The reader should recognize Kepler's equation with τ playing the role of the eccentric anomaly. Thus, the eccentric anomaly (and the true anomaly) is a regularizing variable for the two-body problem. What the regularization process has done is replace the singular differential equation of motion (Eq. 12.3) with a nonsingular one (Eq. 12.8). Hence, if you're going to integrate the two-body equations of motion, a constant ΔE step should be used, not a constant Δt step. The differences as $e \to 1$ will be obvious.

Complications

Most of the N-body integrations ever performed have been for isolated, single-component ($m_n = m \; \forall n \in [1, N]$) systems. The initial conditions are usually generated by a random number generator—either from an arbitrary (and frequently ad hoc) distribution of location and velocity or from the (theoretically) suspected equilibrium distribution. One can complicate the numerical integrations by bringing more physics into the problem. Real stars are not all of the same mass so that the treatment of a multicomponent system is also of interest. Now one deals with N_1 stars of mass m_1, N_2 of mass m_2, and so on, such that

$$N = \sum_n N_n, \qquad M = \sum_n m_n N_n$$

In the limit one deals with a spectrum of masses from some distribution. The observed mass spectrum in galactic clusters was derived by Taff (1975) and is $N(m) \sim m^{-11/4}$.

A second complication is that of binary subsystems within the cluster. These can be present ab initio or formed and accurately kept track of during the course of the cluster's evolution. Hard binary formation between the most massive members of a small cluster is a recurring phenomenon in N-body computations. Such binaries act as energy sinks, tend to come to rest near the cluster center, and speed up the escape rate. The existence of escaping stars means that the N-body system does not conserve mass, energy, or angular momentum, thus limiting the numerical usefulness of these (global, analytical) integrals of the motion. Differential escape appears in multicomponent systems too as do arguments concerning kinetic energy equipartition (e.g., does $m_1 \langle v_1^2 \rangle = m_2 \langle v_2^2 \rangle$ for $m_1 \neq m_2$?) or lack thereof. Multiple-mass groups also allow one to discuss segregation within the cluster (i.e., are the more massive stars more tightly bound to the cluster core than are the less massive ones?). Another common result of N-body computations is the division of the cluster into three regions. The innermost part is the "isothermal" core, meaning that all of the stars residing in the core have the same kinetic energy. Exterior to the core is a less dense, more complicated transition region. Finally, as one continues outward one comes to the lightly populated halo. Halo stars are generally in highly eccentric orbits and are rarely perturbed. Of course, during their infrequent mad dashes through the cluster core they undergo severe perturbations on account of both two-body encounters and the rapidly changing potential of the cluster as a whole (as a function of radial distance from the cluster center—remember that these constructs are highly inhomogenous).

One can inquire as to the outer boundary of a cluster once the cluster has been placed in its appropriate galactic environment. Long ago von Hoerner (1957) and King (1962) showed that the outer boundary of a galactic or globular cluster would be limited by the tidal effects exerted by the remainder of the Galaxy. Thus, one expects real clusters will only extend to $\sim R(M_c/M_g)^{1/3}$ where M_c is the cluster's mass, M_g is the Galaxy's mass, and R is the cluster's galactocentric distance.

There are other constituents in the Galaxy (and in other spiral galaxies) besides star clusters. One of the more massive components are interstellar clouds of gas and dust. The principal constituent of these is atomic hydrogen in the neutral state, or HI. If an interstellar HI cloud passes by a galactic cluster, then it could well exert a significant tidal force. Globular clusters passage through the disc of the Galaxy can provide an external source of heat (because near the disc the gravitational field is constant and normal to the disc—hence passage through such a surface distribution results in a twofold reversal of the gravitational field). Both of these ideas have been advanced by Spitzer (1958; Ostriker et al., 1972). An older general summary of N-body results may be found in van Albada (1968).

Applications to Galaxy Interactions

Galactic cannibalism and tidal stripping (Richstone 1975, 1976) are two different end results of the same process. They differ due to the speed of approach of two galaxies within a rich cluster of galaxies. When the two galaxies rush by each other, their "collision" is fast and the tidal effects one raises on the other could cause plumes of matter (stars, gas, and dust) to be pulled off. Soon after the encounter one might observe a "tail" on one of the galaxies or a "bridge" between them. Toomre and Toomre (1970, 1971, 1972a,b; Toomre 1973) have done computations of this sort. The individual galaxies are modeled as N-body collections of stars and then thrown at each other (numerically). The results, especially when seen in a pseudomotion picture format, are extremely impressive.

If the collision takes place slowly then the two objects might coalesce. The spectrum of coalescence under a variety of conditions was worked out long ago (Field and Saslaw 1965; Field and Hutchings 1968; Taff and Savedoff 1972, 1973) but not applied to clusters of galaxies until Ostriker (1975, 1977, 1978) and White (1975a,b). This process is known as galactic cannibalism and is thought to lead to the formation of the giant elliptical cD galaxies that seem to dominate so many clusters of galaxies. See also Hoessel (1981) and McGlynn and Ostriker (1980).

Another unusual type of galaxy [see Arp's (1966) catalog for pictures] is known as a ring galaxy. Lynds and Toomre (1976) discovered that these could be made if two galaxies hit head on. The ring is a temporary phenomenon, much like the splash in a liquid when something is dropped into it. The self-gravitation of the disturbed galaxy is stronger than the cohesive forces or surface tension of a liquid. The results of many of these computations look so right that one wants to believe in them. The same can be said for the N-body calculations that try to generate spiral structure (at least for a few revolutions). As an example, Toomre (1977) discussing spiral structure, stated:*

> The old puzzle of the spiral arms of galaxies continues to taunt theorists. The more they manage to unravel it, the more obstinate seems the remaining dynamics. Right now, this sense of frustration seems greatest in just that part of the subject which advanced most impressively during the past decade—the idea of Lindlad and Lin that the grand bisymmetric spiral patterns, as in M51 and M81, are basically compression waves felt most intensely by the gas in the disks of those galaxies. Recent observations leave little doubt that such spiral "density waves" exist and indeed are fairly common, but no one still seems to know why.

> To confound matters, not even the n-body experiments conducted on several large computers since the late 1960s have yet yielded any decently long-lived regular spirals. By contrast, quite a few such model disks have developed bar-like or oval structures that have endured almost indefinitely in their interiors. This finding is undoubtedly a major step toward understanding various bar phenomena that complicate the observed spirals, but it also reminds us rudely just how little we really comprehend. To avoid the rapid nonaxisymmetric instabilities associated with this bar-making, the experimental "stars" require (or otherwise they acquire) random velocities proportionately much larger than those observed among stars near the Sun, or than those suggested by edge-on views of other galaxies. It is this dilemma of course, that has fueled much of the recent speculation about massive but largely unseen halos—though the alternative that some very hot inner disks or spheroids might already cure those instabilities seems not yet to have been excluded firmly. Whatever that answer may be, it is clear that the spiral dynamics gets no easier when even the most rapid but least

---

* Reproduced, with permission, from the Annual Review of Astronomy and Astrophysics, Vol. 15 © 1977 by Annual Reviews Inc.

visible members of a galaxy need to be suspected of helping distort the axial symmetry.

The massive halos he refers to were postulated by Ostriker and Peebles (1973; see also Ostriker and Thuan 1975) to account for the fact that collisionless disk systems seemed to require that the ratio of the kinetic energy of their organized motion to the absolute value of their potential energy be less than $\frac{1}{7}$. In order to satisfy this inequality, it seemed as if a massive spheroidal component (the halo) was required to provide the contribution to the potential energy. The radio astronomers have long known that galaxies do not end where the visible light trails off into the background on photographic plates, but there is no convincing evidence for a stellar component of the required mass.

Validity

The reader who is not computationally sophisticated might think that because one uses a 128-bit, eighth-order Runge–Kutta scheme coupled with a predictor corrector to solve Eq. 12.1 the computer is necessarily solving the same problem. Miller (1972), who always warned us of problems in N-body calculations (1964, 1971a,b,c; Butterworth and Miller 1971), has written (Copyright © 1972 by D. Reidel Publishing Company, Dordrecht, Holland):

> In this report, I want to take a different tack: this is supposed to be a meeting of experts in N-body calculations, so I want to concentrate on difficulties with the calculations. In particular, the emphasis will be on attempts to convince you that the bits running around inside those nice, big computers bear some relationship to the physics of stellar systems. It is not ipso facto evident that they do: the mere fact that the experimenter intends his calculation to relate to some kind of system in the sky does not assure any similarity.

To the dismay of, for example, the hydrodynamicists and neutron diffusion experts, it is trivially easy to construct numerical difference schemes for partial differential equations whose solutions bear absolutely no relationship to the original differential equations (Collatz 1966; Richtmyer and Morton 1957). As an example consider

$$\sigma \frac{\partial^2 u}{\partial x^2} = \frac{\partial u}{\partial t}$$

and the difference scheme at the grid $u(x, t) = u(x_m, t_n) = u(m \, \Delta x, n \, \Delta t) = u_m^n$,

$$\sigma \frac{u_{m+1}^n - 2u_m^n + u_{m-1}^n}{(\Delta x)^2} = \frac{u_m^{n+1} - u_m^n}{\Delta t}$$

The formula for $\partial^2 u / \partial x^2$ is just the central difference approximation to a second derivative (with t kept constant, as a partial derivative implies), and the formula for $\partial u / \partial t$ is a simple forward difference (with x held constant). It turns out that whenever $2\sigma \Delta t/(\Delta x)^2 > 1$, one can *analytically* demonstrate that the difference scheme is unstable and that the solution of the difference equation has no relationship to the solution of the differential equation. Moreover, the nature of this difficulty has nothing to do with arithmetic, truncation error, roundoff error, and so on.

The capstone of demonstrations of N-body systems was presented by Lecar (1968). In this study several researchers in the field each integrated the same 25-body system in their own fashion and then discussed the results. Since they could not tell from their results that they had solved the same problem, I would argue that it is explicitly incumbent on purveyors of N-body calculations to show that they are in fact solving Eq. 12.1. Truncation errors, pair correlations, and runaway stars all plague N-body calculations. See the Miller references already given and Standish (1968), Hayli (1971), Allen and Poveda (1971), or Smith (1974). This author's (unpublished) computations on a very simple N-body system explicitly demonstrate a lack of time reversibility (using 102-bit arithmetic!).

KINETIC THEORY

We could jump into a discussion of the N-body approach to stellar dynamics because the groundwork was laid long ago in Chapters One and Two. We cannot do this for the other methods of attack that astrophysicists use on N-body systems because their foundations are deeper, varied, and possibly new to the reader. It is especially important to comprehend the basis in order to understand the nature of the flaws. Thus, the first section [which follows, with permission, Reif (1965) very closely] begins with a discussion of the Boltzmann equation (also known as the Vlasov equation when collisionless in plasma physics and the Fokker–Planck equation in stellar dynamics). After deriving it, we look at the simplest of equilibrium solutions (the isothermal sphere). Sidelights of importance, such as the virial expansion, the relationship with thermodynamics and hydrodynamics, the nonsaturation of gravitational forces, and the BBGKY hierarchy are also discussed.

Once this is under our belt, we can formulate the basis of the Monte Carlo calculations of Hénon and Spitzer et al. These build directly on two-body scattering due to self-gravitation and the various kinetic hypotheses already presented. Then the Hénon and Spitzer work will be briefly discussed. It appears to be circular, involving at best only self-consistency demonstrations. At the worst the situation is similar to the standard 25-body problem Lecar (1968) discussed, since the results of the two sets of computations do not agree.

The ingenuity of astrophysicists has not yet been exhausted. A complete retreat from individual stars can be made and one can dwell in the realm of thermodynamics and hydrodynamics. As these are *derived* from statistical mechanics and kinetic theory (this connection will be briefly illustrated), their lack of validity is evident. The evolutionary computations of Larson (1970a,b) and the stability calculations of Yabushita (1968) and of Lynden-Bell and Wood (1968) will then be examined. The latter two incorrect analyses have been partially corrected in Taff and Van Horn (1974, 1975). Other work of a more rigorous nature in this area will also be mentioned. Finally, this author's presentation (Taff 1977) of the N-body system will be reiterated, which can provide the unequivocal counterexample to the entire field of stellar dynamics.

The Boltzmann Equation

Suppose that, for a given situation which is in general not an equilibrium one, the distribution function $f(\mathbf{r}, \mathbf{v}, t)$ is known. This quantity is defined such that $f(\mathbf{r}, \mathbf{v}, t)\, d\mathbf{r}\, d\mathbf{v}$ is equal to the number of particles whose center of mass at time t is located between $\mathbf{r}$ and $\mathbf{r} + d\mathbf{r}$ and which also have a velocity between $\mathbf{v}$ and $\mathbf{v} + d\mathbf{v}$. One might think that the single-particle distribution function provides a complete description of the macroscopic state of a "gas" of stars and therefore should permit the computation of all quantities of physical interest. This will be true when the system is dilute and lacking in particle-particle correlation. Real clusters of stars are highly correlated.

As an example of the use of f, let $n(\mathbf{r}, t)\, d\mathbf{r}$ be the number of particles which at time t are located between $\mathbf{r}$ and $\mathbf{r} + d\mathbf{r}$. Then

$$n(\mathbf{r}, t) = \int f(\mathbf{r}, \mathbf{v}, t)\, d\mathbf{v}$$

where the integration is over all possible velocities. Furthermore, let $P(\mathbf{r}, \mathbf{v}, t)$ be any function that denotes a property of a star located near $\mathbf{r}$ with a velocity near $\mathbf{v}$ at time t. The mean value of P at the instant of time t, at the location $\mathbf{r}$, is defined by

$$\langle P(\mathbf{r}, \mathbf{v}, t) \rangle = \frac{1}{n(\mathbf{r}, t)} \int f(\mathbf{r}, \mathbf{v}, t) P(\mathbf{r}, \mathbf{v}, t)\, d\mathbf{v}$$

In particular, the mean velocity $\mathbf{u}(\mathbf{r}, t)$ of a star at location $\mathbf{r}$ and time t is

$$\mathbf{u}(\mathbf{r}, t) = \frac{1}{n(\mathbf{r}, t)} \int \mathbf{v} f(\mathbf{r}, \mathbf{v}, t)\, d\mathbf{v}$$

This velocity represents the average flow of the system at a given point. (This is just the hydrodynamic velocity of the fluid.) The "peculiar velocity" $\mathbf{U}$ of a particle is given by

$$\mathbf{U} \equiv \mathbf{v} - \mathbf{u}$$

The Boltzmann Equation Without Collisions

In order to calculate the distribution function $f(\mathbf{r}, \mathbf{v}, t)$, one needs to know what relations this function satisfies. We now turn to deducing these. Assume that each star (of mass m) is subject to an external force $\mathbf{F}(\mathbf{r}, t)$. (For simplicity assume that $\mathbf{F}$ does not depend on the velocity $\mathbf{v}$.) Consider first the particularly simple situation when all of the interactions between stars (i.e., collisions) can be completely neglected. Then, all of the stars that at time t have locations and velocities in the ranges $d\mathbf{r}$ and $d\mathbf{v}$ about $\mathbf{r}$ and $\mathbf{v}$, will an infinitesimally later time $t' = t + dt$ (as a result of their motion under the influence of the force $\mathbf{F}$) have locations and velocities in the ranges $d\mathbf{r}'$ and $d\mathbf{v}'$ near $\mathbf{r}'$ and $\mathbf{v}'$. The appropriate relationships are

$$\mathbf{r}' = \mathbf{r} + \dot{\mathbf{r}}\, dt = \mathbf{r} + \mathbf{v}\, dt$$

$$\mathbf{v}' = \mathbf{v} + \dot{\mathbf{v}}\, dt = \mathbf{v} + \mathbf{F}\frac{dt}{m} \tag{12.9}$$

In the absence of collisions this is all that happens, all stars in the range $d\mathbf{r}\, d\mathbf{v}$ near $\mathbf{r}$ and $\mathbf{v}$ will, after the time interval dt, be found in the range $d\mathbf{r}'\, d\mathbf{v}'$ about $\mathbf{r}'$ and $\mathbf{v}'$. Thus, Eqs. 12.9 imply that

$$f(\mathbf{r}', \mathbf{v}', t)\, d\mathbf{r}'\, d\mathbf{v}' = f(\mathbf{r}, \mathbf{v}, t)\, d\mathbf{r}\, d\mathbf{v} \tag{12.10}$$

The element of volume $d\mathbf{r}\, d\mathbf{v}$ in the six-dimensional $\mathbf{r}, \mathbf{v}$ space may become distorted in shape as a result of this motion. Its new volume is related to the old one by

$$d\mathbf{r}'\, d\mathbf{v}' = |J|\, d\mathbf{r}\, d\mathbf{v}$$

where J is the Jacobian of the transformation in Eq. 12.9:

$$J = \frac{\partial(x', y', z', v_x', v_y', v_z')}{\partial(x, y, z, v_x, v_y, v_z)} = \begin{vmatrix} 1 & 0 & 0 & dt & 0 & 0 \\ 0 & 1 & 0 & 0 & dt & 0 \\ 0 & 0 & 1 & 0 & 0 & dt \\ \dfrac{1}{m}\dfrac{\partial F_x}{\partial x}dt & & & 1 & 0 & 0 \\ \cdot & & & 0 & 1 & 0 \\ \cdot & & & 0 & 0 & 1 \end{vmatrix}$$

Note that all nine terms in the lower left corner of the determinant are proportional to dt. Hence,

$$J = 1 + \mathcal{O}(dt)$$

so that $J = 1$ is correct up to and including first-order terms in the infinitesimal time interval dt. Therefore, it follows that

$$d\mathbf{r}'\, d\mathbf{v}' = d\mathbf{r}\, d\mathbf{v}$$

so,* using these results in Eq. 12.10,

$$f(\mathbf{r}', \mathbf{v}', t') = f(\mathbf{r}, \mathbf{v}, t) \tag{12.11}$$

or

$$f(\mathbf{r} + \dot{\mathbf{r}}\, dt, \mathbf{v} + \dot{\mathbf{v}}\, dt, t + dt) - f(\mathbf{r}, \mathbf{v}, t) = 0$$

* Since J differs from unity only by terms of the second order it follows that Eq. 12.11 actually holds for all times.

Expanding in powers of dt leads to

$$\frac{\partial f}{\partial x}\dot{x}+\frac{\partial f}{\partial y}\dot{y}+\frac{\partial f}{\partial z}\dot{z}+\frac{\partial f}{\partial v_x}\dot{v}_x+\frac{\partial f}{\partial v_y}\dot{v}_y+\frac{\partial f}{\partial v_z}\dot{v}_z+\frac{\partial f}{\partial t}=0$$

This can be written more compactly as

$$\frac{\partial f}{\partial t}+\dot{\mathbf{r}}\cdot\frac{\partial f}{\partial \mathbf{r}}+\dot{\mathbf{v}}\cdot\frac{\partial f}{\partial \mathbf{v}}=\frac{\partial f}{\partial t}+\mathbf{v}\cdot\frac{\partial f}{\partial \mathbf{r}}+\left(\frac{\mathbf{F}}{m}\right)\cdot\frac{\partial f}{\partial \mathbf{v}}=0 \qquad (12.12)$$

Here $\partial f/\partial\mathbf{r}$ denotes the gradient of f with respect to $\mathbf{r}$. Similarly, $\partial f/\partial\mathbf{v}$ denotes the gradient with respect to $\mathbf{v}$, that is, the vector with components $\partial f/\partial v_x$, $\partial f/\partial v_y$, and $\partial f/\partial v_z$. Equation 12.12 is the linear partial differential equation satisfied by f. This relationship asserts that f remains unchanged if one moves along with the stars in phase space. This is a special case of Liouville's theorem and Eq. 12.12 is known as the Boltzmann equation without collisions. (In plasma physics it is sometimes called the Vlasov equation.)

The Boltzmann Equation With Collisions

Suppose now that collisions do take place. Then the number of stars in the range $d\mathbf{r}\,d\mathbf{v}$ can also change by virtue of them. The reason is that stars originally with locations and velocities not in the range $d\mathbf{r}\,d\mathbf{v}$ can be scattered into this range by collisions; conversely, stars originally in this range can be scattered out of it. Let $D_c f\,d\mathbf{r}\,d\mathbf{v}$ denote the net increase per unit time in the number of stars in the range $d\mathbf{r}\,d\mathbf{v}$ as a result of such collisions. Then one can assert, instead of Eq. 12.10, that the number of stars which at the instant $t+dt$ are in the range near $\mathbf{r}+\dot{\mathbf{r}}\,dt$ and $\mathbf{v}+\dot{\mathbf{v}}\,dt$ must be equal to the number of stars which were, at instant t, in the range near $\mathbf{r}$ and $\mathbf{v}$ and moved to $\mathbf{r}+\dot{\mathbf{r}}\,dt$, $\mathbf{v}+\dot{\mathbf{v}}\,dt$ as a result of the external force plus the net change in the number of stars in this range caused by collisions in the time interval dt. In symbols,

$$f(\mathbf{r}+\dot{\mathbf{r}}\,dt,\mathbf{v}+\dot{\mathbf{v}}\,dt,t+dt)\,d\mathbf{r}'\,d\mathbf{v}'=f(\mathbf{r},\mathbf{v},t)\,d\mathbf{r}\,d\mathbf{v}+D_c f\,d\mathbf{r}\,d\mathbf{v}\,dt$$

Using the fact that $d\mathbf{r}'\,d\mathbf{v}'=d\mathbf{r}\,d\mathbf{v}$, this becomes (cf. Eq. 12.12)

$$\frac{\partial f}{\partial t}+\dot{\mathbf{r}}\cdot\nabla_r f+\left(\frac{\mathbf{F}}{m}\right)\cdot\nabla_v f=D_c f \qquad (12.13)$$

This is called the Boltzmann equation.

The compact form in Eq. 12.13 hides a complicated expression. We shall write an explicit expression for $D_c f$ in terms of integrals involving f and describing properly the rate at which particles enter and leave $d\mathbf{r}\,d\mathbf{v}$ as a result of collisions. When this is done, the Boltzmann equation is then an integro-differential equation, that is, both the partial derivatives of f and the integrals of f are involved.

The treatment of the rate of change of f above did not deal with effects of collisions in a realistic way. In particular, correlations between stellar velocities during a collision were neglected. We shall now reformulate the problem in a more rigorous fashion by writing an equation for the distribution function directly in terms of the scattering cross section σ for binary collisions between the stars. The solution of this more accurate equation provides, in principle, a solution of the stellar dynamics problem within the approximations made. Because the equation is complex, the task of solving it is not easy and approximation methods must be used. Nevertheless, despite the increased complexity, there is an important advantage in formulating the problem (more) correctly. The reason is that we will have an equation that is fairly rigorous. Hence, general theorems can be proved and systematic approximation procedures developed. On the other hand, if we start from the simpler formulation above, it is difficult to estimate the errors committed (which are systematic in nature) and it is impossible to know how to correct for them.

Description of Two-Particle Collisions. We commence the discussion by considering in detail collisions between two stars. The two stars under consideration will be treated as simple particles with masses m_1 and m_2, location vectors $\mathbf{r}_1$ and $\mathbf{r}_2$, and velocities $\mathbf{v}_1$ and $\mathbf{v}_2$. The interaction between these particles depends in some way on their relative locations and velocities. We of course know what it is, but for the moment we want to keep the discussion as general as possible.

Conservation of the total linear momentum implies that

$$m_1\mathbf{v}_1 + m_2\mathbf{v}_2 = \mathbf{P} = \text{constant vector}$$

Another quantity of interest is the relative velocity $\mathbf{v}$,

$$\mathbf{v} = \mathbf{v}_1 - \mathbf{v}_2$$

One can use these two equations to express $\mathbf{v}_1$ and $\mathbf{v}_2$ in terms of $\mathbf{P} = M\mathbf{V} = (m_1 + m_2)\mathbf{V}$ and $\mathbf{v}$. The result is

$$(m_1 + m_2)\mathbf{v}_1 = \mathbf{P} + m_2\mathbf{v}, \qquad (m_1 + m_2)\mathbf{v}_2 = \mathbf{P} - m_1\mathbf{v}$$

or

$$\mathbf{v}_1 = \mathbf{V} + \frac{\mu}{m_1}\mathbf{v}, \qquad \mathbf{v}_2 = \mathbf{V} - \frac{\mu}{m_2}\mathbf{v}$$

where

$$\mathbf{V} \equiv \frac{\mathbf{P}}{m_1 + m_2} = \frac{m_1\mathbf{v}_1 + m_2\mathbf{v}_2}{M}$$

is the constant velocity of the center of mass. In addition, we have introduced the quantity

$$\mu \equiv \frac{m_1 m_2}{m_1 + m_2} = \frac{m_1 m_2}{M}$$

which is called the reduced mass. The total kinetic energy K of the particles is

$$K = \tfrac{1}{2}m_1|\mathbf{v}_1|^2 + \tfrac{1}{2}m_2|\mathbf{v}_2|^2 = \tfrac{1}{2}M|\mathbf{V}|^2 + \tfrac{1}{2}\mu|\mathbf{v}|^2$$

Now consider a collision. Denote the velocities of the two stars before they interact by $\mathbf{v}_1$ and $\mathbf{v}_2$. Denote their respective velocities after the collision by $\mathbf{v}_1'$ and $\mathbf{v}_2'$. In terms of the center of mass and relative variables the situation can be simply described. The center of mass velocity $\mathbf{V}$ remains unchanged because there is no external force acting. The relative velocity changes from the value $\mathbf{v}$ before the collision to some value $\mathbf{v}'$ after the collision. We assume that the collision is elastic so that the kinetic energy K remains unchanged. It follows, therefore, that $|\mathbf{v}|^2$ also remains unchanged and that $|\mathbf{v}'| = |\mathbf{v}|$. The only effect of a collision is to change the direction of the relative velocity. The collision process can thus be described by specifying the polar angle θ' and azimuthal angle ϕ' of the final relative velocity $\mathbf{v}'$ with respect to the initial relative velocity $\mathbf{v}$. It is simplest to visualize the relationship between the velocities before and after the collision by considering the corresponding momenta $\mathbf{p}_1 = m_1\mathbf{V} + \mu\mathbf{v}$, $\mathbf{p}_2 = m_2\mathbf{V} - \mu\mathbf{v}$ before the collision and $\mathbf{p}_1'$, $\mathbf{p}_2'$ after the collision. The geometrical relationship are shown in Fig. 48.

Scattering Cross Sections and Symmetry Properties. Stars initially having velocities $\mathbf{v}_1$ and $\mathbf{v}_2$ can get scattered through various angles θ' and ϕ' (depending on the value of the impact parameter D). If the only information available consists of the initial velocities, then the outcome of a scattering process must be described in statistical terms. This can be done by utilizing the quantity σ' defined such that $\sigma'(\mathbf{v}_1, \mathbf{v}_2 \to \mathbf{v}_1', \mathbf{v}_2')\, d\mathbf{v}_1'\, d\mathbf{v}_2'$ is equal to the number of stars per unit time, per unit flux of type 1 stars incident with relative velocity $\mathbf{v}$ upon a

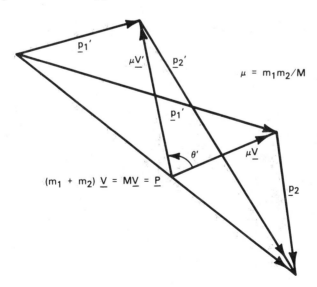

Figure 48. Elastic collision of two particles with initial momenta $\mathbf{p}_1$, $\mathbf{p}_2$ and final momenta $\mathbf{p}_1'$, $\mathbf{p}_2'$.

type 2 star, emerging after scattering with respective final velocities between v_1' and $v_1' + dv_1'$ and between v_2' and $v_2' + dv_2'$. Now

$$\mathbf{v}_1' = \mathbf{V}' + \frac{\mu}{m_1}\mathbf{v}' \qquad \text{and} \qquad \mathbf{v}_2' = \mathbf{V}' - \frac{\mu}{m_2}\mathbf{v}'$$

where, as discussed above $|\mathbf{V}'| = |\mathbf{V}|$ and $|\mathbf{v}'| = |\mathbf{v}|$. The angle σ' must vanish unless v_1' and v_2' are such that these conditions are satisfied. Moreover, the scattering process could be completely described in terms of the equivalent one-body problem of relative motion wherein the direction of $\mathbf{v}'$ is specified in terms of the angles θ' and ϕ'. Hence, we can define a simpler but less symmetrical quantity, the differential scattering cross section σ, with the statement that $\sigma(\mathbf{v}') \, d\Omega'$ is equal to the number of stars per unit time, per unit flux of type 1 stars incident with relative velocity $\mathbf{v}$ upon a type 2 star, emerging after scattering with relative final velocity $\mathbf{v}'$ in a solid angle range $d\Omega'$ about the angles θ' and ϕ'. The differential scattering cross section σ depends in general on the relative speed and on the angles θ' and ϕ'; σ is related to σ' by

$$\sigma(\mathbf{v}') \, d\Omega' = \iint \sigma'(\mathbf{v}_1, \mathbf{v}_2 \rightarrow \mathbf{v}_1', \mathbf{v}_2') \, dv_1' \, dv_2'$$

The integration is over all values of $\mathbf{V}'$ and of $|\mathbf{v}'|$. It is useful to express the velocity range $dv_1' \, dv_2'$ in terms of the variables $\mathbf{V}$ and $\mathbf{v}$. One deduces

$$dv_1 \, dv_2 = dV \, dv \qquad \text{and} \qquad dv_1' \, dv_2' = dV' \, dv'$$

so that

$$dv_1' \, dv_2' = dv_1 \, dv_2$$

The probability σ' has various symmetry properties. These follow because the equations of motion of the particles are invariant under time reversal $(t \rightarrow -t)$. Under such a reversal one obtains the "reverse" collision in which the particles retrace their paths in time. Thus, the relationship

$$\sigma'(\mathbf{v}_1, \mathbf{v}_2 \rightarrow \mathbf{v}_1', \mathbf{v}_2') \, dv_1' \, dv_2' = \sigma'(-\mathbf{v}_1', -\mathbf{v}_2' \rightarrow -\mathbf{v}_1, -\mathbf{v}_2) \, dv_1 \, dv_2$$

or, since $dv_1' \, dv_2' = dv_1 \, dv_2$,

$$\sigma'(\mathbf{v}_1, \mathbf{v}_2 \rightarrow \mathbf{v}_1', \mathbf{v}_2') = \sigma'(-\mathbf{v}_1', -\mathbf{v}_2' \rightarrow -\mathbf{v}_1, -\mathbf{v}_2)$$

must be true. The equations of motion are also invariant under a reflection $(\mathbf{r} \rightarrow -\mathbf{r})$. Again the signs of all of the velocities are changed, but the time order is not. whence

$$\sigma'(\mathbf{v}_1, \mathbf{v}_2 \rightarrow \mathbf{v}_1', \mathbf{v}_2') = \sigma'(-\mathbf{v}_1, -\mathbf{v}_2 \rightarrow -\mathbf{v}_1', -\mathbf{v}_2')$$

It is of particular interest to consider the "inverse" collision, namely that collision obtained from the original one by interchanging the initial and final states. (The inverse collision can also be obtained from the original collision by the operation of time reversal followed by the operation of space reflection.) Successive application of the above properties shows that the collision probabilities for the original and inverse collisions are also equal; namely,

$$\sigma'(\mathbf{v}_1, \mathbf{v}_2 \rightarrow \mathbf{v}_1', \mathbf{v}_2') = \sigma'(\mathbf{v}_1', \mathbf{v}_2' \rightarrow \mathbf{v}_1, \mathbf{v}_2)$$

Derivation of the Boltzmann Equation. We are now finally in a position to make use of this knowledge of collisions to derive an explicit expression for $D_c f$ in the Boltzmann equation (Eq. 12.13). In order to calculate $D_c f$, the rate of change of f caused by collisions, we shall make the following additional assumptions: (1) The star cluster is sufficiently dilute that only two-particle collisions need be taken into account (this is *known* to be a poor assumption). (2) Any possible effects of an external force $\mathbf{F}$ on the magnitude of the collision cross section can be ignored (this too is false in the stellar problem; $\mathbf{F}$ is just the background gravitational force of the cluster). (3) The distribution function $f(\mathbf{r}, \mathbf{v}, t)$ does not vary appreciably during a time interval of the order of the duration of a stellar collision, nor does it vary appreciably over a spatial distance of the order of the range of interstellar forces (again a poor assumption—gravitational $1/r^2$ forces have infinite range). (4) When considering a collision between two stars one can neglect possible correlations between their initial velocities prior to the collision. This is the fundamental approximation in kinetic theory and is known as the assumption of "molecular chaos." It is justified when the density is sufficiently low, for the mean free path will then be much larger than the range of interparticle forces. It is manifestly not true in the stellar dynamics case.

Ignore all of the caveats and focus attention on stars located in the volume element $d\mathbf{r}$ between $\mathbf{r}$ and $\mathbf{r} + d\mathbf{r}$. Consider the collisions that occur there in the time between t and $t + dt$. (Here $d\mathbf{r}$ is taken to be large compared to the range of interparticle forces and dt to be large compared to the duration of a collision. However, they can still be considered infinitesimally small with respect to variations in f by assumption 3.) We are interested in calculating how collisions cause a net change $D_c f(\mathbf{r}, \mathbf{v}, t)\, d\mathbf{r}\, d\mathbf{v}\, dt$ in the number of such particles with velocity between $\mathbf{v}$ and $\mathbf{v} + d\mathbf{v}$. First, the particles in $d\mathbf{r}$ can be scattered out of this range because of collisions with other particles; the resulting decrease is denoted by $D_c^- f(\mathbf{r}, \mathbf{v}, t)\, d\mathbf{r}\, d\mathbf{v}\, dt$. Second, particles in $d\mathbf{r}$ whose velocity is originally not in the range between $\mathbf{v}$ and $\mathbf{v} + d\mathbf{v}$ can be scattered into this range due to collisions with other particles. The resulting increase is denoted by $D_c^+ f(\mathbf{r}, \mathbf{v}, t)\, d\mathbf{r}\, d\mathbf{v}\, dt$. Hence, one can write

$$D_c f = D_c^+ f - D_c^- f \tag{12.14}$$

To calculate $D_c^- f$ consider the volume element $d\mathbf{r}_1$ and those particles in it with velocity near $\mathbf{v}_1$ which are scattered out of this velocity range due to collisions with other particles which are themselves in the same spatial volume element and which have some velocity near $\mathbf{v}_2$. The probability of occurrence of such a collision (wherein particle 1 changes its velocity from $\mathbf{v}_1$ to $\mathbf{v}_1'$, while particle 2 simultaneously changes its velocity from $\mathbf{v}_2$ to $\mathbf{v}_2'$) is described by the scattering probability $\sigma'(\mathbf{v}_1, \mathbf{v}_2 \rightarrow \mathbf{v}_1', \mathbf{v}_2')\, d\mathbf{v}_1'\, d\mathbf{v}_2'$. To obtain the total decrease $D_c^- f_1\, d\mathbf{r}_1\, d\mathbf{v}_1\, dt$ during dt of the number of particles located in $d\mathbf{r}_1$ with velocity between $\mathbf{v}_1$ and $\mathbf{v}_1 + d\mathbf{v}_1$, we must first multiply $\sigma'\, d\mathbf{v}_1'\, d\mathbf{v}_2'$ by the relative flux $|\mathbf{v}_1 - \mathbf{v}_2| f(\mathbf{r}_1, \mathbf{v}_1, t)\, d\mathbf{v}_1$ of type 1 particles incident upon the type 2 particles, and then multiply this by the number $f(\mathbf{r}_1, \mathbf{v}_2, t)\, d\mathbf{r}_1\, d\mathbf{v}_2$ of type 2 particles which can do such scattering. Then we have to sum the result over all possible initial velocities $\mathbf{v}_2$ for type 2 particles and then over all possible final velocities $\mathbf{v}_1'$ and $\mathbf{v}_2'$ of the scattered particles. Whence

$$D_c^- f(\mathbf{r}_1, \mathbf{v}_1, t)\, d\mathbf{r}_1\, d\mathbf{v}_1\, dt = \int_{\mathbf{v}_2'} \int_{\mathbf{v}_1'} \int_{\mathbf{v}_2} [|\mathbf{v}_1 - \mathbf{v}_2| f(\mathbf{r}_1, \mathbf{v}_1, t)\, d\mathbf{v}_1]$$

$$\times [f(\mathbf{r}_1, \mathbf{v}_2, t)\, d\mathbf{r}_1\, d\mathbf{v}_2][\sigma'(\mathbf{v}_1, \mathbf{v}_2 \rightarrow \mathbf{v}_1', \mathbf{v}_2')\, d\mathbf{v}_1'\, d\mathbf{v}_2']$$

Assumption 4 has been used in writing for the probability of the simultaneous presence in $d\mathbf{r}_1$ of particles with respective velocities near $\mathbf{v}_1$ and $\mathbf{v}_2$ an expression proportional to the product

$$f(\mathbf{r}_1, \mathbf{v}_1, t)\, d\mathbf{v}_1 \cdot f(\mathbf{r}_1, \mathbf{v}_2, t)\, d\mathbf{v}_2$$

This explicitly assumes the absence of any correlations between the initial velocities $\mathbf{v}_1$ and $\mathbf{v}_2$, and hence their statistical independence.

We now turn to the calculation of $D_c^+ f$. Considering again the same volume element $d\mathbf{r}_1$, we ask how many particles will end up after a collision with a velocity in the range between $\mathbf{v}_1$ and $\mathbf{v}_1 + d\mathbf{v}_1$. This involves a consideration of the appropriate inverse collisions. Namely, we want to consider all particles in $d\mathbf{r}_1$ with arbitrary initial velocities $\mathbf{v}_1'$ and $\mathbf{v}_2'$ which are such that, after a collision, one particle acquires a velocity in the range $\mathbf{v}_1$ to $\mathbf{v}_1 + d\mathbf{v}_1$, while the other particle acquires some velocity between $\mathbf{v}_2$ and $\mathbf{v}_2 + d\mathbf{v}_2$. This scattering process is described by $\sigma'(\mathbf{v}_1', \mathbf{v}_2' \rightarrow \mathbf{v}_1, \mathbf{v}_2)$ and the relative flux of particles with initial velocity near $\mathbf{v}_1'$ is $|\mathbf{v}_1' - \mathbf{v}_2'| f(\mathbf{r}_1, \mathbf{v}_1', t)\, d\mathbf{v}_1'$. These particles get scattered by the $f(\mathbf{r}_1, \mathbf{v}_2', t)\, d\mathbf{r}_1\, d\mathbf{v}_2'$ particles with velocity near $\mathbf{v}_2'$. Hence, the expression for the total increase in time dt of the number of particles located in $d\mathbf{r}_1$ with velocity between $\mathbf{v}_1$ and $\mathbf{v}_1 + d\mathbf{v}_1$ can be written as

$$D_c^+ f(\mathbf{r}_1, \mathbf{v}_1, t)\, d\mathbf{r}_1\, d\mathbf{v}_1\, dt = \int_{\mathbf{v}_2} \int_{\mathbf{v}_2'} \int_{\mathbf{v}_1'} [|\mathbf{v}_1' - \mathbf{v}_2'| f(\mathbf{r}_1, \mathbf{v}_1', t)\, d\mathbf{v}_1']$$

$$\times [f(\mathbf{r}_1, \mathbf{v}_2', t)\, d\mathbf{r}_1\, d\mathbf{v}_2'][\sigma'(\mathbf{v}_1', \mathbf{v}_2' \rightarrow \mathbf{v}_1, \mathbf{v}_2)\, d\mathbf{v}_1\, d\mathbf{v}_2]$$

Note the following simplifying features; the probabilities for inverse collisions are equal, so that

$$\sigma'(\mathbf{v}_1', \mathbf{v}_2' \rightarrow \mathbf{v}_1, \mathbf{v}_2) = \sigma'(\mathbf{v}_1, \mathbf{v}_2 \rightarrow \mathbf{v}_1', \mathbf{v}_2')$$

Furthermore, we can introduce the relative velocities

$$\mathbf{V} = \mathbf{v}_1 - \mathbf{v}_2, \qquad \mathbf{V}' = \mathbf{v}_1' - \mathbf{v}_2'$$

and remember that the conservation of kinetic energy for elastic collisions implies that

$$|\mathbf{V}'| = |\mathbf{V}| = V$$

It is also convenient to introduce the abbreviations

$$f_1 \equiv f(\mathbf{r}_1, \mathbf{v}_1, t), \qquad f_2 \equiv f(\mathbf{r}_2, \mathbf{v}_2, t)$$

$$f_1' \equiv f(\mathbf{r}_1, \mathbf{v}_1', t), \qquad f_2' \equiv f(\mathbf{r}_2, \mathbf{v}_2, t)$$

Finally, Eq. 12.14 can be rewritten as

$$D_c f_1 = \int \int \int (f_1' f_2' - f_1 f_2) V \sigma'(\mathbf{v}_1, \mathbf{v}_2 \to \mathbf{v}_1', \mathbf{v}_2') \, d\mathbf{v}_2 \, d\mathbf{v}_1' \, d\mathbf{v}_2'$$

We can express this result in terms of $\mathbf{V}'$ and the solid-angle range $d\Omega'$ about this vector. The Boltzmann equation 12.13 for $f(\mathbf{r}_1, \mathbf{v}_1, t)$ can then be written in the explicit form

$$\frac{\partial f_1}{\partial t} + \mathbf{v}_1 \cdot \frac{\partial f_1}{\partial \mathbf{r}_1} + \left(\frac{\mathbf{F}}{m}\right) \cdot \frac{\partial f_1}{\partial \mathbf{v}_1} = \int \int (f_1' f_2' - f_1 f_2) V \sigma \, d\Omega' \, d\mathbf{v}_2 \qquad (12.15)$$

where $\sigma = \sigma(\mathbf{V}')$.

The Solution of the Boltzmann Equation

The Maxwell–Boltzmann Velocity Distribution

Suppose that the external force $\mathbf{F}$ is null and that the system is in equilibrium. Then for noninteracting particles (we will get back to stars soon) the single-particle distribution function only depends on $\mathbf{v}$. From Eq. 12.15 it can be seen that a sufficient condition for $f = f_{MB}(\mathbf{v})$ to be a solution is

$$f_{MB}(\mathbf{v}_1') f_{MB}(\mathbf{v}_2') = f_{MB}(\mathbf{v}_1) f_{MB}(\mathbf{v}_2) \qquad (12.16)$$

where $(\mathbf{v}_1, \mathbf{v}_2) \to (\mathbf{v}_1', \mathbf{v}_2')$ in any collision. This condition is also necessary and demonstrates that f_{MB} is independent of the differential scattering cross section (as long as it is nonzero). The proof involves a demonstration of Boltzmann's H theorem (which one can skip upon a first reading).

The H Theorem. Define the functional $H[f]$ by

$$H \equiv \int f(\mathbf{v}, t) \ln f(\mathbf{v}, t) \, d\mathbf{v}$$

where $f(\mathbf{v}, t)$ satisfies

$$\frac{\partial f_1}{\partial t} = \int \int (f_1' f_2' - f_1 f_2) V \sigma(\mathbf{V}') \, d\Omega' \, d\mathbf{v}_2 \qquad (12.17)$$

Form

$$\frac{dH}{dt} = \int \frac{\partial f(\mathbf{v}, t)}{\partial t}[1 + \ln f(\mathbf{v}, t)] \, d\mathbf{v} \qquad (12.18)$$

Hence if $\partial f/\partial t = 0$, $dH/dt = 0$. Thus, a necessary condition for the vanishing of $\partial f/\partial t$ is $dH/dt = 0$. The next step is to demonstrate the equivalence between $dH/dt = 0$ and Eq. 12.16. This will be accomplished when we show that if f satisfies the Boltzmann equation 12.15, then $dH/dt < 0$ (this result is known as Boltzmann's H theorem).

The expression for $\partial f_1/\partial t$ in Eq. 12.17 is substituted into Eq. 12.18:

$$\frac{dH}{dt} = \iiint \sigma(\mathbf{V}') V(f_2'f_1' - f_2 f_1)[1 + \ln f_1] \, d\Omega' \, d\mathbf{v}_1 \, d\mathbf{v}_2$$

We can interchange $\mathbf{v}_1$ with $\mathbf{v}_2$ in the integral since σ is invariant to such a procedure. Do this, add it to the above expression, and then divide by 2 to consequently express $\dot{H}$ as

$$\frac{dH}{dt} = \frac{1}{2} \iiint \sigma(\mathbf{V}') V(f_1'f_2' - f_1 f_2)[2 + \ln(f_1 f_2)] \, d\Omega' \, d\mathbf{v}_1 \, d\mathbf{v}_2$$

Similarly, the inverse collision symmetry of σ means that we can interchange the $\mathbf{v}_1, \mathbf{v}_2$ pair for the $\mathbf{v}_1', \mathbf{v}_2'$ pair. Thus, we also have the form

$$\frac{dH}{dt} = \frac{1}{2} \iiint \sigma(\mathbf{V}) V'(f_1 f_2 - f_1'f_2')[2 + \ln(f_1'f_2')] \, d\Omega \, d\mathbf{v}_1' \, d\mathbf{v}_2'$$

for $\dot{H}$. We have already seen that $d\mathbf{v}_1' \, d\mathbf{v}_2' = d\mathbf{v}_1 \, d\mathbf{v}_2$, $V = V'$, and $\sigma(\mathbf{V}) = \sigma(\mathbf{V}')$. Take half the sum of the last two expressions for $\dot{H}$ to obtain

$$\frac{dH}{dt} = \frac{1}{4} \iiint \sigma(\mathbf{V}) V(f_1'f_2' - f_1 f_2)[\ln(f_1 f_2) - \ln(f_1'f_2')] \, d\Omega \, d\mathbf{v}_1 \, d\mathbf{v}_2$$

Since $\sigma V > 0$ it follows from the properties of the natural logarithm that the integrand is never positive. Thus, $dH/dt = 0$ if and only if the integrand above vanishes identically—but that implies Eq. 12.16. Note too that this shows that $\lim_{t \to \infty} f(\mathbf{v}, t) = f_{MB}(\mathbf{v})$, independent of the value of $f(\mathbf{v}, t_0)$ for any finite t_0.

Equation 12.16 expresses a conservation law; for every collision $\ln f_{MB}(\mathbf{v})$ is conserved. The conserved quantities for a particle are (1) its mass, (2) its linear momentum, and (3) its kinetic energy (remember there are no forces yet). Thus, $f_{MB}(\mathbf{v})$ must have the form

$$f_{MB}(\mathbf{v}) = C \exp\left[\frac{-m\beta|\mathbf{v} - \mathbf{v}_0|^2}{2}\right]$$

where C, β, and $\mathbf{v}_0$ comprise the five constants. But f_{MB} is the single-particle distribution function and the density is a constant (no forces!). Thus, if the total number of particles is N and they occupy a volume V,

$$\frac{N}{V} = \int f_{MB}(\mathbf{v}) \, d\mathbf{v} = C\left(\frac{2\pi}{m\beta}\right)^{3/2}$$

Also, since there are no forces, the system as a whole must be at rest. Thus, the average velocity **u** must vanish. But direct calculation shows that

$$\mathbf{u} = \frac{V}{N} \int f_{\mathrm{MB}}(\mathbf{v})\mathbf{v}\, d\mathbf{v} = \mathbf{v}_0 = \mathbf{0}$$

Finally, the average energy of a particle is computed:

$$\frac{1}{n} \int \tfrac{1}{2} m v^2 f_{\mathrm{MB}}(\mathbf{v})\, dv$$

where the number density $n = N/V$. With $\mathbf{v}_0 = \mathbf{0}$, this is $3/2\beta$. We need to relate the average energy to a physical quantity. This will be done immediately below with the result that the average energy is $3PV/(2N)$ where P is the pressure that the particles exert through collisions. But the thermodynamic temperature T is defined by the ideal gas law, $PV = NkT$ (k is Boltzmann's constant). Thus, $\beta = 1/kT$ and the Maxwell–Boltzmann velocity distribution, for a dilute system of particles not in an external force field, is

$$f_{\mathrm{MB}}(\mathbf{v}) = \frac{N}{V}\left(\frac{m}{2\pi kT}\right)^{3/2} \exp\left(\frac{-mv^2}{2kT}\right) \tag{12.19}$$

Note too the deep connection between kinetic theory and thermodynamics.

The Computation of the Pressure.   Suppose that we wish to calculate the average force **F** exerted by the gas on a small element of area dA of the container wall. Then we must calculate the mean net momentum delivered to this wall element per unit time by the impinging particles. If attention is focused on an element of area dA lying inside the system an infinitesimal distance in front of the wall, then the above calculation is equivalent to finding the average net momentum transported per unit time across this surface as the particles cross this surface from both directions.* Denote by $\mathbf{\Pi}^+$ the average momentum crossing this surface dA per unit time from left to right, and by $\mathbf{\Pi}^-$ the average momentum crossing this surface dA per unit time from right to left. Then

$$\mathbf{F} = \mathbf{\Pi}^+ - \mathbf{\Pi}^- \tag{12.20}$$

In order to compute $\mathbf{\Pi}^+$ consider an element of surface dA and focus attention on those particles with velocity between **v** and $\mathbf{v} + d\mathbf{v}$ (see Fig. 49). The average number of such particles that cross this area in an infinitesimal time interval dt is the average number of such particles contained in the cylinder of volume $|dA\, v\, dt \cos\theta|$; that is, it is equal to $f(\mathbf{v})\, d\mathbf{v}|dA\, v\, dt \cos\theta|$. By multiplying this number by the momentum ($m\mathbf{v}$) of each particle and then dividing by the time interval (dt), we obtain the mean momentum transported across the area dA

---

* Similarly, and equivalently, we could consider an element of area anywhere inside the gas and ask for the average force which the gas on one side exerts on the gas on the other side. Again this is the same as asking what is the net transport of momentum across the area.

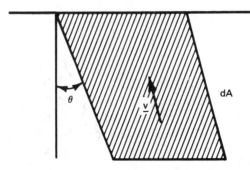

Figure 49. Illustration of momentum transfer through area dA in time dt due to particles with velocity **v**.

per unit time by particles with velocity near **v**. If we sum over all particles that cross this area from left to right, then we deduce, for the total momentum $\mathbf{\Pi}^+$ transported across this area, the expression

$$\mathbf{\Pi}^+ = \int_{v_z \geqslant 0} f(\mathbf{v}) |v \, dA \cos \theta| m\mathbf{v} \, d\mathbf{v}$$

or

$$\mathbf{\Pi}^+ = dA \int_{v_z \geqslant 0} f(\mathbf{v}) |v_z| m\mathbf{v} \, d\mathbf{v}$$

We have put $v_z = v \cos \theta$. A similar expression gives the total momentum $\mathbf{\Pi}^-$ transported across this area from right to left except that the integration must now be over all particles for which $v_z \leqslant 0$. Thus,

$$\mathbf{\Pi}^- = dA \int_{v_z \leqslant 0} f(\mathbf{v}) |v_z| m\mathbf{v} \, d\mathbf{v}$$

The force in Eq. 12.20 is given by the net momentum transported across the surface, that is,

$$\mathbf{F} = dA \int f(\mathbf{v}) v_z m\mathbf{v} \, d\mathbf{v} \tag{12.21}$$

Since the gas is in equilibrium, $f(\mathbf{v}) = f_{mB}(\mathbf{v})$ and is only a function of $v \equiv |\mathbf{v}|$. Note first that

$$F_x = m \, dA \int f_{MB}(\mathbf{v}) v_z v_x \, d\mathbf{v} = 0$$

since the integrand is odd. This expresses the fact that there can be no net tangential force on the wall in an equilibrium situation (and is the basis for the preceding footnote). The mean normal force does not vanish; measured per unit area it gives the average pressure. By Eq. 12.21, P is equal to

$$P = \frac{F_z}{dA} = \int f_{MB}(\mathbf{v}) m v_z^2 \, d\mathbf{v}$$

or

$$P = \langle n m v_z^2 \rangle$$

where we have used the definition

$$\langle v_z^2 \rangle \equiv \frac{1}{n} \int f_{MB}(\mathbf{v}) v_z^2 \, d\mathbf{v}$$

By symmetry, $\langle v_x^2 \rangle = \langle v_y^2 \rangle = \langle v_z^2 \rangle$, so that

$$\langle |\mathbf{v}|^2 \rangle = \langle v_x^2 + v_y^2 + v_z^2 \rangle = 3 \langle v_z^2 \rangle$$

Hence, P can be equivalently written as

$$P = \tfrac{1}{3} \langle n m v^2 \rangle \tag{12.22}$$

Since $\langle v^2 \rangle$ is related to the mean kinetic energy K of a particle, Eq. 12.22 implies the general relationship

$$P = \tfrac{2}{3} n \langle \tfrac{1}{2} m v^2 \rangle = \tfrac{2}{3} \langle n K \rangle \tag{12.23}$$

That is, the mean presure is just equal to two-thirds of the mean kinetic energy per unit volume.

Until now we have not made use of the fact that the number density of particles $f_{MB}(\mathbf{v}) \, d\mathbf{v}$ is given by the Maxwell–Boltzmann velocity distribution. This information allows one to explicitly calculate $\langle v^2 \rangle$ and is equivalent to using the equipartition theorem result that $\langle K \rangle = \tfrac{3}{2} kT$. Therefore, Eq. 12.23 becomes

$$P = nkT$$

and we have regained the equation of state of a classical ideal gas.

Solutions With External Forces

Suppose that the force acting on a particle $\mathbf{F}(\mathbf{r}, t)$ is time independent and conservative. Thus, $\mathbf{F} = -\nabla \phi(\mathbf{r})$ for some potential ϕ. Then, the equilibrium solution of the Boltzmann equation is just

$$f_{MB}(\mathbf{v}) \exp\left[\frac{-\phi(\mathbf{r})}{kT} \right]$$

where $f_{MB}(\mathbf{v})$ is the Maxwell–Boltzmann velocity distribution in Eq. 12.19. To see this, observe that the collision term in the Boltzmann equation, Eq. 12.15,

is still satisfied because we assumed (long ago) that σ was not influenced by
F. Obviously $\partial f/\partial t = 0$, and almost as trivially the sum of the other two terms
on the left-hand side can be seen to vanish. Absorbing the factor $\exp(-\phi/kT)$
into the density, the equilibrium single-particle distribution function may be
written as

$$f(\mathbf{r}, \mathbf{v}) = n(\mathbf{r}) f_{MB}(\mathbf{v})$$

with

$$n(\mathbf{r}) = \int f(\mathbf{r}, \mathbf{v}) \, d\mathbf{v} = \left(\frac{N}{V}\right) \exp\left[\frac{-\phi(\mathbf{r})}{kT}\right]$$

Now back to stars and stellar systems.

The Isothermal Sphere. The external force **F** exerted on a single star in a
cluster of stars is the gravitational force due to the other $N-1$ stars in the
system. With this interpretation for **F** and ϕ, we close the system of equations
self-consistently with Poisson's equation,

$$\nabla^2 \phi = 4\pi G m n = \left(\frac{4\pi G M}{V}\right) \exp\left(\frac{-\phi}{kT}\right)$$

Here $M = mN$ is the total mass in the cluster. If we impose the assumption
of spherical symmetry, then we have an ordinary differential equation to solve
for $\phi(r)$,

$$\frac{1}{r^2} \frac{d}{dr}\left(r^2 \frac{d\phi}{dr}\right) = \frac{4\pi G M}{V} \exp\left(\frac{-\phi}{kT}\right)$$

This is the genesis of the isothermal sphere model for stellar systems. The
model is appropriate whenever the Boltzmann equation is. The boundary
conditions are $\phi(0) = $ a constant (related to the central density) and $d\phi/dr = 0$
at $r = 0$ (a symmetry condition).

In stellar dynamics it is usual to use the following change of variables:

$$\frac{n(r)}{n(0)} = e^{-\psi}, \qquad r = \alpha\xi$$

with

$$\psi = \frac{\phi(r) - \phi(0)}{kT}, \qquad \alpha^2 = \frac{kT}{4\pi G m n(0)}$$

The final result is the equation for an isothermal sphere,

$$\frac{d}{d\xi}\left(\xi^2 \frac{d\psi}{d\xi}\right) = \xi^2 e^{-\psi}, \qquad \psi(0) = \left.\frac{d\psi}{d\xi}\right|_0 = 0$$

The density distribution for an isothermal sphere is shown in Fig. 50. The total

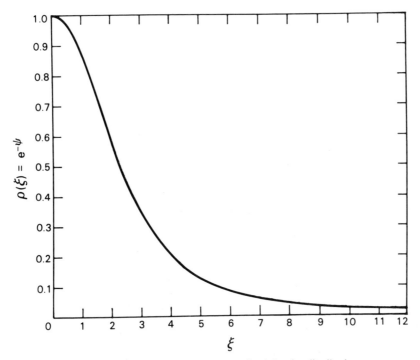

Figure 50. The isothermal sphere normalized density distribution.

mass M inside a radius $R = \alpha \Xi$ is

$$M = 4\pi m \int_0^R n(r) r^2 \, dr = 4\pi\alpha^3 n(0) \left(\xi^2 \frac{d\psi}{d\xi} \right) \Bigg|_{\xi=\Xi}$$

This diverges as $R \to \infty$. The mathematical properties of the isothermal sphere differential equation have been discussed in detail by Chandrasekhar (1939). Multicomponent isothermal spheres were developed by Taff et al. (1975).

More Thermodynamics. We made a connection between kinetic theory and thermodynamics both when the equation of state $P = n/\beta$ was introduced and when β was identified as $1/kT$. The entire thermodynamics of an ideal gas is contained in the work done so far. The total internal energy is just $3N/2\beta = 3NkT/2$. Clearly the specific heat at constant volume $C_V = 3Nk/2$. The work done by the gas when it expands from V to $V + dV$ is $P \, dV$. The first law of thermodynamics defines the heat absorbed,

$$dQ = dU + P \, dV$$

The entropy S is $-VHk$ where H is Boltzmann's functional. From its definition,

$$S = \tfrac{3}{2} Nk \ln(PV^{5/3}) + \text{constant}$$

whence $dS = dQ/T$ showing that $1/T$ is the integrating factor of the heat differential. All of the rest of the thermodynamics of an ideal gas follow from these relationships.

By uncritical analogous reasoning one can discuss the thermodynamics of a self-gravitating (e.g., nonideal) "gas" of stars. Before explicitly discussing the critical physical point, we try to illuminate the nature of the problem within the kinetic theory-thermodynamical context of a nonideal gas.

Alternatives to the Boltzmann Equation

Results From the Virial Theorem

Consider a collection of particles contained in some finite volume. The ith particle is acted upon by a force $\mathbf{F}_i$ composed of contributions from the other particles and from external forces. The x component of the equation of motion for this particle is (suppressing the subscript i for the moment),

$$F_x = \frac{dp_x}{dt}$$

where $p_x = mv_x = m(dx/dt)$ is the x component of the momentum. Similar equations exist for the y and z components as well as for all of the other particles. The x component of the kinetic energy can be written in the form

$$K_x = \frac{p_x^2}{2m} = \frac{1}{2}m\left(\frac{dx}{dt}\right)^2 = \frac{1}{2}m\frac{d}{dt}\left(x\frac{dx}{dt}\right) - \frac{1}{2}mx\frac{d^2x}{dt^2}$$

$$= \frac{1}{2}\frac{d}{dt}(xp_x) - \frac{1}{2}x\frac{dp_x}{dt} = \frac{1}{2}\frac{d}{dt}(xp_x) - \frac{1}{2}xF_x$$

where the last expression has been obtained by a substitution from the equations of motion. A similar result holds for the y and z components, and adding these all together and summing over all of the particles, we find that the total kinetic energy K may be expressed as

$$K = -\frac{1}{2}\sum_i\left[(\mathbf{r}_i \cdot \mathbf{F}_i) + \frac{1}{2}\frac{d}{dt}(\mathbf{r}_i \cdot \mathbf{p}_i)\right]$$

If we now average this expression over some time interval τ, then it can be shown that the last term vanishes for sufficiently large τ. For, in terms of a single rectangular component,

$$\overline{\frac{d}{dt}(xp_x)} = \frac{1}{\tau}\int_0^\tau \frac{d(xp_x)}{dt}\,dt = \frac{1}{\tau}xp_x\Big|_0^\tau \to 0 \qquad \text{as } \tau \to \infty$$

since x and p_x remain finite (the particle is in a box), but we can make τ as large as we please. Thus we obtain the result

$$\bar{K} = -\frac{1}{2}\sum_i(\mathbf{r}_i \cdot \mathbf{F}_i) \tag{12.24}$$

where the bars refer to time averages. This is the famous virial theorem of Clausius, which states that the average kinetic energy of a system of particles is equal to the average of a quantity called the *virial* of the system.

Let us now evaluate the contribution to the virial from the walls of the container. On an element of area of the wall dA the particles exert a time averaged force equal to $P\mathbf{n}\, dA$ where P is the pressure and $\mathbf{n}$ is the outward normal to dA. By Newton's third law of motion, this element of the wall exerts an equal and opposite force on the particles. Integrating over the surface of the container, and transforming the surface integral to a volume integral by the divergence theorem, we find that

$$-\frac{1}{2}\sum_i \overline{(\mathbf{r}_i \cdot \mathbf{F}_i)}_{\text{wall}} = \frac{1}{2}\int P(\mathbf{r} \cdot \mathbf{n})\, dA = \frac{P}{2}\int (\nabla \cdot \mathbf{r})\, dV = \frac{3}{2}PV$$

Substitution of this into Eq. 12.24 yields

$$PV = \frac{2}{3}\bar{K} + \frac{1}{3}\sum_i \overline{(\mathbf{r}_i \cdot \mathbf{F}_i)}$$

where the wall forces are now excluded from the virial. Because most systems in which one is interested are conservative systems without external forces, we can rewrite this as

$$PV = \frac{2}{3}\bar{K} - \frac{1}{3}\sum_i \overline{(\mathbf{r}_i \cdot \nabla_i U)}$$

where U is the potential energy of the system and represents all of the forces between the particles (not restricted to central forces or to additive forces).

This expression can be considerably simplified if the potential energy between pairs of particles, u, is central. For then

$$U = \frac{1}{2}\sum_{i \neq j}\sum u(r_{ij})$$

where $r_{ij} = |\mathbf{r}_i - \mathbf{r}_j|$. Then

$$PV = \frac{2}{3}\bar{K} - \frac{1}{6}\sum_{i \neq j}\sum \overline{\left[r_{ij}\frac{du(r_{ij})}{dr_{ij}}\right]} = \frac{2}{3}\bar{K} - \frac{N(N-1)}{6}\overline{\left[r_{12}\frac{du(r_{12})}{dr_{12}}\right]} \qquad (12.25)$$

If the system is in equilibrium, then we can proceed further and evaluate $\bar{K}$ by replacing the time average with an ensemble average,

$$\bar{K} = \frac{\int K e^{-\mathscr{E}/kT}\, d\mathbf{p}^N\, d\mathbf{r}^N}{\int e^{-\mathscr{E}/kT}\, d\mathbf{p}^N\, d\mathbf{r}^N}$$

where $d\mathbf{p}^N = d\mathbf{p}_1 \cdots d\mathbf{p}_N$, $d\mathbf{r}^N = d\mathbf{r}_1 \cdots d\mathbf{r}_N$ and $\mathscr{E}$ is the total energy of the system. (The potential energy U is time independent for a system in an equilibrium state.) The integration over the momenta is straightforward and yields the well-known result that was derived above, namely,

$$\bar{K} = \tfrac{3}{2}NkT$$

Note that this result is valid for both perfect and imperfect gases. If we substitute this expression for the total kinetic energy into Eq. 12.25 and then replace the time average of the virial by an ensemble average, then we can deduce the expression for the equation of state,

$$PV = NkT - \frac{2\pi N(N-1)}{3V} \int g(r) \frac{du(r)}{dr} r^3 \, dr \qquad (12.26)$$

where $r \equiv r_{12}$ and $g(r)$ is the radial distribution function,

$$g(r) = \frac{V^2 \int e^{-U/kT} d\mathbf{r}_3 \cdots d\mathbf{r}_N}{\int e^{-U/kT} d\mathbf{r}_1 \cdots d\mathbf{r}_N}$$

At this point we can obtain an expression for the second virial coefficient. Expanding U into sums over pairs and neglecting all interactions except those between particles 1 and 2, the first approximation to the radial distribution function will be found to be

$$g(r) = e^{-u(r)/kT} + \cdots$$

Inserting this into the equation of state (Eq. 12.26), we identify

$$B(T) = -\frac{2\pi V}{3kT} \int_0^\infty e^{-u(r)/kT} \frac{du(r)}{dr} r^3 \, dr \qquad (12.27)$$

or

$$B(T) = -2\pi V \int_0^\infty [e^{-u(r)/dT} - 1] r^2 \, dr \qquad (12.28)$$

(as an integration by parts will show) as the second virial coefficient. The integration has been formally extended to infinity because the integrand falls to zero rapidly at large r for *most* $u(r)$. In deriving this we have also made use of the fact that $N \gg 1$. This result is exact, despite my rather hasty derivation. The higher virial coefficients follow from higher approximations to $g(r)$, obtained by expanding the general expression for $g(r)$ as a power series in the density.

An equation of state for a gas merely gives a mathematical relation among the pressure, volume, temperature, and number of molecules at equilibrium. Such a relationship does not have to be expressed in the form of an equation, and in the real world tables are used. The simplest and best known equation of state is that for the perfect gas, $PV = NkT$. Real gases exhibit deviations from this simple equation, but all gases approach it in the limit of very low densities. Of all the modifications proposed to the perfect gas equation of state, the virial equation of state expresses the deviations from the perfect gas equation as an infinite power series in the density; namely,

$$\frac{PV}{NkT} = 1 + BV^{-1} + CV^{-2} + DV^{-3} + \cdots$$

where $B, C, D, \ldots$, are called the second, third, fourth, $\ldots$, virial coefficients. The virial coefficients depend on the temperature and on the particular gas under consideration but are independent of the density or the pressure. The reason for the special importance of the virial equation of state is that it is the only equation of state known that has a sound theoretical foundation. There is a definite interpretation for each virial coefficient in terms of molecular properties. The second virial coefficient represents the deviations from perfection corresponding to interactions between two particles (as we have seen), the third represents the deviations corresponding to interactions among three particles, and so on. Thus, the virial equation of state forms the connection between experimental results and our knowledge of interparticle interactions.

Application to Stellar Systems. The nexus of the problem with the applications of kinetic theory, of statistical mechanics, or of thermodynamics to self-gravitating systems is contained in Eqs. 12.27 and 12.28. Because the interparticle potential u is equal to $-Gm^2/r$, there is trouble at both ends of the domain of integration. First consider $r \to \infty$ and Eq. 12.27. As $r \to \infty$ and $u \to 0, \exp(-u/kT) - 1 \to 1 - u/kT - 1 = -u/kT$. Then one has $B(T) \to -(2\pi VGm^2/kT) \int^{\infty} r \, dr$, which clearly diverges. Of course we were somewhat cavalier in extending the upper range of integration to infinity. For any normal system the interparticle force rapidly goes to zero as the interparticle separation increases—remember that the main area of applicability of kinetic theory was to dilute gases of molecules and atoms, not to dilute gases of stars. Clearly, we can remove this divergence by only extending the integral to the size of the stellar cluster. Exactly where to end the integration (the tidal radius?) and what to do about escaping stars appear to be minor annoyances.

The integral in Eq. 12.28 also diverges as $r \to 0$ (as can be seen by inspection). This difficulty at close approaches is already familiar to us and will reappear when we discuss the two-body scattering cross section in detail (below). It might be argued that it is merely a consequence of the idealization of physical bodies (stars) as particles (e.g., mass points). Hence the divergence can be removed by only integrating down to (say) a stellar diameter for, if two stars approach more closely than that, their identity and form are in question. Indeed, studies of real stellar collisions, stellar coalescence, and so on, have been done (with application to the cores of globular clusters and galactic nuclei). No stellar dynamics computation is interested in such events or descriptions thereof. Exactly where one draws the line and how one handles the resultant correlations (which violate the assumption of molecular chaos) is not clear.

Both of these problems are due to the nonsaturation of gravitational forces (Levy-Leblond 1969). This means that the total energy of any finite collection of self-gravitating mass points does not have a finite, extensive (e.g., proportional to the number of particles) lower bound. Without such a property there can be no rigorous basis for the statistical mechanics of such a system (Fisher and Ruelle 1966). This result is not a consequence of the $1/r^2$ nature of the gravitational force but rather of its unshielded character [cf. Dyson and Lenard (1967) for a discussion of the electrostatic case]. Basically it is that simple. One can ignore the fact that one *knows* that there is no rigorous basis for one's computer manipulations, one can try to improve the situation, or one can look for another job.

The BBGKY Hierarchy

What happens if one lifts all of the assumptions we took on to derive the Boltzmann equation? Basically (and this does *not* deal with the failure of the gravitational force to saturate) one acknowledges the possibility of correlations between particles. Then the probability of two particles being within $d\mathbf{r}$ of $\mathbf{r}$ having velocities within $d\mathbf{v}$ of $\mathbf{v}$ and within $d\mathbf{v}'$ of $\mathbf{v}'$ at time t is not proportional to

$$f(\mathbf{r}, \mathbf{v}, t) \, d\mathbf{v} \cdot f(\mathbf{r}, \mathbf{v}', t) \, d\mathbf{v}'$$

but also depends on the two-particle correlation function, $f_2(\mathbf{r}_1, \mathbf{v}_1, \mathbf{r}_2, \mathbf{v}_2, t)$. The collision term on the right-hand side of Eq. 12.15 also depends on this quantity. Just as one could derive an integro-differential equation for f (the single-particle distribution function) in terms of the two-particle distribution function, one can do the same for f_2 in terms of f_3 (the three-particle distribution function); and f_3 in terms of f_4, f_4 in terms of f_5, ... This hierarchy of coupled equations is known as the Bogoliubov, Born, Green, Kirkwood, and Yvon (BBGKY) hierarchy after the mathematical physicists who developed it. [See Bogoliubov (1946a,b, 1962)]. This development is better than just the Boltzmann equation, but it is still inappropriate for gravitational systems.

Sundmann's Theorem of Total Collapse. Another way of describing the problem with self-gravitating systems is that the systems are inherently unstable to collapse. As there is no shielding of the gravitational force, and it is attractive, the presence of any kind of dissipating mechanisn means that collapse is inevitable. Look at this from another point of view. Let the N-body system consist of N particles, masses m_n, at locations $\mathbf{r}_n$. Form

$$\sum_{n=1}^{N} m_n |\mathbf{r}_n - \mathbf{r}_l|^2 = \sum_{n=1}^{N} m_n r_n^2 - 2\mathbf{r}_l \cdot \sum_{n=1}^{N} m_n \mathbf{r}_n + r_l^2 \sum_{n=1}^{N} m_n \qquad (12.29)$$

Since the center of mass can be located at the origin, this is equal to

$$2I + Mr_l^2$$

where M is the total mass and I is the moment of inertia,

$$I = \frac{1}{2} \sum_{n=1}^{N} m_n r_n^2$$

Now multiply the right-hand side of Eq. 12.29 by m_l and sum over l,

$$\sum_{l,n} m_l m_n |\mathbf{r}_n - \mathbf{r}_l|^2 = 4IM$$

If $l = n$, there is no contribution to the sum, so we can rewrite this as

$$2IM = \sum_{N \geqslant n > l \geqslant 1} m_l m_n |\mathbf{r}_n - \mathbf{r}_l|^2$$

Should this N-body system collapse totally, then $r_{nl} = |\mathbf{r}_n - \mathbf{r}_l| \to 0$ for all n, l combinations so $I \to 0$ too. If this occurs, it will not take forever. To see this, observe that if $r_{nl} \to 0$, then the potential energy of the system,

$$-G \sum_{N \geqslant n > l \geqslant 1} \frac{m_l m_n}{r_{nl}}$$

must approach $-\infty$. But from the virial theorem, Eq. 12.24, we can show that

$$\frac{d^2 I}{dt^2} = 2K + U$$

where K is the kinetic energy. But $\mathscr{E} = K + U$ is a constant, so if $|U| \to \infty$, then $\ddot{I} \to \infty$. Hence there exists some sufficiently large time $t = T \ni \ddot{I} > 1 \; \forall t > T$. Or, for $t > T$, $I > t^2/2 + at + b$ where a and b are constants. Therefore, one concludes that $I \to \infty$ as $t \to \infty$, which contradicts the assumption that $I \to 0$. Sundmann further proved that total collapse cannot occur unless the angular momentum is zero.

Astrophysical Work

Attempts have been made to formulate a statistical approach to stellar systems. Saslaw (1973) contains a summary of the older literature. More recently Katz and Horwitz and their colleagues (Horwitz and Katz 1977, 1978a,b; Katz, Horwitz, and Dekel 1978; Ipser and Horwitz 1979, Ipser 1974; Katz 1980; Lecar and Katz 1981; Katz and Taff 1983) have tried to do so directly from the fundamental concepts of statistical mechanics. It is not clear that they have succeeded. White (1976a,b,c) has done some work in the N-body context on dynamical friction. Heggie (1979a,b) has worked on core collapse. Angeletti and Giannone (1976, 1977a,b, 1980) have extended Larson's work (see below) to multiple-component systems. Lightman and co-workers (1977; Lightman and Fall 1978; Lightman and daCosta 1979) have also investigated collapse, multiple-component systems, and energy diffusion. Without a good physical or numerical basis .... Some work on trying to numerically solve the collisionless Boltzmann equation was done by the author (Taff 1976).

MONTE CARLO CALCULATIONS

The general scenario for life in a globular cluster was given earlier in the chapter. Hénon and Spitzer and his students have attempted to mathematically formulate this process. If this could be done in a correct fashion, then an analysis of many such two-body encounters should explain the evolution of a globular cluster. Since the gravitational force is unshielded and long range, every star in a cluster is always undergoing a "collision" with every other star in the cluster. Thus, the molecular chaos assumption is clearly violated. Its replacement hypothesis in these types of analyses is that all two-body collisions are discrete and independent. This premise allows one to calculate quantities of interest, but its obvious falsity limits my enthusiasm for the results. Finally, the analytical averaging processes used to calculate diffusion coefficients in, for example, velocity require a single-particle distribution function. The standard results are Chandrasekhar's (as summarized in his book, 1942). He used the Maxwell–Boltzmann distribution given in Eq. 12.19. This is not quite circular reasoning—he did not use the Maxwell–Boltzmann velocity distribution multiplied by the isothermal sphere density distribution. The logic is too curved for my taste and I regard the best of the Monte Carlo computations as self-consistency checks.

Clearly, to understand the basis of these calculations, one must consider two-body collisions when the particles interact due to their gravitational attraction. This is dealt with below. From this analysis the differential scattering cross section, the relaxation time, the mean free path, and the velocity diffusion coefficients can all be computed. All of the detail is already in print in Chandrasekhar's text (1942), so we will present the essence only. With this preparation one can then look at the work of Hénon and Spitzer et al.

Two-Body Scattering

The Differential Scattering Cross Section

The differential scattering cross section has already been defined; it is such that $\sigma \, d\Omega$ yields the fraction of particles scattered into the solid angle $d\Omega$ per unit time. If we consider a stationary scattering center subject to a flux of particles with intensity $\mathscr{I}$, then $\mathscr{I}\sigma \, d\Omega$ is the number of particles in the beam scattered into the solid angle $d\Omega$ per unit time. We can characterize each of the beam particles as having a speed at infinity v_∞ and an impact parameter (or distance of closest approach) D. When at infinity the incoming particle's energy $\mathscr{E} = \frac{1}{2}mv_\infty^2$. Also, its angular momentum is simply $L = mDv_\infty = D(2m\mathscr{E})^{1/2}$. The gravitational force exerted by a particle is spherically symmetric, therefore the incident beam of particles must be axially symmetric. Hence,

$$d\Omega = 2\pi \sin\theta \, d\theta$$

where θ is the scattering angle—the angle between the scattered and incident directions.

Now consider a plane perpendicular to the incoming beam of particles. Construct a circular annulus of radius D and thickness dD on it. The number of beam particles crossing through this area is $2\pi\mathscr{I}D\,dD$. All of these particles get scattered, within $d\theta$, into θ. By the definition of σ and the conservation of particles, it must be true that

$$2\pi\mathscr{I}D\,dD = 2\pi\mathscr{I}\sigma \sin\theta \, d\theta$$

Had we been slightly more careful and allowed for negative dD or $d\theta$, absolute value signs would have appeared above since the number of particles is positive. Thus,

$$\sigma = D \csc \theta \left| \frac{dD}{d\theta} \right|$$

The relative hyperbolic orbit can be described as

$$r = \frac{a(e^2 - 1)}{1 + e \cos \phi}$$

where the eccentricity (>1 now) is given by (Chapter Two notation)

$$e^2 = 1 + \frac{2\mathscr{E}L^2}{G^2(M + m)^2}$$

Here M is the mass of the stationary scatterer. In terms of D and v_∞,

$$e^2 = 1 + \frac{D^2 v_\infty^4}{G^2(M + m)^2}$$

For r to be infinite, $1 + e \cos \phi$ must vanish, or $\phi = \pm \cos^{-1}(1/e)$. Thus, the angle between the two asymptotes of the hyperbola, 2ψ, is $\pi - \psi = \cos^{-1}(1/e)$, where

$$\cos \psi = \left[1 + \frac{D^2 v_\infty^4}{G^2(M + m)^2} \right]^{-1/2}$$

Since $\theta = \pi - 2\psi$,

$$\sin\left(\frac{\theta}{2}\right) = \frac{1}{e}$$

whence

$$D = \frac{G(M + m)}{2\mathscr{E}} \cot\left(\frac{\theta}{2}\right)$$

Finally, the expression for σ is

$$\sigma = \frac{1}{4} \frac{G(M + m)}{2\mathscr{E}} \csc^4\left(\frac{\theta}{2}\right)$$

The divergence as $\theta \to 0 (D \to \infty)$ is clear as is the nonconvergence of the total scattering cross section, $2\pi \int_0^\pi \sigma \sin \theta \, d\theta$.

Relaxation Times

Since Chandrasekhar (1942) is the standard reference, we will follow his notation for this discussion. First suppose that the different two-body encounters can each be treated independently and that during such a collision the rest of the cluster can be neglected. Each encounter will result in a deflection $\pi - 2\Psi$ of the star from its original direction of motion and an exchange of energy, ΔE, between the two stars taking part in the encounter. The actual amounts of the deflection and the amount of energy transferred will depend on the initial conditions. The general method is to evaluate the sums

$$\sum \sin^2 2\Psi \qquad \text{and} \qquad \sum (\Delta E)^2$$

for all possible encounters and determine their rates of change with time. When $\sum \sin^2 2\Psi$ becomes on the order of 1, the star would most probably have deviated quite considerably from its original direction of motion. More particularly, if T_D is the time required for $\sum \sin^2 2\Psi$ to become equal to 1, then one may say that by then the star will, on the average, have deviated by $\pi/2$ from its original direction of motion. Similarly, when the root mean square of the exchanged energy, $[\sum (\Delta E)^2]^{1/2}$, becomes of the same order as the initial kinetic energy, then a significant change has occurred. More particularly, if T_E is the time required for $[\sum (\Delta E)^2]^{1/2}$ to become equal to the initial kinetic energy of the star, one may say that by then the star will, on the average, have altered its original energy by about an equal amount. The argument below shows that, in general, T_D and T_E must be of the same order of magnitude.

Let m_1 and $\mathbf{v}_1$ denote the mass and the velocity of a typical cluster star. The parameters defining a collision with a test star of mass m_2 and velocity $\mathbf{v}_2$ are five in number. They are (1) the magnitude v_1, (2) the angle θ between the vectors $\mathbf{v}_1$ and $\mathbf{v}_2$, (3) the azimuthal angle ϕ referred to a system of coordinates the z axis of which coincides with the direction of $\mathbf{v}_2$, (4) the impact parameter D, and (5) the angle Θ between the orbital plane and the fundamental plane containing the vectors $\mathbf{v}_1$ and $\mathbf{v}_2$. For the sake of brevity denote such an encounter by $(v_1, \theta, \phi, D, \Theta)$.

The velocity of the center of mass $\mathbf{V}_g$ remains constant during the encounter. Furthermore, in the orbital plane each star describes a hyperbola about the other, and at the end of the encounter the direction of the relative velocity $\mathbf{V}$ is deflected by an amount $\pi - 2\psi$ (in the orbital plane),

$$\cos \psi = \left[1 + \frac{D^2 V^4}{G^2 (m_1 + m_2)^2} \right]^{-1/2}$$

By definition,

$$\mathbf{V}_g = \frac{1}{m_1 + m_2} (m_1 \mathbf{v}_1 + m_2 \mathbf{v}_2), \qquad \mathbf{V} = \mathbf{v}_2 - \mathbf{v}_1$$

so that

$$\mathbf{v}_2 = \mathbf{V}_g + \frac{m_1}{m_1 + m_2} \mathbf{V}$$

Thus

$$v_2^2 = V_g^2 + 2 \frac{m_1}{m_1 + m_2} V_g V \cos \Phi + \left(\frac{m_1}{m_1 + m_2} \right)^2 V^2$$

where Φ is the angle between $\mathbf{V}_g$ and $\mathbf{V}$. Similarly, at the end of the encounter,

$$v_2'^2 = V_g^2 + 2 \frac{m_1}{m_1 + m_2} V_g V \cos \Phi' + \left(\frac{m_1}{m_1 + m_2} \right)^2 V^2$$

where Φ' is the angle between $\mathbf{V}_g$ and $\mathbf{V}'$. The change in energy is given by

$$\Delta E = \frac{1}{2} m_2 (v_2'^2 - v_2^2)$$

$$= \frac{m_1 m_2}{m_1 + m_2} V_g V (\cos \Phi' - \cos \Phi)$$

Let i be the angle between $\mathbf{V}_g$ and the orbital plane. Also let ϕ, ϕ' be the angles that $\mathbf{V}, \mathbf{V}'$ make with $\mathbf{V}_g$ in the orbital plane. Then

$$\cos \Phi = \cos \phi \cos i, \qquad \cos \Phi' = \cos \phi' \cos i$$

Therefore, ΔE can be rewritten as

$$\Delta E = \frac{m_1 m_2}{m_1 + m_2} V_g V (\cos \phi' - \cos \phi) \cos i$$

or as

$$\Delta E = \frac{2 m_1 m_2}{m_1 + m_2} V_g V \sin \left(\frac{\phi + \phi'}{2} \right) \sin \left(\frac{\phi - \phi'}{2} \right) \cos i$$

But $\phi' - \phi = \pi - 2\psi$, so we can derive

$$\tfrac{1}{2}(\phi + \phi') = \tfrac{1}{2}\pi - \psi + \phi \qquad \text{and} \qquad \tfrac{1}{2}(\phi' - \phi) = \tfrac{1}{2}\pi - \psi$$

Whence

$$\Delta E = -2 \frac{m_1 m_2}{m_1 + m_2} V_g V \cos(\phi - \psi) \cos \psi \cos i$$

The number of $(v_1, \theta, \phi, D, \Theta)$ encounters that take place in an interval of time dt, if $N(v_1, \theta, \phi) \, d\mathbf{v}_1 \, d\theta \, d\phi$ is the number of stars per unit volume with speeds in the range $v_1, v_1 + dv_1$ and in directions confined to the element of solid-angle $\sin \theta \, d\theta \, d\phi$, is

$$N(v_1, \theta, \phi) \, dv_1 \, d\theta \, d\phi \frac{d\Theta}{2\pi} 2\pi D \, dD V \, dt$$

Therefore, the contribution of these encounters to the sum $\sum (\Delta E)^2$ can be written as

$$\sum (\Delta E)^2_{(v_1, \theta, \phi, D, \Theta)} = 2\pi N(v_1, \theta, \phi)(\Delta E)^2 V D \, dD \frac{d\Theta}{2\pi} \, dv_1 \, d\theta \, d\phi \, dt$$

or as

$$\sum (\Delta E)^2_{(v_1, \theta, \phi, D, \Theta)} = 8\pi N(v_1, \theta, \phi) V_g^2 V^3 \left(\frac{m_1 m_2}{m_1 + m_2}\right)^2$$

$$\times \cos^2 i \cos^2(\phi - \psi) \cos^2 \psi D \, dD \frac{d\Theta}{2\pi} \, dv_1 \, d\theta \, d\phi \, dt$$

The remaining problem is to integrate the foregoing expression over the relevant ranges of the five parameters Θ, D, ϕ, θ, and v_1.

As the reader knows, the integration over the impact parameter diverges and some fudge must be performed to save the computation. The integrations over the angles i, θ, and ϕ are relatively straightforward in that there are no divergences to patch up. Finally, one must integrate over the speed distribution. Here Chandrasekhar uses the Maxwell–Boltzmann velocity distribution. Implicit up until now has been the assumption that the spatial density is uniform. Chandrasekhar also shows in a similar analysis that T_D is within a factor of 2 of T_E. It is common today to see the Spitzer and Hart (1971a) version of the formulas evaluated halfway (in mass) out from the cluster center,

$$t_{relax} = \frac{V_m^3}{15.4 G^2 m^2 n \ln(0.4N)} = \frac{0.06 M^{1/2} R_h^{3/2}}{m G^{1/2} \log(0.4N)}$$

where N is the total number of stars of mass $m(M = Nm)$, R_h is the aforementioned halfway radius, n is the mean stellar density interior to R_h, and V_m is the root mean square stellar speed,

$$V_m^2 = \frac{0.4 GM}{R_h}$$

The crossing time is $R_h / V_m = 1.6 R_h^{3/2} / (GM)^{1/2}$ so that

$$\frac{t_{cross}}{t_{relax}} = \frac{26 \log(0.4N)}{N}$$

For a globular cluster with $m = M_\odot$, $N = 10^6$, $R_h = 3$ pc, one finds $t_{cross} = 1.2 \times 10^5$ yr, $t_{relax} = 8.4 \times 10^8$ yr, although one suspects that globular clusters are older than $\sim 5 \times 10^9$ yr.

Velocity Diffusion

Clusters of stars are dynamic entities, so one assumes that their evolution is driven by momentum and energy exchange during collisions. This has come to be called velocity diffusion and is the last piece of analysis necessary to understand the basis of the Monte Carlo calculations. Let $\mathbf{v}_{2g}$ and $\mathbf{v}'_{2g}$ denote the velocity of the star before and after the encounter in the frame of reference in which the center of mass is at rest (still in Chandrasekhar's notation),

$$\mathbf{v}_{2g} = \mathbf{v}_2 - \mathbf{V}_g, \qquad \mathbf{v}'_{2g} = \mathbf{v}'_2 - \mathbf{V}_g$$

with

$$(m_1 + m_2)V_g = m_1 v_1 + m_2 v_2 = m_1 v_1' + m_2 v_2'$$

From these equations we conclude that

$$v_{2g} = \frac{m_1}{m_1 + m_2} V, \qquad v_{2g}' = \frac{m_1}{m_1 + m_2} V'$$

The angle between the vectors v_{2g} and v_{2g}' is the same as that between V and V', that is, $\pi - 2\psi$. Now

$$v_2 \cdot v_2' = v_2 \cdot (v_{2g}' + V_g)$$

or

$$v_2 v_2' \cos(\pi - 2\Psi) = v_2 \cdot v_{2g}' + v_2 \cdot V_g$$

The direction cosines of v_2 with respect to V, a direction in the orbital plane at right angles to V, and a direction perpendicular to the orbital plane are

$$\cos(\Phi - \chi), \qquad -\sin(\Phi - \chi)\cos\Theta, \qquad \sin(\Phi - \chi)\sin\Theta$$

where χ is the angle between v_2 and V_g. Because v_{2g}' is in the same direction as V', the direction cosines of v_{2g}' with respect to the same three directions are

$$\cos(\phi' - \phi), \qquad \sin(\phi' - \phi), \qquad 0$$

Therefore,

$$v_2 \cdot v_{2g}' = v_2 v_{2g}' \{\cos(\phi' - \phi)\cos(\Phi - \chi) - \sin(\phi' - \phi)\sin(\Phi - \chi)\cos\Theta\}$$

We can write, then, that

$$v_2' \cos(\pi - 2\Psi) = v_{2g}'[\cos(\phi' - \phi)\cos(\Phi - \chi)$$
$$- \sin(\phi' - \phi)\sin(\Phi - \chi)\cos\Theta] + V_g \cos\chi$$

Since $\phi' - \phi = \pi - 2\psi$, this can be rewritten as

$$\cos 2\Psi = \frac{1}{v_2'}\{v_{2g}[\cos 2\psi \cos(\Phi - \chi) + \sin 2\psi \sin(\Phi - \chi)\cos\Theta] - V_g \cos\chi\}$$

But

$$V\cos(\Phi - \chi) = v_2 - v_1 \cos\theta \qquad \text{and} \qquad V\sin(\Phi - \chi) = v_1 \sin\theta$$

so

$$V_g \cos\chi = \frac{1}{m_1 + m_2}(m_1 v_1 \cos\theta + m_2 v_2)$$

Substituting the latter expressions into the formula for Φ, and using $v_{2g} = [m_1/(m_1 + m_2)]V$, one can derive the expression

$$\cos 2\Psi = \frac{1}{(m_1 + m_2)v_2'}[m_1(v_2 - v_1 \cos\theta)\cos 2\psi$$
$$+ m_1 v_1 \sin\theta \cos\Theta \sin 2\psi - m_1 v_1 \cos\theta - m_2 v_2]$$

or, alternatively,

$$\cos 2\Psi = \frac{1}{(m_1 + m_2)v_2'}[2m_1(v_2 - v_1 \cos \theta) \cos^2 \psi$$

$$+ 2m_1 v_1 \sin \theta \cos \Theta \sin \psi \cos \psi - (m_1 + m_2)v_2]$$

We can now formulate a relationship for v_2' in terms of v_1, v_2, θ, and ψ,

$$v_2'^2 = \frac{1}{(m_1 + m_2)^2}\{(m_1 + m_2)^2 v_2^2 - 4m_1[(m_2 v_2^2 - m_1 v_1^2)$$

$$+ (m_1 - m_2)v_1 v_2 \cos \theta] \cos^2 \psi$$

$$- 4m_1(m_1 + m_2)v_1 v_2 \sin \theta \cos \Theta \sin \psi \cos \psi\}$$

Substituting for v_2' in the expression for $\cos 2\Phi$ leads to

$$\cos 2\Psi = [2m_1(v_2 - v_1 \cos \theta) \cos^2 \psi +$$

$$\frac{m_1 v_1 \sin \theta \cos \Theta \sin 2\psi - (m_1 + m_2)v_2]}{\{(m_1 + m_2)^2 v_2^2 - 4m_1[(m_2 v_2^2 - m_1 v_1^2)}$$

$$+ (m_1 - m_2)v_1 v_2 \cos \theta] \cos^2 \psi$$

$$- 2m_1(m_1 + m_2)v_1 v_2 \sin \theta \cos \Theta \sin 2\psi\}^{1/2}$$

and

$$\sin^2 2\Psi = 4m_1^2 \cos^2 \psi\{v_1^2 + v_2^2 - 2v_1 v_2 \cos \theta -$$

$$\frac{[(v_2 - v_1 \cos \theta) \cos \psi + v_1 \sin \theta \cos \Theta \sin \psi]^2\}}{\{(m_1 + m_2)^2 v_2^2 - 4m_1[(m_2 v_2^2 - m_1 v_1^2)}$$

$$+ (m_1 - m_2)v_1 v_2 \cos \theta] \cos^2 \psi$$

$$- 2m_1(m_1 + m_2)v_1 v_2 \sin \theta \cos \Theta \sin 2\psi\}$$

The expression for $\sin^2 2\Psi$ simplifies considerably for distant encounters,

$$\sin^2 2\Psi \to \frac{4m_1^2}{(m_1 + m_2)^2 v_2^2}[v_1^2 + v_2^2 - 2v_1 v_2 \cos \theta - v_1^2 \sin^2 \theta \cos^2 \Theta] \cos^2 \psi$$

or

$$\sin^2 2\Psi \to \frac{4m_1^2}{(m_1 + m_2)^2 v_2^2}[(v_2 - v_1 \cos \theta)^2 + v_1^2 \sin^2 \theta \sin^2 \Theta] \cos^2 \psi$$

Let $\Delta v_\parallel$ and $\Delta v_\perp$ denote the changes in the velocity vector parallel to and perpendicular to the original direction of motion of the star,

$$\Delta v_\parallel = v_2' \cos(\pi - 2\Psi) - v_2 \qquad \Delta v_\perp = v_2' \sin 2\Psi$$

Then, from the results deduced above, one can recover the following expressions:

$$\Delta v_{\parallel} = -\frac{2m_1}{m_1 + m_2} [(v_2 - v_1 \cos \theta) \cos \psi + v_1 \sin \theta \cos \Theta \sin \psi] \cos \psi$$

$$\Delta v_{\perp} = \frac{2m_1}{m_1 + m_2} \{v_1^2 + v_2^2 - 2v_1 v_2 \cos \theta$$

$$- [(v_2 - v_1 \cos \theta) \cos \psi + v_1 \sin \theta \cos \Theta \sin \psi]^2\}^{1/2} \cos \psi$$

The sums $\sum (\Delta v_{\parallel})^2$ and $\sum (\Delta v_{\perp})^2$ are simply, if I retain only the dominant terms,

$$\sum (\Delta v_{\parallel})^2 = \frac{1}{m_2^2 v_2^2} \sum (\Delta E)^2, \qquad \sum (\Delta v_{\perp})^2 = v_2^2 \sum \sin^2 2\Psi$$

or

$$\sum (\Delta v_{\parallel})^2 = \frac{1}{4} v_2^2 \frac{dt}{T_E}, \qquad \sum (\Delta v_{\perp})^2 = v_2^2 \frac{dt}{T_D}$$

Hénon and Spitzer's Work

Hénon (1967, 1971a,b, 1973) starts from the general point of view that the above analysis is at least on the right track for large-scale, spherical, self-gravitating systems. In particular, he assumes that the dynamical significance of the relaxation time is that it is also an evolutionary time scale for the system as a whole. In his own words (Copyright © 1971 by D. Reidel Publishing Company, Dordrecht, Holland),

These considerations lead naturally to the basic ideas of the Monte Carlo scheme. First, because N is large, we can divide the gravitational field in two parts: a main smoothed-out field, or mean field, and a small irregular, fluctuating field. We consider the motion of a star during an interval of time Δt, such that

$$t_{\mathrm{cross}} \ll \Delta t \ll t_{\mathrm{relax}} \qquad (2)$$

During that interval, the effects of the fluctuating field can be neglected in a first approximation. The motion of the star is then governed by the mean field. This field is spherically symmetric, and changes only with the time scale t_{relax}, so that it can be taken as time-independent over Δt. The star has then a plane rosette motion, described by simple analytical formulas. The detailed numerical integration along the orbit becomes thus unnecessary: the formulas give at once the motion over the whole interval Δt, which includes a number of revolutions according to (2).

However, the fluctuating field, although small, is not entirely negligible. Its effect is to change slowly and randomly the parameters of the orbit. This effect is small over Δt, but it builds up and becomes significant over a time of the order of t_{relax}; we must therefore take it into account.

It would seem at first view that we must consider the effect of all other stars (which we shall call *field stars*), at all points of the orbit during the interval Δt.

We would then be practically back at the exact N-body integration. To avoid this, we apply here the basic Monte Carlo tactics. First, instead of integrating the perturbations along the orbit, we shall select randomly just one point of the orbit, and we shall compute the perturbation only at this point. Second, instead of considering the effect of all field stars, we shall select randomly just one of them, and compute only the perturbation from that star. Finally, we shall multiply this perturbation by an appropriate fixed factor in order to account for all the time points and all the field stars which have not been considered.

Of course, this procedure will not give the exact perturbation in the motion of the star. But, since this perturbation is a random quantity, it is not its exact value which matters, but its statistical properties; more specifically, its moments of the first and second order. If the procedure is correctly set up, these moments will be correctly reproduced. The evolution of the whole artificial system of stars will be statistically the same as the evolution of the real system, and this is all that is required.

. . .

We have presented the Monte Carlo scheme as an adaptation of the exact N-body integration. But it can also be seen from a different angle, as deriving from the theoretical approach: it can be simply considered as a convenient algorithm for the numerical solution of the Fokker–Planck equation. The sampling effected along the orbit and among the field stars are then merely an application of the classical Monte Carlo trick for the quick evaluation of multiple integrals.

This interpretation makes clear one important point: the assumptions which lie behind the Monte Carlo scheme are essentially the same as the assumptions involved when one writes the Fokker–Planck equation; namely, that a stellar system can be adequately represented by a one-particle distribution function, and that the evolution is only due to binary encounters. Thus, multiple encounters are ignored; in particular the formation and disruption of binaries are not considered.

Clearly I take his caution more seriously than he did.

Spitzer and students (Spitzer and Hart 1971a,b; Spitzer and Shapiro 1972; Spitzer and Thuan 1972; Spitzer and Chevalier 1973; Spitzer and Shull 1975a,b; Spitzer and Mathieu 1980) pursued a slightly different Monte Carlo technique. Unlike Hénon they integrated the individual orbits of a test star. Spitzer et al. also used different velocity diffusion coefficients [based on Spitzer (1962)]. In particular, stars in the halo can escape due to a finite velocity change acquired in a single encounter whereas Hénon allows diffusion only to vanishing (not positive) energies. Both techniques, of course, ignore binaries and very close encounters. Again it is best to quote from Spitzer directly (reprinted courtesy of L. Spitzer and *The Astrophysical Journal*, published by The University of Chicago Press; © 1971 The American Astronomical Society)

The basic procedure followed in this study is to compute numerically the orbits of $\mathcal{N}$ stars in a spherically symmetric system containing a much greater number, N, of stars.

This procedure rests on two basic approaches. First in a system containing N stars, the stellar orbits are calculated by numerical integration on the assumption that the potential field is spherically symmetric; i.e., with this assumption the velocity and the radius of each star are computed as functions of the time. Second, the velocities of all these $\mathcal{N}$ stars are perturbed slightly at intervals to give statistically the same results as would be anticipated from random encounters with neighboring stars.

The numerical integrations of the stellar orbits follow the method applied by Hénon (1964), based on earlier work by Campbell. In this approach, the N stars are divided into $\mathcal{N}$ subgroups, each of which has a number of stars all characterized by the same mass m, the same radial velocity v_r, the same transverse velocity v_t normal to the radius vector, and all situated at the same distance r from the center. Thus each subgroup is essentially a shell, with stars uniformly distributed over the spherical area. Not only is each shell assumed to be spherically symmetric but in addition the distribution of the direction of the transverse velocities is isotropic; i.e., at each point in the shell, the directions of the transverse velocities v_t for the different stars in the shell are uniformly distributed in the plane perpendicular to the radius vector.

Let us now arrange the $\mathcal{N}$ shells in the order of increasing distance r_j from center, and number them from $j = 1$ up to $j = \mathcal{N}$. Then the equations of motion become

$$\frac{dr_j}{dt} = v_{rj}, \tag{7}$$

$$\frac{dv_{rj}}{dt} = \frac{J_j^2}{r_j^3} - \frac{G(j - \frac{1}{2})\mathcal{M}}{r_j^2}, \tag{8}$$

where J_j is the angular momentum per unit mass of the stars in shell j, and $\mathcal{M}$ is the shell mass, taken to be identical for all shells; thus the number of stars in each shell is inversely proportional to the stellar mass m_j. In the numerical work we choose units such that $G\mathcal{M}$ equals unity. The quantity $\frac{1}{2}$ in the last term in equation (2) takes into account the self-attraction of each shell. In the absence of encounters, J_j, is constant for each shell, and equal to $r_j v_{tj}$.

We pass on now to a consideration of collisional perturbations in the velocities v_{rj} and v_{tj}. As usual, we make the familiar general assumptions that only two-body encounters need be considered, that these are random in space and time, and that they occur locally, with each encounter confined to a time interval short compared with the orbital period; this last assumption is not followed consistently since the maximum impact parameter is set equal to the radius of the system, but the error introduced in this way is *believed* [my emphasis] to be small.

Let us consider a gravitational encountered between two stars, denoted by A and B, with velocities $\mathbf{v}_A$ and $\mathbf{v}_B$ before the encounter. The velocity changes, $\Delta \mathbf{v}_A$ and $\Delta \mathbf{v}_B$, resulting from the encounter will be a function of three quantities: the relative velocity $\mathbf{v}_A - \mathbf{v}_B$, the angle Θ between the orbital plane and the plane containing $\mathbf{v}_A$ and $\mathbf{v}_B$, and the impact parameter p, defined as the distance of closest approach in the absence of collisions (Chandrasekhar 1942). In a full Monte Carlo approach to the problem, one might consider individual encounters with v_A, v_B, Θ, and p all chosen at random. This approach would be subject to the objection that an enormous number of encounters would have to be considered, since the cumulative effect of many very small changes in velocity is responsible for collisional relaxation.

A modified Monte Carlo method has been adopted by Hénon (1967), who selects v_A, v_B and Θ at random, but chooses p so that the mean values of Δv and $(\Delta v)^2$ per unit time, integrated over all p, are correctly given. This approach does not give the correct distributions of Δv and $(\Delta v)^2$; hence this method does not give the occasional large changes in velocity that result from occasional close encounters. However, the effect of small encounters is taken into account precisely by this method, which should become exact as $\ln N$ becomes large.

The approach followed in this paper is an even simpler Monte Carlo method, designed to give approximately correct results with a minimum of computations. The averaging procedure introduced by Hénon is carried further, and for a star of velocity v_A, the values of Δv and of $(\Delta v)^2$ are averaged not only over the impact parameter p but over Θ and v_B as well. This procedure requires that specific velocity distributions be assumed; in most situations the assumption of a Maxwellian distribution is a good approximation. Finally, the mean values obtained for Δv and $(\Delta v)^2$ are simplified somewhat to permit more rapid computations.

We now proceed to give the detailed equations used. Let us consider a particular star, of velocity v, whose change of velocity is to be considered in successive interactions with many other stars in the system; such a star is called a "test star" by Chandrasekhar (1942). We denote by $\Delta v_\parallel$ and $\Delta v_\perp$ the collisionally induced changes of v in directions parallel and perpendicular, respectively to the initial value of v. The mean values of $\Delta v_\parallel$, $(\Delta v_\parallel)^2$, and $(\Delta v_\perp)^2$, summed over all collisions in a unit of time, are called diffusion coefficients (Spitzer 1962) and are denoted by $\langle \Delta v_\parallel \rangle$, $\langle (\Delta v_\parallel)^2 \rangle$, and $\langle (\Delta v_\perp)^2 \rangle$, respectively. The stars which a test star encounter will be called "field stars," following Chandrasekhar (1942), and their properties will be denoted with a subscript f. If we assume that the distribution of v_f is Maxwellian. . . .

He goes on to make the (by now) usual excuses for large- and small-impact parameters. Later in this paper and the next one (Spitzer and Hart 1971b), they give a very detailed discussion of the implementation of the method including the various fudge factors they introduce. These serve to speed up the evolution or make the results come out "right." They also compare their results to Hénon's and point out the order of magnitude difference between the two rates of evolution.

There is a paper by Aarseth, Hénon, and Wielen (1974) that discusses, in general, the results of N-body integrations, the two types of Monte Carlo calculations, and Larson's models (see below). While acknowledging that none of these agreed until a posteriori "refinements" were made, these three seem to think that everything has been juggled enough to conform to a "standard" result.

HYDRODYNAMICS AND THERMODYNAMICS

Hydrodynamics and thermodynamics are both derivatives of kinetic theory (statistical mechanics in general, kinetic theory is the special name for the statistical mechanics of dilute gases). Just as the Monte Carlo calculations (as distinct from the principles of a Monte-Carlo-type calculation) relied on the Maxwell–Boltzmann distribution to bridge the gap between the Boltzmann equation and stellar dynamics, so too the hydrodynamical (or fluid mechanical) computations for stellar systems come out of the Boltzmann equation via the Maxwell–Boltzmann distribution. This is exhibited below, closely (and with permission) following the presentation given by Huang (1963). We have already seen the intimate connection between the thermodynamics of dilute gases and kinetic theory. See Huang's text for a more elaborate demonstration of the general relationship between thermodynamics and statistical mechanics.

The fluid mechanical approach is the last major technique used to try to derive the evolutionary paths of stellar systems. Less detailed but global information about the stability of such systems has been sought by thermodynamic arguments. Clearly, if by now tiring to read, since the physical basis for hydrodynamics and thermodynamics does not exist for stellar systems, neither does a meaningful hydrodynamics or thermodynamics. Actually, the thermodynamical computations that gave rise to the concept of the gravothermal catastrophe are self-inconsistent even *if* one grants the applicability of thermodynamics in the stellar dynamical context a priori. We shall consider each of these topics in turn.

Derivation of Hydrodynamics

In order to investigate nonequilibrium phenomena, one must solve the Boltzmann equation. Rigorous properties of solutions to the Boltzmann equation may be obtained from the fact that during a collision there are quantities that are conserved. Let $\chi(\mathbf{r}, \mathbf{v})$ be a property associated with a particle of velocity $\mathbf{v}$ located at $\mathbf{r}$ such that in the collision $\{\mathbf{v}_1, \mathbf{v}_2\} \rightarrow \{\mathbf{v}_1', \mathbf{v}_2'\}$,

$$\chi_1 + \chi_2 = \chi_1' + \chi_2'$$

where $\chi_1 = \chi(\mathbf{r}_1, \mathbf{v}_1)$, and so forth. Then χ is a conserved quantity and the following formula is true:

$$\int \chi(\mathbf{r}, \mathbf{v}) \left[\frac{\partial f(\mathbf{r}, \mathbf{v}, t)}{\partial t} \right]_{\text{coll}} d\mathbf{v} = 0$$

As before $(\partial f / \partial t)_{\text{coll}}$ is the right-hand side of Eq. 12.15.

By the definition of $(\partial f / \partial t)_{\text{coll}}$

$$\int \chi_1 \left(\frac{\partial f_1}{\partial t} \right)_{\text{coll}} d\mathbf{v}_1 = \int \int \int \sigma |\mathbf{v}_2 - \mathbf{v}_1| \chi_1 (f_1' f_2' - f_1 f_2) \, d\mathbf{v}_1 \, d\mathbf{v}_2 \, d\Omega$$

Making use of the properties of the cross section σ and proceeding in a manner similar to that utilized in the proof of the H theorem, we make each of the following changes of variables in the integrals:

$$\mathbf{v}_1 \leftrightarrow \mathbf{v}_2; \mathbf{v}_1 \leftrightarrow \mathbf{v}_1' \quad \text{and} \quad \mathbf{v}_2 \leftrightarrow \mathbf{v}_2'; \quad \text{and} \quad \mathbf{v}_1 \leftrightarrow \mathbf{v}_2' \quad \text{with} \quad \mathbf{v}_2 \leftrightarrow \mathbf{v}_1'$$

In each instance we obtain a different form for the same integral; adding the new formulas to the original one and then dividing the sum by 4, we deduce that

$$\int \chi_1 \left(\frac{\partial f_1}{\partial t} \right)_{\text{coll}} d\mathbf{v}_1 = \frac{1}{4} \int \int \int \sigma |\mathbf{v}_1 - \mathbf{v}_2| (f_2' f_1' - f_2 f_1)$$

$$\times (\chi_1 + \chi_2 - \chi_1' - \chi_2') \, d\mathbf{v}_1 \, d\mathbf{v}_2 \, d\Omega \equiv 0$$

The generalized conservation theorem can be deduced by multiplying the Boltzmann equation on both sides by χ and then integrating over the velocity distribution. The collision term is zero (as we just proved), whence

$$\int \chi(\mathbf{r}, \mathbf{v}) \left[\frac{\partial}{\partial t} + \mathbf{v} \cdot \nabla + \left(\frac{\mathbf{F}}{m} \right) \cdot \frac{\partial}{\partial \mathbf{v}} \right] f(\mathbf{r}, \mathbf{v}, t) \, d\mathbf{v} = 0$$

If $f(\mathbf{r}, \mathbf{v}, t)$ is assumed to vanish when $|\mathbf{v}| \to \infty$, then the last term in the integrand vanishes. Also, remembering the definition of an average value, the remainder reduces to

$$\frac{\partial}{\partial t} \langle n\chi \rangle + \nabla \cdot \langle n\mathbf{v}\chi \rangle - n\langle \mathbf{v} \cdot \nabla \chi \rangle - \frac{n}{m} \langle \mathbf{F} \cdot \nabla_\mathbf{v}\chi \rangle - \frac{n}{m} \langle \chi \nabla_\mathbf{v} \cdot \mathbf{F} \rangle = 0$$

As already mentioned, for particles the conserved quantities are its mass, its linear momentum, and its energy. Hence, we set successively $\chi = m$, $\chi = mv_i$ $(i = 1, 2, 3)$, and $\chi = \frac{1}{2} m |\mathbf{v} - \mathbf{u}(\mathbf{r}, t)|^2$ (where $\mathbf{u}(\mathbf{r}, t) \equiv \langle \mathbf{v} \rangle$). After we do so, we will have three conservation theorems. For $\chi = m$ the result is

$$\frac{\partial}{\partial t} (mn) + \frac{\partial}{\partial \mathbf{r}} \cdot \langle mn\mathbf{v} \rangle = 0$$

or, in terms of the *mass density* $\rho(\mathbf{r}, t) \equiv mn(r, t)$,

$$\frac{\partial \rho}{\partial t} + \nabla \cdot (\rho \mathbf{u}) = 0$$

Next for $\chi = mv_i$,

$$\frac{\partial}{\partial t} \langle \rho v_i \rangle + \sum_{j=1}^{3} \frac{\partial}{\partial r_j} \langle \rho v_i v_j \rangle - \frac{1}{m} \rho F_i = 0$$

To reduce this further observe that

$$\langle v_i v_j \rangle = \langle (v_i - u_i)(v_j - u_j) \rangle + \langle v_i \rangle u_j + u_i \langle v_j \rangle - u_i u_j$$

$$= \langle (v_i - u_i)(v_j - u_j) \rangle + u_i u_j$$

Substituting this result for $\langle v_i v_j \rangle$ and using the fact that mass is conserved, allows us to deduce

$$\rho \left(\frac{\partial u_i}{\partial t} + \sum_{j=1}^{3} u_j \frac{\partial u_i}{\partial r_j} \right) = \frac{1}{m} \rho F_i - \sum_{j=1}^{3} \frac{\partial}{\partial r_j} \langle \rho (v_i - u_i)(v_j - u_j) \rangle$$

Now we define the *pressure tensor* by

$$P_{ij} \equiv \rho \langle (v_i - u_i)(v_j - u_j) \rangle$$

and derive

$$\left(\frac{\partial}{\partial t} + \mathbf{u} \cdot \nabla \right) \mathbf{u} = \frac{\mathbf{F}}{m} - \left(\frac{1}{\rho} \right) \nabla \cdot P$$

Lastly, we set $\chi = \frac{1}{2} m |\mathbf{v} - \mathbf{u}|^2$. Then, because $\langle v_i \rangle = u_i$ and $\partial \mathbf{F} / \partial \mathbf{v}$ is null,

$$\frac{1}{2} \frac{\partial}{\partial t} \langle \rho |\mathbf{v} - \mathbf{u}|^2 \rangle + \frac{1}{2} \nabla \cdot \langle \rho \mathbf{v} |\mathbf{v} - \mathbf{u}|^2 \rangle - \frac{1}{2} \rho \langle \mathbf{v} \cdot \nabla (|\mathbf{v} - \mathbf{u}|^2) \rangle = 0$$

We define *temperature* T by

$$kT \equiv \frac{1}{3} m \langle |\mathbf{v} - \mathbf{u}|^2 \rangle$$

and *heat flux* by

$$\mathbf{q} \equiv \frac{1}{2} m \rho \langle (\mathbf{v} - \mathbf{u}) |\mathbf{v} - \mathbf{u}|^2 \rangle$$

Then this can be written as

$$\frac{1}{2} m \rho \langle \mathbf{v} |\mathbf{v} - \mathbf{u}|^2 \rangle = \frac{1}{2} m \rho \langle (\mathbf{v} - \mathbf{u}) |\mathbf{v} - \mathbf{u}|^2 \rangle + \frac{1}{2} m \rho \mathbf{u} \langle |\mathbf{v} - \mathbf{u}|^2 \rangle$$
$$= \mathbf{q} + \frac{3}{2} \rho kT \mathbf{u}$$

with

$$\rho \langle v_i (v_j - u_j) \rangle = \rho \langle (v_i - u_i)(v_j - u_j) \rangle + \rho u_i \langle v_j - u_j \rangle = P_{ij}$$

The thermal energy conservation equation can be cast in the form

$$\frac{3}{2} \frac{\partial}{\partial t} (\rho kT) + \nabla \cdot \mathbf{q} + \frac{3}{2} \nabla (\rho kT \mathbf{u}) + m P \cdot \nabla \mathbf{u} = 0$$

The pressure tensor is symmetric, so the last term on the left-hand side is equivalent to

$$\sum_{i,j=1}^{3} m P_{ij} \frac{\partial u_j}{\partial r_i} = \sum_{i,j=1}^{3} P_{ij} \frac{m}{2} \left(\frac{\partial u_j}{\partial r_i} + \frac{\partial u_i}{\partial r_j} \right) \equiv \sum_{i,j=1}^{3} P_{ij} \Lambda_{ij}$$

This leads to the final form of the conservation of energy equation,

$$\rho \left(\frac{\partial}{\partial t} + \mathbf{u} \cdot \nabla \right) kT + \frac{2}{3} \nabla \cdot \mathbf{q} = \frac{2}{3} \Lambda \cdot P$$

The Zeroth-Order Approximation

We shall assume that we have been dealing with a gas that, although not in equilibrium, is in near-equilibrium. In particular, in the neighborhood of any locale within the gas the distribution function is Maxwellian. In addition, the density, temperature, and average velocity are presumed to vary slowly. It is, therefore, natural to try the approximation

$$f(\mathbf{r}, \mathbf{v}, t) \approx f'_{MB}(\mathbf{r}, \mathbf{v}, t)$$

where

$$f'_{MB}(\mathbf{r}, \mathbf{v}, t) = n \left(\frac{m}{2\pi kT} \right)^{3/2} \exp\left[-\frac{m}{2kT} (\mathbf{v} - \mathbf{u})^2 \right]$$

The number density n, the temperature T, and the mean motion $\mathbf{u}$ are all slowly varying functions (of $\mathbf{r}$ and t) in f'_{MB}. This cannot represent an exact solution of the Boltzmann equation as

$$\left(\frac{\partial f'_{MB}}{\partial t} \right)_{coll} = 0$$

It should also be clear that, in general,

$$\left(\frac{\partial}{\partial t} + \mathbf{v} \cdot \nabla + \left(\frac{F}{m} \right) \cdot \nabla_v \right) f'_{MB}(\mathbf{r}, \mathbf{v}, t) \neq 0$$

Nonetheless, for the moment let us assume that f'_{MB} is a good approximation and go on to discuss the physical consequences.

If it is a good approximation, then the left-hand side of the above must be approximately equal to zero. This in turn would mean that n, kT, and $\mathbf{u}$ approximately satisfy the conservation theorems. By enforcing this constraint, we can derive equations that restrict the variations of n, kT, and $\mathbf{u}$. To see what they are, we must calculate $\mathbf{q}$ and P_{ij} in lowest order. The results are denoted by $\mathbf{q}'$ and P'_{ij}. Let $C(\mathbf{r}, t) = n(m/2\pi kT)^{3/2}$ and $A(\mathbf{r}, t) = m/2kT$. Then

$$\mathbf{q}' = \frac{1}{2} \frac{m\rho}{n} \int (\mathbf{v} - \mathbf{u}) |\mathbf{v} - \mathbf{u}|^2 C(\mathbf{r}, t) \, e^{-A(\mathbf{r},t)|\mathbf{v}-\mathbf{u}|^2} \, d\mathbf{v}$$

or, with $\mathbf{U} = \mathbf{v} - \mathbf{u}$,

$$\mathbf{q}' = \tfrac{1}{2} m^2 C(\mathbf{r}, t) \int \mathbf{U} U^2 \, e^{-A(\mathbf{r},t) U^2} \, d\mathbf{U} = 0$$

Similarly,

$$P'_{ij} = \frac{\rho}{n} C(\mathbf{r}, t) \int (v_i - u_i)(v_j - u_j) \, e^{-A(\mathbf{r},t)|\mathbf{v}-\mathbf{u}|^2} \, d\mathbf{v}$$

$$= m C(\mathbf{r}, t) \int U_i U_j \, e^{-A(\mathbf{r},t) U^2} \, d\mathbf{U} = \delta_{ij} P$$

where

$$P = \frac{\rho}{3} \left(\frac{m}{2\pi kT} \right)^{3/2} \int U^2 e^{-A(\mathbf{r},t)U^2} \, d\mathbf{U} = nkT$$

(the local hydrostatic pressure) and δ_{ij} is the Kronecker delta function. Substituting these results into the conservation theorems, and noting that

$$\nabla \cdot P' = \nabla P, \qquad P' \cdot \Lambda = P \sum_{i=1}^{3} \Lambda_{ii} = mP\nabla \cdot \mathbf{u}$$

we obtain the equations of motion

$$\frac{\partial \rho}{\partial t} + \nabla \cdot (\rho \mathbf{u}) = 0$$

$$\left(\frac{\partial}{\partial t} + \mathbf{u} \cdot \nabla \right) \mathbf{u} + \frac{1}{\rho} \nabla P = \frac{\mathbf{F}}{m}$$

$$\left(\frac{\partial}{\partial t} + \mathbf{u} \cdot \nabla \right) kT + \frac{1}{c_v} (\nabla \cdot \mathbf{u}) kT = 0$$

where $c_v = \frac{3}{2}$ is the specific heat per particle at constant volume. These are the hydrodynamic equations for the nonviscous flow of a gas. Huang (1963) goes on to derive an error estimate for this approximation, the next approximation, and so on. For our purposes this is enough background, so we turn next to what Larson did.

Larson's Models

Larson's (1970a,b, 1974) starting point is the Boltzmann equation. He assumes (as is usual) spherical spatial symmetry and a near-Maxwellian velocity distribution. He takes moments of the Boltzmann equation up to order 4. Heat flow is governed by third-order moments, and the fourth-order moments are necessary to treat the third-order terms correctly. Larson closed the system at the fifth order. Talpaert and Lefévre (1972) have gone even further. Reading Larson's papers it is clear that he is cognizant of the approximations that he makes. In this section we shall adopt his notation and follow his arguments.

Consider a system with spherical symmetry (r, θ, and ϕ for coordinates) and denote by u, v, and w the corresponding velocity components. Let $\rho(r)$ be the mass density of stars as a function of r, and let $\langle u \rangle$ be the mean velocity in the radial direction. For a system with spherical symmetry the mean tangential velocity components $\langle v \rangle$ and $\langle w \rangle$ must vanish, as must all other moments of odd order in v or w. Moreover, as the tangential directions are equivalent, the value of any moment is unaffected by interchanging v and w. Also define the following higher moments of the velocity distribution:

$$\alpha \equiv \langle (u - \langle u \rangle)^2 \rangle, \qquad \beta \equiv \langle v^2 \rangle = \langle w^2 \rangle$$

$$\varepsilon \equiv \langle (u - \langle u \rangle)^3 \rangle, \qquad \zeta \equiv \langle (u - \langle u \rangle)^4 \rangle$$

(12.30)

Larson argued that instead of using the fourth moment ζ, it would be convenient to write it in terms of a variable ξ, defined as the difference between ζ and the value that ζ would have for a Maxwellian velocity distribution, namely, $3\alpha^2$;

$$\xi \equiv \zeta - 3\alpha^2$$

In Eq. 12.30, α and β are the mean squared random velocities in the radial and transverse directions. The third moment, ε, represents an energy transport or "heat flow" in the radial direction. The quantity ξ may be thought of as representing an excess (or deficiency) of high-velocity stars relative to a Maxwellian velocity distribution.

The moment equations for ρ, $\langle u \rangle$, α, β, ε, and ξ are derived by the usual procedure of multiplying the Boltzmann equation by successively higher powers of the velocity components u, v, w and then integrating over the velocity distribution. When this is done, in addition to those moments already defined, we need $\langle (u - \langle u \rangle) v^2 \rangle$, $\langle (u - \langle u \rangle)^3 v^2 \rangle$, $\langle (u - \langle u \rangle)^5 \rangle$, $\langle (u - \langle u \rangle)^2 v^2 \rangle$, and the corresponding moments with w^2 in place of v^2. To close the system of equations, some way must be found of relating the higher-order moments to the lower-order ones. Larson did this by using an approximate representation of the velocity distribution. He also used this approximation to evaluate relaxation effects caused by collisions.

The Assumed Form of the Velocity Distribution

Let V denote the magnitude of the velocity vector and μ the cosine of the angle between this vector and the radial direction:

$$u - \langle u \rangle = V\mu, \qquad v^2 + w^2 = V^2(1 - \mu^2)$$

Let $f(V, \mu)$ be the distribution function of velocities; the normalization condition is

$$2\pi \int_0^\infty V^2 \, dV \int_{-1}^{+1} f(V, \mu) \, d\mu = 1$$

Because by definition $\langle (u - \langle u \rangle) \rangle = 0$, $f(V, \mu)$ must satisfy the further constraint

$$\langle V\mu \rangle = 2\pi \int_0^\infty V^3 \, dV \int_{-1}^{+1} f(V, \mu) \mu \, d\mu = 0$$

In statistical equilibrium the velocity distribution approaches a Maxwellian distribution,

$$g(V) = (2\pi b)^{-3/2} \, e^{-V^2/2b}$$

The actual velocity distribution $f(V, \mu)$ is expected to be approximated by a Maxwellian distribution, at least in the central part of a cluster where the relaxation effects are most important, so Larson adopted the Maxwellian distribution as a zeroth-order approximation to $f(V, \mu)$. He chose the value of b such that $g(V)$ has the same kinetic energy of random motion as does the actual velocity distribution $F(V, \mu)$ or

$$b = \frac{\alpha + 2\beta}{3}$$

He further assumed that the deviation of $f(V, \mu)$ from a Maxwellian distribution was small and that it could be adequately represented by an expansion in Legendre polynomials. He kept only the first three terms in such a series,

$$f(V, \mu) = g(V) + \sum_{n=0}^{2} a_n(V) P_n(\mu)$$

[The $P_n(\mu)$ are Legendre polynomials and the coefficients $a_n(V)$ are imagined to be small corrections to the Maxwellian distribution $g(V)$.] The term $a_0(V)$ is related to the quantity ξ, $a_1(V)$ to the energy flux parameter ε, and $a_2(V)$ to the anisotropy parameter $\alpha - \beta$. These three terms are the minimum number required to adequately represent the basic features of the perturbed velocity distribution.

Further assume that the functions $\{a_n(V)\}$ may be approximated as the product of $g(V)$ by a power series in V and that only the first few terms in the power series need be retained. Because the velocity distribution must be of the form

$$f(V, \mu) = f'(u - \langle u \rangle, v^2 + w^2) = f'[V\mu, V^2(1 - \mu^2)]$$

that is, it must be a function of V^2 and $V\mu$ only, it is clear that $a_0(V)$ and $a_2(V)$ must contain only even powers of V, whereas $a_1(V)$ must contain only odd powers of V. The simplest polynomials that give nonzero values for ξ, ε, and $\alpha - \beta$ and still satisfy the normalization conditions are

$$a_0(V) = c_0 \left(1 - \frac{2V^2}{3b} + \frac{V^4}{15b^2} \right) g(V)$$

$$a_1(V) = c_1 \frac{V}{b^{1/12}} \left(-1 + \frac{V^2}{5b} \right) g(V)$$

$$a_2(V) = c_2 \frac{V^2}{b} g(V)$$

Evaluate the moments α, β, ε, utilizing this velocity distribution. This allows one to express c_0, c_1, and c_2 in terms of α, β, ε, and ξ:

$$c_0 = \frac{5}{8} \frac{[\xi + \frac{4}{3}(\alpha - \beta)^2]}{b^2}, \qquad c_1 = \frac{5}{6} \frac{\varepsilon}{b^{3/2}}, \qquad c_2 = \frac{\alpha - \beta}{3b}$$

From the expressions for a_0, a_1, and a_2 one can now evaluate the unknown moments required to close the system of equations. The results are

$$\langle (u - \langle u \rangle) v^2 \rangle = \frac{\varepsilon}{3}$$

$$\langle (u - \langle u \rangle)^2 v^2 \rangle = \frac{\xi}{3} - b(\alpha - \beta)$$

$$\langle (u - \langle u \rangle)^5 \rangle = 10 b \varepsilon$$

$$\langle (u - \langle u \rangle)^3 v^2 \rangle = 2 b \varepsilon$$

These equations are expected to be valid if $|\varepsilon|$ and $|\alpha - \beta|$ are small.

The Moment Equations. For a system with spherical symmetry the Boltzmann equation may be written as

$$\frac{\partial f}{\partial t} + u \frac{\partial f}{\partial r} + \dot{u} \frac{\partial f}{\partial u} + \dot{v} \frac{\partial f}{\partial v} + \dot{w} \frac{\partial f}{\partial w} = \left(\frac{\partial f}{\partial t} \right)_{\text{coll}}$$

where

$$\dot{u} = -\frac{\partial \Phi}{\partial r} + \frac{v^2 + w^2}{r}$$

$$\dot{v} = -\frac{uv}{r} + \frac{w^2}{r \tan \theta}$$

$$\dot{w} = -\frac{uw}{r} - \frac{vw}{r \tan \theta}$$

In this equation the term $(\partial f / \partial t)_{\text{coll}}$ represents the rate of change of f on account of encounters between the stars. The moment equations turn out to be

$$\frac{\partial \rho}{\partial t} + \frac{1}{r^2} \frac{\partial}{\partial r} (r^2 \rho \langle u \rangle) = 0$$

$$\frac{\partial \langle u \rangle}{\partial t} + \langle u \rangle \frac{\partial \langle u \rangle}{\partial r} + \frac{1}{\rho} \frac{\partial (\rho \alpha)}{\partial r} + \frac{2}{r} (\alpha - \beta) + \frac{\partial \Phi}{\partial r} = 0$$

$$\frac{\partial \alpha}{\partial t} + \langle u \rangle \frac{\partial \alpha}{\partial r} + 2 \alpha \frac{\partial \langle u \rangle}{\partial r} + \frac{1}{\rho} \frac{\partial (\rho \varepsilon)}{\partial r} + \frac{2 \varepsilon}{r} \left(1 - \frac{2}{3} \frac{\beta}{\alpha} \right) = \left(\frac{\partial \alpha}{\partial t} \right)_{\text{coll}}$$

$$\frac{\partial \beta}{\partial t} + \langle u \rangle \frac{\partial \beta}{\partial r} + 2 \beta \frac{\partial \langle u \rangle}{\partial r} + \frac{1}{3\rho} \frac{\partial}{\partial r} \left(\frac{\beta \rho \varepsilon}{\alpha} \right) + \frac{4}{3} \frac{\beta}{\alpha} \frac{\varepsilon}{r} = \left(\frac{\partial \beta}{\partial t} \right)_{\text{coll}}$$

$$\frac{\partial \varepsilon}{\partial t} + \langle u \rangle \frac{\partial \varepsilon}{\partial r} + 3 \varepsilon \frac{\partial \langle u \rangle}{\partial r} + 3 \alpha \frac{\partial \alpha}{\partial r} + \frac{1}{\rho} \frac{\partial (\rho \xi)}{\partial r} + \frac{2 \xi}{r} \left(1 - \frac{\beta}{\alpha} \right) = \left(\frac{\partial \varepsilon}{\partial t} \right)_{\text{coll}}$$

$$\frac{\partial \xi}{\partial t} + \langle u \rangle \frac{\partial \xi}{\partial r} + 4 \xi \frac{\partial \langle u \rangle}{\partial r} + 6 \varepsilon \frac{\partial \alpha}{\partial r} + 4 \alpha \frac{\partial \varepsilon}{\partial r} = \left(\frac{\partial \xi}{\partial t} \right)_{\text{coll}}$$

The terms $(\partial \alpha / \partial t)_{\text{coll}}$, and so on, represent the effects of encounters.

The Collision Terms. Larson goes on to follow a recommendation of Rosenbluth, MacDonald, and Judd (1957) in order to evaluate the collision terms. He uses the two-body differential scattering cross section already derived and the approximation to the velocity distribution just calculated. Constant spatial density is still assumed. He winds up with

$$\left(\frac{\partial \alpha}{\partial t}\right)_{\text{coll}} = \frac{-4}{5}\frac{(\alpha - \beta)}{T}$$

$$\left(\frac{\partial \beta}{\partial t}\right)_{\text{coll}} = \frac{+2}{5}\frac{(\alpha - \beta)}{T}$$

$$\left(\frac{\partial \varepsilon}{\partial t}\right)_{\text{coll}} = \frac{-87}{160}\frac{\varepsilon}{T}$$

$$\left(\frac{\partial \xi}{\partial t}\right)_{\text{coll}} = \frac{-3}{35}\frac{7\xi - 15(\alpha - \beta)\alpha}{T}$$

where T is the usual relaxation time,

$$T = \frac{1}{16}\left(\frac{3}{\pi}\right)^{1/2}\frac{\langle V \rangle^{3/2}}{G^2 m^2 N \ln[D\langle V^2 \rangle /(2GM)]}$$

with D the cutoff value of the impact parameter ($\langle V^2 \rangle = \alpha + 2\beta$).

Discussion

Larson uses this construct to perform evolutionary computations on the structure of large-scale, self-gravitating systems. The initial conditions he uses are relatively smooth and, as most other computations of this type, the result is an increasingly rapid contraction of the core coupled with a slow diffusion toward zero energy for the halo stars. This is also associated with a corresponding physical expansion of the cluster (unless bounded by a tidal radius). The evolutionary pace in the core quickens as it becomes denser. Herein one has a mechanism for hard binary formation, stellar collisions, massive supernovas, black holes, and so forth. This type of scenario has been replayed by a variety of computational methods in sundry computers.

Applications of Thermodynamics

We saw above how to make the connection between kinetic theory and thermodynamics. The more general derivation of thermodynamics follows a similar prescription. See, for example, Huang's (1963) seventh chapter. Thermodynamic arguments were advanced by Bonnor (1956) and especially by Lynden-Bell and Wood (1968) to discuss the stability of isothermal spheres. As must be clear by now, these analyses are without a rigorous physical basis relating to the nonsaturation of the gravitational force. Moreover, within their own context they are incorrect. Similarly, the hydrodynamic stability analysis of Yabushita (1968) is also wrong. These have been corrected in Taff and Van Horn (1974, 1975). The Lynden–Bell and Wood analysis brought "gravothermal catastrophe" into the vernacular, so the point is worth retelling here.

The Gravothermal Catastrophe

The use of thermodynamics on self-gravitating systems yields "gravothermody-namics." The connection to kinetic theory is through the identification of Boltzmann's functional H as $-S/kV$ where S is the entropy, V is the volume, and k is Boltzmann's constant. To do gravothermodynamics, one uses for the single-particle distribution the product of the isothermal sphere density distri-bution with the Maxwell–Boltzmann velocity distribution. One avoids the embarrassment of an infinite system by arbitrarily truncating the isothermal sphere at some finite radius R. At this point one constructs a spherical wall that exerts just the right pressure to counteract the pressure due to the isother-mal sphere. One can now do work on the system, supply heat to it, change its mass, and so on. Then, as with any thermodynamic analysis, one looks to unusual events in the run of various quantities for hints of phase changes or instabilities. Thus, a zero for a specific heat or a minimum in the enthalpy attracts one's attention.

Lynden-Bell and Wood, within the context of a thermodynamic analysis, showed that there exists a size for an isothermal sphere (constrained by an external wall) at which (1) the entropy is a maximum for a given volume, (2) the energy is least for this volume (greatest binding), (3) the volume is a maximum for a given internal energy, and (4) the adiabat has a vertical tangent. All of these hold simultaneously for an equilibrium configuration. The system is already larger than an isothermal sphere with a positive specific heat at constant volume, so it is unstable to collapse. For as the core collapses its kinetic energy increases, but the negative specific heat prevents the dissipation of this heat, which increases the kinetic energy, and so forth.

Now, how is all of this calculated? Instead of dealing with functions of the volume V, the total energy E, and the number of stars N—the variables in which statistical mechanics is formulated and the physical variable set one uses to describe thermodynamics—it turns out to be more convenient to deal with the dimensionless potential ψ, the dimensionless radius ξ, and the central density $n(0)$. We saw this earlier in the discussion of the isothermal sphere. In terms of $\psi(\xi)$ and its derivatives one can write expressions for the entropy $S(E, V, N)$, the Helmholtz free energy, and so on. The other parameters that appear are $n(0)$ and the temperature T [actually it is better to use $\beta = 1/(kT)$]. So one derives all of the results that Lynden-Bell and Wood did in this nonfundamental representation because of the obvious fact that it is much more convenient to deal with $\psi(\xi)$, $n(0)$, and β than it is to deal with E, V, and N.

All of this is fine, *provided* that the Jacobian of the transformation $J = \partial(E, V, N)/\partial[\xi, n(0), \beta]$ does not vanish. It turns out that the analytical condi-tion for the gravothermal catastrophe is identical to

$$J = 0$$

Thus, *within* the framework of a thermodynamic stability analysis, the conclusion of a gravothermal catastrophe is not warranted. Just as clearly, any discussion of larger isothermal spheres [e.g., those with $\xi > \xi_J$ where $J(\xi_J) = 0$] requires a specific justification if it is to be performed in the ξ, $n(0)$, β representation instead of in the fundamental variable set E, V, and N. No such argument has ever been advanced.

Hydrodynamic Stability Analysis. Yabushita (1968) looked at a hydrodynamical representation of the isothermal sphere and performed a linearized stability analysis of it to look for instabilities. Again, given the correctness of the fluid context for a self-gravitating system, he made a purely mathematical error. The error was minor and in the outer boundary condition. It is corrected in Taff and Van Horn (1974).

RESOLUTION

Raison d'Être

The interesting large-scale, self-gravitating, many-body systems in astronomy are galactic clusters, globular clusters, galaxies, and clusters of galaxies. In each of these instances the system consists of "particles" interacting via Newtonian gravity in three dimensions. Also, for each one, the N-body problem is insoluble analytically, escape from the system is not prohibited, and the close approach of two or more particles requires special computational procedures. In addition to these difficulties, the theoretical basis for the equilibrium solution obtained from statistical mechanics or hydrodynamics (i.e., that the single-particle distribution function is the product of the isothermal sphere density distribution and the Maxwell–Boltzmann velocity distribution) is suspect, the numerical solution of the N-body problem is plagued by the exponential growth of roundoff errors, and the Monte Carlo or hydrodynamical models used to study the system's evolution involve circular reasoning. Finally, the use of thermodynamics or hydrodynamics to investigate the stability of these systems lacks a theoretical basis and suffers from internal self-inconsistencies. Hence, astronomers can only *believe* that they have any understanding of the structure, evolution, or final state of such systems. It is, therefore, imperative that a successful numerical and analytical attack be launched on *some* large-scale, self-gravitating system. Moreover, if the above-mentioned difficulties can be overcome, and the N-body results shown to disagree (for program and machine-independent computations) with the predictions of an exact statistical mechanics, then the astronomical community must squarely face the fact that our understanding of the general N-body problem ends at $N = 2$.

What kind of system should be investigated? Some requirements are clear:

1.   The force between the particles must be attractive Newtonian gravity.

2. The force between the particles must not become unbounded at large particle–particle separations.

3. The force between the particles must not diverge at small particle–particle separations.

4. There must be no external forces.

5. The virial theorem must be satisfied, without a surface term, for both the N-body system and its continuum limit.

6. The statistical mechanics (and hence thermodynamics and hydrodynamics) of the system must be on a sound theoretical basis and be available as an explicit function of N.

7. The limiting case when $N \to \infty$ be available for the statistical mechanics of the system and that this must agree with both the solution of the time-independent collisionless Boltzmann equation and the coupled time-independent hydrodynamical/Poisson equations.

8. The escape of particles must be theoretically impossible.

Such a system is described below. It was first suggested (for some, if not all, of these reasons) to Myron Lecar by Kevin Prendergast (Lecar 1968). Below all of its aspects will be combined and its evolution computed for a large number of initial conditions (as well as for various values of N). We will also investigate whether or not numerical roundoff error is totally insignificant and whether or not the computations are time reversible. Finally, we will test the results in an objective manner. This approach represents a unique perspective on the large-scale, self-gravitational problem. Because of the constraints imposed above and the nature of the comparisons we will make between the results of the N-body integrations and the predictions of statistical mechanics, it is clear that if agreement cannot be obtained for this system, one cannot even *believe* in any other (less severely restricted or fully understood) models.

The Model

The "particles" are identical uniform sheets of mass initially aligned parallel to the yz plane. The initial velocity vectors only have x components. The sheets have infinite extent in the yz plane and zero thickness in the x direction. Each sheet has a mass surface density equal to σ. Because each sheet produces a constant force directed along the x axis (toward the sheet), the motion is always one-dimensional. The magnitude of this force is $2\pi G\sigma$ and, by hypothesis, there are no external forces. Hence, the total energy, the location of the center of mass, and the velocity of the center of mass are all constants. Without loss of generality we choose the latter two to be zero. Thus, requirements 1–4 are satisfied. Below we demonstrate that both the continuum (fluid) and discrete (N-body) systems satisfy the remaining four requirements.

The Continuum System

Rybicki (1971), in a beautiful paper, has shown that in the canonical ensemble the number of sheets per unit length within dx of x is given by

$$n(x)\,dx = N^2\beta\gamma \sum_{n=1}^{N-1} A_n^N \exp(-N\beta\gamma n|x|)\,dx$$

$$= (N^2\beta\gamma\,dx/4)\frac{\int_{-\pi/2}^{\pi/2}\mathrm{sech}^2(N\beta\gamma|x|/2+it)\cos^{2(N-1)}t\,dt}{\int_{-\pi/2}^{\pi/2}\cos^{2(N-1)}t\,dt}$$

where $\gamma = 2\pi G\sigma^2$ and β is the multiplier of the total energy $\mathscr{E}$ in the canonical ensemble weight function [i.e., $\exp(-\beta\mathscr{E})$]. The quantity A_n^N is given by [$\Gamma(z) =$ gamma function]

$$A_n^N = \frac{(-1)^{n-1}n\Gamma^2(N)}{\Gamma(N+n)\Gamma(N-n)}$$

The normalization is such that

$$\int_{-\infty}^{\infty} n(x)\,dx = N$$

and

$$n(0) = \frac{N^2(N-1)\beta\gamma}{2(2N-3)}$$

Rybicki (1971) has also shown that the percentage of sheets with a linear momentum within dp of p is

$$\Pi(p)\,dp = \left[\frac{N\beta}{2\pi\sigma(N-1)}\right]^{1/2}\exp\left[\frac{-NBp^2}{2\sigma(N-1)}\right]dp$$

and that

$$\int_{-\infty}^{\infty}\Pi(p)\,dp = 1$$

Furthermore, the single-particle distribution function is $f(x,p) = n(x)\Pi(p)$. The system is isothermal.

To investigate the limits as $N \to \infty$ while the total mass, $M = N\sigma$, and the total energy, $\mathscr{E} = 3(N-1)/2\beta$, remain fixed, we introduce the dimensionless location and velocity

$$\chi = \frac{x}{L}, \qquad \nu = \frac{p}{\sigma V}$$

wherein

$$L = \frac{2\mathscr{E}}{3\pi GM^2}, \qquad V^2 = \frac{2\mathscr{E}}{3M}$$

We also introduce a dimensionless time,

$$\tau = \frac{t}{T}, \qquad T = \frac{L}{V}$$

As Rybicki has shown (note that his V^2 is twice mine), the new distribution functions are given by

$$n(x)\, dx \rightarrow \left(\frac{N}{2}\right) \mathrm{sech}^2 \chi\, d\chi$$

$$\Pi(p)\, dp \rightarrow \left[\frac{1}{(2\pi)^{1/2}}\right] \exp\left(\frac{-\nu^2}{2}\right) d\nu$$

This does agree with Camm's (1950) separable solution of the time-independent collisionless Boltzmann equation. It also satisfies the coupled, time-independent equations of Poisson and hydrodynamics (given the assumption of isothermality). Thus, requirements 6 and 7 are satisfied. A straightforward computation also shows that the virial theorem is satisfied without a surface term (half of requirement 5).

The Discrete System

It is advantageous to number the sheets such that $x_1 \leqslant x_2 \leqslant \cdots \leqslant x_N$. This is assumed to be the case from now on. To formulate the equations of motion, we first calculate the acceleration on the nth sheet, a_n (α_n in dimensionless units). As each sheet produces a constant force, a_n is merely the number of sheets to the right of the nth sheet minus the number of sheets to its left multiplied by the force per unit mass produced by a single sheet:

$$a_n = 2\pi G\sigma[N - 2n + 1]$$

The equations of motion are

$$\frac{d^2 x_n}{dt^2} = a_n, \qquad n = 1, 2, \ldots, N$$

until there is a collision. A collision between sheets n and $n+1$ occurs when $x_n = x_{n+1}$ at some time t. (The sheets are permeable to each other.) In the dimensionless variables,

$$\frac{d^2 \chi_n}{d\tau^2} = 2\alpha_n, \qquad \alpha_n = \frac{N - 2n + 1}{N}$$

$$\frac{d\chi_n}{d\tau} = \nu_n$$

If the locations and velocities of all the sheets are known at some time $\tau = \tau_1$ and the next collision occurs at some later time $\tau = \tau_2$, then the complete, exact solution of the N-body problem is given by

$$\chi_n(\tau) = \chi_n(\tau_1) + \nu_n(\tau_1)\tau + \alpha_n\tau^2$$
$$\nu_n(\tau) = \nu_n(\tau_1) + 2\alpha_n\tau, \qquad n \in [1, N] \text{ and } \tau_1 \leqslant \tau \leqslant \tau_2$$

τ_2 is computed from

$$\tau_2 = \min_n \left\{ \Delta\nu_n + \left[(\Delta\nu_n)^2 + \left(\frac{8}{N}\right)\Delta\chi_n \right]^{1/2} \right\}$$

with the quantities $\Delta\chi_n$ and $\Delta\nu_n$ given by

$$\Delta\chi_n = \chi_{n+1}(\tau_1) - \chi_n(\tau_1) \geqslant 0$$
$$\Delta\nu_n = \nu_{n+1}(\tau_1) - \nu_n(\tau_1) \qquad \forall n \in [1, N-1]$$

Because escape is obviously impossible, this demonstrates requirement 8.
   The location and velocity of the center of mass are defined by

$$\chi_{cm} = \sum_{n=1}^{N} \frac{\chi_n}{N}, \qquad \nu_{cm} = \sum_{n=1}^{N} \frac{\nu_n}{N}$$

The kinetic and potential energies may be expressed as

$$k = \sum_{n=1}^{N} \frac{\nu_n^2}{2}, \qquad p = -2 \sum_{n=1}^{N} \alpha_n\chi_n$$

This expression for p depends on the equality of the surface densities and the ordering in x. The total energy is $\varepsilon = k + p$ and the moment of inertia is

$$\iota = \sum_{n=1}^{N} \chi_n^2$$

From the equations of motion and the definitions, one can show that

$$\frac{d^2\iota}{d\tau^2} = 2(2k - p)$$

Again, because escape is impossible the virial theorem follows (the remainder of requirement 5).
   The equilibrium values of k, p, ε, and ι calculated from the continuum single-particle distribution function are

$$k = \frac{N}{2}, \qquad p = N, \qquad \varepsilon = \frac{3N}{2}, \qquad \iota = \frac{N\pi^2}{12}$$

Time Scales

All of the numerical computations will be performed using T as the unit of time. The crossing time $(= L/V)$ is unity, the free-fall collapse time $= [4\pi G\sigma n(0)]^{-1/2}$ is $2^{-1/2}$, and the period of oscillation for a sheet placed at rest at $\chi = \chi_0$, P, is

$$P = 2 \int_0^{\chi_0} \{\ln[\text{sech } \chi \cosh \chi_0]\}^{-1/2} \, d\chi$$

For small values of χ_0,

$$P \simeq \pi\sqrt{2}\left[1 + \left(\frac{\chi_0^2}{8}\right)\left(1 - \frac{13\chi_0^2}{96}\right)\right]$$

whereas for large values of χ_0,

$$P \to 4\chi_0^{1/2}$$

If χ_0 is the radius of gyration ($\simeq 0.906997$), then $P \simeq 4.85587$ (obtained using a 32-point Gaussian quadrature formula).

A simplistic estimate of the mean collision time for the system is (mean spacing)/(mean speed) $\simeq (L/N)/V = T/N$. Therefore, the mean collision time for a sheet is $\tau \simeq 1/N^2$. Numerical computations (using a random sample from the equilibrium solution as initial conditions) indicate that this is an underestimate by a factor of 5.

Objective Testing

The predictions of equilibrium statistical mechanics, the time-independent collisionless Boltzmann equation, and the time-independent Poisson and hydrodynamical equations agree and are specific. Namely, not only does the single-particle distribution function separate (i.e., the χ and ν distributions are not correlated) but the distribution function is as given above. (Rybicki has shown, based on the central density, that the approach of the N-body density distribution to the continuum density distribution is of order $1/N$. Only for $N < 100$ would a statistical test be biased.) To test these properties for the numerically determined distribution function I follow the procedures recommended by Kendall and Stuart (1969).

Independence. To test the separability of the single-particle distribution function, we use the null hypothesis,

$$H_0(1): \quad f(\chi, \nu) = g(\chi)h(\nu) \qquad \forall \chi \text{ and } \nu$$

The test statistic is

$$T_{\chi,\nu} = \left[\frac{(N-2)r^2}{(1-r^2)}\right]^{1/2}$$

where r is the correlation coefficient

$$r = \frac{N\sum_{n=1}^{N}\chi_n\nu_n - (\sum_{n=1}^{N}\chi_n)(\sum_{m=1}^{N}\nu_m)}{\{[N\sum_{n=1}^{N}\chi_n^2 - (\sum_{m=1}^{N}\chi_m)^2][N\sum_{n=1}^{N}\nu_n^2 - (\sum_{m=1}^{N}\nu_m)^2]\}^{1/2}}$$

$$= \frac{\sum_{n=1}^{N}\chi_n\nu_n - \chi_{cm}\nu_{cm}}{[\iota/N - \chi_{cm}^2]^{1/2}[2\kappa N - \nu_{cm}^2]^{1/2}}$$

$$= \frac{1}{(8\kappa\iota)^{1/2}}\frac{d\iota}{d\tau}$$

$T_{\chi,\nu}$ has a Student's t distribution with $N-2$ degrees of freedom. Critical values for nonseparability are large values of $|r|$ and, therefore, large values of $T_{\chi\nu}$.

Equality. To test the density and velocity distributions, we use the null hypotheses,

$$H_0(2): \quad g(\chi) = \frac{1}{2}\mathrm{sech}^2\,\chi$$

$$H_0(3): \quad h(\nu) = \frac{e^{-\nu^2/2}}{(2\pi)^{1/2}}$$

The test statistics are

$$X_\chi^2 = \sum_{j=1}^{N_b} \frac{N_b[N_\chi^j - N/N_b]^2}{N}$$

and

$$X_\nu^2 = \sum_{k=1}^{N_b} \frac{N_b[N_\nu^k - N/N_b]^2}{N}$$

where $N_b(=3.28\,N^{2/5})$ is the number of equal probability bins on the interval $(-\infty, \infty)$ and $N_\chi^j(N_\nu^k)$ is the number of sheets in the jth χ(kth ν) bin. The bin boundary points, $\{\chi^j\}, \{\nu^k\}$, are determined from (with $\chi^0 = \nu^0 = -\infty$)

$$1 = N_b \int_{\chi^{j-1}}^{\chi^j} \frac{1}{2}\mathrm{sech}^2\,\chi\,d\chi$$

$$1 = N_b \int_{\nu^{k-1}}^{\nu^k} \left[\frac{1}{(2\pi)^{1/2}}\right] e^{-\nu^2/2}\,d\nu$$

The solution for $\chi^j(\nu^k)$ is

$$\chi^j = \ln\left[\frac{j}{N_b - j}\right]^2$$

$$\nu^k = \mathrm{Pr}^{-1}\left(\frac{k}{N_b}\right)$$

where $\mathrm{Pr}(z) = [1/(2\pi)^{1/2}]\int_{-\infty}^z e^{-u^2/2}\,du$. Both X_χ^2 and X_ν^2 are asymptotically distributed as a chi-squared statistic is with $N_b - 1$ degree of freedom. They are also consistent statistics, asymptotically unbiased, and they maximize the power of the test. Large values of $X_\chi^2(X_\nu^2)$ form the critical region.

(All that is left to do is write efficient code, fill up my computing budget, and let her rip. Watch the journals!)

PROBLEMS

1. Consider a regularization of the form $dt = cr^p\,ds$ where t is the time and r is the radial separation. Show in the two-body problem that if $p = 1$,

$c^2 = a/GM$, then $s = E$. Also show that if $p = 2$, $1/c^2 = GMa(1 - e^2) = L^2$, then $s = v$ (the true anomaly).

2. Continue the above to the case when $p = 3/2$ and $1/c^2 = GM$. Let s now be called the intermediate anomaly (Nacozy 1977). Show that

$$s\sqrt{e} = kF\left(\frac{v}{2}, k\right), \qquad k^2 = \frac{2e}{1+e}$$

where F is the incomplete elliptic integral of the first kind.

3. Show that for an N-particle system

$$\frac{d^2 I}{dt^2} = 2K - \sum_{n=1}^{N} \mathbf{r}_n \cdot \nabla_{\mathbf{r}_n} U$$

from the virial theorem. Whence, using the homogeneity of U in the gravitational case,

$$\frac{d^2 I}{dt^2} = 2K + U$$

and, therefore, there does not exist an arrangement of N particles acted upon by their mutual attractive forces such that $\dot{\mathbf{r}}_n = 0 \, \forall \, n \in [1, N]$.

4. Show that the zeroth-order equations of hydrodynamics possess flow patterns that persist indefinitely.

5. Prove Jeans's theorem: In a steady state the distribution function is a function of the integrals of the motion (assume the validity of the Boltzmann equation).

chapter thirteen

Binary Stars

This chapter closes the circle on celestial mechanics in two senses. The preceding chapter may well have given one the impression that gravity plays a poorly understood role beyond the boundaries of the solar system. Such a supposition is false, which is why Newton's law of gravitation is usually referred to as the *universal* law of gravitation and G is known as the *universal* gravitational constant. For within the realm of the *binary stars* ["a real double star—the union of two stars that are formed together in one system, by the law of attraction" (Herschel 1802)], one can calculate orbits and predict positions in the same manner as for spacecraft and comets. It is the complete success of this enterprise for the many hundreds of binary stars, distributed at random over the celestial sphere, that allows us to stretch

$$\mathbf{F}_{12} = \frac{-Gm_1 m_2 (\mathbf{r}_1 - \mathbf{r}_2)}{|\mathbf{r}_1 - \mathbf{r}_2|^3}$$

beyond the boundaries of the solar system. Hence the adjective "universal." But, alas, even Newton did not get it all right. Newtonian space and time were wedded by Einstein into space–time. First came their union and a flat space–time (the Minkowski metric and special relativity). Next came a beautiful geometrification of space–time, the introduction of curvature, the formal elimination of gravity, and a method of computing the evolution of universes (e.g., the general theory of relativity).

This chapter at least mentions everything one could want to know about binary stars—from cataclysmic variables to the binary pulsars, from the Algol paradox to eclipsing variables, from symbiotic stars to multiple systems, and from accretion discs and x-rays to runaway stars and supernovae. Recurrent novae, astrometric companions, hot spots, Z Camelopardalis variables, atmospheric eclipses, and so on, will be introduced, defined, and very briefly analyzed. The problem with all of this is that, while wonderfully interesting in their own right, a complete discussion of these systems rapidly leaves celestial mechanics and touches on hydrodynamics, stellar evolution, photometry, general relativity, and the rest of astrophysics (to a greater or lesser extent).

Hence, we will continuously break off a discussion to remain within the bounds of this text or commence a discussion with a (sometimes insufficient) minimum amount of preparation.

This chapter offers several advantages. (1) It will provide a concise treatment of the analysis and interpretation of binary star observations and orbits within the classical Newtonian context. (2) We can at least define all of the wondrous species of binary stars mentioned above. We also will outline their dynamical state within the context of the restricted three-body problem and the conservation theorems of energy and angular momentum. (3) We will refocus on the solar system and provide a brief overview of planetary rings and tidal evolution (especially as it applies to the Earth–Moon system). (4) Finally, without developing tensor calculus or the general relativistic field equations, we can show how general relativity makes a difference and what the size of the effects are. Some texts that further expound on the subjects of this chapter are those by Aitken (1935), Batten (1973), Kopal (1959, 1978), and Sahade and Wood (1978).

CLASSICAL BINARY STAR ORBITS

By classical binary star orbits we mean the cut-and-dried discussion of visual binaries and spectroscopic binaries. We shall approach the subject both historically and mathematically.

Historical Overview

As mentioned above, the term *binary star* is used to designate a real double star. The term *double star* is of much earlier origin however. It was used by Ptolemy to describe the appearance of ν Sagittarii (two fifth-magnitude stars whose angular separation is about $14'$). Not every double star defined in this sense constitutes a "binary system." Most are optical pairs that have been accidentally projected onto the line of sight. The first real double star to be found was ξ Ursae Majoris ($=$Mizar; in 1650 by Riccioli). Its principal component (it is really a triple system) was recognized by Pickering as the first spectroscopic binary in 1889. Huyghens saw θ Orionis resolved into the three principal stars of the Trapezium in 1656, and in 1664 Hooke noted that γ Arietis was double. Two additional pairs were discovered before the end of the seventeenth century—α Crucis by Fontenay in 1685 and α Centauri by Richaud in 1689.

These discoveries were accidentally made in the course of observations taken for other purposes. The proximity of these stars seems to have aroused no curiosity. By the mid-1700s several dozen pairs had been recorded (e.g., γ Virginis, Castor, and 61 Cygni). The first scientific argument that at least some of these double stars were the result of physical rather than optical association was Michell's (1767). He pointed out that the frequency distribution of the angular separations of double stars departed from the one that would be expected because of the merely random association of stars (uniformly distributed in space). In particular, there appeared to be too many close pairs. According to Michell, "the natural conclusion from hence is, that is highly probable, and next to a certainty in general, that such double stars as appear to consist of two or more stars placed very near together, do really consist of stars placed nearly together, and under the influence of some general law ... to whatever cause this may be owing, whether to their mutual gravitation, or to some other law or appointment." Herschel (1782) thought that the components of double stars of unequal brightness must be at very different distances and, therefore, particularly suitable for measurements of the relative parallax of their brighter (i.e., nearer) components. He therefore embarked on a systematic search for such pairs. The results of his efforts were gathered in the *First and Second Catalogues of Double Stars* (published in 1782 and 1785). It was not until 1803 that he admitted most of these must be true binary systems (Herschel 1803).

Herschel's continued observations of such systems demonstrated that the relative motion of the fainter component around the brighter (in projection) is an ellipse with the brighter component situated at one of its foci. Additional observations have shown that the apparent motion in these systems is such that the radius vector drawn from the primary to the secondary sweeps out equal areas in equal times. It was, therefore, assumed (from the beginning) that the attractive force in a binary star system is identical with gravitation.

We need to inquire whether or not this assumption can be justified. Although it is supported by all of the available evidence, a rigorous mathematical proof of its validity is difficult because the motion observed is not the true motion but its projection upon a plane perpendicular to the line of sight. Therefore, the primary may lie at any point within the ellipse described by the secondary. Hence, in Leuschner's (1916) words, "mathematical difficulties are encountered in establishing a law of force which is independent of the angle θ, the orientation." He then says that in binary systems, "when the law is arbitrarily assumed to be independent of the orientation, as was found to be the case in the solar system, two possibilities arise, namely, either that the force is in direct proportion to the distance r between the two stars or that the Newtonian law applies. It can be shown, however, that when, in the case of an elliptic orbit, the force is proportional to r, the primary star must be in the center of the ellipse. As this has never been found to be the case, the only alternative is the Newtonian law" (see Chapter Six). It should be clearly understood that the difficulty in demonstrating the universality of the law of gravitation is a purely mathematical one.

Visual and photographic double-star astronomy has made great strides since Herschel's time. It has been enriched by the discovery of thousands of new pairs. Furthermore, those binaries (less than 100) for which absolute orbits as well as parallaxes have been determined have become the source of our knowledge of stellar masses. There is another type of binary system in which the components are very close together. Therefore, the speed of their orbital motions is increased. In fact, the radial component may produce measurable Doppler shift(s) in the line spectra of such a system. Many hundreds of close binaries of this nature—called *spectroscopic binaries*—have been discovered by this effect. Algol was first recognized as one by Vogel in 1889. Observations of spectroscopic binary stars have provided invaluable data on the physical properties of a much wider class of stars than those that appear (to us) as visual binaries. Spectroscopic observations are capable of providing information on the absolute properties of stars. However, they cannot furnish masses or absolute dimensions unless the inclination of the orbit (to the celestial sphere) is known. This piece of information can be deduced if the inclination is large enough for the two stars to eclipse. This will produce the characteristic variation of light by which *eclipsing variables* are known.

Eclipsing binaries represent the oldest known close binary systems. Montanari appears to be the first to have noted (in 1670) that the star Algol did not always shine with a constant light. Twenty-five years later the variability of Algol was confirmed by Maraldi. However, neither he nor Montanari seems to have noticed the periodicity of Algol's light changes; this was discovered by Goodricke in 1782. Goodricke not only reported the discovery of the first known short periodic variable star (those known before were long periodic or irregular in nature), accompanied by an excellent estimate of its period, but near the end of his paper (Goodricke 1783) he proposed two explanations for this phenomenon. One of them was "... if it were not perhaps too early to hazard even a conjecture on the cause of this variation, I should imagine it could hardly be accounted for otherwise than ... by the interposition of a large body revolving around Algol...." He also found the variability of β Lyrae as well as that of δ Cephei. It was not until 1889, when Vogel (1890) recognized Algol as a spectroscopic binary whose conjunctions coincided with the minima of light, that the binary nature of Algol and other similar eclipsing variables was established. The first eclipsing variables were discovered in 1782 (Algol) and 1783 (β Lyrae), but it was not until 1880 that Pickering attempted to interpret the light changes of the former (Pickering 1880). In 1912 Russell and Shapley developed the first general method for an analysis of the light changes of eclipsing variables (Russell 1912a,b,c; Russell and Shapley 1912a,b). They also provided techniques for preliminary solutions for the elements. In order to do so, however, several reasonable but simplistic assumptions had to be made concerning the form of components (spheres or similar ellipsoids) or the distribution of brightness over their discs. The principal data that can be extracted from the light curves of eclipsing variables are the fractional radii of the constituents and the inclination of the orbital plane; spectroscopic observations provide a calibration of the masses and the dimensions of the system in absolute units. It is a combination of both that provides the larger part of what is known of the absolute properties of close binary systems and of their components.

An important theoretical development was the realization that if the stars are distorted and moving in elliptical orbits about their common center of mass, then the major axis of the ellipse will rotate at an angular speed determined by the degree of central condensation of the components. In many cases this will be essentially determined by one component alone. Thus, this offers an observational test of an extremely important stellar property. The original paper calling attention to this was also by Russell (1928). However, he neglected to allow for the changing shape of the stars with their varying separation at different parts of the elliptical orbit. Cowling (1938) called attention to this and derived a correct formula. Later work took into account certain higher-order terms in the apsidal motion.

Various researchers tried to apply these ideas to individual systems, often with conflicting results. Three factors have contributed to this. The rate of rotation of the line of apsides (relative to the orbital period) had to be determined from the observational data in order for the central condensation to be computed. Because the apsidal period was usually of the order of decades, the observational data normally did not have a sufficiently long time span to permit an accurate determination. Hence, estimates of this by various workers (for the same system) often varied widely. Furthermore, the values of the relative radii were needed to a high order of precision and a good value of the mass ratio was required. Both of these are difficult to acquire.

Another problem was resolved with the study of ε Aurigae (Kuiper, Struve, and Strömgren 1937) at Yerkes Observatory. This was followed a few years later by an investigation of β Lyrae (Struve 1941; Kuiper 1941; Gill 1941; Greenstein and Page 1941). These investigations led Struve to postulate the existence of gaseous streams between the components of close binaries and of an outer expanding gaseous envelope. This hypothesis helped to interpret some of the unusual spectral features that these stars displayed. Understanding this system represented a breakthrough. Earlier Barr (1908) had performed an analysis of the available orbits and velocity curves of spectroscopic binaries. He had found that often the velocity curves were unsymmetrical. These asymmetries, with only a few exceptions, were similar in the sense that the ascending branch of the curve was of greater length than the descending branch. A logical consequence of this was that the values of the longitude of periastron did not show a uniform distribution but were concentrated in the first quadrant. This peculiar distribution, known as the Barr effect, already suggested (as Barr himself pointed out) that the observed radial velocities were contaminated by some extraneous effect. The gas streams were the solution.

The success that had been attained, the further discovery by Joy (1942) that the double hydrogen emissions displayed by the eclipsing system RW Tauri undergo eclipses as well (and therefore must arise from a gaseous ring around the primary component) broke open a rather staid field. T CrB (recurrent Nova Coronae Borealis 1866 and 1946) was recognized as an eclipsing variable with a period of 227.6 days (Sanford 1949; cf. Kraft 1958). Then came the discovery by Walker (1956) that Nova Herculis 1934 is an eclipsing variable (DQ Her). It consists of a close pair of subdwarfs (one of which is the nova). Subsequent searches among stars of this type led to the discovery of the binary nature of several other postnovae—such as GK Per (Nova Persei 1901), V603 Aql (Nova Aquilae 1918), WZ Sge (Nova Sagittae 1913 and 1946), and T Aur (Nova Aurigae 1891). The last two of these turned out to be eclipsing variables. The orbital periods of these systems range from ~1.5 hr for WZ Sge to ~5 hr for T Aur; only GK Per has a period longer than a day (1.904 days). These data strongly support the contention that nova outbursts occur only among the components of close binary systems and that the binary nature is a *prerequisite* for such phenomena. The same appears to be true not only for triggering instabilities that produce nova outbursts but also the (more frequent) outbursts of light exhibited by cataclysmic variables of the SS Cygni and U Geminorum types. Both of these systems are binaries consisting of close pairs of subdwarfs. All other known stars of this type (such as RX And, SS Aur, RU Peg, etc.) proved to be binaries with orbital periods of less than 9 hr.

Even greater surprises have appeared in the more massive systems. In 1940 Gaposchkin discovered the Wolf–Rayet star HD 193576 to be an eclipsing variable (V444 Cyg) in which the star exhibiting the Wolf–Rayet spectrum (of mass $9.8 M_\odot$) is the less massive component. In subsequent years the Wolf–Rayet stars HD 214419 (CQ Cep), HD 168206 (CV Ser), and XC Cep were shown also to be eclipsing variables. If one takes account of observational effects (i.e., the random distribution of orbital inclinations), it is probable that all known Wolf–Rayet stars are components of close binary systems.

In 1971 the Uhuru satellite discovered that two known x-ray sources [Cen X-3 and Her X-1 (HZ Her)] are binaries with periods 2.087 and 1.700 days and, moreover, that they exhibit intensity oscillations with a period of 4.84 and 1.24 sec. Walker (1956) found the system DQ Her (Nova Her 1934) to exhibit optical oscillations with a period of 70 sec. Therefore, the possibility suggested itself that in Cen X-3 and Her X-1 one may have to deal with components of close binaries similar to the postnova DQ Her but (because of the short period of oscillation) smaller size and (presumably) higher density—akin to neutron stars rather than to white dwarfs.

The Centaurus X-3 source proved to be considerably more interesting because of its very large mass. The x-ray source underwent a total eclipse each 2.087 days. The same is also true of the Cygnus X-1 source (period 5.60 days) whose massive secondary component (of approximately $10 M_\odot$) appears to be the most promising candidate for a stellar black hole.

Another recent development was the discovery in 1974 of the fact that the pulsar PSR 1913+16 appears to be the component of a close binary system (cf. Hulse and Taylor 1975). The binary nature of such a system was not detected by optical means—rather it was indicated by the fact that the time interval of 0.05903 sec between successive radio pulses of PSR 1913+16 did (and does) not remain constant (or increase, as for all single objects of this type, uniformly with time), but fluctuates with a period of 0.3230 day. This is a result of the varying distance of the object during the course of the pulsar's revolution around the center of mass. The time interval between successive pulses fluctuates periodically because of the light equation in the system. Such an orbit is of great interest and its existence provides an experimental laboratory for theories of general relativity.

Selection Effects

It should be clear that visual binaries can be discovered only when they are relatively close. It is pointless to search for them beyond a few hundred parsecs. Although distance is the most important selection factor in the discovery of visual binaries, an almost equally important factor is the difference in brightness between the components. A faint companion is harder to detect by visual methods than one nearly equal in brightness to the primary, especially if their angular separation is small. Nonetheless, the observations of visual systems have furnished astronomy with much valuable data which include the masses and luminosities of their components, the form and orientation of their orbits in space, and so on. The fact that the relative orbits are ellipses, with one component located at its focus, also provided the first observational proof of the validity of Newton's law of gravitation beyond the confines of the solar system.

Turning to close binaries, a completely different situation is met. The binary nature of such systems can be recognized by a periodic variation in radial velocity or (for systems with suitable orientation of the orbital plane with respect to the line of sight) by a periodic variation in their brightness due to eclipses. For close binaries the probability of discovery will increase with diminishing distance between the components and also with an increase of their mass or luminosity. The closer (or more massive) the system, the faster the orbital motion and, therefore, the larger the Doppler shifts. Also, the closer the system, the greater the probability that the components will mutually eclipse each other.

Visual Binary Star Orbits

Preliminaries

Before embarking on a discussion of visual star orbits it might be well to recall how the path of a single star appears on the sky. The displacement due to annual parallax can be described by the parallax factors in right ascension and declination P_α and P_δ,

$$P_\alpha = R_\odot(\cos \varepsilon \cos \alpha \sin \Lambda_\odot - \sin \alpha \cos \Lambda_\odot)$$

$$P_\delta = R_\odot[(\sin \varepsilon \cos \delta - \cos \varepsilon \sin \alpha \sin \delta) \sin \Lambda_\odot$$

$$- \cos \alpha \sin \delta \cos \Lambda_\odot]$$

The star may also have a proper motion relative to the Sun so that the full time dependence of the geocentric path can be represented as

$$\xi = c_x + \mu_x t + \pi_r P_\alpha, \qquad \eta = c_y + \mu_y t + \pi_r P_\delta \qquad (13.1)$$

ξ and η are known as standard coordinates (see Chapter Seven), traditionally used to analyze stellar motions in photographic astrometry. The proper motion components are μ_x and μ_y, the c's are arbitrary zero-point constants, and t is the time in tropical years. Finally, since the star of interest (or program star) has its position reduced relative to the background (or reference) stars, the parallax so deduced is relative—hence the use of the symbol π_r. Only if the reference stars were infinitely distance (and therefore infinitely bright!) would the parallax be an absolute one. The relative to absolute reduction need not concern us here.

The superposition of the parallactic motion and the proper motion results in an equidistant spiral across the celestial sphere (see Fig. 51). Any systematic departure from such a path would be ascribed to revolution about the center of mass of a multiple system. It would be the center of mass that moved through interstellar space uniformly and whose motion upon the parallactic ellipse reflected the Earth's (actually the Earth–Moon barycenter's) annual tour of the solar system.

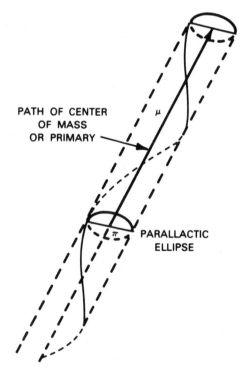

Figure 51.   The path, across the celestial sphere, of a single star.

In the case of visual binaries the motion is obvious. One of the pair, usually the brighter star, is designated as the *primary*. The other component is called the *secondary*. The observational datum consists of the measurement of the angular distance (or distance ρ) between them and the position angle θ of the secondary relative to the primary (measured from the North through the East). Usually, ρ is given in seconds of arc and θ in degrees. Only when the linear distance to the binary is known (i.e., the parallax has been determined) can the linear extent of the orbit be deduced.

The true revolution of the two components about their center of mass is projected onto the plane of the sky. This immediately yields an indeterminacy as two orbits, symmetrical with respect to the plane of the sky, will satisfy the observations. Only if one has information concerning the motion along the line of sight (e.g., radial velocities) can this ambiguity be resolved. (Its practical result is a slight change in the meaning of the orbital element set.) The nature of this projection is to preserve the location of the center of the true ellipse, but the primary no longer appears to be at the true orbit's focus. However, as the diameter of the apparent ellipse through the primary is the projection of the major axis of the true ellipse, the locations of periastron and apoastron are unambiguous. Also, a line parallel to the true line of nodes remains parallel and unshortened; a similar line perpendicular to the true line of nodes remains orthogonal but foreshortened by the cosine of the inclination.

If one plots the values of ρ and θ, one can see that the secondary traverses an apparent orbit whose equation may be written as

$$Ax^2 + 2Hxy + By^2 + 2Gx + 2Fy + 1 = 0 \qquad (13.2)$$

($A, B > 0$ and $AB - H^2 > 0$). Clearly only five numbers are necessary to compute values for the unknowns. Were one to proceed in such a fashion, the most accurately known quantity (the time of observation) would be totally ignored. Hence such a technique is never used. (In writing Eq. 13.2 I assumed that the primary is at the origin of the xy coordinate system; $x = \rho \cos \theta$, $y = \rho \sin \theta$.)

Before presenting the most commonly used orbit determination method for visual binaries (that of Thiele–Innes), we want to define the orbital elements as is peculiarly done in this field. The period P is expressed in years and means the same as it has throughout this book. Similarly, the time of periastron passage T, the eccentricity e, and the semi-major axis a (expressed in seconds of arc unless the parallax is known) retain their standard meanings. Because the orbital elements are defined for the relative orbit of the secondary (mass = M_s) about the primary (mass = M_p), the sizes of their individual orbits about their common center of mass (a_p, a_s) are related to a by

$$a_p = \frac{aM_s}{M}, \qquad a_s = \frac{aM_p}{M}$$

where $M = M_p + M_s$. The ambiguity concerning the tilt of the orbital plane is artificially resolved by only using inclinations i in the first quadrant. Consequently, the longitude of the ascending node may be that of the descending node. Hence, longitude of the node (or even better, position angle of the node) is a preferred terminology; $\Omega \in [0, 180°]$. The argument of periastron ω is the angle in the plane of the true orbit measured between the line of nodes and the major axis. In order to accommodate the uncertainty of the nodal point, ω is measured from the nodal point to periastron in the direction of the secondary's motion; $\omega \in [0, 360°]$. Observers of eclipsing and spectroscopic systems frequently quote the complement of ω as ω.

The Thiele–Innes Method

The Thiele–Innes method (Thiele 1883; Innes's work of 1926 can be found in van der Bos 1926, 1932) is based on certain combinations of the orbital elements known as the Thiele–Innes constants. In the orbital plane the secondary's location is given by

$$X = \frac{r}{a}\cos v, \qquad Y = \frac{r}{a}\sin v, \qquad Z = 0$$

When this is projected onto the plane of the sky, one obtains for the distance ρ and the position angle θ

$$\rho \cos(\theta - \Omega) = r \cos(v + \omega)$$

$$\rho \sin(\theta - \Omega) = r \sin(v + \omega) \cos i$$

$$z = r \sin(v + \omega) \sin i$$

where z is the projection along the line of sight. Alternatively,

$$
\begin{aligned}
x = \rho \cos \theta &= r(\cos \omega \cos \Omega - \sin \omega \sin \Omega \cos i) \cos v \\
&\quad - r(\sin \omega \cos \Omega + \cos \omega \sin \Omega \cos i) \sin v \\
y = \rho \sin \theta &= r(\cos \omega \sin \Omega + \sin \omega \cos \Omega \cos i) \cos v \\
&\quad - r(\sin \omega \sin \Omega - \cos \omega \cos \Omega \cos i) \sin v \\
z = r \sin \omega \sin i \cos v &+ r \cos \omega \sin i \sin v
\end{aligned}
\tag{13.3}
$$

The obvious combinations of trigonometric functions of the Euler angles of the orbit lead to the definition of the Thiele–Innes constants A, B, C, and F, G, H;

$$A = a(\cos \omega \cos \Omega - \sin \omega \sin \Omega \cos i)$$

$$B = a(\cos \omega \sin \Omega + \sin \omega \cos \Omega \cos i)$$

$$C = a \sin \omega \sin i$$

$$F = -a(\sin \omega \cos \Omega + \cos \omega \sin \Omega \cos i)$$

$$G = -a(\sin \omega \sin \Omega - \cos \omega \cos \Omega \cos i)$$

$$H = a \cos \omega \sin i$$

The projection from the orbital plane onto the plane of the sky can now be written as (since $Z \equiv 0$)

$$x = AX + FY, \quad y = BX + GY, \quad z = CX + HY$$

Note that the Thiele-Innes constants depend on the angular quantities (a is in seconds of arc!) and that X and Y are time dependent.

The exercise of going from the Thiele-Innes constants back to the orbital elements is straightforward:

$$A + G = a(1 + \cos i) \cos(\omega + \Omega)$$

$$A - G = a(1 - \cos i) \cos(\omega - \Omega)$$

$$B - F = a(1 + \cos i) \sin(\omega + \Omega)$$

$$B + F = -a(1 - \cos i) \sin(\omega - \Omega)$$

Whence

$$\tan(\omega + \Omega) = \frac{B - F}{A + G}, \quad \tan(\omega - \Omega) = \frac{B + F}{G - A}$$

$$a(1 + \cos i) = (A + G) \sec(\omega + \Omega) = (B - F) \csc(\omega + \Omega)$$

$$a(1 - \cos i) = (A - G) \sec(\omega - \Omega) = -(B + F) \csc(\omega - \Omega)$$

A geometrical interpretation of the Thiele-Innes constants is given in Problem 3. The next step is to demonstrate that the orbit can be determined using the data set $\{\rho, \theta, t\}$ and the Thiele-Innes constants.

Orbit Determination

The key is to represent the area swept out from t_1 and t_2, and then to t_3, in terms of the Thiele-Innes constants. Twice the areal constant of the apparent orbit (e.g., $\rho^2 \dot{\theta}$) is equal to $x\dot{y} - y\dot{x}$. It is also equal to twice the area of the apparent ellipse divided by the orbital period. For ellipses, area is just the product of π and their semi-principal axes (e.g., area $= \pi \times$ semi-minor axis $\times$ semi-major axis). Therefore, in terms of Thiele-Innes constants,

$$\rho^2 \dot{\theta} = ne(AG - BF)$$

where n is the mean motion ($= 2\pi/\text{period}$) and e is the eccentricity. We define the quantity Δ_{pq} by

$$\Delta_{pq} = \rho_p \rho_q \sin(\theta_q - \theta_p)$$

ρ_p and θ_p are the distance and position angle at $t = t_p$. Since $x = \rho \cos \theta$ and $y = \rho \sin \theta$, we can write Δ_{pq} as

$$\Delta_{pq} = x_p y_q - x_q y_p = (AG - BF)(X_p Y_q - X_q Y_p)$$

The second form follows from the definition of Thiele-Innes constants. Now, combining these two results with Kepler's equation, we deduce that

$$\frac{\Delta_{pq}}{\rho^2 \dot{\theta}} = \frac{\sin(E_q - E_p) - e(\sin E_q - \sin E_p)}{n}$$

and then that

$$t_q - t_p - \frac{\Delta_{pq}}{\rho^2 \dot{\theta}} = \frac{(E_q - E_p) - \sin(E_q - E_p)}{n}$$

where E_p is the value of the eccentric anomaly corresponding to the instant $t = t_p$. The last result is Thiele's fundamental formula.

Define the auxiliary angles $E_{21} = E_2 - E_1$, $E_{32} = E_3 - E_2$, and $E_{31} = E_3 - E_1$. Then

$$t_2 - t_1 - \frac{\Delta_{12}}{\rho^2 \dot{\theta}} = \frac{E_{21} - \sin E_{21}}{n}$$

$$t_3 - t_2 - \frac{\Delta_{23}}{\rho^2 \dot{\theta}} = \frac{E_{32} - \sin E_{32}}{n}$$

$$t_3 - t_1 - \frac{\Delta_{13}}{\rho^2 \dot{\theta}} = \frac{(E_{32} + E_{21}) - \sin(E_{32} + E_{21})}{n}$$

The times of observation (t_1, t_2, and t_3) are known and the projected areas Δ_{12}, Δ_{23}, and Δ_{13} can be calculated by plotting the orbit. Similarly, $\rho^2 \dot{\theta}$ can be computed from such a graph. Therefore, all three left-hand sides are known and one has three equations in two unknowns. An explicit algebraic solution for, say, the sine and cosine of E_2 can be obtained without difficulty:

$$e \sin E_2 = \frac{\Delta_{23} \sin E_{21} - \Delta_{12} \sin E_{32}}{\Delta_{12} + \Delta_{23} - \Delta_{13}}$$

$$e \cos E_2 = \frac{\Delta_{23} \cos E_{21} + \Delta_{12} \cos E_{32} - \Delta_{13}}{\Delta_{12} + \Delta_{23} - \Delta_{13}}$$

Whence come e, E_{21} and $E_1 = E_2 - E_{21}$, and $E_3 = E_2 + E_{23}$. Next, via Kepler's equation, we deduce the corresponding mean anomalies and then the time of periastron passage. Finally, we calculate $\{X_j, Y_j\} = \{\cos E_j - e, (1 - e^2)^{1/2} \sin E_j\}$ and A, B, F, and G. Clearly, with a surfeit of observations the appropriate statistical procedures should be employed.

Spectroscopic Binary Star Orbits

Spectroscopic binaries are, in general, those binaries further away from the Earth. Hence, the semi-major axis a (in seconds of arc) is much smaller than the seeing disc. Therefore, these systems cannot usually be resolved into their components. However, since the stars are members of a true binary system, they revolve about their center of mass. This revolution produces periodic changes in their velocity components. In particular, such changes are noticeable along (almost) any line of sight. It follows that the spectral lines of such stars will exhibit a periodic fluctuation as a consequence of their Doppler shifts. If only one set of lines is visible (because the primary is much brighter than the secondary), it is referred to as a *single-lined spectroscopic binary.* This is the more common situation. With this introduction let me turn to analyzing the orbit.

Orbital Analysis

Let the XY plane be tangent to the celestial sphere at the center of mass and let the Z axis (perpendicular to the XY plane) be parallel to the line of sight (along which the radial velocities are measured). The radial velocities are considered positive when the star is receding from the observer and negative when it is approaching the observer. The orientation of the X and Y axes remains unknown. When the star is at any point in its orbit, its distance z from the XY plane will be (see Eq. 13.3)*

$$z = r \sin i \sin(v + \omega)$$

The radial velocity is equal to dz/dt

$$\frac{dz}{dt} = \sin i \sin(v+\omega)\frac{dr}{dt} + r \sin i \cos(v+\omega)\frac{dv}{dt}$$

From the laws of motion

$$r\frac{dv}{dt} = \frac{na(1+e \cos v)}{(1-e^2)^{1/2}}, \qquad \frac{dr}{dt} = \frac{nae \sin v}{(1-e^2)^{1/2}}$$

Therefore

$$\frac{dz}{dt} = \frac{na \sin i}{(1-e^2)^{1/2}}[e \cos \omega + \cos(v+\omega)] \tag{13.4}$$

This is the fundamental equation connecting the radial velocities with the elements of orbit. The observed velocities contain the velocity of the center of mass (V). This is a constant. V must therefore be subtracted from the observed values to make them purely periodic. Alternatively, the velocity curve is periodic only with respect to a line representing the velocity of the system as a whole. This line is called the V axis. Equation 13.4 applies only to the velocities counted from the V axis. If $d\zeta/dt$ represents the velocity as actually observed (i.e., the velocity referred to the zero axis), then one has

$$\frac{d\zeta}{dt} = V + \frac{dz}{dt}$$

---

*Remember, we have assumed that the spectrum of only one component is visible; when both spectra are measurable, then the relative orbit of one with respect to the other may be found using the same formulas but changing the value of the effective mass. The relative and absolute orbits are similar.

Five constants enter the right-hand member of Eq. 13.4: $a \sin i, e, \mu$ (through n), ω, and T. These are the five orbital elements that can be determined by measures of the radial velocity. However, because the inclination of the orbital plane cannot be determined, the value of the semi-major axis remains unknown. It is customary, therefore, to regard the combination $a \sin i$ as an "element." Furthermore, it should be clear that the position of the line of nodes cannot be determined (though one can find the times when the star passes through each of the nodal points). The various elements have the same definitions as in the case of visual binary star orbits except that the angle ω in spectroscopic binary orbits is always measured from the ascending node, the node at which the star is moving away from the observer. The radial velocity has its maximum positive value at this node and its minimum positive value (or maximum negative value) at the descending node. It should also be noted that the unit of time for n (and therefore for P) is the day, not the year, as it is in visual binary orbits.

We take A and B as the magnitudes of the radial velocity curve ordinates at the points of maximum reckoned from the V axis, regarding B as a positive quantity and writing $K = a \sin i/(i - e^2)^{1/2}$. Then

$$A = K(1 + e \cos \omega), \qquad B = K(1 - e \cos \omega)$$

Or, equivalently,

$$K = \frac{A + B}{2}, \qquad eK \cos \omega = \frac{A - B}{2}, \qquad e \cos \omega = \frac{A - B}{A + B}$$

Hence, we may write Eq. 13.4 in the form

$$\frac{dz}{dt} = K[e \cos \omega + \cos(v + \omega)] = \frac{A - B}{2} + \frac{A + B}{2} \cos(v + \omega)$$

or, including V,

$$\frac{d\zeta}{dt} = V + \frac{A - B}{2} + \frac{A + B}{2} \cos(v + \omega)$$

K is the half-amplitude of the radial velocity curve.

To this point practically all methods of spectroscopic orbit determination are identical. When the fundamental relations are given as above and the radial velocity curve has been drawn, various methods are available for computing orbital elements (other than the period which is assumed to be known). Of these, the method devised by Lehmann-Filhés (1894) will be presented.

Method of Lehmann-Filhés

Given the observations and the radial velocity curve drawn with the value of P assumed, the first step is to fix the V axis. This is found by using the condition that the integral of dz/dt (the area under the radial velocity curve) must be the same for those portions above and below the V axis. This may be accomplished by graphical means. Having found the V axis, one may determine the points of maximum and minimum velocity. When dz/dt vanishes,

$$\cos(v+\omega) = -\frac{A-B}{A+B} = -e\cos\omega$$

If v_1 is the true anomaly corresponding to the point traversed by the star on the way from the ascending node to the descending node, and v_2 is the true anomaly for the other null point, then $\sin(v_1+\omega)$ will be positive, $\sin(v_2+\omega)$ negative, and

$$\cos(v_1+\omega) = -\frac{A-B}{A+B}, \qquad \cos(v_2+\omega) = -\frac{A-B}{A+B}$$

$$\sin(v_1+\omega) = \frac{2\sqrt{AB}}{A+B}, \qquad \sin(v_2+\omega) = -\frac{2\sqrt{AB}}{A+B}$$

Let Z_1 and Z_2 (they are simply expressible in terms of areas under the radial velocity curve) be defined by

$$Z_1 = r_1 \sin i \sin(v_1+\omega)$$

$$Z_2 = r_2 \sin i \sin(v_2+\omega) = -r_2 \sin i \sin(v_1+\omega)$$

so that

$$-\frac{Z_1}{Z_2} = \frac{r_1}{r_2} = \frac{1+e\cos v_2}{1+e\cos v_1}$$

since $r = [a(1-e^2)]/(1+e\cos v)$. We write $v+\omega-\omega$ for v, expand and reduce, and then, with the aid of the above relationships, deduce that

$$-\frac{Z_1}{Z_2} = \frac{\sin(v_1+\omega)-e\sin\omega}{\sin(v_1+\omega)+e\sin\omega}$$

Whence

$$e\sin\omega = \frac{Z_2+Z_1}{Z_2-Z_1}\sin(v_1+\omega) = \frac{2\sqrt{AB}}{A+B}\cdot\frac{Z_2+Z_1}{Z_2-Z_1}$$

But $e\cos\omega = (A-B)/(A+B)$, so both e and ω are known. (The values of A and B were taken from the radial velocity curve too.)

At the time of periastron passage $v = 0°$, hence

$$\left.\frac{dz}{dt}\right|_{t=T} = K(1+e)\cos\omega$$

This gives the ordinate corresponding to the point of periastron passage. The abscissa of this point, properly combined with the epoch chosen for the beginning of the curve, defines T, the time of periastron passage.

By definition

$$K = \frac{na \sin i}{(1 - e^2)^{1/2}}$$

and thence

$$n = \frac{K(1 - e^2)^{1/2}}{a \sin i} = \frac{A + B}{2} \cdot \frac{(1 - e^2)^{1/2}}{a \sin i}$$

from which one may deduce $a \sin i$. The unit of time for A and B is the second and that for n is the day, therefore the factor 86,400 must be introduced. The result then becomes

$$a \sin i = 86,400 \frac{K}{n}(1 - e^2)^{1/2}$$

In terms of the Thiele–Innes constants the relative radial velocity of the secondary with respect to the primary is (M = mean anomaly)

$$\frac{dz}{dt} = C\frac{dX}{dt} + H\frac{dY}{dt} = nC\frac{dX}{dM} + nH\frac{dY}{dM}$$

To reduce from seconds of arc per year to kilometers per second requires the introduction of the absolute parallax π and the factor 4.737 (to change A.U./yr to km/sec). The relative radial velocity, in kilometers per second, is

$$V = L\frac{dX}{dM} + N\frac{dY}{dM}$$

where

$$\pi L = 4.737 nC, \qquad \pi N = 4.737 nH$$

Because of the 180° uncertainty in ω, the constants C, H, πL, and πN are uncertain as to sign. Finally, we compute the mass function. Since $M_1 a_1 = M_2 a_2$, it follows that

$$\frac{M_2}{M_1} = \frac{a_1}{a_2} = \frac{a_1 \sin i}{a_2 \sin i}, \qquad \frac{M_2}{M_1 + M_2} = \frac{a_1}{a_1 + a_2} = \frac{a_1}{a}$$

In the usual units Kepler's third law takes the form (i.e., $M_\odot$, A.U., year)

$$M_1 + M_2 = \frac{a^3}{P^2}$$

Thus,

$$\left(\frac{M_2}{M_1 + M_2}\right)^3 = \frac{a_1^3}{a^3} = \frac{a_1^3}{(M_1 + M_2)P^2}$$

$$\frac{M_2^3}{(M_1 + M_2)^2} = \frac{a_1^3}{P^2}$$

We still have to multiply by $\sin^3 i$ because only the combination $a \sin i$ can be deduced. The mass function is

$$f(M) = \frac{(M_2 \sin i)^3}{(M_1 + M_2)^2} = \frac{(a_1 \sin i)^3}{P^2}$$

Complications

Rotation: Other stars rotate about an axis just as the Sun does. This causes a broadening of the observed spectral lines and leads to an asymmetrical radial velocity curve.

Gas Streams: The Barr effect mentioned earlier was discovered because of the nonuniform distribution of argument of periastron values. In particular there was an excess of ω values lying in the first quadrant. Struve deduced the cause of this phenomenon by studying a spectroscopic binary that also underwent eclipses. The systematic error is caused by gaseous streams or rings of material streaming from one of the binary components to the other. More will be said about such close binary systems below.

Reflection: If both stars are of comparable luminosity, then not only may both sets of spectral lines be visible but the apparent light from one star may result from both its intrinsic luminosity and the reflection of its companion's light. Such phenomena are known as "hot spots." Obviously, in the deciphering of the light curves, these types of complications could wreak havoc.

Double-Lined Binaries

Another result of comparable luminosities is the visibility of both sets of spectral lines. They will be periodically shifted about the V axis out of phase with one another (barring complicating factors). Now

$$dz_1 / dt = K_1 [e \cos \omega + \cos(v + \omega)]$$
$$dz_2 / dt = K_2 [e \cos(\omega + \pi) + \cos(v + \omega + \pi)]$$
$$= -K_2 [e \cos \omega + \cos(v + \omega)]$$
$$= -\frac{K_2}{K_1} \frac{dz_1}{dt}$$

The relative radial velocity dz/dt is given by

$$dz / dt = K [e \cos \omega + \cos(v + \omega)]$$
$$K = K_1 + K_2 = \frac{n(a_1 + a_2) \sin i}{(1 - e^2)^{1/2}}$$

so that the expression for $a \sin i$ now takes the form

$$a \sin i = (K_1 + K_2) \frac{P}{2\pi} (1 - e^2)^{1/2}$$

With a in kilometers and P in days the numerical constant is equal to 13,751.
    The mass ratio is given by

$$\frac{M_2}{M_1} = \frac{a_1}{a_2} = \frac{a_1 \sin i}{a_2 \sin i} = \frac{K_1}{K_2}$$

As usual

$$M_1 + M_2 = \frac{a^3}{P^2}$$

so, with $a = a_1 + a_2$,

$$M_1 \sin^3 i = \frac{(a_2 \sin i)(a \sin i)^2}{P^2}$$

$$M_2 \sin^3 i = \frac{(a_1 \sin i)(a \sin i)^2}{P^2}$$

Therefore, the mass ratio can be computed. If the binary is also eclipsing, then
one can determine i and, thence M_1 and M_2 (as opposed to M_1/M_2 and
$M_1 \sin^3 i$, etc.).

Eclipsing Binaries

We mentioned that one of the first spectroscopic binary systems to be discovered was Algol (β Persei) and that it had also been known as a variable
star. There are other stars whose light varies in the same manner as does that
of Algol—that is, although it remains constant most of the time, at regular
intervals it fades to a minimum. It may remain constant at the minimum for
a short duration and then recover full brightness or the change may be
continuous. The hypothesis that the star, as viewed from the Earth, undergoes
a total, annular, or partial eclipse (the eclipsing body being a relatively dark
star revolving about the common center of mass) accounts for observations
of this type. Unless the darker star is nonluminous, there should be a second
minimum when the bright star passes between it and the Earth. The relative
depth of the two minima depends on the relative intensities of the two stars
and on their relative sizes.
    In the years 1912–1915 Russell and Shapley (cf. references given in the
Historical Overview) made a thorough investigation of this problem. Russell
developed a general analytical method and Shapley applied it to ~100 systems.
In the general case the problem is an extremely complicated one, for not only
should the orbits be regarded as elliptical with planes inclined to the line of
sight, but the stars should be assumed to be ellipsoids with the longest diameter
of each being directed toward the other. Moreover, the discs may not be
uniformly bright (they may be darker toward the limb as the Sun is), or the
nearer side (which receives radiation from the other) may be brighter than the
farther side. The complete specification of an eclipsing binary system requires

Table 13.   Generalized Binary Elements

Orbital Elements		Eclipse Elements	
Semi-major axis	a	Radius of larger star	R_1
Eccentricity	e	Radius of smaller star	R_2
Longitude of periastron	ω	Light of larger star	L_1
Inclination	i	Light of smaller star and at least three	
Period	P	constants defining amount of	
Epoch of principal conjunction	t_0	elongation, of darkening at the	
		limb, and of brightening of one star	
		by radiation of the other	L_2

a knowledge of at least 13 quantities which, in Russell's notation, are shown in Table 13.

In Russell's (1912c) words:

> The longitude of the node must remain unknown, as there is no hope of telescopic separation of any eclipsing pair.

> The value of a in absolute units can be found only from spectroscopic data. In the absence of these, it is desirable to take a as an unknown but definite unit of length, and express all other linear dimensions in terms of it. Similarly, the absolute values of L_1 and L_2 can be determined only if the parallax of the system is known. But in all cases the combined light of the pair, $L_1 + L_2$, can be taken as the unit of light and the apparent brightness at any time expressed in terms of this. This leaves the problem with eleven unknown quantities to be determined from the photometric measures. Of these, the period is invariably known with a degree of accuracy greatly surpassing that attainable for any of the other elements, and the epoch of principal minimum can be determined, almost independently of the other elements, by inspection of the light-curve. Of the remaining elements, the constants expressing ellipticity and reflection may be derived from the observed brightness between eclipses. These effects are often so small as to be detected only by the most refined observations. The question of darkening toward the limb may well be set aside until the problem is solved for the case of stars that appear as uniformly illuminated disks.

> This leaves us with six unknowns. Fortunately, systems of such short periods as those of the majority of eclipsing variables, usually have nearly circular orbits (as is shown both by spectroscopic data and by the position of the secondary minimum). The assumption of a circular orbit is therefore usually a good approximation to the facts and often requires no subsequent modification.

Russell first discussed the simplified problem:

> *Two spherical stars, appearing as uniformly illuminated disks, and revolving about their common center of gravity in circular orbits, mutually eclipse one another.*

It is required to find the relative dimensions and brightness of the two stars, and the inclination of the orbit, from the observed light curve.

The determination of the orbit can be made by simple geometrical methods, but their practical application demands the tabulation and use of rather complicated functions.

We may assume P and t_0 as already known. If the radius of the relative orbit is taken as the unit of length, and the combined light of the two stars as the unit of light, we have to determine four unknown quantities. Of the various possible sets of unknowns, we select the following:

Radius of the larger star	R_1
Ratio of radii of the two stars	k
Light of the larger star	L_1
Inclination of the orbit	i

The radius of the smaller is then $R_2 = kR_1$, and its light $L_2 = 1 - L_1$. It should be noticed that, with the above definitions, k can never exceed unity, but L_2 will exceed L_1 whenever the smaller star is the brighter (which seems to be the fact in the majority of observed cases).

We shall not further pursue the subject here. It is a complicated exercise in transcendental curve fitting (e.g., it involves elliptic functions unavoidably).

Stellar Masses

Visual, spectroscopic, and eclipsing binaries make up the bulk of objects on which extensive observations and orbit fitting have been performed. An additional class of object, wherein the companion is unseen, is called an *astrometric binary*. It is briefly treated below. Before going on to consider the discovery and analysis of these types of stellar systems, we summarize our knowledge of stellar masses as deduced from the analysis of binary systems.

Multiple systems in general, and binary systems in particular, are the only means we have of directly ascertaining the masses of extraterrestrial objects. Even this is not *direct*, but through the intermediary of celestial mechanics. For visual binaries the semi-major axis is known only in arc seconds. Hence, without a determination of the parallax, the masses of the binary remain unknown. For the single-lined spectroscopic binaries only the mass function $M_2^3 \sin^3 i / (M_1 + M_2)^2$ can be determined. In the double-lined spectroscopic binaries one can deduce $M_1 \sin^3 i$ and $M_2 \sin^3 i$ separately. Neither the semi-major axis nor the masses can be computed for an eclipsing binary system. Thus, only if we are lucky enough to find a binary exhibiting more than one of these indicators of multiplicity can the entire problem be solved. Table 14, adapted from A. H. Batten, *Binary and Multiple Systems of Stars*, © 1973, Pergamon Press Ltd. (with permission), illuminates the situation nicely.

Astronomers are not content to leave things in so poor a state of affairs. They have used statistics and large samples in order to try to tease out as

Table 14. Summary of Information Obtainable from Binary Systems

Element		Visual Binary	Spectroscopic Binary		Eclipsing Binary
			One Spectrum	Two Spectra	
P		Yes	Yes	Yes	Yes
A		Apparent a''	$a_1 \sin i$	$a \sin i$	No
e		Yes	Yes	Yes	Yes
ω		Yes	Yes	Yes	Yes
T		Yes	Yes	Yes	Yes
i		Yes	No	No	No
Ω		Yes[a]	No	No	No
M_1		If parallax	$f(M)$	$M_1 \sin^3 i$	No
M_2		Known		$M_2 \sin^3 i$	No
Radii	R_1	No	Can be estimated from knowledge of spectrum and luminosity		$r_1 = (R_1/a)$ r_2
Fractional luminosity	L_1 L_2	Yes Yes	Can be estimated from knowledge of spectrum		Yes Yes
Spectral types		Yes	Yes	Yes	If several colors available
Limb darkening	u_1 u_2	No No	No No	No No	In principle In principle
Ellipticity		No	No	No	In principle

[a] Ambiguous without radial-velocity observations.

much additional information as possible. For example, unless there is some galactic-scale correlation unknown to us, the inclinations of binary systems' orbital planes should be randomly distributed with respect to the line of sight. Therefore, the probability of an inclination between i and $i + di$ is $\sin i \, di$. From this distribution the average value of $\sin^3 i$ can be obtained, $\langle \sin^3 i \rangle = 3\pi/16$. Using this one can disentangle a from $a \sin i$.

A second astrophysical correlate is between the *spectral type* of a star and its luminosity. The spectral type is a surface-temperature-based descriptor defined in terms of the strength of certain spectral lines. Furthermore, for most *main-sequence* stars there is a correlation between mass and luminosity known as the *mass–luminosity relationship*. A main sequence star is an ordinary star burning hydrogen into helium in a quasi-stable mode. This phase of a star's life lasts tens of millions to tens of billions of years (and is inversely correlated with its mass). Hot, bright, massive, early-spectral-type stars live a short life whereas cool, dim, less massive, late-spectral-type stars live a long one. The existence of these correlations mean that from the spectral type one can infer a mass. If this is done for a visual binary system, then the semi-major axis so derived is by the use of *dynamical parallax*.

Astrometric Binaries

An *astrometric binary* is a stellar system whose photographic image is that of a single star but whose orbital motion is reflected in systematic residuals from Eqs. 13.1. In some cases the system may be resolved visually and the analysis for resolved astrometric binaries and unresolved astrometric binaries is slightly different. Moreover, if the stars are of comparable luminosity the weighted center of light intensity (or photocenter) will trace out the consequences of the orbital motion.

Consider first resolved astrometric binaries. Let the brighter component (the primary) be designated as 1, the fainter component (the secondary) be designated as 2. If $\mu = M_2/(M_1 + M_2)$, then the new forms due to orbital motion of Eqs. 13.1 (for the primary) can be written as

$$\xi = c_x + \mu_x t + \pi_r P_\alpha - \mu \, \Delta\xi, \qquad \eta = c_y + \mu_y t + \pi_r P_\delta - \mu \, \Delta\eta$$

These are for the primary; for the secondary the latter factors would be $(1 - \mu) \, \Delta\xi$ and $(1 - \mu) \, \Delta\eta$. One can now introduce Thiele–Innes constants as usual.

Another approach is to express the semi-major axis of the relative orbit (a) in terms of the semi-major axis of the primary about the center of mass (a_1),

$$a_1 = \mu a$$

Then, using the Thiele–Innes representation of the relative orbit, the orbital displacements in x and y are given by

$$-\mu \, \Delta x = -\frac{a_1}{a}(Bx + Gy)$$

$$-\mu \, \Delta y = -\frac{a_1}{a}(Ax + Fy)$$

or

$$-\mu \, \Delta x = a_1 Q_\alpha, \qquad -\mu \, \Delta y = a_1 Q_\delta$$

The Q's are called *orbital factors* and one writes

$$\xi = c_x + \mu_x t + \pi_r P_\alpha + a_1 Q_\alpha, \qquad \eta = c_y + \mu_y t + \pi_r P_\delta + a_1 Q_\delta$$

For unresolved astrometric binaries the fractional distance of the primary to the photocenter is given by

$$\beta = \frac{L_2}{L_1 + L_2}$$

where L_1 and L_2 are the luminosities of the primary and secondary. The fractional separation of the photocenter to the barycenter is $\mu - \beta$ and the semi-major axis of the photocentric orbit is equal to $(\mu - \beta)a$. Another way of looking at these motions is to remember that, referred to the center of mass of the binary system, the orbits of the primary, the secondary, and the photocenter are all similar to the relative orbit and proportional to μ, $1 - \mu$, and $\mu - \beta$. Should $(\mu - \beta)a < 0$, then the photocenter and the secondary are on the same side of the barycenter; otherwise they are on opposite sides. Should the less massive star be the more luminous, then $\beta - \mu > \mu$ and the photocentric orbit will be larger than the primary's. This situation is abnormal, however. Finally, the terms in ξ and η due to photocenter displacement are

$$-(\mu - \beta)\,\Delta x, \qquad -(\mu - \beta)\,\Delta y$$

CLOSE BINARY SYSTEMS

A definition of a *close binary* system or an interacting binary system is a two-component stellar system so close together in space that the evolutionary history of each of the components departs appreciably from the ordinary development of a comparable single star. Before the central unifying dynamical idea was invented, each close binary system (almost) was designated as a uniquely peculiar object. As this tendency was compounded, elucidation of the basic phenomena was further confounded because many of these objects were not even recognized as binaries! So, one learning about this field is immediately overwhelmed by names or phrases: SS 433, cataclysmic variables, HZ Hercules, contact binaries, HR 1099, binary pulsar, Cygnus X-1, Roche lobe, U Germinorum variables, symbiotic stars, RS Canum Venaticorum systems, dwarf novae, UW Canis Majoris, apsidal motion, PSR 1913 + 16, x-ray binary, Hercules X-1, semi-detached systems, W Ursa Majoris, starspots and hotspots, AM Hercules, recurrent novae, interacting, close, wide binaries, and mass exchange. I can't lead you through the details of this astrophysical thicket, but I can briefly describe the dynamics and define the principal types of close binary systems.

The Roche Model

In the discussion of the three-body problem (Chapter Six) five equilibrium solutions (in the rotating reference frame) were given for a particle of infinitesimal mass. These five points are collectively referred to as the Lagrangian points. Those along the line joining the two components were unstable. Now consider a binary system with components of unequal mass. The (originally) more massive component evolves faster. Once hydrogen is expended by conversion into helium, the star has lost its principal energy source. It can no longer maintain hydrodynamic equilibrium and it expands (considerably). Stars that complete this expansion process are known as red giants for they become cool, huge, reddish stars. However, our hypothetical star is a member of a binary system. Once it expands far enough, it will fill the *Roche surface* or extend until its *Roche lobe* is reached—which is exactly when its atmospheric radius crosses the unstable Lagrangian point between it and the secondary. This part of its atmosphere is no longer bound. Hence, the material spilling through this point becomes detached from the primary and moves within the confines of the total potential of the restricted three-body system. It may form a disc in the orbital plane and accrete even more material. It may stream onto the surface of the secondary thereby creating a hotspot, x-ray or γ-ray radiation, or surface nucleosynthesis (with corresponding elemental abundance anomalies in the secondary's spectrum). It may eventually transfer so much mass, energy, and angular momentum that the secondary becomes the more massive component of the system and the geometrical character of the orbit is completely altered.

The *conservative model* for close binary systems involves a number of simplifying assumptions. It includes the assumptions that the stars are spherically symmetrical and that each is in hydrostatic equilibrium. The first is reasonable, even in the presence of rotational and tidal distortions, because most stars are highly centrally condensed. The second is a necessary simplification to deal with an immensely complicated hydrodynamical problem. It is further assumed that the orbit is circular with synchronous rotation. This results from tidal friction and typically occurs over a longer time scale. Fortunately nuclear burning time scales are even longer ($\sim 10^9$ years) while the mass transfer itself occurs on a gravitational contraction time scale ($\sim 10^6$ years; also known as the Kelvin–Helmholtz time scale). One further assumes that the Roche lobe is a spherically symmetric surface centered on the more massive component. Finally, the essence of the conservative model is that the total mass, energy, and angular momentum of the *system* are conserved (e.g., two stars plus gas streams).

Suppose that the two stars have masses M_1 and M_2, separation a, period P, and orbital angular momentum L. Then, from Kepler's laws and the definition of L,

$$4\pi^2 a^3 = G(M_1 + M_2)P^2$$

$$L^2 = \frac{G(M_1 M_2)^2 a}{M_1 + M_2}$$

Hence, if L is fixed and $M = M_1 + M_2$ is fixed, then both M_1/M_2 and a may change:

$$a = \frac{ML^2}{G(M_1 M_2)^2} = \frac{\text{const}}{M_1^2(M - M_1)^2}$$

The separation is a minimum when $M_1 = M_2$.

The Algol Paradox

An Algol-type system is a binary system consisting of late-spectral-type main-sequence primary (spectral-type late B or early A*) and an even later spectral-type subgiant secondary (G or K). As mentioned above, a main-sequence star is one undergoing its long-lived, quasi-stable, nuclear conversion of hydrogen to helium. A subgiant is a star that has already passed through this stage. Hence, assuming the coeval formation of the binary, the subgiant must be the more massive component because the rate of stellar evolution is inversely proportional to the initial mass. More than this, Algol-type systems (and in addition to Algol, U Cephei is a good example‡) have distorted light curves and radial velocity curves. The paradox is that, given reasonable hypotheses concerning, for example, masses, the larger subgiant is the less massive component. The resolution, developed by Crawford (1955), Morton (1960), and Smak (1962, 1964), was the realization that mass exchange must have occurred via the Roche lobe mechanism. Furthermore, once mass transfer commences, the radius of the Roche lobe shrinks (because of the changing mass ratio). But, while shrinking, it enters denser layers of the primary's atmosphere thereby speeding up the mass transfer process. When corrections to the orbital elements are included in the process, it becomes further reinforced (the period and semi-major axis decrease). Hence, this positive feedback system continues until the mass ratio is unity. Kippenhahn and Weigert (1967) and Paczynski (1966, 1967a–d) completed the first picture of the evolutionary history of these systems.

Orbital Dynamics

A review of the change in orbital elements in close binary systems due to the exchange of mass between the components was given by Kruszewski (1966). The exercise is straightforward. As usual, we have

$$\mathscr{E} = -\frac{GM}{2a}, \qquad L^2 = GMa(1 - e^2)$$

where $\mathscr{E}$ is the total energy per unit mass, M is the sum of the masses, and L is the magnitude of the total orbital angular momentum per unit mass. Because many of these stars are in nearly circular orbits it is better to use

$$h = e \sin v, \qquad k = e \cos v$$

as variables than it is to use e and ω (see Chapter Nine). If instead of the

---

* The spectral-type sequence, from early to late, is OBAFGKMS. It is also a temperature sequence (O hottest, B cooler, etc.). Furthermore, each letter is subdivided into subclasses O8, O9, B0, B1, and so on. Late B is B5 or higher; early A is A5 or lower.

‡ Names of variable stars depend on the method of discovery. If discovered optically they are named for the constellation they occur in starting with the letter R as in R Coronae Borealis. Next comes S, T, ..., Z, RR, RS, and so on, up until ZZ. Next is AA, AB, ..., QZ (with no J's). This allows for 334 names. The next variable is known as V335, V336, and so on. This procedure is followed unless the star already has a proper name (e.g., Betelgeuse), a name in the Bayer system (α Ori), or in the Flamsteed system (58 Ori). Sources discovered in other wavelength bands are usually named for the catalog in which their existence is announced, with numbers that specify approximate position in right ascension and declination (e.g., PSR 1916+34 is near $\alpha = 19^h 16^m$, $\delta = +34°$) or in galactic coordinates.

true anomaly and eccentricity, we use the radial velocity ($\dot{r}$) and the separation (r), then

$$h = \frac{L\dot{r}}{GM}, \qquad k = \frac{L^2}{GMr} - 1$$

A change in M of δM, in, for example, $\mathscr{E}$ of $\delta\mathscr{E}$ results in changes in a, P, h, and k of

$$\frac{\delta a}{a} = \delta M - \frac{\delta\mathscr{E}}{\mathscr{E}}$$

$$\frac{\delta P}{P} = \frac{\delta M}{M} - \frac{3\delta\mathscr{E}}{2\mathscr{E}}$$

$$\delta h = -\frac{\delta M}{M} + \frac{\delta L}{L} + \frac{L\delta\dot{r}}{GM}$$

$$\frac{\delta k}{1+k} = -\frac{\delta M}{M} + \frac{2\delta L}{L} - \frac{\delta r}{r}$$

Assuming that the infinitesimal parcel of material follows a planar restricted three-body trajectory simplifies the computation. A further reduction of difficulty occurs when the mass loss (gain) is assumed to be isotropic. In the simplest case,

$$\delta e = 0$$

$$\frac{\delta a}{a} = -\frac{\delta M}{M} + \frac{2\delta L}{L}$$

$$\frac{\delta P}{P} = -\frac{2\delta M}{M} + \frac{3\delta L}{L}$$

Principal Types of Close Binary Systems

The assemblage of types of close binary systems will be discussed in alphabetical order. The descriptive phrase may refer to the basic phenomena, the long-standing name, or the definitive example type.

Atmospheric Eclipses.   When one element of the eclipsing binary is a supergiant (with a greatly extended atmosphere), then when the larger star passes in front of the other component (typically an early main-sequence star), one sees light diminution due to an atmospheric eclipse. These are also known as VV Cephei systems. The four best known examples are ε Aurigae, 31 Cygni, 32 Cygni, and VV Cephei.

Cataclysmic (or Explosive) Variables.   This type of binary system is the result of mass exchange via Roche lobe overflow onto a compact companion, usually a white dwarf. A white dwarf is an evolved star, totally devoid of nuclear (or appreciable gravitational) energy sources held up by quantum mechanical effects—namely, electron degeneracy at the Fermi surface. There is a maximum mass for such objects of $\sim 1.44 M_\odot$. Usually, because of the time needed to get to the white dwarf stage, this is not the first episode of mass exchange in the system. Some of these binaries develop unstable accretion discs. When the instability manifests itself there is a very rapid dumping of material onto the compact object. Because the companion is compact, this means that the in-falling material has fallen down a deep potential energy well. Hence, it arrives with considerable kinetic energy. This leads to a (short-lived) thermonuclear runaway on the compact object's surface. Nova Herculis 1936 = DQ Herculis is an example of this. The system is also a $P = 4.7$-hr eclipsing binary.

Within the class of cataclysmic variables there are four subclasses: (1) classical novae (DQ Herculis, GK Persae), (2) dwarf novae (U Geminorum, Z Camelopardalis), (3) novalike variables, and (4) recurrent novae (T Coronae Borealis, T Pyxidis, RS Ophiuchi).

Classical Nova. Stars whose brightness rapidly increases (by a factor of 10^{3-4} in luminosity) due to an explosion. The ejected matter typically moves at speeds of ~ 1000 km/sec. It usually takes years for the system to settle back down.

RS CVn Systems. A cooler subgiant (K0) coupled with a hotter F or G main sequence star with a mass ratio of unity. The periods are ~ 1–15 days, so these systems are semi-detached or detached. The orbital periods of these systems undergo unpredictable changes. This is rare for binary systems whose components are so widely separated. Moreover, outside of the eclipses there is strong emission from ionized calcium (the H and K lines in particular).

Symbiotic Stars. The prototypical symbiotic star is Z Andromedae. These stars' spectra show low-temperature absorption lines and emission features associated with high-excitation environments. Both light variation and spectral features show quasi-periodic variations. They appear to be binaries consisting of a late-type (M0) giant star and an early type (B9) subdwarf with orbital periods measured in months.

Wolf–Rayet Stars. Very early spectral type, very hot, very luminous, and very massive stars in a binary system (named after the two French astronomers who discovered the prototypes in 1867). Their high luminosities exert an appreciable radiation pressure on the outermost (and least strongly bound) parts of their atmospheres. This leads to measurable mass loss ($\sim 10^{-6} M_\odot$/yr) at considerable relative velocities ($\sim 10^3$ km/sec). The result is broad emission lines and a very extended atmosphere. Examples are λ_2 Velorum, θ Muscae, CV Serpentis, and V444 Cygni.

W Ursae Majoris Stars.   The highest number density of all types of binary stars are W UMa stars. They are short-period (<1 day), middle-spectral-type (F or G), eclipsing variables. The mass ratio is ~2 and their continuously curving light curves imply that their atmospheres are extended. Other abnormalities show in their mass-to-luminosity ratios, mass-to-radius ratios, and in the fact that both components are at the same effective temperature. Lucy (1968a,b) suggested that both stars overflow their Roche lobes and, as a result, share a common convective envelope that is optically thick. This model appears to be qualitatively correct, and a consequence is that there is no entropy difference between the stars' atmospheres.

RELATED ODDS AND ENDS

This last section very briefly treats a variety of odds and ends related to close binary systems and then brings in some new physics—general relativity. Tidal friction, apsidal motion, the Earth–Moon system and synchronous rotation, planetary rings, accretion discs, multiple systems, and the binary pulsar are the final topics. This discussion is divided into systems without extra material, systems with it, and nonclassical systems.

Systems Without Extra Material

Apsidal Motion

From the discussion of realistic approximations to the geopotential (Chapter One), we know that rotation causes departures from spherical symmetry in a fluid body. The resulting equilibrium figure (in the simplest case of uniform rotation about a fixed axis) has an axially symmetric mass distribution. Therefore, the external gravitational field of a rotationally distorted fluid mass will not be spherically symmetric. As most stars rotate, even for single stars the assumption of spherical symmetry is not strictly true. The presence of a neighbor can cause a nonrotating object to depart from spherical symmetry. Such a nearby external mass will raise tides. These tides will not be spherically symmetrical. When two rotating fluid masses in close proximity revolve about their common center of mass, then there are both rotational deformations and tide-generated deformations. If both of these are small effects, then a linear perturbation analysis should suffice. Furthermore, there will be no explicit coupling between the perturbing influences. As seen in Chapters Nine and Ten, departures from spherical symmetry produce secular changes in some of the orbital elements—in particular, the argument of periastron advances. This allows a probe of the interior structure of stars for the rate of increase of ω depends on the oblateness.

The analysis can be found in Kopal (1959, 1978) or in Batten (1973). The secular advance in ω, $\Delta\omega$, is given by

$$\frac{\Delta\omega}{2\pi} = k_1\left[g + (15f + g)\left(\frac{M_2}{M_1}\right)\right]\left(\frac{R_1}{a}\right)^5$$
$$+ k_2\left[g + (15f + g)\left(\frac{M_1}{M_2}\right)\right]\left(\frac{R_2}{a}\right)^5$$

where M_1, M_2 are the two stellar masses (with corresponding radii R_1, R_2), a is the semi-major axis, k_1 and k_2 are constants dependent on the internal structure of the two stars, and f, g are functions of the orbital eccentricity given by

$$f = \left(1 + \frac{3e^2}{2} + \frac{e^4}{8}\right)(1 - e^2)^{-5}, \qquad g = (1 - e^2)^{-2}$$

The ratio of the period of apsidal advance, P_ω, to the orbital period P is given by

$$\frac{P_\omega}{P} = \frac{2\pi}{\Delta\omega}$$

All of this sounds good, but the best efforts of the double-star observers and stellar interiors model builders have not been able to elucidate much about the internal distribution of matter in the binary system stars.

Tidal Friction

The tides raised upon the other member of a binary system would be aligned along the instantaneous relative radius vector *if* the two bodies were perfectly elastic. Stars are not and terrestrial planets obviously have a certain degree of rigidity. Hence, tidal bulges are not instantaneously aligned but, as it takes the deformed material a certain amount of time to react, the tidal bulges will lag. The effects of this are that the body raising the tides now exerts a couple on the asymmetric body. This in turn slows down the rotation rate. Eventually, such dissipative processes result in a synchronized state of rotation–revolution as is the case for the Earth–Moon system. The period of rotation of the Moon about its axis is the same as its period of revolution about the Earth. That is why the Moon always keeps the same face toward us. This state is the lowest energy-stable equilibrium state (see Goldreich and Tremaine 1982). A similar type of phenomenon not only occurs elsewhere in the solar system (i.e., Sun–Mercury, Pluto–Charon, etc.) but occurs in binary star systems too.

In the case of the Earth–Moon system there is a considerable amount of empirical evidence to support this type of dissipative evolution. Until recently the Earth's rotation rate defined our time system; if it were slowing down, then Earth rotation time would lag behind a uniform time system. Of course the Moon is moving and would therefore appear to have a secular acceleration.

If we take the Earth–Moon system to be an isolated one, then its total (orbital plus spin) angular momentum must be conserved. Since neither the mass of the Earth nor that of the Moon is changing, their rotational periods and mutual revolution periods are being altered in a fashion consistent with the conservation of angular momentum—namely the Moon is receding from the Earth. The principal cause of energy dissipation appears to be overcoming friction between the oceans and the sea floors during the tide raising and lowering process. The main observable effect is that past eclipses (which have been recorded for thousands of years) did not occur when they "should have" because the time scale is changing. The Earth's rotation rate is decreasing by $\simeq 1200''/\text{century}^2$. See Szebehely and McKenzie (1977) for a discussion of the stability of the system.

Systems With Extra Material

As is surely known, the solar system is filled with all sorts of additional material besides the Sun, a few (major) planets, and their moons. Planetary rings, minor planets, dust, and assorted other debris abound. Stellar systems other than ours also have rings (called accretion discs) and other bodies (called secondaries in triple, quadruple, etc., systems). The basic generalities of these additional objects appear to be understood, but a detailed understanding is lacking. Accretion discs have recently been reviewed by Pringle (1981) and Verbunt (1982), whereas planetary rings were covered by Goldreich and Tremaine (1982). Hence, the discussion of these two subjects will be very limited. Multiple ($N > 2$) stellar systems are usually hierarchical in structure—a close binary with a distant companion or two close binaries separated by a much larger distance (forming a wide quartet). Therefore, these objects are treated as if they were disturbed by a third body, and the results already given in Chapters Nine and Ten will suffice.

Planetary Rings

The discovery of the rings of Uranus sparked modern observers and theoreticians to try to delineate the features of these objects and to attempt to understand their evolution. Spacecraft flybys of Jupiter (and the discovery of its ring) and Saturn further fueled these efforts. Saturn's rings surely exhibit unusual, bizarre, and fascinating structure which was totally unexpected. One can see, especially in the Uranian system, the interplay between theory and experiment in a series of papers published in *The Astronomical Journal* (Elliot et al. 1978; Millis and Wasserman 1978; Nicholson et al. 1978; Elliot et al. 1981a,b; Nicholson, Matthews, and Goldreich 1982). The spokes in Saturn's rings appear to be a magnetohydrodynamical effect (Porco and Danielson 1982). It is not clear why Greenberg (1981) had to publish a paper explaining the results of elementary perturbation theory (see the problems).

Saturn's rings, except for Cassini's division, appear in ground-based telescopes to be smooth, uniform discs of material. The fact that they are discontinuous in the extreme, being composed of thousands of individual ringlets, requires an evolutionary explanation and a stability mechanism. Uranus has ~10 ringlets, but they are widely spread apart. Some of them are even elliptical or inclined to the Uranian equator. Jupiter has a single faint ring. Neptune's rings have been searched for, but none have been found (so far). The Poynting-Robertson effect (drag due to radiation), plasma drag, electromagnetic interactions, synchronous satellites, resonances, density waves, and so on, all have been invoked to explain some part of the multiplicity of observed features. Planetary rings are about as well understood as are the Kirkwood gaps.

Accretion Discs

As outlined in the Roche Model section of this chapter, once hydrogen burning in a star's core ceases (for lack of fuel—the star still has plenty of primordial hydrogen but it is not in the hot, dense, core where nuclear reactions can occur), the radiative and thermal pressures holding the star up against its self-gravity diminish. This causes a contraction of the core (which will eventually become dense and hot enough to burn the transmutated hydrogen—now in the form of helium—into carbon) and an expansion of the outer parts. Once, when a member of a binary system, the outer radius exceeds the distance to the in-between, unstable, collinear Lagrangian point, the material is free of the parent star. It can now wander about the binary as a third body of infinitesimal mass. Due to radiative losses, viscous effects, and other damping mechanisms this material can condense upon the secondary, especially if it is a compact object (=white dwarf, neutron star, or black hole). If it does so in the form of a disc of material (conservation of angular momentum requires discs rather than spherical shells), this material will become heated as more matter accretes. This collisional kinetic energy will be radiated away. Finally, if the disc is inherently unstable, then periodically the disc will dump material directly onto the surface of the compact object. After falling down this deep potential energy well, there will be an energetic radiation of the converted kinetic energy and possibly thermonuclear reactions on the surface of the compact object.

Objects such as SMC X-1, Vela XR-1, Centaurus X-3, Cygnus X-1, X-2, and X-3, Scorpius X-1, and Hercules X-1 are all x-ray binaries. Their periods are short (~1 day), the spectral type of the visible star is that of an early giant, and they exhibit other unusual features such as pulsed emission, x-ray eclipses, or radio emission. Their luminosities are $\sim 10^{37}$ erg/sec $(=10^4 L_\odot)$ and their temperatures are $\sim 10^8$ K. The luminosity (L) is related to the compact object's mass (M) and radius (R) by

$$L \sim \frac{GM\dot{m}}{R} \sim L_\odot \left(\frac{M}{M_\odot}\right)\left(\frac{R_\odot}{R}\right)(\dot{m}/10^{-8} M_\odot/\text{yr})$$

where $\dot{m}$ is the rate at which the primary is losing mass. The resulting temperature (T) is given by

$$T \sim \frac{GMH}{kR} \sim 10^7 \left(\frac{M}{M_\odot}\right)\left(\frac{R_\odot}{R}\right) \text{ K}$$

where H is the hydrogen atom mass and k is Boltzmann's constant. See Blumenthal and Tucker (1974) and Shu and Lubow (1981). Both the temperature requirements and especially the coherence in the pulsed x-ray emission demand that the secondary be collapsed (e.g., size < speed of light × coherence time; coherence times are ~1 sec). The dynamics of accretion and the state and structure of the accretion disc are really hydrodynamical in nature.

General Relativity

The first equation in this book expresses the global invariance of classical mechanics (in the nonrelativistic limit). It is nonrelativistic because the speeds of all bodies are small compared with the speed of light (c) and all gravitational fields are weak (i.e., $GM/Rc^2 \ll 1$, where M is the total mass contained in a sphere of radius R). It is global because the parameters of the transformation law, $\mathbf{r}_0$, $\mathbf{v}_0$, the rotation matrix R, and t_0, are constants over all space and time. We know that when speeds approach c, this transformation fails and another, the Lorentz transformation, must be used. It too is global. Einstein invented a theory of gravitation that is locally invariant. The parameters of the coordinate transformation are not constants over all space and time but vary with their space-time location. By extending the concept of invariance of the laws of physics to include local invariance, Einstein showed how to geometrize the gravitational field. That is, he explicitly showed how one could keep the apple that hit Newton on the head moving along a geodesic at a constant speed. In order to do so, he gave up flat space-time and introduced the concept of curved space-time. This generalization is known as general relativity.

Research in general relativity lay dormant for decades. It has undergone a resurgence because of improved experimental techniques, the development of geometrifications of space-time different than Einstein's, and the discovery of general relativistic laboratories among the stars. The desire to unify the four basic forces of nature and to invent a quantum theory of gravity has also played a role. What can one say in a few pages? Not much, since general relativity requires its own mathematical superstructure (tensor calculus). We shall show where the effects of general relativity make measurable predictions in celestial mechanics and shall indicate the size of these relativistic perturbations.

Classical Tests

Three classical methods exist for testing general relativity. As with all such tests, they rely on the fact that general relativity and classical mechanics make different quantitative predictions in the occurrence of certain phenomena. One prediction is concerned with a secular increase in the argument of periastron of an orbit. In the weak field limit $(GM/Rc^2 \ll 1)$, the general relativistic equations for the space–time path of a geodesic can be recast into an equation of motion for a particle's trajectory. It has the form

$$\ddot{r} = -\frac{GM}{r^2} + \frac{K}{r^3}$$

The solution (Problem 17, Chapter Two) is a precessing ellipse, whence the argument of periapse advances. The result is, when the proper value of K is used,

$$\Delta\omega = \frac{6\pi GM}{ac^2(1 - e^2)}$$

This is about 43″/cent for Mercury's orbit. This is also the unexplained residual in Mercury's orbit deduced from hundreds of years of observation.

There is another observable solar system effect. Once we accept that light rays travel on the geodesics of a curved space–time, where the source of the curvature is due to the presence of massive bodies, it follows that a photon passing close by any massive object will be deflected from a straight-line path. We need an infinitely distant source, such as a star and a massive body—say the Sun. The one set of circumstances that allows the observations of stars that are very close to the solar disc are those of a total solar eclipse. The first one, after Einstein published his work, was in 1919. For a star θ radians away from the solar center the deflection is predicted to be $1″.75(1 + \cos\theta)R_\odot/(2D)$, where D is the light ray's distance of closest approach. This result was verified in 1919 and is known to be good to better than 1% today (using radio astronomy techniques).

The last classical test is that of the gravitational red shift. The prediction is that a light ray emerging from a deep gravitational field will be shifted to the red. The amount is $\sim 2\Phi/c^2$, where Φ is the Newtonian potential. Although the effect can be observed on the Earth (using the Mössbauer effect), it was first detected in neutron stars and white dwarfs. These collapsed, compact objects have large values of GM/Rc^2. In fact, white dwarf theory, precision astrometry, and general relativity have all been combined to do these tests.

Other Tests

We will slough over the difference between tests of the principle of equivalence, metrical theory tests, and tests of general relativity per se. A test of the former is the equality of gravitational and inertial mass referred to in Chapter One. This has been demonstrated to 1 part in 10^{12} for laboratory-sized objects, and to a 1 part in 10^2 for planet-sized objects. The physics of the gravitational red shift effect can be reformulated into a time delay effect for clocks in different gravitational fields. Clocks both on spacecraft and on aircraft flown around the world have been utilized (with results ranging from 1 part in 10^2 to 1 part in 10^4).

One of the most precise tests of general relativity was proposed by Shapiro in 1964. He realized that the transit time of a light ray between two fixed points would be increased if a massive body was interposed. The cause is the bending of the geodesic into a path of longer arc length. The amount is (for the Sun)

$$\Delta t = \left(\frac{4GM_\odot}{c^3}\right)\ln\left(\frac{r_\oplus + r_p + r}{r_\oplus + r_p - r}\right)$$

where $r_\oplus$ and r_p are the heliocentric distances of the Earth and another planet (the two "fixed points") and r is their separation. The maximum value of Δt is $\sim 250\ \mu\text{sec}$. This test is extremely sensitive when using spacecraft as the other "planet," especially if it has transmitters at two frequencies (to correct for interplanetary plasma effects).

Finally, before moving on to the binary pulsar, there is the Lense–Thirring effect. Its solar system manifestation is in the secular advance of the longitude of the ascending node at a rate equal to

$$\frac{2GI\omega}{a^3 c^2 (1 - e^2)^{3/2}}$$

Herein I is the moment of inertia of the central body and ω is its angular speed of rotation.

The Binary Pulsar

A pulsar is a star that is extremely condensed, spinning, possessing a magnetic field, and surrounded by a relativistic plasma. The star got into this dense ($\sim 10^{14-15}\ \text{gm/cm}^3$), small ($\sim 1\ \text{km}$) state as the result of a supernova explosion. The few solar masses of material at the core of the (former) O or early B star imploded when the instability that triggers the supernovae phenomena could no longer be suppressed. The remnant is a neutron star, so called because the matter density is high enough that all atoms are disrupted and the free electrons are (literally) squeezed into the nuclear protons to produce neutrons. When such an object, with an associated magnetic field, spins in an ionized, high-temperature medium, it produces a beamed pulse of electromagnetic radiation known as synchrotron radiation. If we are in the path of the beam, we will see it once per pulsar revolution period. If this is occurring optically, then we would see a pulsed, periodic flashing—and for a few pulsars this is exactly what we observe. It turns out that it is easier to see the pulsing at radio frequencies. So, even though the technology for optical detection existed in the 1890s (as well as some now well-known pulsars, especially the Crab nebula), these objects are discovered as a result of radio observations.

A binary pulsar is a pulsar occurring as one component of a binary system. Given the violence of the supernovae event, the survival of a binary through such a process must be considered rare. At least three such systems are now known: PSR 1913 + 16, PSR 0820 + 02, and PSR 0655 + 64. The first system in the list was discovered in 1974 by Hulse and Taylor (1975). The binary system has a 0.323-day period (=7.75 hr) and an eccentricity of 0.617. The mass function is $0.13 M_\odot$ with $i \simeq 47°$. Here is a fully relativistic system, with a clock (the pulsar rotation period) good to 1 part in 10^{10}, to play with. The argument of periastron shift is 4° per year, not 43″ per century. That's what the excitement is about!

We will not discuss orbital analysis here—basically, the observed variations in the pulse period are interpreted as being caused by binary motion. The situation is analogous to that of a spectroscopic binary, and a total of five parameters of the orbit may be fixed (the other two pulsars are less well studied, less massive, and in larger, nearly circular orbits—see Taylor 1981).

From the advance of periastron one can derive the combined mass of the system, $2.8 M_\odot$. The gravitational red shift and the (special relativistic) transverse Doppler shift yield a different function of the component's masses than their sum, whence, after measuring these effects, one deduces the masses of the two components (pulsar = $1.43 M_\odot$, companion = $1.40 M_\odot$). From the Keplerian orbital analysis one knows the mass function, so the inclination is known too.

One of the unique phenomena predicted by the theory of general relativity is that of gravitational radiation. Like electromagnetic radiation, gravitational waves propagate along the geodesics of space–time at the speed of light. The first-order secular rate of decrease in a binary system's energy due to such radiative damping is given by

$$\frac{\langle \Delta \mathscr{E} \rangle}{P} = \Delta \mathscr{E})_{\text{sec}} = -\frac{32 G^4 M_1^2 M_2^2 (M_1 + M_2)}{5 a^5 c^5 (1 - e^2)^{7/2}} \left(1 + \frac{73}{24} e^2 + \frac{37}{96} e^4 \right)$$

The corresponding $\mathscr{E} = -\mu/2a$ decrease in revolution period is

$$\dot{P} = -\frac{12 \pi^2 c^4}{G^2 M_1 M_2 M P} \Delta \mathscr{E})_{\text{sec}}$$

or -2.4×10^{-12} sec/sec for PSR 1913+16. The observed value is -2.1×10^{-12} sec/sec. A general relativistic binary has the potential for other tests of the theory, but they are too small to be observed. So it is not believed that exotic systems are needed to test relativity for macroscopic bodies, special relativistic effects have already been measured in Earth-bound artificial satellites.

This last, all-to-brief section is included as a tease. Celestial mechanics is not an old staid, closed field of study. Among the formula manipulators on the newest computers, the dozens of artificial satellites traversing the solar system, the asteroid belt around Vega, and the general relativistic laboratories of black holes, pulsars, and x-ray bursters in globular clusters, the Universe is just as wonderful, strange, and beautiful as it was to Newcomb or Laplace or Galileo or Ptolemy, or even a smart *Australopethicene africanus*. Look up and be entranced.

PROBLEMS

1. Consider the following two different methods of calculating the perturbing effects of a close binary pair on a distant third body. First, let the equal mass components of the binary pair revolve about their common center of mass in circular orbits of radius a. Then compute the time-averaged value of the force they produce (only do the computation in their orbital plane). Show that the result is

$$\langle \mathbf{F}(\mathbf{r}) \rangle = -\frac{2Gm}{\pi r^2 (r^2 - a^2)}[(r+a)E(k) + (r-a)K(k)]\mathbf{r}$$

where $\mathbf{r}$ is the vector from the center of mass of the binary to the third body and $k^2 = 4ar/(a+r)^2$. Second, show that ($m = $ a binary component's mass)

$$\langle U(\mathbf{r}) \rangle = -\frac{4Gm}{\pi(r+a)}K(k)$$

and since $\langle \mathbf{F} \rangle$ is invariant to $a \to -a$, $-\nabla \langle U \rangle = \langle \mathbf{F} \rangle$. Can you think of other equivalent formulations? (Hint: regard the binary as two hoops of matter of appropriate density.)

2. Show that if a spectroscopic binary is really a triple system, then there is another term in the expression for z,

$$z = z_{\text{usual}} + v_r t$$

where v_r is the radial velocity of the center of mass of the triple system. This term is called the light equation (correction). Explain how this would affect the mean anomaly and specifically justify the result

$$M = M_{\text{usual}} + \frac{v_r t}{c}$$

Finally, deduce the real-to-apparent period ratio

$$P_{\text{true}} = P_{\text{app}}\left(\frac{1 - v_r}{c}\right)$$

3. Prove that a geometrical interpretation of the Thiele–Innes constants is that A and B are the coordinates of the projected periastron, referred to the center of the apparent orbit as the origin, while F and G are the coordinates of the points where the projection of the minor axis intersects the auxiliary ellipse in the geometrical interpretation of Kepler's equation (again refined to the center of the apparent orbit). Thus, deduce that the direction cosines of the true major and minor axes are

$$(A, B, C)/a \qquad \text{and} \qquad (F, G, H)/a$$

4. Deduce

$$A \pm G = 2a \cos(\omega \pm \Omega)^{\cos^2}_{\sin^2}(i/2)$$

$$\pm B - F = 2a \sin(\omega \pm \Omega)^{\cos^2}_{\sin^2}(i/2)$$

and the inverse relationships

$$\tan(\omega \pm \Omega) = a \pm \frac{B - F}{A \pm G}$$

$$\tan^2(i/2) = \frac{A - G}{A + G} \frac{\cos(\omega + \Omega)}{\cos(\omega - \Omega)} = -\frac{B + F}{B - F} \frac{\sin(\omega + \Omega)}{\sin(\omega - \Omega)}$$

5. Let a binary's distance d, parallax π, semi-major axis a, and radial velocity v_r be denoted by a subscript zero at epoch t_0. Show that under the joint assumption of uniform rectilinear motion through space for the center of mass of the binary for the solar system barycenter that their values at some other time t are approximately given by

$$d = d_0 + \frac{v_r}{4.737}(t - t_0), \qquad p = \left(\frac{d_0}{d}\right)p_0, \qquad a = \left(\frac{d_0}{d}\right)a_0$$

Further show that if the proper motion of the binary's center of mass is μ with a position angle P, then the changes in i, ω, and Ω are approximately given by ($\Delta t = t - t_0$, etc.)

$$\Delta i = \mu_0 \, \Delta t \, \sin(P_0 - \Omega_0)$$

$$\Delta \omega = \mu_0 \, \Delta t \, \csc i_0 \cos(P_0 - \Omega_0)$$

$$\Delta \Omega = \mu_0 \, \Delta t [\sin P_0 \tan \delta_0 - \cot i_0 \cos(P_0 - \Omega_0)]$$

where $\delta_0 =$ declination of center of mass at t_0. Finally, demonstrate that the corresponding changes in the Thiele-Innes parameters are

$$\Delta A = \left(\frac{A_0}{a_0}\right) \Delta a + F_0 \, \Delta \omega - B_0 \, \Delta \Omega + C_0 \, \Delta i \sin \Omega_0$$

$$\Delta B = \left(\frac{B_0}{a_0}\right) \Delta a + G_0 \, \Delta \omega + A_0 \, \Delta \Omega - C_0 \, \Delta i \cos \Omega_0$$

$$\Delta C = \left(\frac{C_0}{a_0}\right) \Delta a + H_0 \, \Delta \omega + C_0 \, \Delta i \cot i_0$$

$$\Delta F = \left(\frac{F_0}{a_0}\right) \Delta a - A_0 \, \Delta \omega - G_0 \, \Delta \Omega + H_0 \, \Delta i \sin \Omega_0$$

$$\Delta G = \left(\frac{G_0}{a_0}\right) \Delta a - B_0 \, \Delta \omega + F_0 \, \Delta \Omega - H_0 \, \Delta i \cos \Omega_0$$

$$\Delta H = \left(\frac{H_0}{a_0}\right) \Delta a - C_0 \, \Delta \omega + H_0 \, \Delta i \cot i_0$$

6. Show that the first-order secular rate of change of $\pi = \omega + \Omega$ due to J_2 and J_4, for an elliptical ring of material of radius a, eccentricity e, is given by

$$\dot{\pi} = \left(\frac{GM}{a^3}\right)^{1/2} \left[\frac{3J_2}{2a^2(1-e^2)^2} - \frac{15J_4}{4a^4(1-e^2)^4}\right]$$

where M is the mass of the primary.

Appendix

IAU (1976) System of Astronomical Constants

The astronomical unit of time is 1 day (86,400 sec). An interval of 36,525 days is 1 Julian century. The astronomical unit of mass is the mass of the Sun ($M_\odot$). The astronomical unit of length (or unit distance $= A$) is that length for which the Gaussian gravitational constant (k) takes the value 0.01720209895 when the units of measurement are the astronomical units of length, mass, and time. The dimensions of k^2 are those of the constant of gravitation (G).

Gaussian gravitational constant	$k = 0.01720209895$
Speed of light	$c = 299{,}792{,}458 \text{ m/sec}$
Light time for unit distance	$\tau_A = 499.004782 \text{ sec}$
Equatorial radius for the Earth	$a_e = 6{,}378{,}140 \text{ m}$
Flattening factor for the Earth	$f = 1/298.257$
Geocentric gravitational constant	$GM_\oplus = 3.986005 \times 10^{14} \text{ m}^3/\text{sec}^2$
Constant of gravitation	$G = 6.672 \times 10^{-11} \text{ m}^3/\text{kg} \cdot \text{sec}^2$
Ratio of mass of Moon to that of Earth	$\mu = 0.01230002$
General precession in longitude, per Julian century, at J2000.0	$p = 5029\rlap{.}''0966$
Obliquity of the ecliptic at J2000.0	$\varepsilon = 23°26'21\rlap{.}''448$
Constant of nutation at J2000.0	$N = 9\rlap{.}''2025$
Unit distance	$A = c\tau_A = 1.49597870 \times 10^{11} \text{ m}$
Solar parallax	$\pi_\odot = \sin^{-1}(a_e/A) = 8\rlap{.}''794148$
Constant of aberration at J2000.0	$\kappa = 20\rlap{.}''49552$
Heliocentric gravitational constant	$GM_\odot = 1.32712438 \times 10^{20} \text{ m}^3/\text{sec}^2$
Mass of the Sun	$M_\odot = 1.9891 \times 10^{30} \text{ kg}$

Inverse ratios of the masses of the planets to the Sun's:

Mercury	6,023,600	Jupiter	1,047.355
Venus	408,523.5	Saturn	3,498.5
Earth	332,946.0	Uranus	22,869
Earth + Moon	328,900.5	Neptune	19,314
Mars	3,098,710	Pluto	3,000,000

Gravity field of Earth:

$$J_2 = +0.00108263, \qquad J_3 = -0.254 \times 10^{-5}, \qquad J_4 = -0.161 \times 10^{-5}$$

The new standard epoch (designated J2000.0) will be 2000 January $1\overset{d}{.}5$, which is JD2451545.0, and the new standard equinox will correspond to this instant.

References

Aarseth, S. J., M. Hénon, and R. Wielen, *Astron. Astrophys.* **37**, 183 (1974).

Aarseth, S. J., and M. Lecar, *Ann. Rev. Astron. Astrophys.* **13**, 1 (1975).

Aitken, R. G., *The Binary Stars*, McGraw-Hill, New York, 1935 (reprinted by Dover, New York, 1964).

Aksnes, K., *Astrophys. Norvegica* **10**, 69 (1965); *Astron. J.* **75**, 1066 (1970).

Allen, C., and A. Poveda, *Astrophys. Sp. Sci.* **13**, 350 (1971).

Angeletti, L., and P. Giannone, *Mem. Soc. Astron. Italiana* **47**, 245 (1976); *Astrophys. Sp. Sci.* **46**, 205 (1977a); *Astrophys. Sp. Sci.* **50**, 311 (1977b); *Astron. Astrophys.* **85**, 113 (1980).

Anthony, M. L., and G. F. Fosdick, *J. Aero. Sci.* **28**, 789 (1961).

Anthony, M. L., and L. M. Perko, *ARS J.* **31**, 1413 (1961).

Arnold, V. I., *Mathematical Methods of Classical Mechanics*, Springer-Verlag, New York, 1978.

Arp, H., *Astrophys. J. Suppl.* **14**, 1 (1966).

Bachman, C. G., *Laser Radar Systems and Techniques*, ARTECH House, Dedham, MA, 1979.

Ball, J. A., M.I.T. Lincoln Laboratory Technical Note 1969-42, 1969.

Barr, J. M., *J. Roy. Astr. Soc. Can.* **2**, 70 (1908).

Bate, R. A., D. D. Mueller, and J. E. White, *Fundamentals of Astrodynamics*, Dover, New York, 1971.

Batten, A. H., *Binary and Multiple Systems of Stars*, Pergamon, Oxford, 1973.

Battin, R. H., *Astronautical Guidance*, McGraw-Hill, New York, 1964.

Bean, B. R., and E. J. Dutton, "Radio Meteorology," Natl. Bur. Stand. Monograph No. 92, U.S. Government Printing Office, Washington, D.C., 1966.

Bertrand, J., *Comptes Rendus* **77**, 849 (1873).

Bettis, D. G., and V. Szebehely, *Astrophys. Sp. Sci.* **14**, 133 (1971).

Bisnovaty-Kogan, G. S., I. V. Estulin, N. G. Havenson, V. G. Kurt, G. A. Mersov, and I. D. Novikov, *Astrophys. Sp. Sci.* **75**, 219 (1981).

Black, W., in B. D. Tapley and V. Szebehely, Eds., *Recent Advances in Dynamical Astronomy*, D. Reidel Publ., Dordrecht, Holland, 1973, p. 61.

Blitzer, L., *ARS J.* **32**, 1102 (1962).

Blumenthal, G. R., and W. H. Tucker, *Ann. Rev. Astron. Astrophys.* **12**, 23 (1974).

Bogoliubov, N. N., *J. Phys. (USSR)* **10**, 256 (1946a); *J. Phys. (USSR)* **10**, 265 (1946b); *Stud. in Stat. Mech.* **1**, 1 (1962).

Bogoliubov, N. N., and Y. A. Mitropolsky, *Asymptotic Methods In The Theory of Nonlinear Oscillations*, Hindustan Publ. Corp., New Delhi, 1961.

509

Bonnor, W. B., *Mon. Not. R. astr. Soc.* **116**, 351 (1956).

Boss, B., *General Catalogue of 33342 Stars for the Epoch 1950,* Carnegie Institution of Washington Publication No. 468, Washington, D.C., 1936.

Broucke, R., *Astrophys. Sp. Sci.* **72**, 33 (1980).

Broucke, R., and P. Cefola, *Cel. Mech.* **7**, 388 (1973).

Brouwer, D., *Astron. J.* **52**, 223 (1946).

Brouwer, D., and G. M. Clemence, *Methods of Celestial Mechanics,* Academic, New York, 1961.

Brown, E. W., *An Introductory Treatise on the Lunar Theory,* Cambridge University Press, London, 1896 (reprinted by Dover, New York, 1960).

Brown, E. W., and C. A. Shook, *Planetary Theory,* Cambridge University Press, London, 1933 (reprinted by Dover, New York, 1964).

Bruns, H. *Acta. Math.* **11**, 25 (1887).

Butterworth, E. M., and R. H. Miller, *Astrophys. J.* **170**, 275 (1971).

Cajori, F., *Principia,* Vols. 1 and 2, University of California Press, Berkeley, 1934.

Camm, G-L., *Mon. Not. R. astr. Soc.* **110**, 305 (1950).

Chandrasekhar, S., *An Introduction to the Study of Stellar Structure,* University of Chicago Press, Chicago, 1939 (reprinted by Dover, New York, 1967); *Principles of Stellar Dynamics,* University of Chicago Press, Chicago, 1942 (reprinted by Dover, New York, 1960).

Chew, M. C., *Ann. Math. Stat.* **38**, 494 (1967).

Collatz, L., *The Numerical Treatment of Differential Equations,* Springer-Verlag, New York, 1966.

Cook, G. E., *Geophys. J. RAS* **6**, 271 (1962).

Cowling, T. G., *Mon. Not. R. astr. Soc.* **98**, 734 (1938).

Crawford, R., *Astrophys. J.* **121**, 71 (1955).

Dallas, S. S., Jet Propulsion Laboratory Technical Report No. 32-1267 (1970).

Danby, J. M. A., *Fundamentals of Celestial Mechanics,* Macmillan, New York, 1962.

Davis, M. S., in B. D. Tapley and V. Szebehely, Eds., *Recent Advances in Dynamical Astronomy,* D. Reidel Publ., Dordrecht, Holland, 1973, p. 351.

Deutsch, R., *Orbital Dynamics of Space Vehicles,* Prentice-Hall, Englewood Cliffs, NJ, 1963.

Dicke, R. H., *Sci. Amer.* **205**, 84 (1961).

Dyson, F. S., and A. Lenard, *J. Math. Phys.* **8**, 423 (1967).

Ehricke, K. A., *Space Flight,* Vols. I and II, Van Nostrand, Princeton, NJ, 1960, 1962.

Eichhorn, H., *Astron. Astrophys.* **102**, 35 (1981).

Eichhorn, H., and A. Rust, *Astr. Nach.* **292**, 37 (1970).

Elliot, J. L., E. Dunham, L. H. Wasserman, R. L. Millis, and J. Churms, *Astron. J.* **83**, 980 (1978).

Elliot, J. L., J. A. Frogel, J. H. Elias, I. S. Glass, R. G. French, D. J. Mink, and W. Liller, *Astron. J.* **86**, 127 (1981a).

Elliot, J. L., R. G. French, J. A. Frogel, J. H. Elias, D. J. Mink, and W. Liller, *Astron. J.* **86**, 444 (1981b).

Eötvös, R., *Math. Natur. Berichte Ungarn.* **8**, 65 (1891); *Ann. Phys. Chemie* **59**, 354 (1896).

Escobal, P. R., *Methods of Orbit Determination,* Wiley, New York, 1965; *Methods of Astrodynamics,* Wiley, New York, 1968.

Escobal, P. R., and D. A. Affatali, *J. Astronaut. Sci.* **19**, 179 (1967).

Euler, L., *Mém. de Berlin,* 228 (1760); also see *Nov. Comm. Petrop* **10**, 207 (1764) and **11**, 152 (1765).

Field, G. B., and J. Hutchings, *Astrophys. J.* **153**, 737 (1968).

Field, G. B., and W. C. Saslaw, *Astrophys. J.* **142**, 568 (1965).

Fisher, M. E., and D. Ruelle, *J. Math. Phys.* **7**, 260 (1966).

Fricke, W., and A. Kopff, *Veroff. des Astron. Rechen-Inst. Heidelberg* No. 10 (1963).

Froeschlé, C., and H. Scholl, in R. L. Duncombe, Ed., *Dynamics of the Solar System*, D. Reidel Publ., Dordrecht, Holland, 1979, p. 223.

Gaposchkin, E. M., Smithson. Astrophys. Obs. Spec. Rep. No. 353 (1973); *Phil. Trans. R. Soc. Lond.* **284A**, 515 (1973); *J. Geophys. Res.* **79**, 5377 (1974).

Gaposchkin, E. M., and K. Lambeck, Smithson. Astrophys. Obs. Spec. Rep. No. 315 (1970); *J. Geophys. Res.* **76**, 4855 (1971).

Garfinkel, B., *Astron. J.* **63**, 88 (1958); *Astron. J.* **64**, 353 (1959); *Astron. J.* **69**, 223 (1964).

Garfinkel, B., and K. Aksnes, *Astron. J.* **75**, 85 (1970).

Gauss, K. F., *Theoria Motus*, 1809 (trans. C. H. Davis), Little, Brown, Boston, 1857 (reprinted by Dover, New York, 1963).

Gehrels, T., and E. F. Tedesco, *Astron. J.* **84**, 1079 (1979).

Geyling, F. T., and H. R. Westerman, *Introduction to Orbital Mechanics*, Addison-Wesley, Reading, MA, 1971.

Gibbs, W. J., *Mem. Nat'l. Acad. Sci.* **4** (1888).

Gill, J. R., *Astrophys. J.* **93**, 118 (1941).

Goldreich, P., and S. Tremaine, *Ann. Rev. Astron. Astrophys.* **20**, 249 (1982).

Goldstein, H., *Classical Mechanics*, 2nd ed., Addison-Wesley, Reading, MA, 1980.

Goodricke, J., *Phil. Trans. R. Soc. Lond.* **73**, 474 (1783).

Greenberg, R., *Astron. J.* **86**, 912 (1981).

Greenstein, J. L., and T. L. Page, *Astrophys. J.* **93**, 128 (1941).

Hayli, A., *Astrophys. Sp. Sci.* **13**, 309 (1971).

Heggie, D. C., *Mon. Not. R. astr. Soc.* **186**, 155 (1979a); *Mon. Not. R. astr. Soc.* **188**, 525 (1979b).

Hénon, M., *Bull. Astron.* **2**, 91 (1967); *Astrophys. Sp. Sci.* **13**, 284 (1971a); *Astrophys. Sp. Sci.* **14**, 151 (1971b); in L. Martinet and L. Mayor, Eds., *Dynamical Structure and Evolution of Stellar Systems*, Observatory, Geneva, 1973, p. 183.

Henrard, J., in O. Calame, Ed., *High Precision Earth Rotation and Earth-Moon Dynamics*, D. Reidel Publ., Dordrecht, Holland, 1982, p. 227.

Henrici, P., *Discrete Variable Methods In Ordinary Differential Equations*, Wiley, New York, 1962; *Error Propagation for Difference Methods*, Wiley, New York, 1963; *Elements of Numerical Analysis*, Wiley, New York, 1964.

Herget, P., *The Computation of Orbits*, Edwards Bros. Inc., Ann Arbor, MI, 1948; *Astron. J.* **70**, 1 (1965).

Herschel, W., *Phil. Trans. R. Soc. Lond.* **72**, 82 (1782); *Phil. Trans. R. Soc. Lond. for 1802* **477** (1802); *Phil. Trans. R. Soc. Lond. for 1803* **339** (1803).

Hoelker, R. F., and R. Silber, Army Ballistic Missile Agency–Redstone Arsenal Report DA-TM-2-59, 1959; also see *Adv. in Ball. Miss. Sp. Tech.* **3**, 164 (1961).

Hoessel, J. G., *Astrophys. J.* **241**, 293 (1981).

Hohmann, W., *Die Erreichbarkeit der Himmelskörper*, Oldenburg, Munich, 1925 (NASA Tech. Trans. F.44, 1960).

Hori, G.-I., and Y. Kozai, in G. E. O. Giacaglia, Ed., *Satellite Dynamics*, Springer-Verlag, Berlin, 1975, p. 1.

Horwitz, G. and J. Katz, *Astrophys. J.* **211**, 226 (1977); *Astrophys. J.* **222**, 941 (1978a); *Astrophys. J.* **223**, 311 (1978b).

Huang, K., *Statistical Mechanics*, Wiley, New York, 1963.

Hulse, R. A., and J. H. Taylor, *Astrophys. J.* **195**, L51 (1975).

Ince, E. L., *Ordinary Differential Equations*, Longmans, Green and Co. Ltd., London, 1927 (reprinted by Dover, New York, 1956).

Ipser, J. R., *Astrophys. J.* **193**, 463 (1974).

Ipser, J. R., and G. Horwitz, *Astrophys. J.* **232**, 863 (1979).

Jackson, J. D., *Classical Electrodynamics*, 2nd ed., Wiley, New York, 1975.

Jacobi, C. G. J., *Comptes Rendus* **3**, 59 (1836).

Jeffreys, W. H., *Comm. Assoc. Comp. Mach.* **14**, 538 (1971).

Jeffreys, W. H., and V. G. Szebehely, *Comm. Astrophys.* **8**, 9 (1978).

Joy, A. H., *Publ. Astron. Soc. Pac.* **54**, 35 (1942).

Kaplan, G. H., Ed., *U.S.N.O. Circ.* No. 163 (1981).

Katz, J., *Mon. Not. R. astr. Soc.* **190**, 497 (1980).

Katz, J., and G. Horwitz, *Astrophys. J. Suppl.* **33**, 251 (1977).

Katz, J., G. Horwitz, and A. Dekel, *Astrophys. J.* **223**, 299 (1978).

Katz, J., and L. G. Taff, *Astrophys. J.* **264**, 476 (1983).

Kaula, W. M., *Theory of Satellite Geodesy*, Blaisdell, Waltham, MA, 1966.

Kendall, M. G., and A. Stuart, *The Advanced Theory of Statistics*, 3rd ed., Hafner, NY, 1969.

Kent, S. T., and H. R. Betz, *Cel. Mech.* **1**, 91 (1969).

King, I. R., *Astron. J.* **67**, 471 (1962).

King, I. R., and J. E. Hesser, Ed., *Star Clusters*, D. Reidel Publ., Dordrecht, Holland, 1980, p. 139.

Kingston, R. H., *Detection of Optical and Infrared Radiation*, Springer-Verlag, New York, 1978.

Kinoshita, H., *Cel. Mech.* **15**, 501 (1977).

Kippenhahn, R., and A. Weigert, *Zeit. f. Astrophysik* **66**, 58 (1967).

Kirkwood, D., *Meteoric Astronomy*, Lippincott, Philadelphia, 1867, Ch. 13.

Kiselev, A. A., and O. P. Bykov, *Sov. Ast. A. J.* **17**, 816 (1974); *Sov. Ast. A. J.* **20**, 496 (1976).

Klebesadel, R. W., I. B. Strong, and R. A. Olson, *Astrophys. J.* **182**, L85 (1973).

Kopal, Z., *Close Binary Systems*, Wiley, New York, 1959; *Dynamics of Close Binary Systems*, D. Reidel Publ., Dordrecht, Holland, 1978.

Kowal, C. J., W. Liller, and B. G. Marsden, in R. L. Duncombe, Ed., *Dynamics of the Solar System*, D. Reidel Publ., Dordrecht, Holland, 1979, p. 245.

Kozai, Y., *Astron. J.* **64**, 367 (1959).

Kraft, R. P., *Astrophys. J.* **127**, 625 (1958).

Kruszewski, A., *Adv. Astron. Astrophys.* **4**, 233 (1966).

Kryloff, N. M., and N. N. Bogoliubov, *Introduction to Non-linear Mechanics*, Publ. Acad. Sci. Ukr. S.S.R., 1937.

Kuiper, G. P., *Astrophys. J.* **93**, 133 (1941).

Kuiper, G. P., O. Struve, and B. Strömgren, *Astrophys. J.* **86**, 570 (1937).

Kurth, R., *An Introduction to the Mechanics of the Solar System*, Pergamon, New York, 1959.

Lagrange, J. L., *Collected Works, Vol. VI*, Gauthier-Villars, Paris, 1873, p. 229.

Langebartel, R. G., *Astrophys. Sp. Sci.* **75**, 437 (1981).

Larson, R. B., *Mon. Not. R. astr. Soc.* **147**, 323 (1970a); *Mon. Not. R. astr. Soc.* **150**, 93 (1970b); *Mon. Not. R. astr. Soc.* **166**, 585 (1974).

Lecar, M., *Bull. Astron.* **3**, 91 (1968).

Lecar, M., and J. Katz, *Astrophys. J.* **243**, 983 (1981).

Lehmann-Filhés, R., *Astr. Nach.* **136**, 17 (1894).

Lerch, F. J., S. M. Klosko, R. E. Laubscher, and C. A. Wagner, Goddard Space Flight Center Doc. X-921-77-246, 1977.

Leuschner, A. O., *Univ. Cal. Chronicle* **18**, No. 2 (1916).

Levy-Leblond, J.-M., *J. Math. Phys.* **10**, 806 (1969).

Lighthill, M. J., *Introduction to Fourier Analysis and Generalized Functions*, Cambridge University Press, London, 1958.

Lightman, A. P., *Astrophys. J.* **215**, 914 (1977).

Lightman, A. P., and L. N. daCosta, *Astrophys. J.* **228**, 543 (1979).

Lightman, A. P., and S. M. Fall, *Astrophys. J.* **221**, 567 (1978).

Lin, C. C., and F. H. Shu, *Astrophys. J.* **140**, 646 (1964).

Lindstedt, A., *K. Akad. Wiss. St. Petersburg* **31**, 4 (1882).

Liu, J. J. F., *AIAA J.* **12**, 1511 (1974).

Long, S. A. T., *Cel. Mech.* **12**, 225 (1975).

Lorell, J., J. D. Anderson, and H. Lass, Jet Propulsion Laboratory Technical Report No. 32-482, 1964.

Lorell, J., and A. Liu, Jet Propulsion Laboratory Technical Report No. 32-1513, 1971.

Lucy, L. B., *Astrophys. J.* **151**, 1123 (1968a); *Astrophys. J.* **153**, 877 (1968b).

Lynden-Bell, D., and R. Wood, *Mon. Not. R. astr. Soc.* **138**, 495 (1968).

Lynds, R., and A. Toomre, *Astrophys. J.* **209**, 382 (1976).

Maxwell, A. D., *Astron. J.* **42**, 13 (1932).

McClain, W. D., Computer Sciences Corp. Tech. Rep. TR-77/6010, 1977.

McCuskey, S. W., *Introduction to Celestial Mechanics*, Addison-Wesley, Reading, MA, 1963.

McGlynn, T. A., and J. P. Ostriker, *Astrophys. J.*, **241**, 915 (1980).

Meeks, M. L., Ed., *Methods of Experimental Physics*, Vol. 12B, Academic, New York, 1976.

Meeus, J., *Die Sterne* **45**, 116 (1969).

Michell, J., *Phil. Trans. R. Soc. Lond.* **57**, 234 (1767).

Michie, R. W., *Ann. Rev. Astron. Astrophys.* **2**, 49 (1964).

Miller, R. H., *Astrophys. J.* **140**, 250 (1964); *J. Comp. Phys.* **8**, 449 (1971a); *J. Comp. Phys.* **8**, 464 (1971b); *Astrophys. J.* **165**, 391 (1971c); in M. Lecar, Ed., *Gravitational N-Body Problem*, D. Reidel Publ., Dordrecht, Holland, 1972, p. 124.

Millis, R. L., and L. H. Wasserman, *Astron. J.* **83**, 993 (1978).

Morgan, H. R., *Catalog of 5268 Standard Star, 1950.0 Based on the Normal System N30*, Astron. Pap. Amer. Eph. 13, Pt. III, 1952.

Morse, P. M., and H. Feshbach, *Methods of Theoretical Physics*, Vols. 1 and 2, McGraw-Hill, New York, 1953.

Morton, D. C., *Astrophys. J.* **132**, 146 (1960).

Moser, J. K., *Comm. Pure Appl. Math.* **23**, 609 (1970).

Moulton, F. R., *Celestial Mechanics*, Macmillan, New York, 1902 (reprinted by Dover, New York, 1970); *Astron. J.* **23**, 93 (1903).

Moyer, T. D., *Cel. Mech.* **23**, 33 (1981a); *Cel. Mech.* **23**, 53 (1981b).

Mulholland, J. D., *Publ. Astron. Soc. Pac.* **84**, 357 (1972).

Murdoch, D. C., *Linear Algebra for Undergraduates*, Wiley, New York, 1957.

Nacozy, P. E., *Astron. J.* **81**, 787 (1976); *Cel. Mech.* **16**, 309 (1977).

Neutsch, W., *Astron. Astrophys.* **102**, 59 (1981).

Nicholson, P. D., K. Matthews, and P. Goldreich, *Astron. J.* **87**, 433 (1982).

Nicholson, P. D., S. E. Persson, K. Matthews, P. Goldreich, and G. Neugebauer, *Astron. J.* **83**, 1240 (1978).

Ostriker, J., *Astrophys. J.* **202**, L113 (1975); *Astrophys. J.* **217**, L125 (1977); *Astrophys. J.* **224**, 320 (1978).

Ostriker, J. P., and P. J. E. Peebles, *Astrophys. J.* **186**, 467 (1973).

Ostriker, J. P., L. Spitzer, Jr., and R. A. Chevalier, *Astrophys. J.* **176**, L51 (1972).

Ostriker, J. P., and T. X. Thuan, *Astrophys. J.* **202**, 353 (1975).

Paczynski, B., *Acta Astron.* **16**, 231 (1966); **17**, 1 (1967a); **17**, 193 (1967b); **17**, 287 (1967c); **17**, 355 (1967d).

Pavelle, R., M. Rothstein, and J. Fitch, *Sci. Am.* **245**, 136 (1981).

Peale, S. J., *Ann. Rev. Astron. Astrophys.* **14**, 215 (1976).

Pickering, E. C., *Proc. Am. Acad. Sci.* **16**, 1 (1880).

Pierpont, J., *The Theory of Functions of Real Variables*, Vol. II, Ginn & Co., Boston, 1912 (reprinted by Dover, New York, 1959).

Plummer, H. C., *An Introductory Treatise on Dynamical Astronomy*, Cambridge University Press, London, 1918 (reprinted by Dover, New York, 1960).

Poincaré, H., *Comptes Rendus* **123**, 1224 (1896).

Porco, C. G., and G. E. Danielson, *Astron. J.* **87**, 826 (1982).

Pringle, J. E., *Ann. Rev. Astron. Astrophys.* **19**, 137 (1981).

Reif, F., *Foundations of Statistical and Thermal Physics*, McGraw-Hill, New York, 1965.

Richstone, D. O., *Astrophys. J.* **200**, 535 (1975); *Astrophys. J.* **204**, 642 (1976).

Richtmyer, R. D., and K. W. Morton, *Difference Methods for Initial Value Problems*, Interscience Publishers, New York, 1957.

Rosenbluth, M. N., W. M. MacDonald, and D. L. Judd, *Phys. Rev.* **107**, 1 (1957).

Russell, H. N., *Astrophys. J.* **35**, 315 (1912a); *Astrophys. J.* **36**, 54 (1912b); *Astrophys. J.* **36**, 133 (1912c); *Mon. Not. R. astr. Soc.* **88**, 641 (1928).

Russell, H. N., and H. Shapley, *Astrophys. J.* **36**, 239 (1912a); *Astrophys. J.* **36**, 385 (1912b).

Rybicki, G., *Astrophys. Sp. Sci.* **14**, 56 (1971).

Sahade, J., and F. B. Wood, *Interacting Binary Stars*, Pergamon, Oxford, 1978.

Sammet, J. E., *Adv. Comp.* **8**, 47 (1966).

Sanford, R. F., *Astrophys. J.* **109**, 81 (1949).

Saslaw, W. C., *Publ. Astron. Soc. Pac.* **85**, 5 (1973).

Scholl, H., in R. L. Duncombe, Ed., *Dynamics of the Solar System*, D. Reidel Publ., Dordrecht, Holland, 1979, p. 217.

Scott, F. P., and J. A. Hughes, *Astron. J.* **69**, 368 (1964).

Serafin, R. A., *Cel. Mech.* **21**, 351 (1980).

Shu, F. H., and S. H. Lubow, *Ann. Rev. Astron. Astrophys.* **19**, 277 (1981).

Siegel, C. L., and J. K. Moser, *Lectures on Celestial Mechanics*, Springer-Verlag, New York, 1971.

Skolnik, M. I., *Radar Handbook*, McGraw-Hill, New York, 1970.

Smak, J., *Acta Astron.* **12**, 28 (1962); *Publ. Astron. Soc. Pac.* **76**, 210 (1964).

Smart, W. M., *Celestial Mechanics*, Cambridge University Press, London, 1953.

Smith, G. R., *Cel. Mech.* **19**, 163 (1979).

Smith, H., in D. G. Bettis, Ed., *Lecture Notes in Mathematics*, Springer-Verlag, New York, 1974, p. 360.

Spitzer, L., Jr., *Astrophys. J.* **127**, 17 (1958); *Physics of Fully Ionized Gases*, Wiley, New York, 1962.

Spitzer, L., Jr., and R. A. Chevalier, *Astrophys. J.* **183**, 565 (1973).

Spitzer, L., Jr., and M. H. Hart, *Astrophys. J.* **164**, 399 (1971a); **164**, 483 (1971b).

Spitzer, L., Jr., and R. D. Mathieu, *Astrophys. J.* **241**, 618 (1980).

Spitzer, L., Jr., and S. L. Shapiro, *Astrophys. J.* **173**, 529 (1972).

Spitzer, L., Jr., and J. M. Shull, *Astrophys. J.* **200**, 339 (1975a); *Astrophys. J.* **201**, 773 (1975b).

Spitzer, L., Jr., and T. X. Thuan, *Astrophys. J.* **175**, 31 (1972).

Sridharan, R., and M. L. Renard, *Cel. Mech.* **2**, 179 (1975).

Sridharan, R., and W. P. Seniw, M. I. T. Lincoln Laboratory Technical Note 1979-25, 1979; M.I.T. Lincoln Laboratory Technical Note 1980-1, Vol. 1, 1980.

Staff of the SAO, *Star Catalog of Positions and Proper Motions of 258,997 Stars for the Epoch and Equinox of 1950.0*, Smithsonian Institute Publication 4652, Washington, D.C., 1966.

Standish, E. M., *Bull. Astron* **3**, 135 (1968).

Sterne, T. E., *Astron. J.* **62**, 96 (1957); *Astron. J.* **63**, 28 (1958); *An Introduction to Celestial Mechanics*, Interscience Publishers, New York, 1960.

Stone, L. D., *Theory of Optimal Search*, Academic, New York, 1975.

Struve, O., *Astrophys. J.* **93**, 104 (1941); *Astrophys. J.* **99**, 222 (1944).

Stumpff, P., *Astron. Astrophys.* **56**, 13 (1977).

Sundmann, K. F., *Acta. Math.* **35**, 105 (1912).

Szebehely, V., *Theory of Orbits*, Academic, New York, 1967; in B. D. Tapley and V. Szebehely, Eds., *Recent Advances in Dynamical Astronomy*, D. Reidel Publ., Dordrecht, Holland, 1973, p. 75.

Szebehely, V., and R. McKenzie, *Astron. J.* **82**, 303 (1977).

Taff, L. G., *Icarus* **20**, 21 (1973); *Astron. J.* **79**, 1280 (1975); *J. Comp. Phys.* **20**, 160 (1976); *Astron. Quart.* **1**, 183 (1977); M.I.T. Lincoln Laboratory Technical Note 1979-49, 1979a; M.I.T. Lincoln Laboratory Project Report ETS-44, 1979b; M.I.T. Lincoln Laboratory Technical Note 1980-24, 1980; *Computational Spherical Astronomy*, Wiley, New York, 1981.

Taff, L. G., and D. L. Hall, *Cel. Mech.* **16**, 481 (1977); *Cel. Mech.* **21**, 281 (1980).

Taff, L. G., P. M. S. Randall, and S. A. Stansfield, M.I.T. Lincoln Laboratory Technical Report 618, 1984.

Taff, L. G., and M. P. Savedoff, *Mon. Not. R. astr. Soc.* **160**, 89 (1972); *Mon. Not. R. astr. Soc.* **164**, 357 (1973).

Taff, L. G., and J. M. Sorvari, M.I.T. Lincoln Laboratory Technical Note 1979-38, 1979; *Cel. Mech.* **26**, 423 (1982).

Taff, L. G., and H. M. Van Horn, *Mon. Not. R. astr. Soc.* **167**, 427 (1974); *Astrophys. J.* **197**, L23 (1975).

Taff, L. G., H. M. Van Horn, C. H. Hansen, and R. R. Ross, *Astrophys. J.* **197**, 651 (1975).

Talpaert, Y. and F. Lefévre, *Bull. Cl. Sci. Acad. R. Belg.* **58**, 759 (1972).

Taylor, J. H., in W. Sieber and R. Wielebinski, Eds., *Pulsars*, D. Reidel Publ., Dordrecht, Holland, 1981, p. 361.

Thiele, T. N., *Astron. Nach.* **104**, 245 (1883).

Tisserand, F., *Traite de Mecanique Celeste*, Vol. I, 1889.

Tobey, R. G., IBM Technical Report 00.1365, 1965.

Tobey, R. G., R. J. Bobrow, and S. N. Zilles, *Proc. Amer. Fed. Info. Proc. Soc.* **27**, 37 (1965).

Toomre, A., *Sci. Am.* **229**, 38 (1973); *Ann. Rev. Astron. Astrophys.* **15**, 437 (1977).

Toomre, A., and J. Toomre, *Bull. Amer. Astron. Soc.* **2**, 350 (1970); *Bull. Amer. Astron. Soc.* **3**, 390 (1971); *Bull. Amer. Astron. Soc.* **4**, 214 (1972a); *Astrophys. J.* **178**, 623 (1972b).

van Albada, T. S., *Bull. Astr. Neth.* **19**, 479 (1968).

van der Bos, W. H., *Union Obs. Circ.* **68**, 354 (1926).

van Houten, C. J., I. van Houten-Groeneveld, P. Herget, and T. Gehrels, *Astron. Astrophys. Suppl.* **2**, 339 (1970).

Verbunt, F., *Sp. Sci. Rev.* **32**, 379 (1982).

Vogel, H. C., *Astron. Nach.* **123**, 289 (1890).

von Hoerner, S., *Astrophys. J.* **125**, 451 (1957); *Zeit. f. Astrophys.* **50**, 184 (1960).

Walker, M., *Astrophys. J.* **123**, 68 (1956).

Watson, G. N., *A Treatise on the Theory of Bessel Functions*, Cambridge University Press, London, 1922.

White, S. D. M., *Mon. Not. R. astr. Soc.* **174**, 19 (1975a); *Mon. Not. R. astr. Soc.* **174**, 467 (1975b); *Mon. Not. R. astr. Soc.* **174**, 19 (1976a); *Mon. Not. R. astr. Soc.* **174**, 467 (1976b); *Mon. Not. R. astr. Soc.* **177**, 717 (1976c).

Whittaker, E. T., *A Treatise on the Analytical Dynamics of Particles and Rigid Bodies*, 4th ed., Cambridge University Press, London, 1937 (reprinted by Dover, New York, 1944).

Wielen, R., *Veröff. des Astrons. Rechen-Inst. Heidelberg*, No. 19 (1967).

Woolard, E. W., and G. M. Clemence, *Spherical Astronomy*, Academic, New York, 1966.

Yabushita, S., *Mon. Not. R. astr. Soc.* **140**, 109 (1968).

Index